Proceedings of the International Conference on Security, Surveillance and Artificial Intelligence (ICSSAI-2023), Dec 1–2, 2023, Kolkata, India

Editor biographies

Editor 1: Dr. Debasis Chaudhuri

Dr. Debasis Chaudhuri presently working as a Professor in the Department of Computer Science & Engineering, Techno India University, West Bengal. Before that He was the senior scientist and Deputy General Manager at DRDO Integration Centre, Panagarh, West Bengal in the Rank of Joint Secretary, Ministry of Defense, Govt. of India. He was Project scientist at Indian Statistical Institute, Kolkata. He was involved in ISI-ADRIN Project, Dept. of Space, Govt. of India in the year of 1994 to June 1996. He worked as professor at Defense Institute of Advance Technology (Deemed University) Pune forInformation Classification: Generalone and half years during 2001–July 2004. He was visiting Professor at University of Nebraska, USA during Aug 2003–2004. He Worked at Integrated Test Range (ITR), Chandipur for one year during January 2005—December 2005. He was the senior scientist and DGM at DRDO Integration Centre, Panagarh from July 2014 to 31st October 2021.

Editor 2: Dr. Jan-Harm Pretorius

Dr. Jan-Harm Pretorius obtained his BSc Hons (Electrotechnics) (1980), MIng (1982) and DIng (1997) degrees in Electrical and Electronic Engineering at the Rand Afrikaans University and an MSc (Laser Engineering and Pulse Power) at the University of St Andrews in Scotland (1989), the latter cum laude. He is a trained Baldrige (USA) and South African Excellence Foundation (SAEF) assessor. He worked at the South African Atomic Energy Corporation (AEC) as a Senior Consulting Engineer for 15 years. He also worked as the Technology Manager at the Satellite Applications Centre (SAC) of the Council for Scientific and Industrial Research (CSIR). He is currently a Professor and Head of School: Postgraduate School of Engineering Management in the Faculty of Engineering and the Built Environment. He has co-authored over 240 research papers and supervised 50 PhD and over 260 Master's students. He is a registered professional engineer, professional Measurement and Verification (M&V) practitioner, senior member of the Institute of Electrical and Electronic Engineering (IEEE), fellow of the South African Institute of Electrical Engineers (SAIEE) and a fellow of the South African Academy of Engineering.

Editor 3: Dr. Debashis Das

Dr. Debashis Das is currently working as an Associate Professor and Head of the Department in the department of Computer Science & Engineering at Techno India University, West Bengal. He did his PhD from IIT(ISM) Dhanbad in 2017. His research interest is in Image Processing, Biometric Security and Steganography. He is associated with a project on Vision based Surveillance funded by DRDO, Govt. of India. He also served as a reviewer for a number of peer reviewed journals and conferences. He is the life member of Indian Society for Technical Education (ISTE) and member of IEEE.

Editor 4: Mr. Sauvik Bal

Mr. Sauvik Bal currently working as an Assistant Professor in the Department of Computer Science & Engineering, Techno India University, West Bengal. Prior to that he worked in Haldia Institute of Technology, West Bengal and University of Engineering & Management, Jaipur. He received his Master degree in Computer Science & Engineering from Maulana Abul Kalam Azad University of Technology (formerly known as West Bengal University of Technology), in 2015. In both the UG and PG levels, he has Eight years of teaching and research experience. His research interest is in Natural Language Processing, Machine Learning, Deep Learning. He is a professional Member of ACM, Life Member of Indian Society for Technical Education (ISTE), Member of Computer Science Teachers Association, and International Association of Engineers. He attached with various international conferences.

International Conference on Security, Surveillance and Artificial Intelligence (ICSSAI-2023)

Proceedings of the International Conference on Security, Surveillance and Artificial Intelligence (ICSSAI-2023), Dec 1–2, 2023, Kolkata, India

Edited by
Debasis Chaudhuri
Jan-Harm Pretorius
Debashis Das
Sauvik Bal

CRC Press
Taylor & Francis Group
Boca Raton London New York

CRC Press is an imprint of the
Taylor & Francis Group, an **informa** business

First edition published 2024
by CRC Press
4 Park Square, Milton Park, Abingdon, Oxon, OX14 4RN

and by CRC Press
2385 NW Executive Center Drive, Suite 320, Boca Raton FL 33431

CRC Press is an imprint of Informa UK Limited

British Library Cataloguing-in-Publication Data
A catalogue record for this book is available from the British Library

ISBN: 9781032549972 (pbk)
ISBN: 9781003428459 (ebk)

DOI: 10.1201/9781003428459

Typeset in Sabon LT Std
by HBK Digital

Printed and bound in India

Contents

List of figures

List of tables

Preface

The International Conference on Security, Surveillance & Artificial Intelligence (ICSSAI2023) was held in West Bengal, India during December 1–2, 2023. The conference was organized by the Techno India University, one of the renowned universities in the state of West Bengal which is committed for generating, disseminating and preserving knowledge.

ICSSAI 2023 was a two-day conference that aims to gather students, academicians, researchers and industry practitioners from all over the country and the globe to present advances in the field of computer science and foster creativity through knowledge sharing. The conference primarily focused on the broad domain of information security, surveillance and its real life applications, artificial intelligence and industrial applications.

ICSSAI2023 received 128 research articles from the various parts of the country and the globe. All the submitted articles have undergone double blind peer review process made by the domain experts selected from different countries. The proceeding editors have finally accepted 46 (acceptance rate is 35.9%) articles for the presentation and publication at the conference after a rigorous analysis of the technical review comments. We would like to thank the authors for their contribution in the conference. We would like to show our sincere gratitude to all the reviewers, advisory committee members and technical program committee members for their continuous support, suggestions and well wishes.

ICSSAI2023 provided a common forum to exchange and share the experiences and technical knowledge of the mixed research minds participated in the technical event. We would like to thank all the eminent speakers for enlightening and sharing thoughts on the core domain and advanced research perspective. We would like to thank the management of Techno India University, to facilitate appropriate arrangement and continuous support to organize such a large scale event. A sincere thank to all the internal committee members and the steering committee members for their tireless work to make the conference a successful one. We would also like to thank Taylor & Francis for their fruitful association and collaboration in preparing and publishing the quality research articles in a single volume. We are also grateful to the esteemed journal editors for their collaboration and commitment to publish some of the promising research works in the special issue volumes.

Hope the conference proceeding will attract, help and be useful for the entire research fraternity to a good extent.

Dr. Jan Ham C Pretorius
Dr. Debasis Chaudhuri
Dr. Debashis Das
Mr. Sauvik Bal

Note of Appreciation

डा. एन. राजेश पिल्लै
उत्कृष्ट वैज्ञानिक एवं निदेशक

Dr. N. Rajesh Pillai
OS & Director

भारत सरकार
रक्षा मंत्रालय
रक्षा अनुसंधान तथा विकास संगठन
वैज्ञानिक विश्लेषण समूह
मेटकॉफ भवन, दिल्ली - 110 054

Government of India
Ministry of Defence
Defence R&D Organisation
Scientific Analysis Group
Metcalfe House, Delhi - 110 054

MESSAGE

I am happy to note that Techno India University, Salt Lake, Kolkata is organising the First International Conference on Security, Surveillance and Artificial Intelligence (ICSSAI - 2023) on 1st & 2nd December, 2023. I congratulate each and every member of this conference for their commitment and dedication towards the success of the conference.

The Artificial Intelligence (AI) techniques like deep learning, machine learning, pattern recognition, Natural language processing (NLP), and computer vision play a major role in several automation applications, both in industries and day-to-day life. AI techniques enable us to provide several smart solutions for human society, not only for urban areas but also for rural areas. AI can be applied to various domains, such as, smart city design, green energy developments, efficient banking solutions, IoT applications, expert medical systems, surveillance for civil as well as defence industries etc. AI and in particular Machine Learning provides many techniques for security and privacy assessment. AI is also widely used in the area of medical science which has a major impact on human lives. Thus, a smart human society with a good quality of life can be thought of from the novel ideas/concepts generated from the conference. Surveillance in today's society is required to remove crime and troubles, not only in smart cities and villages in rural areas, but also to handle many challenging defence applications. To generate solutions for such problems, research in the related areas is seriously required. Situations bring research opportunities to scholars, faculties and practitioners. Conferences in this domain provide wide scope for interdisciplinary research to different communities.

Scientists and engineers are essential to a nation's progress. They promote initiatives that advance economic development, encourage improvements that raise our standard of living, and contribute to such programmes. This Conference is both timely and an excellent platform for the scientists and engineers to establish meaningful collaborations around the world.

I wish the conference and all the participants a great success.

डॉ. एन. राजेश पिल्लै / Dr. N. RAJESH PILLAI
उत्कृष्ट वैज्ञानिक एवं निदेशक / Outstanding Scientist & Director
वैज्ञानिक विश्लेषण समूह / Scientific Analysis Group
रक्षा मंत्रालय, डीआरडीओ / Ministry of Defence, DRDO
मेटकॉफ भवन परिसर, दिल्ली-110054
Metcalfe House Complex, Delhi-110054

दूरभाष/Telephone: 011-23819828, 23812695   फैक्स/Fax: 011-23812683   ई-मेल/E-mail: director.sag@gov.in

Introductory Note

Security surveillance has recorded a new path with artificial intelligence (AI) integration. AI used in monitoring systems has reconfigured how we monitor and protect our environment. These new systems can automatically and quickly analyze vast data from video footage, identifying suspicious behaviors or irregularities in real time. AI can differentiate between normal activities and potential security threats using advanced object and face recognition and communication analysis. Therefore, it can be used to improve the efficiency of security personnel. Furthermore, the security surveillance potential of AI can adapt and learn from existing situations to continuously improve the system's accuracy. This technology has the advantage of reducing false alarms. Cooperation between security surveillance and AI enhances our ability to prevent and respond to security breaches and offers a proactive approach to ensuring the safety of our environments.

Moreover, one of the most exciting features of AI in security surveillance is its potential to predict and prevent incidents before they occur. Recently, a French start-up developed security surveillance for agriculture that permits radar to map the whole area, and whenever there is an intrusion, to release a drone that will go closer to the intruder to record video of the scene. This system also uses image recognition to identify any intruder, and based on its algorithm, provides an early warning to the operators. Thus, security surveillance capabilities are currently implemented in some airports to strengthen security at the boarding gate and improve the flow of passengers, such as in China. It is also used in some public spaces to detect abandoned bags to act against dangerous packages. This proactive approach empowers security teams to take preventive measures, reducing the likelihood of security breaches and enhancing overall safety.

Prof. Jan Harm C Pretorius
Organizing Chair, ICSSAI-2023
Professor, Postgraduate School of Engineering Management
University of Johannesburg, Cape Town, South Africa

The primary purpose of surveillance is to keep an eye on public areas or the perimeter of safe places in order to keep them safe from burglary, intrusion, fire, and other potential threats. The human operators are the weakest link in the security system, regardless of how sophisticated the cameras and VMS are. The issue is that it is difficult for humans to continuously monitor several screens without losing focus, which can eventually cause them to miss important occurrences. Additionally, because individuals become tired easily, their productivity and response times unavoidably suffer, increasing the likelihood of human error.

In order to identify people, cars, objects, attributes, and events, artificial intelligence for surveillance imaging uses computer software programmes that assess the sounds and visuals from video surveillance cameras. For the property being protected by camera surveillance, security contractors programme the software to define restricted areas within the camera's field of view (a parking lot, for example, but not the sidewalk or a public street outside the lot), as well as programme for times of day (such as after the close of business). If the artificial intelligence (AI) notices a trespasser violating the "rule" that nobody is authorised in that area at that time of day, it will send an alarm. Governments and technology proponents point to the potential of artificial intelligence (AI) to help secure international borders more effectively and, in some cases, more safely, as one of the benefits of the rise of artificial intelligence (AI), which promises to streamline operations in border management. With much interaction and discussion, a fascinating and educational scientific agenda is being created.

Prof. Debasis Chaudhuri
Organizing Chair, ICSSAI-2023
Professor, Computer Science & Engineering
Techno India University, West Bengal, India

Part 1

Information Security

1 SLASH: A Secure and Lightweight Authentication Scheme for Enhancing the Reliability of IoT-based Smart Healthcare Systems

Srabana Pramanik[1,a], Sandip Roy[2,b], and Soumya Banerjee[2,c]

[1]Department of CSE (SoE) Presidency University, Bengaluru, India

[2]Virginia Modeling, Analysis and Simulation Center Old Dominion University, Suffolk, Virginia, USA

Abstract

With the increasing dependency on Internet of Things (IoT) and its expanding application domain, the transformation of services into smart services is becoming a reality. However, this evolution comes with a range of security challenges due to the limitations of IoT end-node devices, including less computational abilities, limited storage capacity, low power backup, and limited processing power. To address these challenges, a robust authentication method is essential to establish the faith and trust in management of various IoT devices and ensures a trustworthy communication among all entities. In this paper, we propose an authentication technique with dynamic identity tailored for IoT environments, leveraging communication Channel properties, one-way hashing operation, and XOR operations. The software simulation tool Proverif has been utilized to validate the scheme's safety and security against various attacks. Comparative performance study demonstrates the superiority of our proposed protocol over related approaches concerning computational ability, memory storage, and communication expenses.

Keywords: Mutual authentication, Proverif, Internet of Things, Security analysis, Fuzzy extractor.

Introduction

The swift evolution of the IoT has accelerated the growth of the fourth industrial revolution, with various sectors transforming to provide smart services worldwide. IoT brings together products from different manufacturers, connecting them under one umbrella to offer enhanced and intelligent services to end-users [16].It constitutes a dynamic network of physical objects, devices, and sensors embedded with software, actuators, and network connectivity, enabling data sharing and responsiveness through internet connectivity. The number of connected devices in the IoT ecosystem has grown exponentially, reaching 50 billion by 2020 [8]. However, this proliferation of devices and cloud connectivity also opens the door for malicious activities. As technology advances, it becomes increasingly crucial to establish a strong security support system for the IoT environment to ensure

[a]srabana.edu@gmail.com, [b]sroy@odu.edu, [c]s1banerj@odu.edu

end-to-end security. Authentication plays a vital role in enhancing security, enabling secure sessions after a secure booting process in IoT devices. However, traditional authentication systems involving shared secret keys become impractical due to resource constraints in IoT devices, such as limited memory and physical exposure [13]. Therefore, it is a need to design a safe, secure and robust authentication methodology that accommodates these limitations and addresses potential attacks.

In this research paper, a novel authentication scheme is projected for IoT environment using channel characteristics. In IoT deployment, various types of communication channels are established, that includes message communications among various users of the IoT system, the authentication server and the IoT node devices [6]. All these communication channels require mutual authentication between the parties involved [11]. Our focus in this article is on the communication of IoT node devices and the Gateway authentication server.

Motivation and Contribution of the Research

In this work, we aim for enhancing the security of various message communications required among resource constrained IoT nodes and Gateway servers in IoT deployments. To elaborate the methodology of the proposed scheme, the real-time application Tele-care medicine information system is utilized. This approach creates various stake holder's identity in a dynamic way and builds unique session key dynamically for every session, thereby enhancing the overall security of the communication process.

The key contributions and outcomes of the paper can be summarized as follows:

1. Lightweight and Efficient Authentication Scheme: The paper proposes a novel authentication scheme that utilizes a lightweight irreversible hash operation, XOR operation, and communication channel parameters as a factor of authentication among the communicating parties before establishing any transaction.
2. Robust Security Analysis provide a comprehensive evaluation of the Proverif Software Validation. The proposed protocol is rigorously tested using the Proverif tool in the security analysis phase. The simulation results confirm that the protocol is resilient against various adversarial attacks, such as replay, privileged-insider, Man-in-the-Middle (MITM) attacks and ensures a high level of protection for the communication.
3. In the performance analysis, the paper demonstrates that the recommended scheme outperforms previously published protocols on memory usage, computation time, and communication overhead. This demonstrates both the efficiency and scalability of the proposed scheme for resource limited IoT environments.

The proposed scheme is successful validated through Proverif, demonstrates its capability to provide secure communication and protect against potential attacks. Finally, analysis of various performance metrics highlights the superiority of the proposed scheme over existing protocols, affirming its suitability for resource constrained IoT deployments.

The remaining portion of the article is structured as follows. Section 2. elaborates the system models which also include threat model and physical channel characteristics. Section 3. explains the proposed methodology followed by informal security analysis and formal security analysis in section 4 & 5. Section 6. gives details of comparative performance studies. Section 7. proves the proposed protocol's superiority among three protocols.

The system Model

In the Tele-care medicine information system (TMIS), each patient is equipped with one or more medical-IoT devices (Di). These devices are continuously gathering the patients' health related secret data, and at regular intervals, the secret medical health data is transmitted and uploaded to a dedicated centralized medical server (S).

Before initiation of data upload, a secure but efficient authentication process is required to verify the legitimacy of already registered medical-IoT devices. This authentication process ensures that only authorized devices can communicate with the central server, providing an essential layer of security. The proposed scheme is designed by focusing the Single server-based telecare medicine Information System. Figure 1.1 elaborates that [16].

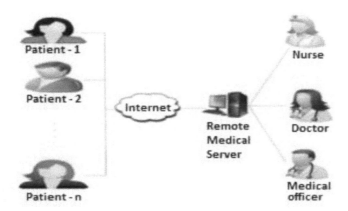

Figure 1.1 Single server-based telecare medicine information system.

Threat Model

In a communication scenario involving two parties over a public channel, the security is analyzed under the Dolev-Yao (DY) threat model, here, an attacker can freely access, modify, and even inject false messages. Additionally, the intruder can leverage techniques like power analysis and side-channel attacks to extract confidential information from stolen IoT devices, which may have been physically captured by the intruder or obtained from the IoT user's mobile device [4]. To address security concerns, most of the authentication schemes in literature were modeled using the CK—adversary model (abbreviated from the adversary model proposed by Canetti and Krawczyk), which encompasses all abilities of the DY adversary and allows the intruder to compromise both session states as well as session specific keys. The CK—Adversary model places significant emphasis on the need to minimize the potential consequences of disclosing ephemeral secrets on the confidentiality of individual session keys during the course of communication [5]. To meet these security requirements, a proposed protocol has been designed. This protocol aims to ensure protection against ephemeral-secret-leakage attacks that could compromise the security of the communication. By implementing measures to safeguard session keys and ephemeral secrets, the proposed protocol intends to maintain the confidentiality and integrity of the communication even in the presence of a powerful adversary.

Physical Channel Characteristics

To enhance the toughness of the projected protocol, an extra factor is introduced to strengthen mutual authentication among communicators. This extra factor leverages as non-cryptographic parameter, like physical channel characteristics, traffic patterns, and energy consumption. Each entity involved in the protocol generates a unique user profile, or fingerprint, by selecting different features or parameters.

For instance, the features of the device's profile or the historical log data are continuously monitored and compared by following the network traffic of each device. If there is any variation in network traffic, such as a sudden increase, in both devices, then it must be re-authenticated each other.

The additional factor integrated into the protocol, utilizes non-cryptographic parameters and user-specific fingerprints are generated from a diverse feature set. The constant monitoring and comparison of network traffic, coupled with the dynamic nature of wireless channels, contribute to increased robustness and authentication accuracy.

The projected protocol heavily relies on the shared communication channel parameters among two devices. However, due to the inherent randomness and dynamic nature of wireless channels, various channel-specific parameters exhibit significant variations over time.

In many existing physical layer security (PLS) techniques, it is presumed that the channel among two communicating entities are reciprocal with each other. Both entities may perceive channel-specific parameter and use them to encrypt data to ensure its confidentiality [7]. But in reality, the assumption may not be always hold true. Usually, the channel-specific parameters and characteristics can vary among the transmitter and receiver, frequently changing over time. Therefore, it is possible that the channel-based nonce derived by the transmitter may not always be identical to the channel-based nonce extracted by the receiver $(CH_{ni}! = CH_{nj})$. The information about channel state, the received signal intensity, and the angle of arrival are widely recognized as significant contributors to the generation of channel-based nonces [3].

To overcome this challenge of non-reciprocity in the channel, the proposed protocol employs fuzzy extractor technique. This technique allows the system to reconcile the differences in channel parameters between users and ensures that both entities can effectively use these parameters for encryption and authentication purposes.

Proposed Scheme

The proposed protocol utilizes two operations, one-way hash operation (h(.)) and XOR operation. To address the challenge of channel non-reciprocity, the fuzzy extractor method is employed.

In this scheme, the fuzzy extractor technique has two key functions: one is Gen(.) and other is Rep(.). By incorporating these two functions within the fuzzy extractor technique, the scheme can effectively handle the variations in channel nonces between two entities. This ensures that both transmitter and receiver can generate and utilize the same channel-based key (α) for authentication and data encryption, even when there are slight differences in their channel parameters. As a result, the projected scheme achieves enhanced security and reliability in a dynamic wireless environment.

Gen(.): The probabilistic function, Gen(.) uses the channel nonce to generates two outputs: one-bit channel-based key (α_i) and a reproduction parameter (γ_i). Where, the nonce is extracted by transmitter (CH_{ni}). The purpose of this function is to produce a uniform string of randomly generated bits from the given input CH_{ni}. Therefore, it can be expressed as follows: Gen $(CH_{ni}) = (\alpha_i, \gamma_i)$.

Rep(.): When the input deviates somewhat from the original, this function is used to reconstruct the uniform string of randomly produced bits. The difference between the inputs is measured using the Hamming distance, and it should be less than or equal to a predefined threshold value (t) for the function to work. Specifically, the inputs to the Rep(.) function are CH_{nj} (the channel nonce extracted by receiver) and γ_i (the reproduction parameter). Using these inputs, the Rep(.) function generates the channel-based key (α_i), which is essential for authentication and data encryption. Therefore, it can be expressed as follows: $\alpha_i = $ Rep (CH_{nj}, γ_i) (Roy et al., 2018b).

Patients equipped with medical IoT devices which periodically collect medical data and uploading that into authoritative server. Before each upload they have to undergo through the authentication procedure. This will ensure the security of the smart healthcare systems.

Table 1.1: Details of used notations.

Notation used	Description
D_i	*IoT node*
S	*Server*
ID_i	*Device identity*
CH_{ni}, CH_{nj}	*Communication channel parameter*
a_i	*Only in registration phase, the pre-shared key is used between S & Di.*
R_{Ni}, R_{Sj}	*Randomly generated variables*
T_1, T_2	*Freshly generated Time stamp*
$h(.)$	*Hash function*
$\oplus$	*XOR function*
$\parallel$	*Concatenation function*
Skn / Sks	*Session specific key for a particular session*

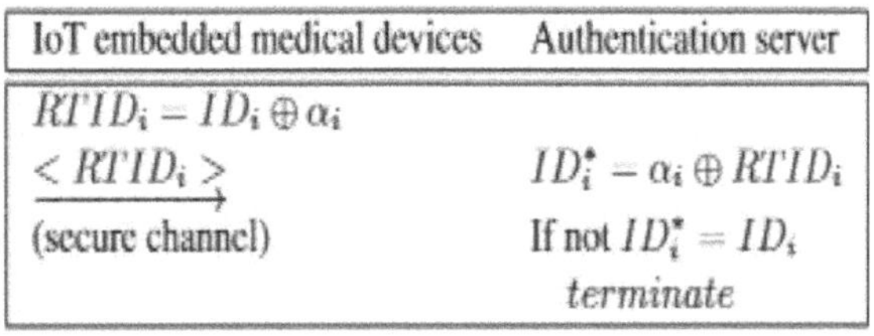

IoT embedded medical devices	Authentication server
$RTID_i = ID_i \oplus \alpha_i$	
$< RTID_i >$	$ID_i^* = \alpha_i \oplus RTID_i$
(secure channel)	If not $ID_i^* = ID_i$
	terminate

Figure 1.2 Detail of the registration of the IoT embedded medical devices with the authentication server.

IoT embedded medical devices	Authentication server
$Gen(CH_{ni}) = \alpha_i, \gamma_i$	
$TID_i = h(ID_i \oplus CH_{ni}) \oplus T1$	
$M_1 = h(ID_i \| T1) \oplus (R_{Ni} \| \gamma_i)$	
$M_2 = h(R_{Ni} \| \alpha_i) \oplus (TID_i)$	$M3 = h(ID_i \| T1)$
$< M1, M2, T1 >$	$(R_{Ni} \| \gamma_i) = M1 \oplus M3$
$\xrightarrow{}$	
(public channel)	$\alpha_i^* = Rep(CH_{nj}, \gamma_i)$
	$TID_i = h(ID_i \oplus CH_{nj}) \oplus T1$
	$M4 = h(R_{Ni} \| \alpha_i^*) \oplus TID_i$
	If not $M2 = M4$
	terminate
	Select $R_{sj} \in \mathbb{Z})_F$
	$M5 = h(ID_i \| T1 \| R_{Ni}) \oplus R_{sj} \oplus T2$
$M7 = h(ID_i \| T1 \| R_{Ni})$	$SK^* = h(ID_i \| T2 \| T1 \| R_{sj} \| R_{Ni} \| \alpha_i^*)$
$R_{sj} = M7 \oplus M5 \oplus T2$	$M6 = h(SK \| R_{sj} \| R_{Ni} \| T2)$
$SK^* = h(ID_i \| T2 \| T1 \| R_{sj} \| R_{Ni} \| \alpha_i)$	$< M5, M6, T2 >$
	$\xleftarrow{}$
$M8 = h(SK^* \| R_{sj} \| R_{Ni} \| T2)$	(public channel)
If not $M8 = M6$	
terminate	
$SK = SK^*$	

Figure 1.3 Detail of the authentication between IoT embedded medical devices and authentication server.

The Pseudocode of the Proposed Work

Before registration the server and the IoT devices both have the identity of the IoT devices and pre-shared variable which will be used till registration phase.

1. **Registration phase (Registration for newly installed IoT embedded medical devices)**
   IoT embedded medical devices constructs the temporary $RTID_i$ and shares it securely with the server, described in Figure 1.2.
2. **Authentication phase**
   Figure 1.3 details the authentication between IoT embedded medical devices and authentication server. T1 and T2 are timestamps and SK is the computed session key.

Security Analysis

This section analyses how the proposed scheme defends various adversarial attacks.

Proposition 1: The projected scheme is secured against Impersonation attack and Reply attack.

Proof: If an intruder wants to send a fake message to IoT node side as server then he has to know the random variables (Rni &Rsi) and the recent timestamps (T1&T2) which is nearly impossible. Not only that it gives protection from reply attack also by combining random variable and timestamp with all the messages.

Proposition 2: The proposed system ensures the mutual authentication.

Proof: Both the IoTend-node and server verify each communication through two key hash data, by comparing M2 = h($R_{Ni} \| \alpha_i$) $\oplus$ TID_i & M4 = h($R_{Ni} \| \alpha_i^*$) $\oplus$ TID_i in server side, it generates random variable, R_{Si} and T2. Then M6 = h($SK \| R_{Sj} \| R_{Ni} \|$ T2) & M8 = h(SK^* $\| R_{Sj} \| R_{Ni} \|$ T2)) is compared in IoT side. Where all hash values are combined with (Rni &Rsi) and α. Thus, the projected scheme provides Mutual Authentication.

Proposition 3: The proposed system maintaining forward secrecy.

Proof: In this case both the entries share session key, $SK = h(ID_i \parallel T2 \parallel T1 \parallel R_{Sj} \parallel R_{Ni} \parallel \alpha_i^*)$, which contains short-term key (R_{Sj}, R_{Ni}) and long-term key (IDi). For each and every session these random nonce changes, for which the dynamic nature of the session key is maintained. If the communication key is captured for a particular session, after that, also it will not be going to affect the next sessions. Hence, the scheme preserves forward secrecy.

Proposition 4: Giving safeguard against Ephemeral-Secret-Leakage (ESL) attack & Known-Key-Session-Specific-Temporary-Information (KSSTI) attack.

Proof: In all communication the raw identity is not communicated. In move on only the hash key values are used. If one of the parameters of session key is hacked, then also it is tough to gather all parameters of session key of a specific session. Getting the long-term secret are very tough. Based on KSSTI attack, if the attacker collects the ephemeral-random-number by observing the transmission, then also it is next to impossible to get the access on long-term secrets at a time. Hence, the projected scheme hinders the ESL attacks and the KSSTI attacks.

Usage of Proverif Simulator for Formal Security Verification

In this section, the authentication protocol undergoes through a rigorous verification using the ProVerif software, an automatic cryptographic formal protocol verifier tool. It utilizes the pi calculation. ProVerif generally explores the tested protocol's complete state space and checks the vulnerabilities related to the session key (Vangala et al., 2022).

By using the ProVerif tool, the security properties of the projected authentication protocol are thoroughly examined (Pramanik et al., 2022b). The verified security properties include:

- Secrecy of the Session Key.
- Provision of strong secrecy property as well as strong anonymity property.
- Resistance to various security attacks including replay, impersonation, and modification attacks.

ProVerif version 2.04 is used to simulate the results which are presented in Figure 1.4. The result shows that the session key is successfully obtained during the authentication procedure and can withstand attacks.

Performance Analysis

This part of the paper focuses on the comparative analysis of various functionalities of the proposed scheme in contrast to previously published protocols. The evaluation includes three key aspects: execution time, communication overhead, and memory use of each IoT node.

To ensure a fair comparison, the registration process for each IoT node is performed only once, and the calculation considers only the login-authentication phases. In the comparative study, related recently published protocols are taken into account [9, 10, 12]. The reason for including these protocols is that they were executed under the same situation as the projected scheme. In the proposed protocol the channel parameter is used as factor of authentication.

Table 1.2: For different operation the rough operational time.

Symbols	Operation	Operational time (in ms)
T_h	Hash operation or PUF operation	0.320
$T_{mac} \approx T_h$	MAC operation	0.320
T_{epm}	ECC (point multiplication)	17.10
$T_{fe} \approx T_{epm}$	Fuzzy extractor	17.10
T_{end}	Encryption or decryption	5.60

Table 1.3: Comparison study of computational time.

Protocols	Total computational complexity	Total overhead (in ms)
Proposed method	$12T_h + 1\ T_{fe}$	20.94
[12]	$9T_h + 8T_{epm}$	47.68
[9]	$10T_h + 1\ T_{fe}$	20.3
[10]	$6T_h + 3T_{end} + 2T_{puf}$	38.08

Table 1.4: Numbers of bit requirement.

Parameter wise	Bit's requirement
For identity (value)	160
Freshly generated time-stamp	160
ECC (For multiplication)	160
Hash value	256
XOR operation	160
Concatenation operation	160

Table 1.5: Comparative investigation of communication expenses.

Related Protocols	Total Transmission overhead(bits)	Number of messages
Proposed method	1050	2
[12]	1504	3
[9]	1760	3

Table 1.6: Comparison of memory usage.

Related Schemes	Comparison (bits)
Proposed method	160
[9]	160
[12]	576

Computational Complexity Analysis

The proposed scheme incorporates several essential operations, such as Hash, MAC (Massage Authentication Code), Fuzzy Extractor, ECC (point multiplication), encryption or decryption, and PUF. Despite including these computationally intensive tasks, the scheme demonstrates remarkably shorter computational time when compared to most existing protocols, with the exception of Melki et al's scheme. However, it is important to note that our proposed scheme offers superior functionality and a higher level of security features. Specifically, during each update of patient health into the medical server, the medical IoT device node and the server efficiently authenticate each other in a mutual manner which is takes 20.94 milliseconds.

Communication Overhead Calculation

In this study, the communication overhead signifies total the total number of bits required for message communication between the IoT device and the gateway server. It is very important to note that the transmission of bits during the registration process is not counted in the calculation. However, during the login-authentication stage, the total number of bits transmitted is taken into account to assess the communication overhead. Table 1.4 presents a comprehensive overview of the number of bits required for each parameter used in our proposed protocol and other relevant schemes [1].

Memory Requirement

In this scheme, the memory usage is calculated as [160] = 160 bits, representing the memory space required to store < IDi> by following Pramanik et al's [15] scheme. During the registration process, each IoT node has to stores only IDi, in its memory. These parameters are securely stored for future communication purposes. By limiting the storage key parameters, the scheme ensures efficient memory utilization while retaining the necessary information required for authentication and secure data transfer. This approach optimizes

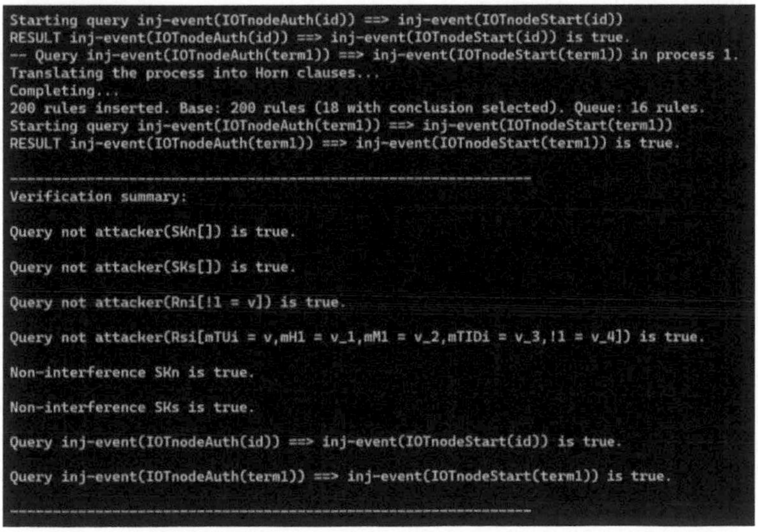

Figure 1.4 Screenshot of simulation output.

memory usage and guarantees that IoT nodes can maintain the essential data needed for seamless communication with the gateway server in the IoT ecosystem.

Conclusion and Future Direction

The research presents a pioneering authenticated key settlement approach using optimized irreversible hash, XOR functions, and fuzzy extractor to ensure the security of IoT environments. By generating dynamic session keys for data transfers, the protocol offers enhanced security, particularly in critical applications such as smart healthcare systems. Rigorous security and accuracy analyses confirm the effectiveness of the scheme, making it a viable choice for securing the ever-growing IoT ecosystem. While this research focuses on securing a specific part of the IoT ecosystem, the next step involves extending the proposed technique to encompass the entire IoT infrastructure. This ambitious future work aims to create a comprehensive and secure authentication mechanism for all aspects of the IoT environment before any communication occurs.

References

1. Alzahrani, B. A. and Mahmood, K. (2021). Provable privacy preserving authentication solution for internet of things environment. *IEEE Access*, 9, 82857–82865.
2. Vangala, S. Roy and Das, A. K. (2022). Blockchain-based lightweight authentication protocol for iot-enabled smart agriculture. *2022 International Conference on Cyber-Physical Social Intelligence (ICCSI)*, Nanjing, China.
3. Badawy, A., Elfouly, T., Khattab, T., Mohamed, A., and Guizani, M. (2016). Unleashing the secure potential of the wireless physical layer: Secret key generation methods. *Physical Communication*, 19, 1–10.
4. Banerjee, S., Odelu, V., Das, A. K., Chattopadhyay, S., Rodrigues, J. J., and Park, Y. (2019). Physically secure lightweight anonymous user authentication protocol for internet of things using physically unclonable functions. *IEEE Access*, 7, 85627–85644.
5. Canetti, R. and Krawczyk, H. (2001). Analysis of key-exchange protocols and their use for building secure channels. *International Conference on the Theory and Applications of Cryptographic Techniques*, 453–474.
6. Chatterjee, S. and Roy, S. (2018). An efficient dynamic access control scheme for distributed wireless sensor networks. *International Journal of Ad Hoc and Ubiquitous Computing*, 27(1), 1.
7. Hamamreh, J. M. and Arslan, H. (2017). Secure orthogonal transform division multiplexing (OTDM) waveform for 5G and beyond. *IEEE Communications Letters*, 21(5), 1191–1194.
8. He, D. and Zeadally, S. (2015). An analysis of RFID authentication schemes for Internet of Things in healthcare environment using elliptic curve cryptography. *IEEE Internet of Things Journal*, 2(1), 72–83.
9. Melki, R., Noura, H. N., and Chehab, A. (2020). Lightweight multi-factor mutual authentication protocol for IoT devices. *International Journal of Information Security*, 19(6), 679–694.
10. Muhal, M. A., Luo, X., Mahmood, Z., and Ullah, A. (2018). *Physical Unclonable Function Based Authentication Scheme for Smart Devices in Internet of Things*. Ic, 160–165.
11. Nandy, T., Idris, M. Y. I. Bin, Md Noor, R., Mat Kiah, M. L., Lun, L. S., Annuar Juma'At, N. B., Ahmedy, I., Abdul Ghani, N., and Bhattacharyya, S. (2019). Review on security of internet of things authentication mechanism. *IEEE Access*, 7, 151054–151089.
12. Panda, P. K. and Chattopadhyay, S. (2020). A secure mutual authentication protocol for IoT environment. *Journal of Reliable Intelligent Environments*, 6(2), 79–94.

13. Pramanik, S., Sakkari, D. S., and Pramanik, S. (2022). Remediation measures to make the insecure internet of things deployment secure. *International Journal of Engineering Trends and Technology*, 70(6), 155–164.
14. Pramanik, S., Sakkari, D. S., and Pramanik, S. (2022). Privacy conserving authenticated key settlement approach for remote users in IoT based Telecare Medicine information system. *Smart Health*, 26, 100355.
15. Pramanik, S., Sakkari, D. S., and Pramanik, S. (2022). Ensure the dynamic identity and PUF based authenticated key settlement approach for the IoT infrastructure. *Proceedings of the Indian National Science Academy*, 88(4), 670–687.
16. Roy, S. (2018). Design and analysis of remote authentication and access control for wireless communications.
17. Roy, S., Chatterjee, S., Chattopadhyay, S., and Gupta, A. K. (2016). A biometrics-based robust and secure user authentication protocol for e-healthcare service. *2016 International Conference on Advances in Computing, Communications and Informatics (ICACCI)*.
18. Roy, S., Chatterjee, S., Das, A. K., Chattopadhyay, S., Kumari, S., and Jo, M. (2018). Chaotic map-based anonymous user authentication scheme with user biometrics and fuzzy extractor for crowdsourcing internet of things. *IEEE Internet of Things Journal*, 5(4), 2884–2895.

2 Will Block Chain be Secured Post-Quantum Computing?

Amrit Mukherjee[a], Pratik Goswami[b], Rudolf Vohnout, Ivo Bukovsky, and Ondre Budik

Department of Computer Science and Engineering, Faculty of Science, University of South Bohemia, Czech Republic

Abstract

The abstract discusses the emergence of quantum computing as a powerful tool for solving complex problems using the principles of quantum mechanics. Quantum algorithms can solve complex problems much faster than traditional computing algorithms. The use of quantum physics concepts such as superposition, interference, and entanglement enable algorithms to solve non-polynomial problems in constant and polynomial time. Blockchain technology is a distributed ledger that records data securely and transparently using public-key cryptography and hashing. Shor's algorithm, a quantum algorithm, possesses the capability to efficiently determine the prime factors of an integer within polynomial time. This efficiency poses a significant challenge to encryption methods like RSA, which rely on the inherent difficulty of integer factorization that requires extensive computational time.

Keywords: Crisis, ownership, variables.

Introduction

In recent years, quantum computing has risen as a potent solution for addressing intricate challenges by harnessing the principles of quantum mechanics to manipulate qubits (quantum bits). These quantum algorithms excel in performing computations significantly faster than their classical counterparts, offering a remarkable advantage in expeditiously solving complex problems. Let us talk about the ability to compute before computers were introduced. We cannot even compare today's computing abilities with those of that era. That will be the same case. We cannot imagine enough scalable quantum computers that can solve many of the most challenging problems in a fraction of a minute. It may take thousands of years to compute with our current generation of computers. It will prove to be the next revolution in the computing field. Various practical computational algorithms and data structures are designed to operate on classical computers. While many of these algorithms have quantum counterparts, the potential needed is already present in their respective quantum versions, which prove to be much more potent due to the unique nature of quantum mechanical information processing. A basic illustration of this is demonstrated through Deutsch's problem, which reveals that quantum computation can be considerably

[a]amukherjee@jcu.cz, [b]rvohnout@jcu.cz

faster than classical computation. In this problem, we are given a function that is either balanced (for a complete set of inputs, the equal number of output bits 0 and 1) or constant (returns the same value regardless of input), and our focus is to resolve whether the function is balanced or constant. We must perform two calculations in classical computing, one for each input value. However, in quantum computing, we only need to execute a single calculation, which enables us to identify whether the function is balanced or constant. However, the algorithm is not capable of identifying output values.

Quantum computing algorithms have been developed for solving problems in constant or polynomial time, which might take exponential or super–polynomial time:

- Deutsch–Jozsa algorithm (for solving black box problems)
- Simon's algorithm (for solving black box problem)
- Shor's algorithm (for solving integer factorization)
- Estimating Gauss sums (for finding Gauss exponential sum)
- Grover's algorithm (to increase quadratic speed in calculations)

These algorithms solve the non-polynomial problem in constant and polynomial time by using the techniques of quantum physics like entanglement, superposition and interference. The concept of cryptocurrency practically introduced blockchain. The first virtual currency that was born along with the blockchain was Bitcoin. As stated by Nakamoto [1], blockchain technology may be characterized as a distributed ledger that securely stores data (with the aid of hashing), which helps in preventing it from unwanted access. Blockchain technology uses a decentralized network and has been demonstrated to be immutable and transparent. Every participant on the network can see the updates made to the blockchain without revealing or leaking any personal information about the network user. Blockchain helps its participants use it securely only through public–key cryptography (or asymmetric cryptography), which ensures that the transaction is authentic or not. As Fernandez-Carames [2] mentioned, hashing also plays a crucial part in blockchain because it produces a unique hash value for each block, which is also used to link the blocks together. Use cases of the blockchain in different fields are healthcare, logistics, e-voting, smart factories, virtual currencies and other applications.

The widely used asymmetric key algorithms are as follows:

- ECDSA (Elliptic) [5]
- RSA (Rivest) [4]
- ECDH (Elliptic Curve) [5]
- DSA (Digital) [7]

These algorithms do not prevent the information from being penetrable but transform them into some other format which can be decoded but will take a couple of thousands of years. These algorithms use some mathematical equations that cannot be solved even with a computer in comparatively less time. The paper's structure is as follows: Section II delves into blockchain terminology and basic concepts, while Section III explores the potential threat to public key security posed by quantum algorithms. In a similar vein, Section IV elucidates the discussion concerning the threat to hash function security caused by quantum algorithms. Furthermore, Sections V and VI provide insights into the ideal blockchain and the conclusion regarding quantum-based cryptography systems.

Blockchain Terminology and Basic Concepts

Before reviewing how quantum computers are a threat to blockchain we have to understand some of the basic terminology and concepts. As the term suggests, the Blockchain is a chain of blocks containing information regarding any digital transaction may or may not be monetary. Briefly each node has its unique hash value generated with the help of different hash functions like RIPEMD-160, MD5, BLAKE2, SHA-256 and many more.

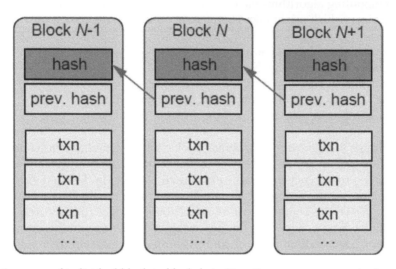

Figure 2.1 Structure of individual block in blockchain ("txn" represents transaction).

Within each block, you'll find the hash value for the current block, the hash value for the previous block, as well as relevant transaction data specific to that block and the broader blockchain. The inclusion of the previous block's hash value creates a chain-like structure, enabling the chronological tracking of transactions. To foster decentralization, a public ledger is maintained, with each blockchain participant holding a copy. This decentralized nature, coupled with consensus protocols, ensures that every participant can verify new transactions, making data alteration and manipulation exceedingly challenging, thereby endowing the network's data with immutability.

The widely used consensus techniques as mentioned by Rajan [8] are:

- Proof-of-work (used by the Bitcoin)
- Practical byzantine fault tolerance (PBFT)
- Proof of stake (PoS)
- Proof of burn (PoB)
- Proof of capacity

In addition to using private and public keys, blockchain also employs the idea of public key cryptography. In a blockchain, parties must verify that the subsequent transaction was completed by them using their own digital signatures before exchanging information or processing it. Here, the digital signature is basically the private key which is used to encrypt the information like real life signatures used to show the confirmation that the transaction has taken placed while the public key shared across the network is used to verify whether the transaction that has taken place is valid or not.

Threat to Public Key Security due to Quantum Algorithms

Shor's Algorithm

The algorithm is named after the inventor Peter Shor. It is an algorithm for locating an integer's prime factors on a quantum computer in polylogarithmic time (meaning polynomial in log N time). Time taken by the parameters are:

- Classical computer: $-O(\sqrt{N})$,
- Quantum computer: $-O((log\ N)^2\ (log\ log\ N)\ (log\ log\ log\ N))$

Here, N denotes the number to find the factors.

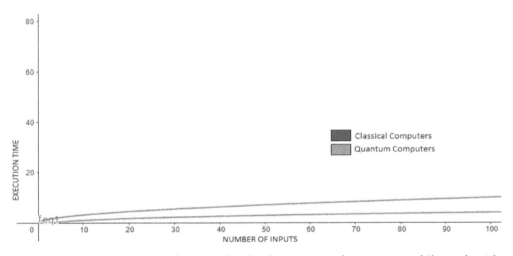

Figure 2.2 Comparison between the time taken by the computers for execution of Shor's algorithm.

The algorithm is useful to breach data encryption created by algorithms such as RSA that rely on the extensive time taken in the process of integer factorization. It is certain that factoring is a specific application of the larger hidden subgroup problem, Shor's approach was originally created to factor huge composite integers produced by multiplying two significant primes. By modifying Shor's algorithm, all such problems can be resolved, including the discrete logarithm problem. Due to these results, algorithms, such as Diffie-Hellman key exchange, elliptic curve cryptography, the Digital Signature algorithm, and ElGamal encryption also become insecure [9].

Table 2.1: Different types of Encryption algorithms that are in danger due to Shor's Algorithm by Berlekamp [9].

Algorithms	Function	Key size/signature size	Pre-quantum security level
ECDSA	Signature	256 bits	128 bits
RSA-1024	Signature, Encryption	1024 bits	80 bits
RSA-2048	Signature, Encryption	2048 bits	112 bits
RSA-3072	Signature, Encryption	3072 bits	128 bits
DSA-3072	Signature	3072 bits	128 bits

Security Challenges on Hash Function due to Quantum Algorithms

Grover's Algorithm

This algorithm provides a specific solution for the problem of searching the input value for a function which is very difficult to reverse engineer. In the blockchain if a person wants to attack then he not only has to change only information regarding that particular block but also has to change the entire chain of blocks as they are connected with their hashes as changing only one hash means that will result in breaking the chain. As this algorithm provides the solution for inverting the solutions from a function into the input values which will make the blockchain systems more prone to attacks as explained by Bernstein [9]. Time taken to crack and invert all the hash values is:

- For classic computer: $O(n)$
- For quantum computer: $O(\sqrt{N})$

As we can see how drastically improvements these quantum algorithms show in time complexity in comparison with the classical algorithm.

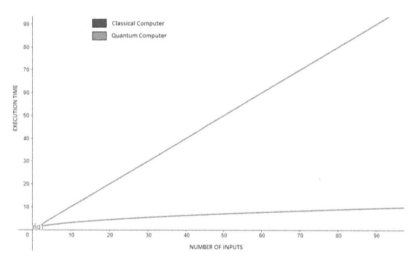

Figure 2.3 Comparison between the time taken by the computers for execution of Grover's algorithm.

Ideal Blockchain Traits

A post-quantum cryptosystem's appropriateness hinges solely on its established efficiency. Primarily, the system must possess diminutive key sizes, thereby curtailing the requisite storage capacity and simplifying computational processes. Special attention should be accorded to the Internet of Things (IoT) end-devices, given their inherent limitations in storage and computational capabilities, as they constitute integral components within blockchain networks.

Moreover, it is imperative to constrain the length of both signatures and hashes to minimize the blockchain's overall footprint. Addressing the issue of protracted execution times per transaction is paramount to enhancing the efficiency of the post-quantum cryptosystem. Resolving this concern will lead to decreased execution times, thereby enabling the blockchain to process a greater number of transactions within stipulated timeframes. This

Table 2.2: Different types of hashing algorithms that are in danger due to Grover's algorithm [9].

Algorithms	Main affected blockchain/ DLTs	Estimated post-quantum security level	Key size/ hash size	Pre-quantum security level	Function
SHA–256	Bitcoin, Ethereum, Dash, Litecoin	128 bits	256 bits	256 bits	Hash function
Ethash	Ethereum	128/256 bits	256/512 bits	256/512 bits	Hash function
Scrypt	Litecoin, NXT	128 bits	256 bits	256 bits	Hash function
RIPEMD160	Bitcoin, Ethereum, Litecoin	80 bits	160 bits	160 bits	Hash function
KECCAK-256	Monero, Bytecoin	128 bits	256 bits	256 bits	Hash function
KECCAK-384	IOTA	192 bits	384 bits	384 bits	Hash function
SHA3-256		128 bits	256 bits	256 bits	Hash function
AES-123		64 bits	128 bits	128 bits	Symmetric Encryption
AES-256		128 bits	ts	256 bits	Symmetric Encryption

enhancement must be coupled with a focus on reduced computational complexity, ensuring the inclusion of resource-constrained devices in blockchain transactions.

Energy conservation assumes critical importance in the realm of blockchain technology, particularly in light of previous criticisms aimed at energy-intensive consensus protocols, exemplified by Bitcoin. This heightened power consumption is not solely attributed to the consensus protocol but is also exacerbated by intricate security mechanisms. Overall, these attributes constitute indispensable prerequisites for an efficient post-quantum cryptosystem suitable for integration into diverse blockchain environments. The essential factors encompassed in crafting an ideal blockchain encompass compact key sizes, streamlined computational processes, concise digital signatures and hash lengths, and expedited transaction execution. The assurance of cryptographic systems' security and compatibility with a variety of blockchain implementations is contingent upon the incorporation of these discussed features.

Conclusion

Quantum Based Cryptography Systems

Quantum computing not only ushers in waves of innovation in technology and algorithms but also unveils new frontiers in encryption. It explores vast possibilities for transforming various cryptographic methods, potentially reshaping blockchain technology. Among these, quantum key distribution (QKD) stands out as the most advanced and mature quantum cryptographic technology, resilient against quantum computers, thanks to its departure from traditional cryptographic reliance on mathematical complexity. Instead, QKD leverages the fundamental principles of quantum mechanics to safeguard cryptographic keys. Nevertheless, it's worth noting that the process of measuring quantum systems in the QKD framework is often intricate, as highlighted by Blömer in 2007 [11]. While QKD doesn't directly transmit message data, it can effectively encrypt messages using any encryption algorithm, thereby ensuring their certified security.

In addition to QKD, an innovative approach introduced by Rajan and Visser presents a quantum counterpart to the blockchain concept, as elucidated by Blömer in 2007 [11]. These technologies are still in the developmental stages, yet they hold the potential to provide robust defenses against future quantum computer attacks, thus offering valuable enhancements to blockchain security. Some of the suggested cryptosystems are:

- Lattice-based cryptosystems
- Multivariate-based cryptosystems
- Code-based cryptosystems
- Super singular elliptic curve isogeny cryptosystems

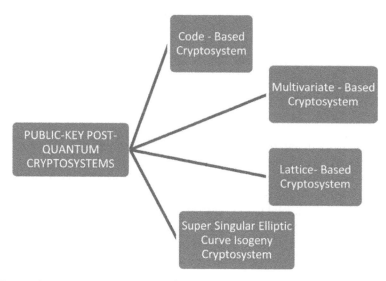

Figure 2.4 Types of post quantum cryptography systems suggested [2].

Code-Based Cryptosystem

These are cryptographic constructs founded upon principles that incorporate error correction codes. An exemplary instance of such a cryptosystem is McEliece's, which, as elucidated by Blömer in 2007 [11], emerged during the 1970s. The primary impetus behind the development of these cryptosystems is the mitigation of vulnerabilities associated with the syndrome decoding problem. Notably, these cryptographic systems exhibit a notable advantage in terms of expeditious encryption and decryption processes relative to their counterparts. However, they necessitate extensive matrices serving as both public and private keys, potentially consuming several megabytes of storage capacity. To address this challenge, a viable avenue entails delving into research on matrix compression techniques amenable to implementation through programming methodologies. It is essential to underscore that these cryptosystems presently offer security levels ranging from 128 to 256 bits of classical security. Given the advent of quantum computing, there exists a pressing need for a substantial augmentation of security measures. Key sizes span the spectrum, encompassing diminutive sizes for private keys are 320 bits of types ROLLO-II and RQC, extending to substantial dimensions like 15.5 KB for public keys of type HQC. This diversification in key sizes has proven instrumental in facilitating swift blockchain transactions, as documented by Fernandez-Carames in 2020 [2].

Multivariate-Based Cryptosystem

While multivariate-based schemes display resilience against quantum attacks, additional research is needed to improve their decryption speed, decrease key sizes, and mitigate ciphertext overhead. The challenges associated with multivariate-based schemes can be divided into two categories, which align with either the NP-hard or NP-complete class, as explained by [12]. Cryptosystems built upon Matsumoto-Imai's algorithm, those employing Hidden Field Equations (HFE), and schemes based on square matrices with random quadratic polynomials all present promising pathways within the domain of multivariate-based techniques.

Lattice-Based Cryptosystem

Lattice-based cryptography is built upon the mathematical framework of lattices, which are structured collections of points in n-dimensional spaces characterized by periodic properties. The security of lattice-based schemes relies on the assumed computational complexity of lattice problems, with a primary focus on the shortest vector problem (SVP). Classified as an NP-hard problem, SVP involves the daunting challenge of identifying the shortest non-zero vector within a lattice. Quantum computers face similarly formidable obstacles when dealing with other lattice-related problems like the closest vector problem (CVP) and the shortest independent vector problem (SIVP). Lattice-based cryptographic systems offer rapid and efficient execution, resulting in enhanced transaction efficiency for blockchain users. However, similar to other post-quantum cryptographic methods, the implementation of lattice-based systems necessitates the use of substantial cryptographic keys and incurs significant ciphertext overhead. For example, well-known lattice-based schemes such as NTRU and New Hope frequently require keys ranging from a few bytes to several kilobytes in size. In addition to polynomial algebra, the most promising lattice-based cryptographic systems are grounded in the learning with errors (LWE) problem and its various derivatives, including LP-LWE and Ring-LWE, which are presently undergoing development and refinement. A lattice-based public-key cryptosystem that advanced to the second round of the NIST competition provides classical security levels ranging from 128 to 368 bits, along with quantum security levels spanning 84 to 300 bits. Notably, IoT round five employs a 128-bit private key, while FrodoKEM-1344 boasts an exceptionally extensive private key of 344,704 bits.

Super Singular Elliptic Curve Isogeny Cryptosystems

In the realm of ordinary elliptic curves, post-quantum cryptosystems employ isogeny protocols as a defense against quantum attacks. This strategy has given rise to several promising post-quantum cryptosystems, each characterized by key sizes in the range of a few thousand bits. Among these, SIKE stands out as the sole public-key encryption scheme grounded in isogeny that successfully met the NIST criteria. To attain a classical security level of 128 bits, SIKE leverages pseudo-random walks within supersingular isogeny graphs, necessitating a 2640-bit public key and a 2992-bit private key, as outlined by Fernandez-Carames [2].

References

1. Nakamoto, S. (2019). Bitcoin: a peer-to-peer electronic cash system. Accessed Nov. 2, 2019. Available from: https://bitcoin.org/bitcoin.pdf
2. Fernandez-Carames, T. M. and Fraga-Lamas, P. (2020). Towards post-quantum blockchain: A review on blockchain cryptography resistant to quantum computing attacks. *IEEE access,* 99(1), 1.
3. Rivest, L., Shamir, A., and Adleman, L. (1983). A method for obtaining digital signatures and public-key cryptosystems. *Communication ACM,* 26(1), 96–99.
4. Koblitz, N. (1987). Elliptic curve cryptosystems. *Mathematics of Computation,* 48(177), 203–209.
5. Dife, W. and Hellman, M. (1976). New directions in cryptography. *IEEE Transactions on Information Theory,* 22(6), 644–654.
6. Digital Signature Standard (DSS), Standard FIPS 186-2, NIST, Jan. 2000.
7. Kiktenko, E., Pozhar, N., Anufriev, M., Trushechkin, A., Yunusov, R., Kurochkin, Y., Lvovsky, A., and Fedorov, A. (2018). Quantum-secured blockchain. *Quantum Science and Technology,* 3(3).
8. Rajan, D. and Visser, M. (2019). Quantum blockchain using entanglement in time. *Quantum Report,* 1(1), 3–11.
9. Berlekamp, E., Mceliece, R., and Van Tilborg, H. (1978). On the inherent intractability of certain coding problems. *IEEE Transactions on Information Theory,* 24(3), 384–386.
10. Buchman, J. and Dahmen, E. (2009). Post-Quantum Cryptography. Berlin, Germany: Springer-Verlag.
11. Blömer, D. and Naewe, S. (2007). Sampling methods for shortest vectors, closest vectors and successive minima. International Colloquium on Automata, Languages and Programming, pp. 65–77.
12. Azarderakhsh, Y. R., Jalali, A., Jao, D., and Soukharev, V. (2017). A postquantum digital signature scheme based on supersingular isogenies. *In International Conference on Financial Cryptography and Data Security,* pp. 163–181.

3 Study and Analysis of the Recent Trends for Security Mechanisms in Mobile Adhoc Network

Bidisha Banerjee[a] and Sarmistha Neogy[b]

Jadavpur University, Jadavpur, Kolkata, India

Abstract

Mobile Ad-Hoc Networks (MANETs) offer dynamic, self-configuring wireless communication among mobile devices without the need for fixed infrastructure. However, their unique features, such as open communication channels, limited resources, and lack of centralized control, make them susceptible to various security threats. This article delves into the manifold security challenges faced by MANETs, like node misbehavior, eavesdropping, packet dropping, and more. It also investigates the vulnerabilities posed by the network's dynamic nature and decentralized architecture. Drawing from recent research, the article evaluates advanced security measures employed to protect MANETs, including encryption, secure routing protocols, trust management, and swarm intelligence-based approaches. The goal is to guide network administrators and researchers in selecting the most suitable security solutions for different scenarios.

Keywords: Cryptography, machine learning, MANET(s), security, swarm intelligence, trust.

Introduction

MANETs have revolutionized the way mobile devices communicate in the absence of a fixed infrastructure. MANETs are characterized by their dynamic and self-configuring nature, enabling devices to form temporary networks on the fly, without relying on centralized control or pre-established network infrastructure. These networks find applications in diverse fields, including military operations, disaster relief scenarios, vehicular networks, and IoT deployments, where traditional fixed infrastructures may be unavailable or impractical. The architecture of MANET is shown in Figure 3.1.

Despite their numerous advantages, MANETs present unique security challenges due to their decentralized and resource-constrained nature [3]. The open and wireless communication medium exposes MANETs to a wide array of security threats, ranging from node misbehavior and data interception to malicious attacks on routing protocols and denial-of-service (DoS) exploits [4, 5, 6]. Such vulnerabilities can severely compromise

[a]bidishab.rs.cse@jadavpuruniversity.in, [b]sarmisthaneogy@gmail.com

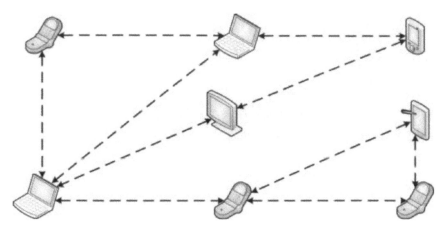

Figure 3.1 Mobile adhoc network.

the confidentiality, integrity, and availability of data being transmitted, making the design and implementation of robust security mechanisms in MANETs of paramount importance [7].

Literature suggests that several studies have addressed specific aspects of MANET security over the years. For instance, Korir et al. [2] conducted an extensive analysis of security threats in MANETs, considering their strengths and challenges, and proposes the need for more efficient and secure routing protocols. Kharaisat et al. [1] evaluated the performance of various intrusion detection system in detecting malicious activities within MANETs, shedding light on the strengths and limitations of each solution.

This paper presents a detailed security analysis of recent trends and solutions in Mobile Ad-Hoc Networks (MANETs), aiming to provide researchers valuable insights for safeguarding communication within these dynamic networks. While previous studies have addressed specific aspects of MANET security, this review consolidates and synthesizes the existing knowledge, analyzing current security mechanisms and exploring emerging trends. Ultimately, this review aims to contribute to the growing body of knowledge on MANET security and promote the development of robust and adaptive security measures for the expanding landscape of MANETs.

The paper is divided into the following sections: Section 2 provides a comprehensive exploration of security mechanisms and techniques. In Section 3, we delve into the crucial discoveries stemming from an extensive literature review. Section 4 provides the future direction of the work. Lastly, Section 5 concludes the summary of the entire article.

Security Mechanisms

Securing MANETs is crucial due to their decentralized, dynamic nature, and vulnerability to various security threats. Researchers and practitioners have developed numerous security mechanisms and techniques to enhance MANETs' resilience against these threats. This section explores the sophisticated security mechanisms and techniques proposed to protect MANETs from potential attacks and vulnerabilities. The latest trends in the security mechanisms of MANET have been categorized mainly into four distinct parts, as depicted in Figure 3.2.

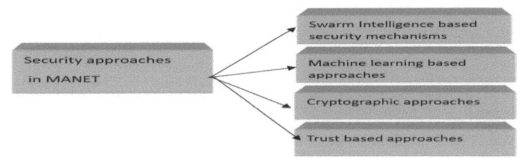

Figure 3.2 Various security approaches for MANET.

Machine Learning-Based Intrusion Detection Mechanisms

Intrusion detection in MANETs identifies and counters unauthorized or malicious activities within the network. These self-configuring networks comprise mobile devices like laptops or cellphones that communicate without centralized infrastructure. Given their dynamic and decentralized nature, intrusion detection is vital for ensuring their security and stability against potential attacks [8].

Ponnusamy et al. examined the intricacies of implementing Intrusion Detection Systems (IDS) in wireless settings, especially in IoT devices, WSN, and MANETs. The study discusses different intrusion detection methods and IDS strategies for wired and wireless networks. It underscores the architectural distinctions that necessitate custom IDS designs for wireless networks. A notable finding is the scarcity of network traces for training machine-learning models for IoT-specific breaches, highlighting a need for more tailored IDS solutions in wireless contexts [9]. Laqtib et al. studied the use of deep learning in Intrusion Detection Systems (IDS) for MANETs. Noting that traditional cryptography might not combat new threats, the paper highlights the potential of deep learning models, such as CNN, Inception-CNN, Bi-LSTM, and GRU, for the dynamic MANET setting. The research compares the efficacy of CNN and RNN for intrusion detection, emphasizing deep learning-based IDS's high accuracy and environmental resilience. The study guides the selection of suitable IDS methods in MANETs based on specific needs [31]. Pandey et al. emphasized the importance of advanced routing in MANETs due to their dynamic nature. They suggested using Artificial Neural Network (ANN) and Support Vector Machine (SVM) to detect black hole attacks, utilizing the AODV protocol. Comparing black hole AODV with their Secure AODV (SAODV) approach, the study shows the latter's improved performance and security against black hole attacks in MANETs [32].

Swarm Intelligence-Based Security Mechanisms

Swarm Intelligence techniques, inspired by collective behavior of social organisms, play a crucial role in MANET security by creating dynamic, adaptive defense mechanisms [30]. These algorithms allow nodes to collaborate and make decentralized decisions, effectively responding to changing network conditions and evolving attack patterns. This flexibility ensures robust and effective security even against new and previously unknown threats [14].

Swarm-based Intrusion Detection Systems (IDS) have demonstrated their efficiency in monitoring network traffic and identifying malicious activities. Leveraging the collective intelligence of nodes, these IDS can detect anomalies and potential attacks with greater accuracy, minimizing false positives and improving threat detection [15]. Swarm Intelligence has a profound impact on MANET security by offering dynamic, adaptive, and efficient solutions. These bio-inspired techniques empower the network to effectively detect and respond to threats, ensure energy-efficient operations, and maintain robust connectivity even in the face of adversarial conditions. As MANETs continue to evolve and face new challenges, Swarm Intelligence remains a promising approach to bolster the security and overall performance of these MANETs [16].

Veeraiah et al. developed a multipath routing protocol using fuzzy clustering and fuzzy Naive Bayes intrusion detection algorithms for energy and security optimization. The Bird Swarm-Whale Optimization Algorithm (BSWOA), combining bird swarm optimization (BSA) and whale optimization algorithm (WOA), implements multipath routing, considering several factors for optimal route selection. Evaluated under various attack scenarios, the proposed BSWOA protocol showed effectiveness in quality of service parameters, outperforming existing methods and proving suitable for MANETs and future wireless networks [10].

Kumar et al. introduced a novel IDS called CEGS-GDBC for multi-attack intrusion detection in MANETs, addressing node mobility and energy optimization challenges. The IDS comprises three main components: 1) Cluster head election using Dual Network centrality, 2) Cluster formation with Epsilon Greedy Swarm optimization, and 3) Multi-attack intrusion detection with a Gradient Deep Belief Network Classifier. Designed to reduce memory and computation while achieving high detection rates for DoS and Zero-Day attacks, the CEGS-GDBC method's efficiency was confirmed through extensive simulations, showing significant improvements in attack detection, memory consumption, and computational time compared to other methods [11].

Trust-Based Security Mechanisms

Trust-based security methods have emerged as effective strategies for enhancing the security of MANETs. These methods leverage the concept of trust to evaluate the reliability and behavior of network nodes. By establishing trust relationships among nodes, trust-based security approaches can identify and isolate malicious or compromised nodes, thereby mitigating potential threats and attacks. These methods enable MANETs to adapt to dynamic environments, promote cooperative behavior, and provide a robust defense mechanism against various security challenges [17]. The prime intension of trust-based security in MANETs is to discern and isolate malicious nodes while fostering cooperation among reliable nodes. Nodes gauge the trustworthiness of their peers based on behavior, granting more privileges to trustworthy nodes and restricting or monitoring suspicious ones. However, these methods can be fooled by colluding nodes or adept attackers. Hence, there's ongoing research to enhance the efficacy of trust-based security in MANETs [18]. Shukla et al. discussed the significance of MANETs in various sectors and highlighted their vulnerability to security threats, specifically the wormhole attack, which creates wormhole links between nodes. The paper proposes a trust-based approach to detect and mitigate the wormhole attack using parameters like receiving time and data rate. The evaluation shows that the wormhole attack negatively impacts the network's performance, but the proposed trusted approach improves the Packet Delivery Ratio (PDR) and throughput, while the end-to-end delay remains similar to an unaffected network [20]. In 2021, Shirajuddin et al. introduced the TBSMR trust-based multipath routing protocol for MANETs, aiming for secure, efficient data transmission and enhanced Quality of Service (QoS). TBSMR considers factors like congestion control and malicious node detection. It operates in three stages: route discovery, malicious node detection, and data forwarding. Simulations revealed TB-SMR's superiority over existing methods in metrics like Packet Delivery Ratio, Packet Loss Rate, and throughput, ensuring better QoS and security in MANETs [21].

Cryptography-Based Security Mechanisms

Cryptographic approaches are pivotal in addressing security challenges in MANET by utilizing encryption, authentication, and key management techniques to establish a strong framework for safeguarding sensitive data and preventing unauthorized access. These methods guarantee confidentiality, integrity, authentication, and non-repudiation of information exchanged within the network.

Singh et al. proposed a city-wide communication network for smart cities using wireless sensor networks. They addressed uneven node distribution and high message volume by combining a trust-based method with cryptography. This technique identifies suspicious nodes and finds alternative paths, enhancing network security. Cryptography, especially cyclic shift and bitwise XOR operations, are key components of this trust-centric approach [23]. In this paper, Bondada et al. proposed a secure and energy-efficient routing protocol for MANETs using group key management and asymmetric key cryptography. The protocol involves specialized nodes, CK and DK, for key generation and distribution, reducing energy consumption for other nodes. Extensive experiments demonstrate its superiority over existing protocols. The proposed approach enhances the security and performance of MANETs by efficiently managing secret keys and mitigating security threats [24]. Sanmuganathan et al. proposed a cluster-based group key management system for MANETs, utilizing elliptic curve cryptography and the sailfish optimization algorithm for optimal private key selection. This approach reduces computational overhead during re-keying and improves computational overhead, latency, packet delivery rate, and network lifetime compared to existing methods. It provides two-level security with reduced computational complexity and better performance in MANET group key management [25].

Comparative Analysis

Based on the survey presented in section 2 (Tables 3.1 and 3.2), we have summarized the following strengths and weaknesses of each approach:

- **Trust-Based Intrusion Detection for MANETs:**
  - Resilience to unknown Attacks: Trust-based intrusion detection can struggle to detect unknown or zero-day attacks as they may not exhibit patterns of previously seen malicious behaviour.
  - Overhead and Scalability: The trust management overhead can become significant in large MANETs, impacting network performance and scalability.
  - Self-Healing Capability: Trust-based systems can adapt and self-heal by isolating compromised nodes, contributing to network resilience.
- **Cryptographic Approaches for Securing MANETs:**
  - Computational Overhead: Cryptographic operations can be computationally expensive, especially for resource-constrained MANET nodes, impacting network performance.
  - Key Management Complexity: Effective key management is crucial for cryptographic security, and it can be complex and challenging to implement in dynamic MANET environments.

Table 3.1: Comparative analysis of security techniques for MANET.

Authors	Based on	Algorithms	Simulator	Performance metrics	Strengths	Weaknesses	Year
Sangeetha et al. [8]	Machine learning	BHD	Matlab	Attacker processing percentage, Defender processing percentage	Can detect multiple form of attacks	No real-time application	2021
Veeraiah et al. [10]	Machine learning and Swarm intelligence	BSWOA	NS2	Minimal energy, throughput, detection rate, and minimal delay	Achieve effective multipath routing while considering energy-efficiency and security in the network	The increased complexity might make it harder to debug and maintain the system	2020
Dilipkumar et al. [11]	Swarm intelligence	CEGSGDBC	-	Memory consumption, computational time, true positive rate	Compact cluster formation, decreased computational time and memory consumption	Dependence of the parameters like topology and node density; lack of real-world testing	2021
Einy et al. [12]	swarm intelligence and machine learning	MOPSOFLN	-	Classification rate, false positive rate, precision, recall, f-measures	Improved performance, selective feature subset, flexibility and adaptability	Computational complexity, generalization to unknown attacks	2021
Rajeshkumar et al. [13]	swarm intelligence	CTAAMPSO	-	Packet delivery ratio, end-to-end delay, control packet overhead	Trust worthy identification of malfunctioning nodes	The performance vary significantly at the levels of malware	2022
Zhang et al. [19]	Trust based	DATEA	Matlab	Trust value, residual energy	Combined trust values gives better result	Scalability, overhead, or the impact of network dynamics are not discussed	2019
Shukla et al. [20]	Trust based	Trusted AODV	NS3	PDR, end-to-end delay, throughput	Works well in the context of wormhole attack	Only one type of attack is considered	2021
Shirajuddin et al. [21]	Trust based	TBSMR	NS2	throughput, PDR, PLR, end-to-end delay	Guarantees enhancement of Qos of Manet, ensures secure communication	No attack scenarios have been experimented	2021
Pari et al. [22]	Trust based and cryptographic approach	HTRM	NS3	End-to-end delay, packet delivery ratio, performance, power consumption and keycomputing time	Security are provided by asymmetric key cryptography, improved performance metric	Computational overhead, key management complexity	2022

Table 3.2: Comparative analysis of security techniques for MANET.

Authors	Based on	Algorithms	Simulator	performance metrics	Strengths	Weaknesses	Year
Bondada et al. [24]	Cryptographic approach	EERP-	NS2	PDR, throughput, End-to-end delay, communication overhead, key computa- tional time	Centralized key management, reduced energy dissemination	overhead from centralization, implementation complexity	2022
Shanmuganathan et al. [25]	Cryptograp approach and swarm intelligence	ECCSFOAGKM-Manets	Python, NS3	PDR, throughput, latency, network lifetime, computational overhead and security level	Ensures security with cryptographic approach, Computational overhead decreased.	Implementation challenges in choosing private key	2023
Soni et al. [26]	Crypto graphic approach	EL- CKP		avalanche effect, lightweight speed	High security, lightweight speed	Chaotic function complexity, key management challenges	2023
N.Prakash et al. [27]	Swarm intelligence	Bee-IP	NS2	End-to-end delay, PDR, Throughput	Path quality is maintained	Computational overhead	2023
Arunmozhi et al. [28]	Swarm intelligence	energy efficient defense scheme with SI	NS2	PDR, routing overhead and end-to-end delay	Ensure high security using SI approach	Only one type of attack is taken into account	2023
Popli et al. [29]	swarm intelligence	Hybrid method using EPSO and ACO	Matlab	Delay, throughput, PDR and PLR	provide accurate identification of malicious node	Only one type of attack is considered.	2020

- Key Distribution and Revocation: Cryptographic systems may face challenges in distributing and revoking keys efficiently, particularly in large and constantly changing MANETs.
- Machine Learning-Based Intrusion Detection for MANETs:
  - Training Data Availability: Machine learning-based approaches require extensive and representative training data to build accurate intrusion detection models.
  - False Positives and False Negatives: These approaches may produce false positives and false negatives, leading to misclassification of legitimate nodes as malicious or vice versa.
  - Adaptability and Evolution: Machine learning models need continuous updates to keep up with new attack patterns and network changes.
- Swarm Intelligence-Based Approaches for Securing MANETs:
  - Scalability: The performance of swarm intelligence-based approaches may decline with the increasing size and complexity of the MANET due to the overhead associated with swarm coordination.
  - Decentralization: Swarm intelligence-based systems can be inherently decentralized, making them suitable for autonomous MANETs without centralized control.
  - Adaptability to Network Topology Changes: Swarm intelligence algorithms can dynamically adapt to changes in network topology, improving network efficiency and resilience.

In summary, selecting a security method for MANETs depends on specific deployment conditions and needs. Cryptographic systems ensure robust security but can be resource-intensive and face key management challenges. Trust-based detection offers resilience but may not tackle new threats. Machine learning techniques detect new attacks but require extensive data and updates. Swarm intelligence-based methods adapt to the network but may not scale well. For optimal protection, a holistic approach combining multiple strategies is recommended for MANETs.

Future Direction

In order to improve intrusion detection, anomaly recognition, and network resilience, machine learning approaches and swarm intelligence-based processes must be combined. Utilizing swarm intelligence for node authentication and trust building will foster strong cooperation, while swarm-based routing protocols can enable efficient packet routing in dynamic topologies. The ability to detect anomalies will be improved by combining swarm intelligence with machine learning, and self-adaptive security measures based on swarm intelligence will react dynamically to new threats. Furthermore, investigating the potential of swarm intelligence for MANET IoT device security offers exciting potential. However, by combining energy-saving techniques, privacy protection, and cross-layer coordination with trust-based and cryptographic solutions, MANETs can enhance their security, adaptability, and resilience in a variety of dynamic network contexts.

Conclusion

Each MANET security strategy has its own pros and disadvantages, and the acceptability of each relies on the particular deployment scenarios and requirements. Despite its durability and self-healing abilities, trust-based intrusion detection may have trouble defending against unidentified threats. While providing strong security protections, cryptographic techniques can be computationally taxing and difficult to manage keys. Machine learning-based techniques are excellent at spotting new threats, but they require a large training set and frequent updates. Although swarm intelligence-based solutions are flexible to network changes, scaling issues may arise. To conclude, we can argue that a thorough security plan may comprise the hybridization of several approaches, exploiting their strengths and successfully managing their shortcomings, to guarantee the greatest possible protection for MANETs. MANETs can be protected with the highest level of security in a variety of operational scenarios by carefully weighing the distinctive characteristics of each method.

References

1. Khraisat, A., Gondal, I., Vamplew, P., and Kamruzzaman, J. (2019). Survey of intrusion detection systems: techniques, datasets and challenges. *Cybersecurity*, 2(1), 1–22.
2. Korir, F., and Cheruiyot, W. (2022). A survey on security challenges in the current MANET routing protocols. *Global Journal of Engineering and Technology Advances*, 12(01), 078–091.
3. Sivapriya, N., and Mohandas, R. (2022). Analysis on essential challenges and attacks on MANET security appraisal. *Journal of Algebraic Statistics*, 13(3), 2578-2589.
4. Gurung, S., and Chauhan, S. (2020). A survey of black-hole attack mitigation techniques in MANET: merits, drawbacks, and suitability. *Wireless Networks*, 26, 1981–2011.
5. Liu, X., Han, J., Ni, G., Zhang, C., and Liu, Y. (2019). A multipath redundant transmission algorithm for MANET. In Communications, Signal Processing, and Systems: Proceedings of the 2017 International Conference on Communications, Signal Processing, and Systems (pp. 518–524). Springer, Singapore.
6. Hadi, R. M., Abdullah, S. H., and Abedi, W. M. S. (2022). Proposed neural intrusion detection system to detect denial of service attacks in MANETs. *Periodicals of Engineering and Natural Sciences*, 10(3), 70–78.
7. Deryabin, M., Babenko, M., Nazarov, A., Kucherov, N., Karachevtsev, A., Glotov, A., and Vashchenko, I. (2019). Protocol for secure and reliable data transmission in MANET based on modular arithmetic. In 2019 International Conference on Engineering and Telecommunication (EnT), (pp. 1–5). IEEE.
8. Sangeetha, V., Vaneeta, M., Kumar, S. S., Pareek, P. K., and Dixit, S. (2021). Efficient Intrusion detection of malicious node using bayesian hybrid detection in MANET. *IOP Conference Series: Materials Science and Engineering*, 1022(1), 012077. IOP publishing.
9. Ponnusamy, V., Humayun, M., Jhanjhi, N. Z., Yichiet, A., and Almufareh, M. F. (2022). Intrusion detection systems in internet of things and mobile ad-hoc networks. *Computer Systems Science Engineering*, 40(3), 1199–1215.
10. Veeraiah, N., and Krishna, B. T. (2020). An approach for optimal-secure multi-path routing and intrusion detection in MANET. *Evolutionary Intelligence*, 1–15.
11. Dilipkumar, S., and Durairaj, M. (2021). Epilson swarm optimized cluster gradient and deep belief classifier for multi-attack intrusion detection in MANET. *Journal of Ambient Intelligence and Humanized Computing*, 1–16.
12. Einy, S., Oz, C., and Navaei, Y. D. (2021). Network intrusion detection system based on the combination of multiobjective particle swarm algorithm-based feature selection and fast-learning network. *Wireless Communications and Mobile Computing*, 1–12.
13. Rajeshkumar, G., Kumar, M. V., Kumar, K. S., Bhatia, S., Mashat, A., and Dadheech, P. (2023). An improved multi-objective particle swarm optimization routing on MANET. *Computer Systems Science Engineering*, 44(2), 1187–1200.
14. Banerjee, B., and Neogy, S. (2023). An efficient swarm based technique for securing MANET transmission. In Proceedings of the 24th International Conference on Distributed Computing and Networking, (pp. 300–304).

15. Nasir, M. H., Khan, S. A., Khan, M. M., and Fatima, M. (2022). Swarm intelligence inspired intrusion detection systems—a systematic literature review. *Computer Networks*, 205, 108708.
16. Srivastava, A. (2023). Swarm intelligence for network security: a new approach to user behavior analysis. *International Research Journal of Engineering and Technology*, 10(2), 379–383.
17. Khanna, N., and Sachdeva, M. (2019). Study of trust-based mechanism and its component model in MANET: current research state, issues, and future recommendation. *International Journal of Communication Systems*, 32(12), e4012.
18. Muzammal, S. M., Murugesan, R. K., and Jhanjhi, N. Z. (2020). A comprehensive review on secure routing in internet of things: mitigation methods and trust-based approaches. *IEEE Internet of Things Journal*, 8(6), 4186–4210.
19. Zhang, D. G., Gao, J. X., Liu, X. H., Zhang, T., and Zhao, D. X. (2019). Novel approach of distributed adaptive trust metrics for MANET. *Wireless Networks*, 25, 3587–3603.
20. Shukla, M., and Joshi, B. K. (2021). A trust based approach to mitigate wormhole attacks in mobile adhoc networks. In 2021 10th IEEE international conference on communication systems and network technologies (CSNT), (pp. 776–782). IEEE.
21. Sirajuddin, M., Rupa, Ch., Iwendi, C., and Biamba, C. (2021). TBSMR: A trust-based secure multipath routing protocol for enhancing the QoS of the mobile ad hoc network. *Security and Communication Networks*, 1–9.
22. Pari, S. N., and Sudharson, K. (2023). Hybrid trust based reputation mechanism for discovering malevolent node in MANET. *Computer Systems Science Engineering*, 44(3), 2775–2789.
23. Singh, S., Pise, A., Alfarraj, O., Tolba, A., and Yoon, B. (2022). A cryptographic approach to prevent network incursion for enhancement of QoS in sustainable smart city using MANET. *Sustainable Cities and Society*, 79, 103483.
24. Bondada, P., Samanta, D., Kaur, M., and Lee, H. N. (2022). Data security-based routing in MANETs using key management mechanism. *Applied Sciences*, 12(3), 1041.
25. Shanmuganathan, C., Boopalan, K., Elangovan, G., and Sathish Kumar, P. J. (2023). Enabling security in MANETs using an efficient cluster based group key management with elliptical curve cryptography in consort with sail fish optimization algorithm. *Transactions on Emerging Telecommunications Technologies*, 34(3), e4717.
26. Soni, A. K., Gupta R., and Khare, A., (2023). An efficient and lightweight chaotic function with key exchange protection for man in the middle attack in mobile ad-hoc networks (MANET). 10.21203/rs.3.rs-3128165/v1
27. Prakash, M.N. and Sasieswaran, C., Path Selection Based Honey Bee Based QOS Routing for MANETS. *IJMRSET* (2023), 6(2). DOI:10.15680/IJMRSET.2023.0602015.
28. Arunmozhi, S. A., Rajeswari, S., and Venkataramani, Y. (2023). Swarm intelligence based routing with black hole attack detection in MANET. *Computer Systems Science Engineering*, 44(3), 2337–2347.
29. Popli, L. D. R. (2020). Sinkhole attack detection in manet using swarm intelligence techniques. *International Journal of Engineering Applied Sciences and Technology*, 5(4), 155–160.
30. Prasath, A. R. (2021). Bi-fitness swarm optimizer: blockchain assisted secure swarm intelligence routing protocol for MANET. *Indian Journal of Computer Science and Engineering*, 12(5), 1442–1458.
31. Laqtib, S., Yassini, K. E., and Hasnaoui, M. L. (2019). A deep learning methods for intrusion detection systems based machine learning in MANET. In Proceedings of the 4th International Conference on Smart City Applications, (pp. 1–8).
32. Pandey, S., and Singh, V. (2020). Blackhole attack detection using machine learning approach on MANET. In 2020 International Conference on Electronics and Sustainable Communication Systems (ICESC), (pp. 797–802). IEEE.

4 Adaptive Threshold Based Robust Reversible Watermarking Technique

Alina Dash[a], Kshiramani Naik[b], and Priyanka Priyadarshini[c]
Department of IT, VSSUT, Burla, Sambalpur, Odisha, India

Abstract

Reversible watermarking technique ensures image ownership verification through the insertion of a unique watermark signal into the host image. The suggested method utilizes a blend of discrete cosine transform (DCT) and discrete wavelet transform (DWT) for the secure embedding of the watermark. The process starts by breaking down the image into distinct 8x8 blocks, followed by the application of a 3-level DWT to the watermark. An adaptive thresholding method is employed to select the necessary DC coefficient, Subsequently, this modified watermark is employed to be embedded within the DCT coefficients of the host image. The retrieval process for both the watermark and the original image is uncomplicated and involves the application of inverse DWT and inverse DCT. To assess the technique's robustness, various image processing attacks are conducted, demonstrating its ability to maintain high image quality and resilience against common attacks. The proposed reversible watermarking scheme offers an efficient and reliable solution for protecting digital content and ensuring secure ownership verification.

Keywords: Adaptive thresholding, DC coefficients, discrete cosine transform (DCT), discrete wavelet transform (DWT), Reversible watermarking.

Introduction

With the proliferation of digital content and the ease of its distribution, ensuring the authenticity and ownership of multimedia assets has become a critical concern in today's interconnected world. Watermarking has emerged as a powerful technique to address these challenges, enabling content owners to assert their rights and protect their intellectual property. Watermarking allows content creators to embed a unique identifier into their content that can be used to prove ownership or detect unauthorized use of the content. Reversible watermarking is especially useful as it allows the original content to be recovered without any loss of information, which can be important in applications such as forensics and legal evidence.

The research presents a promising approach for reversible watermarking using DCT, DWT, and adaptive thresholding.

---

[a]alinadash_cse@vssut.ac.in, [b]kshiramaninaik_it@vssut.ac.in, [c]ppriyadarshini398@gmail.com

This technique empowers content owners to protect their multimedia assets by seamlessly embedding and extracting watermarks without any loss of the original data. The scheme's robustness is demonstrated through extensive experimentation, showcasing high image quality and resilience against various attacks. This method provides a valuable contribution to the field of multimedia security and content protection, enabling efficient ownership verification and safeguarding digital assets in today's interconnected world.

The paper adheres to a meticulously structured format. It initiates with Section 2, where a contextual analysis and pertinent prior research in the domain are outlined. Moving to Section 3, an all-encompassing review of the existing literature is presented. Section 4 furnishes an elaborate depiction of the proposed scheme. The outcomes of the experiments are exhaustively examined in Section 5. Ultimately, in Section 6, the paper concludes by succinctly summarizing the principal discoveries and contributions of the study.

Literature Survey

In previous times many research works have been conducted in the field of reversible watermarking to increase the watermark's sensitivity, which makes it more exposed to minor alterations. The process of various methods used for embedding an image are done to preserve the integrity and extractability of the image. Some of such algorithms proposed recently are mentioned below.

Leng et al. [1] introduced Digital Image Zero-watermarking Technology, which reviews and presents various techniques like DCT, DWT, and SVD. Zero-watermarking overcomes limitations of traditional methods by directly embedding watermark information without changing pixel values. The paper addresses challenges of attacks and false detection, offering potential solutions. It serves as a valuable reference for researchers exploring zero-watermarking approaches.

The utilization of the DWT-SVD Algorithm for Digital Watermarking has been successfully executed by Malik et al. [2]. The research paper thoroughly assesses diverse digital watermarking methods as well as their practical uses. The newly introduced DWT-SVD algorithm showcases a proficient approach to seamlessly implant watermarks into host images. Furthermore, the technique for extracting watermarks based on the inverse transform underscores its efficiency. This investigation presents a significant advancement in the realm of digital watermarking and offers promising potential across numerous applications focused on safeguarding digital images.

Anand and Singh [3] developed an improved watermarking technique to protect patient data in medical images. They used DWT-SVD domain to embed multi-watermarks and applied Hamming code to reduce channel noise distortion. The watermarked image was encrypted and compressed, with Chaotic-LZW showing the best performance. However, HyperChaotic-LZW proved more robust against various attacks. The method achieved high robustness, imperceptibility, security, and compression ratio for medical images.

Hasan et al. [4] presented an Encryption-Based Image Watermarking Algorithm in 2DWT-DCT Domains. It is based on reviews of existing image watermarking and encryption-based methods, emphasizing their strengths and weaknesses. Specifically focusing on 2DWT-DCT-based watermarking, the authors discuss the significance of coefficient selection and limitations of existing methods. Their proposed encryption-based watermarking algorithm demonstrates superiority over other existing techniques, validated through various performance metrics.

Ernawan et al. [5] introduced an Improved Image Watermarking approach where they focused on modified selected DWT-DCT coefficients, emphasizing the need for an enhanced technique with suitable coefficient selection. Their proposed method outperforms existing watermarking techniques, as demonstrated by various performance metrics.

Ariatmanto and Ernawan [6] proposed an adaptive scaling factor and DCT coefficients based watermarking. They emphasize the significance of appropriate coefficient selection for embedding and introduce a method to calculate adaptive scaling factors. The technique demonstrates superior performance in robustness and imperceptibility, making it valuable for image watermarking applications.

Proposed Work

The proposed scheme is reversible watermarking scheme which means that the host image and provided watermark can be extracted without any distortions after the embedding process. The phases of the proposed watermarking framework described in below section. The process of embedding and extraction phases are illustrated in Figures 4.1 and 4.2 respectively.

Embedding Algorithm

Input:
   Original host image H
   Secret watermark image W

Output:
   Watermarked image W_{ed}

Algorithm:
1. Decompose the original image into non-overlapping blocks of size 8x8.
2. For each block in the original image:
    a. Apply the DCT method to the block, obtaining the DCT coefficients.
3. Subject the watermark to 3-level DWT (Discrete Wavelet Transform).
4. For each DCT coefficient block obtained in Step 2:
    a. Extract the corresponding wavelet coefficient block from the watermark's 3-level DWT.
    b. Perform adaptive thresholding on the extracted wavelet coefficient block.
    c. Embed the threshold wavelet coefficients into the DCT coefficient block.
5. Combine the modified DCT coefficient blocks to form the watermarked image.
6. Apply the inverse DCT on each block of the watermarked image to get final watermarked image.
7. Output the watermarked image.

Extraction Algorithm:

Input:
   Watermarked image W_{ed}
   Original watermark

Output:
   Extracted watermark: The extracted watermark image

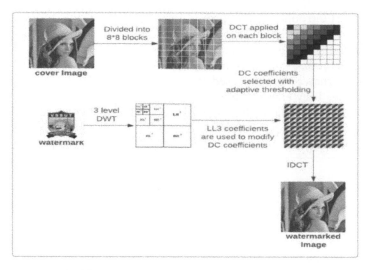

Figure 4.1 Watermark embedding.

Algorithm:
1. The watermarked image is divided into 8x8 size non-overlapping blocks.
2. Per every block in the watermarked image:
   a. Apply the DCT on the block, obtaining the DCT coefficients.
3. Apply inverse DCT along with thresholding technique to each DCT coefficient block:
   a. IDCT is applied to the DCT coefficient block to get the spatial-domain block.
   b. Thresholding technique is applied to the spatial-domain block to extract the cover information.
   c. The extracted cover information is stored as the block.
4. Combine the modified blocks to form the extracted cover image.
5. In order to get back the watermark, IDWT is applied to extracted cover image.
6. Apply the IDWT to the extracted cover image
7. Output the extracted watermark.

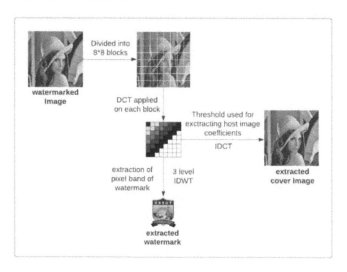

Figure 4.2 Watermark extraction.

Experimental Result

In this study, five grayscale images (Lenna, Pepper, Sailboat, Baboon, House) from the SIPI database were used as input. A 150x150 grayscale logo was selected as the watermark. The watermark embedding and extraction process were carried out using Python, employing OpenCV, NumPy, and Matplotlib libraries for experimental purpose The outputs included watermarked versions of the images, successfully embedding the logo while maintaining visual quality. The scheme demonstrated robustness and imperceptibility, making it suitable for copyright protection and data integrity verification. Figure 4.3 represent the list of Cover images and watermark image taken along with the watermarked image generated and extracted.

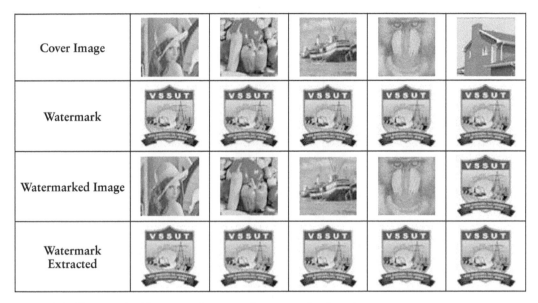

Figure 4.3 Watermarked images and extracted watermark of test images.

Performance of the Model

Tables 4.1, 4.2, and 4.3 display a range of image quality metrics, including Peak Signal to Noise Ratio (PSNR), Structural Similarity Index Measure (SSIM), Normalized Cross-Correlation (NCC), Mean Square Error (MSE), and Bit Error Rate (BER). These metrics were computed by evaluating the provided input image against the resulting watermarked images.

Table 4.1: Quality parameters values (PSNR, SSIM, NCC, MSE and BER) of proposed method.

Image	PSNR	SSIM	NCC	MSE	BER
Lenna	60.771	0.999	0.999	0.054	0.68
Peppers	64.042	0.999	0.999	0.025	0.32
Sailboat	56.428	0.998	0.999	0.147	1.77
Baboon	74.254	0.999	0.999	0.002	0.03
House	53.334	0.999	0.999	0.301	1.75
Average	61.675	0.998	0.999	0.105	0.91

Table 4.2: Presented PSNR and SSIM values, along with a juxtaposition against values from other established papers.

Image Name	Image Size	Lai et al. [12]		Nasrin M. et al. [1]		Ferda Ernawan et al. [2]		Dhani et al. [6]		Suggested Scheme	
		PSNR	SSIM	PSNR	SSIM	PSNR	SSIM	PSNR	SSIM	PSNR	SSIM
Lenna	512×512	48.708	0.992	44.824	0.980	47.176	0.987	45.731	0.994	60.771	0.999
Peppers	512×512	46.613	0.992	43.889	0.981	47.158	0.987	45.953	0.995	64.042	0.999
Sailboat	512×512	45.613	0.986	42.479	0.965	46.918	0.985	43.796	0.990	56.428	0.998
Baboon	512×512	35.596	0.988	43.301	0.986	46.116	0.990	45.682	0.996	74.254	0.999
House	512×512	40.185	0.985	41.944	0.957	47.300	0.985	43.830	0.987	53.334	0.999
Average		43.373	0.988	43.287	0.973	46.934	0.986	44.998	0.992	61.675	0.998

Table 4.3: Comparing the average PSNR values to those of other existing methods.

L.Feng et al. [10]	J. Lang et al. [9]	D.S.Lin et al. [11]	J.Guo et al. [8]	A.Anand et al. [3]	T.H. Hu et al. [7]	proposed
37.72	40.021	36.425	40.321	44.19	40.48	61.765

Results after Attack on Images

Watermark attacks involve manipulating watermarked content to disrupt information transmission or watermark detection. Attackers may use various methods, from simple modifications to advanced cryptographic techniques, to retrieve, alter, or remove watermarked data or insert incorrect data. Types of attacks include basic, cryptographic, legal, geometric, removal, and protocol attacks. These attacks can render existing watermarking techniques ineffective, emphasizing the need for robust methods to withstand them. In our experiment, we tested a watermark attack on Lena and obtained the result of NCC and PSNR values in Table 4.4. The NCC values of different images after going through various attacks are presented in Table 4.5 and graphical representation is illustrated in Figure 4.4.

Table 4.4: NCC and PSNR values of Lenna after being suffering from different attacks.

Attacks	Description	NCC	PSNR
GF-3	Gaussian lowpass filter with kernel: 3×3	0.9956	36.51
GF-5	Gaussian lowpass filter with (kernel: 5×5)	0.9935	35.54
JPEGQ-4	JPEG compression (quality: 40)	0.9956	35.69
JPEGQ-5	JPEG compression (quality: 50)	0.9962	40.77
JPEGQ-9	JPEG compression (quality: 90)	0.9988	40.77
SN-1	Speckle noise (variance:0.001)	0.9460	36.21
SPN-02	Salt & pepper noise (density: 0.002)	0.9913	35.67

Table 4.5: NCC values of different images under various attacks.

IMAGES	GF-3	GF-5	JPEGQ-4	JPEGQ-9	SN-1	SPN-02	MF-3
Lenna	0.995	0.993	0.995	0.998	0.946	0.991	0.996
Pepper	0.995	0.994	0.994	0.998	0.963	0.992	0.994
Sailboat	0.992	0.989	0.996	0.999	0.919	0.992	0.992
Baboon	0.985	0.974	0.990	0.998	0.917	0.988	0.979
House	0.999	0.999	0.999	0.999	0.946	0.994	0.999

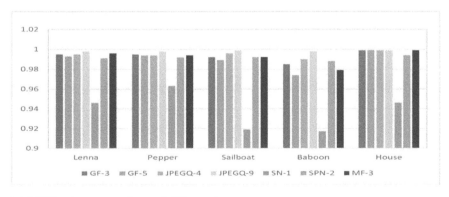

Figure 4.4 NCC value comparison of different images.

Conclusion

Proposed watermarking technique using DWT, DCT, and adaptive thresholding outperforms other methods with higher PSNR, SSIM, and NCC values. It achieves better visual fidelity and numerical accuracy between cover and watermarked images, even under attacks and compressions. The method was evaluated on grayscale test images, showing its robustness and imperceptibility. It effectively ensures digital content ownership and integrity.

References

1. Leng, X. X., Xiao, J., Li, D. Y., and Shen, Z. Y. (2013). Study on the digital image zero-watermarking technology. *Advanced Materials Research, 765*, 1113–1117. (Trans Tech Publications Ltd).
2. Malik, V., Sangwan, N., and Sangwan, S. (2017). Digital watermarking using DWT-SVD algorithm. *Advances in Computational Sciences and Technology*, 10(7), 2161–2171.
3. Anand, A., and Singh, A. K. (2020). An improved DWT-SVD domain watermarking for medical information security. *Computer Communications*, 152, 72–80.
4. Hasan, N., Islam, M. S., Chen, W., Kabir, M. A., and Al-Ahmadi, S. (2021). Encryption based image watermarking algorithm in 2DWT-DCT domains. *Sensors, 21*(16), 5540.
5. Ernawan, F., Ariatmanto, D., and Firdaus, A. (2021). An improved image watermarking by modifying selected DWT-DCT coefficients. *IEEE Access, 9*, 45474–45485.
6. Ariatmanto, D., and Ernawan, F. (2022). Adaptive scaling factors based on the impact of selected DCT coefficients for image watermarking. *Journal of King Saud University-Computer and Information Sciences*, 34(3), 605–614.
7. Hu, H. T., Chang, J. R., and Hsu, L. Y. (2016). Robust blind image watermarking by modulating the mean of partly sign-altered DCT coefficients guided by human visual perception. *AEU-International Journal of Electronics and Communications, 70*(10), 1374–1381.
8. Guo, J., Zheng, P., and Huang, J. (2015). Secure watermarking scheme against watermark attacks in the encrypted domain. *Journal of Visual Communication and Image Representation*, 30, 125–135.
9. Lang, J., and Zhang, Z. G. (2014). Blind digital watermarking method in the fractional Fourier transform domain. *Optics and Lasers in Engineering, 53*, 112–121.
10. Feng, L. P., Zheng, L. B., and Cao, P. (2010). A DWT-DCT based blind watermarking algorithm for copyright protection. In 2010 3rd International Conference on Computer Science and Information Technology, (Vol. 7, pp. 455–458). IEEE.
11. Lin, S. D., Shie, S. C., and Guo, J. Y. (2010). Improving the robustness of DCT-based image watermarking against JPEG compression. *Computer Standards and Interfaces*, 32(1–2), 54–60.
12. Lai, C. C. (2011). An improved SVD-based watermarking scheme using human visual characteristics. *Optics Communications*, 284(4), 938–944.

5 Enhancing Data Security in Cloud Storage: Utilizing Cryptographic Algorithms for Healthcare Industry

Shrabani Sutradhar[1,a], Sunil Karforma[2,b], Rajesh Bose[3,c], and Sandip Roy[3,d]

[1]Assistant Professor, Brainware University, Kolkata, India

[2]Professor and Dean, The University of Burdwan, Burdwan, India

[3]Professor, JIS University, Kolkata, India

Abstract

The data safety and privacy concerns are valid reasons when dealing with sensitive healthcare data and entrusting to third-party cloud providers. In order to address these concerns, various cryptographic algorithms have been developed to secure data in cloud storage frameworks. ElGamal Encryption, Feistel Cipher, and Curve25519 are three examples of cryptographic algorithms that can be utilized to store data in the cloud. These algorithms are commonly used for secure data transmission over networks and for storing data in a non-human-readable format. The proposed system that incorporates these algorithms aims to ensure the secure and efficient transfer, segmentation, encryption, merging, decryption, and recovery of data. By employing these algorithms, the system enhances the security of multi-cloud storage infrastructures.

Keywords: Advanced encryption standard (AES), cloud computing, Cryptographic algorithms, data encryption standard (DES), data security, elGamal encryption, healthcare industry, multi-cloud storage, rivest–shamir–adleman (RSA).

Introduction

The safety and privacy of medical data remain valid concerns. Protecting this data from unauthorized access and ensuring its confidentiality, integrity, and availability are crucial for healthcare organizations. To address these concerns, cryptographic algorithms have been developed to secure data in cloud storage frameworks. This paper proposes a comprehensive system that incorporates three powerful cryptographic algorithms, namely ElGamal Encryption [6], Feistel Cipher [7], and Curve25519 Costello and Smith [4], to enhance data security in cloud storage frameworks Kambatla et al. [8]. These algorithms offer robust protection and efficient processing capabilities, ensuring the secure transfer, segmentation, encryption, merging, decryption, and recovery of data. The use of ElGamal Encryption in the proposed system provides a strong level of data confidentiality. By utilizing public and private key pairs, ElGamal Encryption ensures that only authorized individuals. This algorithm has been widely used for secure data

[a]shrabanidas1989@gmail.com, [b]sunilkarforma@yahoo.com, [c]rajesh.bose@jisuniversity.ac.in, [d]sandip.roy@jisuniversity.ac.in

transmission and storage Sahni et al. [18]. The Feistel Cipher algorithm employed in the system enables efficient data segmentation, encryption, and merging. With multiple rounds of substitution and permutation operations, Feistel Cipher ensures that the data is securely divided into smaller segments, toprevent unauthorized users from data breach Liao et al. [13]. Feistel Cipher has been recognized for its effectiveness in protecting data integrity and confidentiality. The system further incorporates the Curve25519 algorithm for data encryption. Curve25519 is a widely adopted elliptic curve cryptography algorithm known for its strong security and high efficiency Li et al. [14]. By encrypting the data using Curve25519, the system ensures that the stored information is in a non-human-readable format, safeguarding it from unauthorized access Li et al. [15]. To validate the proposed system's effectiveness, a pseudo code representation of the methodology and a detailed analysis of its space and time complexity are provided. The system demonstrates efficient space and time complexity, making it suitable for secure and efficient data storage, retrieval, and processing in cloud storage frameworks.

The proposed system offers several novel features and strengths compared to alternative technologies. It provides robust data security, efficient encryption, segmentation, data recovery mechanisms, user-friendliness, and scalability. Furthermore, potential areas for feature enhancements and future work are discussed, including integration with homomorphic encryption, advanced access control mechanisms, secure data sharing protocols, advanced anomaly detection techniques, and performance optimization.

By incorporating the proposed system into their cloud storage frameworks, healthcare organizations can enhance data security, protect sensitive information, and maintain the trust of patients and healthcare providers. The system provides a comprehensive solution to address the security and privacy aspect, contributing to advancement of secure and efficient healthcare information management.

Data Merging

Data merging is a critical operation that allows fragmented or segmented data to be combined back into its original form. In the context of the proposed system, data merging plays a vital role in reconstructing the original healthcare data after it has been securely segmented and distributed. The Feistel Cipher algorithm employed in the system facilitates efficient data segmentation and encryption. During the data merging process, the system reverses the segmentation process by applying the inverse operations of the Feistel Cipher's substitution and permutation rounds. This ensures that the segmented data pieces are securely combined to reconstruct the original healthcare data.

Data Segmentation

Data segmentation involves dividing large datasets into smaller, more manageable segments. The proposed system leverages the Feistel Cipher algorithm for effective data segmentation. Feistel Cipher employs multiple rounds of substitution and permutation operations to securely divide the data into smaller segments. This segmentation process not only aids in distributing the data across the cloud storage infrastructure but also adds an additional layer of security. Even if an unauthorized entity gains access to one segment, the Feistel Cipher's encryption process makes it exceedingly difficult to decipher the original information without access to the entire data.

Data Recovery

Data recovery is a demanding condition of any secure storage system, as it ensures that data can be retrieved and reconstructed over the event of data loss or corruption. The proposed system incorporates robust data recovery mechanisms to address these concerns. In the context of the ElGamal Encryption and Curve25519 encryption methods, the corresponding decryption processes are used for data recovery. Authorized users with the appropriate private keys can decrypt the encrypted data segments, and the system then applies the reverse Feistel Cipher operations to merge the segments and reconstruct the original data. This recovery process ensures that healthcare organizations can reliably retrieve their sensitive data even in the face of unexpected challenges. The proposed cryptographic system focuseson secure data merging, efficient segmentation, and reliable data recovery. By combining the strengths of ElGamal Encryption, Feistel Cipher, and Curve25519, we get a comprehensive solution to resolve the safety concerns for keeping healthcare data in cloud storage frameworks. This holistic approach contributes to the advancement of secure and efficient healthcare information management, fostering trust among patients, healthcare providers, and organizations utilizing cloud-based.

In the next section reviews the works and the section-3 contain the Systems architecture using proposed methods. Section-4 depicts the result analysis. Secttion-5 discusses the conclusion and future work.

Literature Review

Several studies have focused on addressing data security concerns in cloud storage frameworks, particularly in the healthcare industry. In a paper Zhang and Zhang [23] Authors proposes a cloud storage security model that combines ABE (Attribute-based encryption) and Blockchain technology. The model ensures secure gateway over healthcare data storage in the cloud, protecting data confidentiality and integrity. Khan et al. [10] suggest a hybrid cryptographic solution that combines symmetric and asymmetric encryption algorithms for safe data transfer in cloud computing environment. The proposed technique enhances data confidentiality and integrity, ensuring secure data transfer between healthcare systems and cloud storage. This research focuses on secure multi-keyword ranked search schemes in cloud [21]. The suggested system ensures data privacy and efficient keyword-based search over encrypted healthcare data stored in the cloud. It incorporates cryptographic techniques to protect sensitive information and enable secure search operations. This paper proposes a secure and efficient data sharing framework and a cryptographic approach for cloud-assisted healthcare systems Kumar et al. [12, 11]. The framework utilizes cryptographic algorithms, including homomorphic encryption and ABE (Attribute-based encryption), to facilitate secure data sharing while preserving data privacy and integrity. The proposed approach incorporates a combination of cryptographic algorithms, including hybrid encryption and elliptic curve cryptography, to protect sensitive healthcare data and enable secure sharing among authorized entities. Author R. Kumar and S. Kumar proposed a secure cloud storage framework for healthcare data using a combination of hybrid encryption and attribute-based access control (ABAC) mechanisms Sutradhar et al. [20]. The system utilized symmetric and asymmetric encryption algorithms along with access policies based on user attributes to secure data confidentiality and fine-grained access control. In the study, the authors discussed the reliable storage environment for healthcare data incorporating homomorphic encryption and Blockchain technology Zkik et al. [22]. Chatterjee et al. [3] provide a multi-cloud environment while maintaining data integrity and traceability. A research article by A. Patel and M. J. Patil presented a cloud storage framework for secure healthcare data management, employing a combination of encryption, data deduplication, and secure access control mechanisms Sahana et al. [17]. The system utilized a hybrid encryption approach, combining symmetric and asymmetric encryption algorithms, to ensure data confidentiality, integrity, and availability. X. Wang et al. proposed a secure and efficient data storage framework for cloud-assisted telehealth systems, utilizing advanced encryption and access control techniques. Shen et al. [19] introduced a novel cloud storage system for healthcare data based on combination different techniques. Shifting focus, Alzaidi et al. [1] explored MANET network security against wormhole attacks, while Kavitha et al. [9] proposed a two-hop mechanism for indirect trust computation. Meanwhile, Menon et al. [16] harnessed LDA and ASV-RF with PSO for intelligent data classification in healthcare diagnoses.

Methodology

The primary objective of this proposed system is to bolster the security of healthcare data within cloud storage frameworks. To achieve this, a trio of cryptographic algorithms—ElGamal Encryption, Feistel Cipher, and Curve25519—are integrated into the system's architecture. This integration serves to safeguard the sentient medical data to reserve in the cloud environment, thus reinforcing the healthcare system's data security [2]. The ensuing methodology delineates the stepwise process encompassing data transfer, segmentation, encryption, merging, decryption, and recovery, as depicted in Figure 5.1.

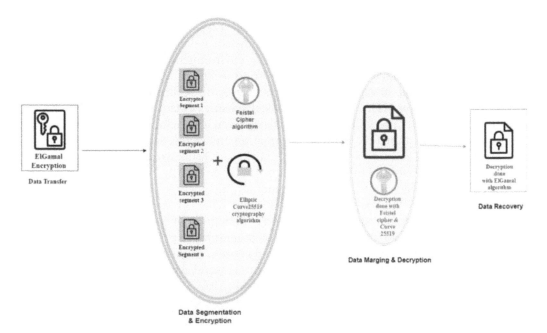

Figure 5.1 Stepwise encryption procedure.

Secure Data Transfer

Facilitating the secure movement of healthcare data from local environments to the cloud storage framework constitutes the initial phase. Prior to transmission to the cloud, the data undergoes local encryption using the ElGamal Encryption algorithm. This algorithm, utilizing public and private key pairs for encryption and decryption Dutta et al. [5], ensures the data's confidentiality.

Data Segmentation

In order to enable efficient storage and processing within the cloud, extensive healthcare datasets are partitioned into smaller, manageable segments. The Feistel Cipher algorithm is then applied to each segment to enhance its security. Leveraging multiple rounds of substitution and permutation operations, the Feistel Cipher renders unauthorized decryption arduous, thereby fortifying the security of healthcare data.

Data Encryption

Upon data retrieval necessity, the system retrieves encrypted data segments from the cloud storage framework. Subsequently, these encrypted segments are decrypted using the Curve25519 algorithm. The ensuing decrypted segments are amalgamated using the Feistel Cipher algorithm, culminating in the reconstruction of the original healthcare data.

Data Decryption

The concluding phase encompasses the application of the ElGamal Encryption algorithm to decrypt the reconstructed healthcare data. The private key, aligned with the public key employed during the data transfer encryption, is instrumental in this decryption process. Authorized personnel can thus access the original healthcare data in a comprehensible.

Data Recovery

To address contingencies such as data loss or corruption, robust data recovery mechanisms are integrated into the system. Backup copies of encrypted healthcare data segments are maintained to facilitate recovery processes. Employing decryption and merging procedures, the system can restore lost or compromised healthcare data effectively. By fusing the ElGamal Encryption, Feistel Cipher, and Curve25519 algorithms into the system, the proposed methodology fortifies the secure transfer, segmentation, encryption, merging, decryption, and recovery of healthcare data within the cloud. Through these algorithmic safeguards, the multi-cloud storage infrastructure's security is substantially elevated, curtailing unauthorized access and safeguarding the sanctity of sensitive healthcare information.

Pseudocode Representation of the Methodology

Here's a pseudocode representation of the methodology described:

// Step 1: Data Transfer

```
encryptedData = ElGamalEncrypt(data, publicKey)
sendToCloud(encryptedData)
```

// Step 2: Data Segmentation

```
segments = segmentData(data)
encryptedSegments = []
for segment in segments:
encryptedSegment = FeistelCipherEncrypt(segment)
encryptedSegments.append(encryptedSegment)
```

// Step 3: Data Encryption

```
for segment in encryptedSegments:
encryptedSegment = Curve25519Encrypt(segment)
storeInCloud(encryptedSegment)
```

// Step 4: Data Merging

```
encryptedSegments = retrieveFromCloud()
decryptedSegments = []
for segment in encryptedSegments:
decryptedSegment = Curve25519Decrypt(segment)
decryptedSegments.append(decryptedSegment)
reconstructedData = FeistelCipherDecrypt(decryptedSegments)
```

// Step 5: Data Decryption

```
decryptedData = ElGamalDecrypt(reconstructedData, privateKey)
```

// Step 6: Data Recovery

```
backupSegments = retrieveBackupFromCloud()
restoredSegments = []
for segment in backupSegments:
decryptedSegment = Curve25519Decrypt(segment)
restoredSegments.append(decryptedSegment)
restoredData = FeistelCipherDecrypt(restoredSegments)
```

Result Analysis

The proposed system incorporates three cryptographic algorithms (ElGamal Encryption [6]), Feistel Cipher [7], and Curve25519 Costello and Smith [4] to enhance data security in cloud storage frameworks. The system ensures the secure transfer, segmentation, encryption, merging, decryption, and recovery of medical data. Through the utilization of these algorithms, the system ensures the privacy, accuracy, and accessibility of valuable information, creating difficulties for unauthorized users attempting to gain entry and decipher the data.

Simulation Setup

Creating a detailed simulation setup involves configuring hardware and software components. On the hardware side, utilize a powerful server for virtualization and allocate ample resources. Set up client machines for data transfer and processing. In the software setup, install a compatible OS, like Ubuntu, and use virtualization/containerization platforms for flexibility. Employ Python, Java, or C++ for programming and integrate suitable cryptography libraries. Simulate cloud storage using solutions like NFS or HDFS, and establish network connections with tools like GNS3. Generate synthetic healthcare data and develop scripts mimicking the proposed system's methodology. Monitor with performance tools and visualize data for analysis. In simulation execution, start instances, transfer and encrypt data, segment and encrypt further, retrieve and decrypt, simulate loss and recovery, and analyze performance metrics. This comprehensive setup enables testing and refining the proposed system's functionality.

Complexity Analysis of Space & Time

Intricacy of the recommended system can be analyzed based on the cryptographic algorithms with Proposed System (PS) with Technology A (TA) and Technology B (TB) and the operations performed in Table 5.1.

Table 5.1: Space and time complexity analysis.

Operations	PS	AES and RSA	DES and Blowfish
Data Transfer	Const. Time	Const. Time	Const. Time
Data Segmentation	LIN. Time	LIN. Time	LIN. Time
Data Encryption	LIN. Time	LIN. Time	LIN. Time
Data Merging	LIN. Time	LIN. Time	LIN. Time
Data Decryption	LIN. Time	LIN. Time	LIN. Time
Data Recovery	LIN. Time	LIN. Time	N/A

(*n represents the size of the data*) (*Constant Time (Const. Time)-O(1)*) (*Linear Time (LIN. Time)-O(n)*)

The proposed system exhibits efficient space and time complexity for the various operations involved. The complexities are generally linear (LIN) time (O(n)), here n indicates the volume of the processed data. A in terms of complexity. However, Technology B may have higher time complexity due to the use of older algorithms.

The use of efficient algorithms like Curve25519 in the proposed system contributes to its high efficiency. Curve25519 offers fast encryption and decryption operations, resulting in reduced processing time. The Feistel Cipher algorithm used for segmentation and merging also has a moderate time complexity, making it suitable for large datasets.

In terms of space complexity, the proposed system requires additional storage for encrypted data segments and backup copies. However, the space complexity is generally manageable and proportional to the size of the data.

Novelty and Strength of the Proposed System

The proposed system incorporates a combination of three cryptographic algorithms (ElGamal Encryption, Feistel Cipher, and Curve25519) to enhance data security in cloud storage frameworks. It provides several novel features and demonstrates strength in terms of security, efficiency, segmentation, data recovery, usability, and scalability. The following (Table 5.2) summarizes the novelty and strength of the proposed system compared to alternative.

Overall, the proposed system demonstrates several advantages over the alternative technologies in terms of data security, encryption strength, efficiency, segmentation, data recovery, usability, and scalability.

It incorporates a combination of strong cryptographic algorithms to ensure robust protection of sensitive healthcare data in the cloud.

Table 5.2: Novelty and strength of the proposed system compared to alternative technologies.

Features	Proposed System	AES and RSA	DES and Blowfish
Data Security	Strong	Moderate	Weak
Encryption	ElGamal, Feistel, Curve25519	AES, RSA	DES, Blowfish
Algorithm Efficiency	High	Moderate	Low
Segmentation	Yes	Yes	No
Data Recovery	Yes	Yes	No
Usability	User-friendly	Complex	Moderate
Scalability	Highly scalable	Moderate	Limited

Conclusion

In conclusion, the proposed system introduces a holistic and tailored approach to bolster data security within cloud storage infrastructures, with a specific focus on the healthcare sector. Through the integration of advanced cryptographic measures such as ElGamal Encryption, Feistel Cipher, and Curve25519 algorithms, the system establishes a robust framework that guarantees the utmost confidentiality, integrity, and accessibility of critical healthcare information stored in the cloud. This innovation enables not only secure data transfer, segmentation, and encryption but also seamless merging, decryption, and recovery processes. By addressing these multifaceted aspects of data protection, the system emerges as a potent solution to empower healthcare institutions in safeguarding their sensitive data assets effectively. With its comprehensive suite of security functionalities, the system not only aligns with the unique demands of the healthcare industry but also paves the way for a secure and efficient future in cloud-based healthcare data management. In an era where data breaches pose significant threats, this system stands as a cornerstone for organizations aiming to fortify their data security measures and ensure the continued trust of patients and stakeholders alike.

Future Work

In future developments, the integration of homomorphic encryption hold the potential to significantly enhance privacy and security by enabling execution on secure healthcare. Additionally, implementing secure data sharing protocols or secure multiparty computation can facilitate safe collaboration among healthcare organizations, preserving data privacy during sharing. Incorporating state-of-the-art anomaly detection techniques, such as machine learning-based algorithms, will bolster the system's capability to identify and counteract security threats. Continual performance optimization remains crucial, involving the reduction of computational overhead and enhancement of scalability for improved efficiency. By pursuing these avenues of development, the proposed healthcare data security system can adapt to evolving challenges in cloud storage frameworks, ensuring the safeguarding of sensitive patient information and maintaining the trust of both patients and healthcare providers.

References

1. Alzaidi, M. S., Shukla, P. K., Sangeetha, V., Pandagre, K. N., Minchula, V. K., Sharma, S., and Prashanth, V. (2023). Applying machine learning enabled myriad fragment empirical modes in 5G communications to detect profile injection attacks. *Wireless Networks*, 2023, 1–14.
2. Chatterjee, P., Bose, R., Banerjee, S., and Roy, S. (2022). Enhancing security of cloud based LMS by deploying secure Loopback Protocol. *International Journal of Mechanical Engineering*, 7(1), 1474–1481.
3. Chatterjee, P., Bose, R., and Roy, S. (2020). A review on architecture of secured cloud based learning management system. *Journal of Xidian University*, 14(7), 365–376.
4. Costello, C., and Smith, B. (2018). Montgomery curves and their arithmetic: The case of large characteristic fields. *Journal of Cryptographic Engineering*, 8, 227–240.
5. Dutta, A., Bose, R., Kumar Chakraborty, S., and Roy, S. (2022). A security provocation in cloud-based computing. In Gupta, D., Goswami, R. S., Banerjee, S., Tanveer, M., and Pachori, R. B., eds. Pattern Recognition and Data Analysis with Applications. Lecture Notes in Electrical Engineering (Vol. 888, pp. 343–355). Springer Nature Singapore
6. ElGamal, T. (1985). A public key cryptosystem and a signature scheme based on discrete logarithms. *IEEE Transactions on Information Theory*, 31(4), 469–472.
7. Feistel, H. (1973). Cryptography and computer privacy. *Scientific American*, 228(5), 15–23.
8. Kambatla, K., Gharachorloo, K., and Barford, P. (2013). The emergence of cloud-based federated electronic health record systems. *IEEE Cloud Computing*, 1(2), 20–27.
9. Kavitha, A., Reddy, V. B., Singh, N., Gunjan, V. K., Lakshmanna, K., Khan, A. A., and Wechtaisong, C. (2022). Security in IoT mesh networks based on trust similarity. *IEEE Access*, 10, 121712–121724.
10. Khan, M. K., Hussain, M., Alazab, M., and Asim, M. (2021). Hybrid cryptographic technique for secure data transmission in cloud computing environment. *Journal of Ambient Intelligence and Humanized Computing*, 12(10), 13985–14000.
11. Kumar, R., Kumar, P., Tripathi, R., Gupta, G. P., Islam, A. N., and Shorfuzzaman, M. (2022). Permissioned blockchain and deep learning for secure and efficient data sharing in industrial healthcare systems. *IEEE Transactions on Industrial Informatics*, 18(11), 8065–8073.
12. Kumar, V., Chaki, N., and Verma, A. (2021). A cryptographic approach for secure healthcare data sharing in the cloud. *Future Generation Computer Systems*, 120, 160–173.
13. Liao, J., Cao, Z., Dong, X., and Chen, G. (2018). A secure data sharing framework for collaborative healthcare systems using attribute-based encryption and blockchain technology. *IEEE Access*, 6, 76608–76616.
14. Li, X., Zheng, G., Zomaya, A. Y., and Zhou, B. B. (2019). Privacy-preserving data sharing in cloud-assisted wearable edge healthcare systems. *IEEE Transactions on Industrial Informatics*, 15(3), 1753–1763.
15. Li, X., Shen, J., Zhang, Q., and Lu, L. (2020). A novel secure multiparty computation framework for privacy-preserving collaborative anomaly detection in IoT-enabled healthcare systems. *IEEE Transactions on Industrial Informatics*, 16(4), 2696–2706.

16. Menon, S. P., Shukla, P. K., Sethi, P., Alasiry, A., Marzougui, M., Alouane, M. T. H., and Khan, A. A. (2023). An intelligent diabetic patient tracking system based on machine learning for E-health applications. *Sensors*, 23(6), 3004.

17. Sahana, S., Bose, R., and Sarddar, D. (2016). Harnessing RAID mechanism for enhancement of data storage and security on cloud. *Brazilian Journal of Science and Technology*, 3, 1–13.

18. Sahni, A., Patel, V. M., Hsu, C. H., and Ma, Y. (2017). Secure cloud computing for healthcare data via homomorphic encryption. *ACM Transactions on Internet Technology*, 18(3), 1–21.

19. Shen, J., Deng, X., and Xu, Z. (2019). Multi-security-level cloud storage system based on improved proxy re-encryption. *EURASIP Journal on Wireless Communications and Networking*, (1), 1–12.

20. Sutradhar, S., Karforma, S., Bose, R., and Roy, S. (2023). A dynamic step-wise tiny encryption algorithm with fruit fly optimization for quality of service improvement in healthcare. *Healthcare Analytics*, 3, 1–15.

21. Yu, Y., and Zhang, Y. (2021). A secure multi-keyword ranked search scheme in cloud computing. *IEEE Access*, 9, 49861–49872.

22. Zkik, K., Orhanou, G., and El Hajji, S. (2017). Secure mobile multi cloud architecture for authentication and data storage. *International Journal of Cloud Applications and Computing (IJCAC)*, 7(2), 62–76.

23. Zhang, X., and Zhang, S. (2022). An efficient cloud storage security model based on attribute encryption and blockchain. *IEEE Access*, 10, 19934–19947.

6 Blockchain as a Security Solution in the Internet of Vehicle: A Survey

Sayan Sen Sarma[a]

Assistant Professor, Mahishadal Girls' College, West Bengal, India

Abstract

For the successful integration between automobile and IoV, vehicular ad hoc network (VANET) plays a major role. In this era of smart cities and smart communications, intelligent transport system (ITS) is being used to provide different interactive services in VANET. This may include various applications like safety concerns of the drivers or the passengers, road guidance, driving assistance, infotainment etc. This continuous advancements in different areas of VANET provides an advantageous and comfortable user experience, however, it also exposes the network to different security challenges. Considering the relevance of the field we first study the different characteristics and security issues of VANET. The next section contains a survey on blockchain based countermeasures of the addressed security issues. Blockchain is a popular distributed ledger and is very well suited to address the challenges in IoVs. This paper provides a details study about the application of blockchain to fill the existing security gaps in IoV.

Keywords: Block chain, IoV, security, VANET.

Introduction

Internet of vehicles (IoV) is a network containing entities like sensor equipped vehicles, software interface and other technologies to provide data connectivity and data exchanging facility according to the existing protocols over the VANET [2]. ITS is being used to provide different interactive services with a goal of improving the performance matrices of VANET [1]. IoV contains vehicle information collecting units called on-board units (OBUs) and those collected data are usually communicated using short range communication protocols like LTE-V or dedicated short-range communication (DSRC) [4]. The data collected through OBUs are usually real time data which helps to take actions like broadcasting event triggered messages, dynamic route planning etc. Data communication in IoV takes part in two phases [2]. Phase one is V2V communication which stands for vehicle-to-vehicle communication and phase two is V2I communication which stands for vehicle-to-infrastructure communication. Working procedure of both V2V and V2I communication phases in conventional IoV infrastructure are shown in Figure 6.1 [3].

[a]sayansensarma@gmail.com

Both the communication phases use OBUs to collect vehicle information and follow the dedicated short-range communication to communicate the collected data (such as signalling information, information about accidents, weather/traffic status etc.) in real-time. This architecture significantly helps to improve the QoS of the network. Unfortunately it also invites some security challenges. For example, in a real life scenario with heterogeneous traffic generated during the process of information collection, malicious vehicles can always convey false information or tamper the shared data with false identity and it may cause several security threats. Sybil attack, DoS attack, impersonation attack, replay attack etc. are the results of such security challenges. Considering the relevance of the field first we study the different characteristics and security issues of VANET. The next section contains a survey on blockchain based countermeasures of the addressed security issues. Blockchain is a popular distributed ledger and is very well suited to address the challenges in IoVs. This paper provides a details study about the application of blockchain to fill the existing security gaps in IoV.

Background

OBU, Road Side unit (RSU), and trusted applications (TA) are the three main constructors of The VANETs architecture and it contains two types of communication technologies as V2V and V2I communication as shown in Figure 6.1. In V2V communication contact vehicles confer with each other and interchange the traffic-linked data particulars within the wireless network range [4]. Whenever any mishap like accidents or congestion occurs, immediately a vehicle within such network alerts other vehicles within the same network to avoid that faulty road or area, by sending an alert signal. RSUs are installed in the side of the road to help the vehicles to share information through V2I communication. With the help of the shared information drivers get notified about the different network conditions in advance.

Prior to the communication, RSUs and OBUs are registered by a third party to keep themselves in trusted list. The RSUs usually position itself on the road and they are responsible for the authentication of the communication between TA and OBU [6]. OBUs are fitted in the vehicles to perform communications with RSUs through DSRC communication mode. In a dynamic adhoc network with the help of OBUs, the vehicles converse with each other. For both fixed and mobile nodes contact less services has been presented with a narrow ingress to the network's infrastructure. In VANET most of the existing nodes are with high mobility features and therefore the network topology varies frequently [7]. This makes the VANET topology more complex and dynamic in nature. The main properties of VANET are **high mobility, driver protection, vibrant network topology, variable network density, no power limits, restriction of transmission power, network strength, extensive scale, extensive computational processing** etc.

Security Issues in VANET

Along with providing better user experience safe and security is one of the primary concerns of VANET. It is vital to fight against different attacks and to provide a secure service to the end users. A details study about different security issues of VANET has been provided in Table 6.1 [5, 11].

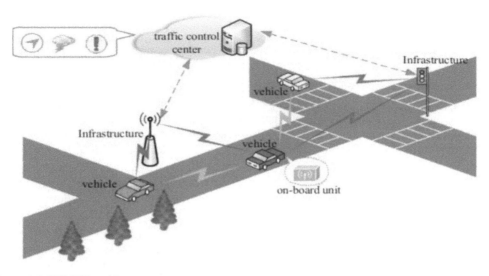

Figure 6.1 VANET architecture.

Table 6.1: Different security issues in VANET.

Security Issue	Requirement
Availability	Heterogeneous highly scalable nature of the network with mobile nodes make it easy to get exposed under denial of service attack-based attempts (Arif et al., 2019).
Authentication	Restricts intruders to send message (Zeadally et al., 2012).
Integrity	Guarantees that no alteration has been done with the date in the middle of the path between the sender and the receiver (Afzal et al., 2020).
Confidentiality	Prevents data from being affected by adversaries (Seikh et al., 2020).
Nonrepudiation	Ensures that the source of the originating message should not deny the fact that it has generated a specific message (Seikh et al., 2020).
Pseudonymity	Helps the participants to hide their original identity (Afzal et al., 2020).
Privacy	Prevents the leak of sensitive Information about the identity and location of the driver (Zeadally et al., 2012).
Scalability	Increase ability of the network to adopt with the dynamically changing requirements and response accordingly (Awang et al., 2017).
Mobility	Communicating nodes changes their position in every second dynamically to make mobility omnipresent (Zeadally et al., 2010).
Data Verification	Gets rid of malicious messages present in the network (Afzal et al., 2020).
Access Control	Preserve the policy rights and to maintain the different roles of all stakeholders of the network (Awang et al., 2017).
Key Management	Required for the encryption and decryption process during the secured communication between the nodes (Chaudhry et al., 2020) [9].
Location Verification	As most of the nodes are mobile hence authentication of location is one of the main requirements (Sari et al., 2015).

Introduction to Blockchain

Blockchain stands to be a suitable tool to resolve the vulnerabilities in the areas of security, privacy and trust management in heterogeneous IoV applications [12]. The main features (Kchao et al., 2019) which makes it eligible to integrate with an IoV applications are, (**1**) **Decentralization:** Every node is independent, (**2**) **Transparency:** The operation procedure of the entire system is kept transparent, (**3**) **Collective maintenance:** Maintenance system of the system has been ran by the equal participation of each node, (**4**) **Reliable database:** To increase the reliability all the nodes stores the same copy of the blockchain ledger and (**5**) **Automation:** Data/resource sharing is done automatically though the use of smart contracts.

A block header Singhand Kim [13] generally contains metadata like, a hash value of point the previous block, timestamp, level of difficulty of date mining, random number (Nonce), root of the Merkle tree etc. To implement the smart contract based services and applications platforms like, Hyperledger, Tron and Ethereum etc., smart contract programmability has been widely used [12]. The mechanism of network access control Li et al. [10] divides the blockchain into three types as: (**1**) **Public blockchain:** This is designed for everyone to access and modify the chain data, (**2**) **Private blockchain:** In this case, there will be limited nodes and every node will not be able to associate with the blockchain and (**3**) **Consortium blockchain:** In this partially decentralized blockchain some nodes are granted with the advance authorization to participate in a chain.

Blockchain Based Implementation of IoV Security

In this section we focus to analyse the integration process of blockchain in different IoVs applications. Table 6.2 classifies and illustrates multiple proposed blockchain applications of recent times. Through the survey of literature, a detail study on differen application areas are given below.

Access Control

With the swift development of the IoV, several vehicular information systems have been created. The increasing use of IoV applications has affected the transportation services drastically [8]. Blockchain applications are by nature decentralized and scalable. Hence, the methods proposed by Sharma and Chakraborty [12] are very much applicable for futuristic IoV applications. It implements the authorized access control by uploading the validated transactions. A three step authentication technique has also been proposed [29]. With the implementation of smart contracts this blockchain based method preserves the security and privacy.

Message Validation

In literature we can find that most of the IoV applications are based on the sharing of information. Different applications like, routing, signalling, weather prediction, accidental information sharing etc. are all based on the shared information which are available in the network [14]. If we do not incorporate a message validation technique then it is always possible that a malicious user can attack the system with forged messages. Authors in (Abassi et al., 2018) have proposed some blockchain based methods for message validation.

Trust Management

In a heterogeneous IoV environment trust is one of the major security concern among the communicating devises. If we take decision based on the assumption that a third party communicator can always be trusted then it can create threat and make the system vulnerable. Authors in [17], proposed a trust-less architecture based on blockchain. Here, each vehicle keeps track of each other's trust value and depending on it a transaction is validated [19].

Table 6.2: Application of blockchain in different security aspects of IoV.

IoV Security	Access Control	(Sharma et al., 2018)	BlockAPP
		(Singh et al., 2018)	Intelligent vehicle trust point (IVTP)
	Message Validation	(Kchaou et al., 2018)	Trust clustering mechanism for VANET (TCMV)
		(Abassi et al., 2018)	Distribute trust clustering mechanism for VANET (DTCMV)
		(Wang et al., 2021)	Blockchain-based traffic event validation (BTEV) framework
Trust Management		(Yang et al., 2019)	Vehicular announcement protocol echo-announcement
		(Li et al., 2018)	Anonymous cloaking region construction scheme
		(Luo et al., 2020)	Blockchain-based trust management with BTCPS scheme
		(Liu et al., 2020)	Blockchain-based privacy preserving authentication (BPPA) scheme
Certificate Management		(Lu et al., 2019)	Decentralized key management mechanism (DBKMM)
		(Ma et al., 2020)	A blockchain-based anonymous reputation system (BARS)
		(Lu et al., 2018)	Semicentralized traffic signal control (SCTSC) mode
Data Management		(Cheng et al., 2019)	Miner selection and block verification solutions
		(kang et al., 2019)	Mobile crowd sensing (MCS) with blockchain
		(Yin et al., 2020)	A DQDA incentive mechanism
Privacy Preserving		(Lu et al., 2020)	A hybrid blockchain-PermiDAG
		(Feng et al., 2020)	Blockchain-based software-defined VANET (block- SDV) framework

Certificate Management

In classical methods used for IoV applications all the transactions are usually kept anonymous through an encryption process based on a public key certificate [18]. The success and failure of such systems depends on the single point which is the circulation and revocation of the certificate. Li et al. [10] proposed some blockchain based methods to increase the system reliability by addressing this problem of single point failure.

Data Management

In traditional IoV based data management systems trust and integrity can anytime be compromised in a heterogeneous environment. In Cheng et al. [23], authors have proposed some blockchain based data management techniques which trust and integrity of the system in heterogeneous environment. Management of data has been done in two phases Kang et al. [22] namely, (1) on-chain data management which deals with preservation and distribution of the data and (2) off-chain data which deals with query processing.

Privacy Preserving

As the intelligent transport systems (ITS) is getting developed, vehicle information is generated in large numbers and with greater accuracy. For example, the driving experience can be improved by camera data, which can record vehicle accidents. However, the preservation of privacy is a vital issue that needs to be addressed. In Luo et al. [16], blockchain based methods has been proposed to resolve the conflict between the availability and privacy preservation of data [20].

Conclusion and Future Direction

In recent times rapid development has been observed in the field of IoV and its applications in automobile industries. For the successful integration between automobile and IoV, VANET plays a major role. In this era of smart cities and smart communications, intelligent transport system (ITS) is being used to provide different interactive services in VANET. This continuous advancements in different areas of VANET provides an advantageous and comfortable user experience, however, it also exposes the network to different security challenges. Sybil attack, DoS attack, impersonation attack, replay attack etc. are the results of such security challenges. Considering the relevance of the field we have studied the different characteristics and security issues of VANET. Next, a survey has been done on block chain based countermeasures of the addressed security issues. Block chain is a popular distributed ledger and is very well suited to address the challenges in IoVs. This paper has covered a detail study about the application of blockchain to fill the existing security gaps in IoV.

It is assumed that in recent future every vehicle will be using some sorts of IoV applications. In such scenarios, blockchain based distributed IoV applications are more suitable than the centralized IoV applications by nature. Blockchain is also supposed to maintain the core information system against a very low cost. Although, we have shown an extensive survey to explain the applications like security, privacy, trust and certificate management through various IoV applications, still there are many unaddressed aspects. Hance, in recent future there is a large scope of addressing many open areas for blockchain based IoV applications.

References

1. Samaras, N. S. (2016). Using basic MANET routing algorithms for data dissemination in vehicular ad hoc networks (VANETs). In 24th Telecommunications Forum (TELFOR), Belgrade, Serbia. (pp. 1–4). doi: 10.1109/TELFOR.2016.7818741.

2. Wantoro, J., and Mustika, I. W. (2014). M-AODV+: an extension of AODV+ routing protocol for supporting vehicle-to-vehicle communication in vehicular ad hoc networks. In IEEE International Conference on Communication, Networks and Satellite (COMNETSAT), Jakarta, Indonesia. (pp. 39–44). doi:10.1109/COMNETSAT.2014.

3. Pathak, C., Shrivastava, A., and Jain, A. (2016). Ad-hoc on demand distance vector routing protocol using Dijkastra's algorithm (AODV-D) for high throughput in VANET (Vehicular Ad-hoc network). In 11th International Conference on Industrial and Information Systems (ICIIS), Roorkee, India, (pp. 355–359). doi: 10.1109/ICIINFS.2016.8262965.

4. Rasool, U., Zikria, Y. B., and Kim, S. W. (2017). A review of wireless access vehicular environment multichannel operational medium access control protocols: Quality-of-service analysis and other related issues. *International Journal of Distributed Sensor Networks*, 13, 1–22. doi:155014771771017.10.1177/1550147717-710174.

5. Arif, M., Wang, G., Bhuiyan, M. Z. A., Wang, T., and Chen, J. (2019). A survey on security attacks in VANETs: communication, applications and challenges. *Vehicular Communications*, 19, 1–36. ISSN 2214-2096.

6. Zeadally, S., Hunt, R., Chen, Y. S., Irwin, A., and Hassan, A. (2010). Vehicular adhoc networks (VANETs): status, results, and challenges. *Telecommunication Systems*, 50, 1–25. doi: 10.1007/s11235-010-9400-5.

7. Sheikh, M. S., Liang, J., and Wang, W. (2020). Security and privacy in vehicular ad hoc network and vehicle cloud computing: a survey. *Wireless Communications and Mobile Computing*, 2020, 1–25. (Article ID 5129620) doi: 10.1155/2020/5129620.

8. Awang, A., Husain, A., Kamel, K., and Aissa, S. (2017). Routing in vehicular adhoc networks: a survey on single and cross-layer design techniques, and perspectives. *IEEE Access*, 5, 59497–59517. doi:10.1109/ACCESS.2017.2692240.

9. Chaudhry, S. A. (2020). PALK: password-based anonymous lightweight key Agreement framework for smart grid. *International Journal of Electrical Power and Energy System*, 121, 106121. ISSN 0142-0615, https://doi.org/10.1016/j.ijepes.2020.

10. Li, X., Han, Y., Gao, J., and Niu, J. (2019). Secure hierarchical authentication protocol in vanet. *IET Information Security*, 4(1), 99–110.

11. Sari, A., Onursal, O., and Akkaya, M. (2015). Review of the security issues in vehicular ad hoc networks (VANET). *International Journal of Communications, Network and System Sciences*, 8, 552–566. doi: 10.4236/ijcns.2015.813050.

12. Sharma, R., and Chakraborty, S. (2018). BlockAPP: using blockchain for authentication and privacy preservation in IoV. In IEEE Globecom Workshops (GC Wkshps), Abu Dhabi, United Arab Emirates, (pp. 1–6). doi: 10.1109/GLOCOMW.20-18.8644389.

13. Singh, M., and Kim, S. (2018). Branch based blockchain technology in intelligent vehicle. *Computer Networks*, 145, 219–231. doi: 10.1016/j.comnet.2018.08.016.

14. Kchaou, A., Abassi, R., and Guemara, S. (2018). Toward a distributed trust management scheme for VANET. In Proceedings of the 13th International Conference on Availability, Reliability and Security (ARES 2018). Association for Computing Machinery, New York, NY, USA, (Vol. 53, pp. 1–6). doi: https://doi.org/10.1145/3230833.32-32824.

15. Li, L., Liu, J., Cheng, L., Qiu, S., Wang, W., Zhang, X., and Zhang, Z. (2017). CreditCoin: a privacy-preserving blockchain-based incentive announcement network for communications of smart vehicles. *IEEE Transactions on Intelligent Transportation Systems*, 19(7), 2204–2220. doi: 10.1109/TITS.2017.27.

16. Luo, B., Li, X., Weng, J., Guo, J., and Ma, J. (2020). Blockchain enabled trust-based location privacy protection scheme in VANET. *IEEE Transactions on Vehicular Technology*, 69(2), 2034–2048. doi: 10.1109/TVT.2019.2957744.

17. Liu, X., Huang, H., Xiao, F., and Ma, Z. (2020). A blockchain-based trust management with conditional privacy-preserving announcement scheme for VANETs. *IEEE Internet of Things Journal*, 7(5), 4101–4112. doi:10.1109/JIOT.2019-.2957421.

18. Lu, Z., Wang, Q., Qu, G., Zhang, H., and Liu, Z. (2019). A blockchain based privacy preserving authentication scheme for VANETs. *IEEE Transactions on Very Large Scale Integration (VLSI) Systems*, 27(12), 2792–2801. doi: 10.1109/TVLSI.2019.2-92-9420.

19. Ma, Z., Zhang, J., Guo, Y., Liu, Y., Liu, X., and He, W. (2020). An efficient decentralized key management mechanism for VANET with blockchain. *IEEE Transactions on Vehicular Technology*, 69(6), 5836–5849. doi: 10.1109/TVT.2020.2-972923.

20. Lu, Z., Liu, W., Wang, Q., Qu, G., and Liu, Z. (2018). A privacy-preserving trust model based on blockchain for VANETs. *IEEE Access*, 6, 45655–45664. doi: 10.1109/ACCESS.2018.2864189.

21. Cheng, L., Liu, J., Xu, G., Zhang, Z., and Wang, W. (2019). SCTSC: a semicentralized traffic signal control mode with attribute-based blockchain in IoVs. *IEEE Transactions on Computational Social Systems*, 6(6), 1373–1385. doi:10.1109/TCSS.2019.2904633.

22. Kang, J., Xiong, Z., Niyato, D., Ye, D., Kim, D. I., and Zhao, J. (2019). Toward secure blockchain-enabled internet of vehicles: optimizing consensus management using reputation and contract theory. *IEEE Transactions on Vehicular Technology*, 68(3), 2906–2920. doi: 10.1109/TVT.2019.2894944.

23. Chen, C., Wu, J., Lin, H., Chen, W., and Zheng, Z. (2019). A secure and efficient blockchain-based data trading approach for internet of vehicles. *IEEE Transactions on Vehicular Technology*, 68(9), 9110–9121. doi: 10.1109/TVT.2019.2-927533.

24. Lin, X., Wu, J., Mumtaz, S., Garg, S., Li, J., and Guizani, M. (2021). Blockchain-based on-demand computing resource trading in IoV-assisted smart city. *IEEE Transactions on Emerging Topics in Computing*, 9(3), 1373–1385. doi:10.1109/TETC.2020.2971831.

25. Lu, Y., Huang, X., Zhang, K., Maharjan, S., and Zhang, Y. (2020). Blockchain empowered asynchronous federated learning for secure data sharing in internet of vehicles. *IEEE Transactions on Vehicular Technology*, 69(4), 4298–4311. doi: 10.1109/TVT.2020.2973651.

26. Zhang, D., Yu, F. R., and Yang, R. (2019). Blockchain-based distributed software-defined vehicular networks: a dueling deep Q-learning approach. *IEEE Transactions on Cognitive Communications and Networking*, 5(4), 1086–1100. doi:10.1109/TCCN.2019.2944399.

27. Kchaou, A., Abassi, R. and El Fatmi, S. G., Towards a Secured Clustering Mechanism for Messages Exchange in VANET, 32nd International Conference on Advanced Information Networking and Applications Workshops (WAINA), Krakow, Poland, 2018, pp. 88–93, doi:10.1109/WAINA.2018.00068.

7 A Novel Secure and Reliable 12T SRAM Memory Design for Biometric Information Processing Systems

Priyanka Sharma[1,a], Vaibhav Neema[1,b], and Nitesh Kumar Soni[2,c]

[1]Institute of Engineering and Technology, Devi Ahilya University, Indore, MP, India
[2]Medi-Caps University, Indore, MP, India

Abstract

This research addresses the security and reliability concerns associated with Static Random Access Memory (SRAM) in biometric information processing systems. SRAM is commonly used to store sensitive biometric data, making it susceptible to non-invasive side-channel attacks, such as the Leakage Power Analysis (LPA) attack which compromise its data security. Additionally, SRAM's susceptibility to soft errors poses significant challenge to hold the data correctly. In this context, a novel 12T SRAM cell design is proposed that mitigates these vulnerabilities while maintaining proper functionality. The proposed 12T SRAM cell incorporates various protective features, including sleep cells to equalize leakage current during hold operations, a feedback path to prevent soft errors, and PMOS stacking to improve read stability. Monte Carlo simulations are performed to evaluate the 12T cell's resistance to LPA attacks, soft errors, and read stability. The results demonstrate that the proposed 12T cell and 6T cell exhibits 98% and 0.1% correlation respectively in leakage current indicating high resistance of 12T cell and vulnerability of 6T cell from LPA attacks. Also, 12T cell provide improved read failure probability as compared to 6T cell with 50% failure probability when simulated at 500mV for FS corner. Furthermore, the image quality of fingerprint templates when stored in proposed 12T cell is 34.71dB PSNR while it is only 6 dB PSNR when stored in 6T cell. This outcome underscores the 12T cell's capability to offer high-quality image storage.

Keywords: LPA attack, soft error, SRAM, stability.

Introduction

SRAM is a memory element that is commonly used in modern biometric information processing systems. These systems store sensitive biometric data in the form of images, requiring secure and reliable storage solutions. However, SRAM faces a significant threat from non-invasive side-channel attacks (SCAs) [1] that can extract secret image data through software and physical methods [2]. LPA attack has escalated into a significant worry within cryptographic systems due to its inherent threat to data security [3], as it compromises data security by revealing important memory information through power consumption analysis. This represents a substantial security risk to the cryptographic modules employed in the storage of biometric images [4, 5]. Additionally, it is crucial for the

---

[a]psharma@ietdavv.edu.in, [b]vneema@ietdavv.edu.in, [c]nitesh@medicaps.ac.in

storage memory to maintain high reliability and prevent any loss or corruption of sensitive information during read, write, and hold operations. The SRAM is usually used in biometric finger print authentication system in which fingerprint template of a user is stored for future reference during registration. When verification of that user is required, the SRAM memory temporary retrieves the stored database and compares it with the fingerprint template provided as input to the device at that instance as shown in Figure 7.1. While the fingerprint template is temporarily stored in SRAM the data is at high risk to get extracted by any unauthorized person. Therefore, it is crucial for the SRAM memory to be secure enough to prevent any unauthorised access of this sensitive data. Additionally, the SRAM memory should also be reliable to ensure the accurate matching of input with the stored database. Conventional 6T SRAM memory is used in such devices due to its fast access and low power consumption. However, at the same time, it suffers from low reliability and a high risk of LPA attack [6–9]. Moreover, since this type of system stores sensitive information in the form of an image, if SRAM has low reliability, it will suffer from low image quality [10].

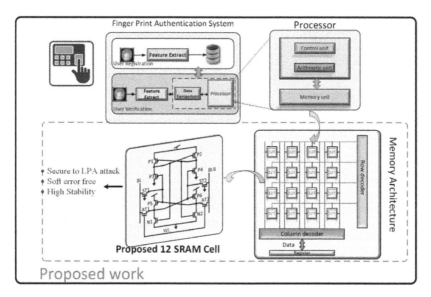

Figure 7.1 Block diagram for finger print authentication system.

This research makes several significant contributions:

1. Investigating leakage current in the 6T SRAM cell and its relationship with stored data through correlation analysis.
2. Conducting an analysis of soft errors in the 6T SRAM cell and creating a scatter plot to visualize critical voltage patterns.
3. Employing Monte Carlo (MC) simulations to evaluate RSNM for the 6T cell, culminating in the determination of its probability of failure.
4. Developing a 12T SRAM cell design that is both resistant to LPA attack and free from soft errors, ensuring heightened stability.
5. Comparing image quality when finger print image stored in the proposed 12T cell and stored in 6T SRAM using Peak Signal-to-Noise Ratio (PSNR).

The paper's subsequent structure unfolds as follows: Section II delves into an elaborate explanation of the proposed 12T cell. Section III offers a comparative analysis between the 6T and 12T cells. Comprehensive insights from detailed simulation results are in Section IV, with Section V encapsulating a thorough summary of the paper's contents.

Proposed Memory cell for Finger Print Authentication System

The proposed 12T cell incorporates various features aimed at protecting against LPA attacks and preventing soft errors (shown in Figure 7.1). During hold operation, the use of sleep cells S1 and S2 ensures that the leakage current in the bit-lines (BL and BLB) is similar for both stored values (0 and 1), making it challenging for an attacker to determine the actual stored value by LPA because of high correlation between bit-lines leakage current. To, prevent soft errors, a large feedback path is implemented in the design. This helps in safeguarding the storage node value from flipping during radiation events, enhancing the cell's reliability. During write "0" operation, the Word Line (WL) at high, while BL is set to zero and BLB is at VDD. This configuration turns on specific transistors (P1, P4, P6, AT1, AT2 and N2) and turns off others (P2, P3, P5, AT1, AT2 and N1), allowing for proper write "0" operation with Node A and D obtaining a value of 0, and Node B and C at VDD. During read, the WL control signal at high, and both BL and BLB are pre-charged with VDD. Activation of transistors N3 and N4 is performed to read operation. If the cell stores a value of 1, nodes A and D will be at VDD, while nodes B and C will be at 0. The BL voltage remains unchanged, and BLB is discharged via ON transistor N1 which creating a discharge path between BLB and GND. With the help of sense amplifier difference between bit-lines sense, to completing the read operation. The use of a PMOS stack arrangement enhances noise immunity during the read operation, ensuring that the correct value is read from the cell even in noisy environments. Overall, the proposed 12T SRAM cell incorporates protective features against LPA attacks and soft errors, while maintaining proper functionality during read and write operations.

Comparative Analysis of 6T SRAM and Proposed Cell

Figure 7.2 shows standard 6T SRAM design challenges and the methods used to overcome them in the proposed 12T SRAM cell as to make the memory secure and reliable.

(i) LPA Analysis

The 6T SRAM cell, depicted in Figure 7.2 A1, includes two storage nodes, Q and QB, and experiences power consumption during the hold state due to data-dependent leakage components which is susceptible to a leakage power analysis attack [6, 7]. The half-select [10–12] case of 6T where WL is 0 and the column is selected as shown in Figure 7.3, the shared bit-line (BL) connects to VDD, while BLB connects to ground as illustrated in Figure 7.2 A1. The leakage behaviour of the 6T cell is asymmetrical depending on the paths. These leakage currents contribute to the steady-state information leakage. In 6T order to account for variations within the die, 1000 MC simulations were conducted in Figure 7.4(a). The mean leakages were found to be significantly different for a "1" and a "0" with values of 2.99pA and 570fA, respectively, under the BL at VDD and BLB at 0 which shows data dependency of leakage current on stored data and increases the risk of LPA attack [4, 5] because of low correlation (only 0.1%) between bit-lines leakage current. The leakage

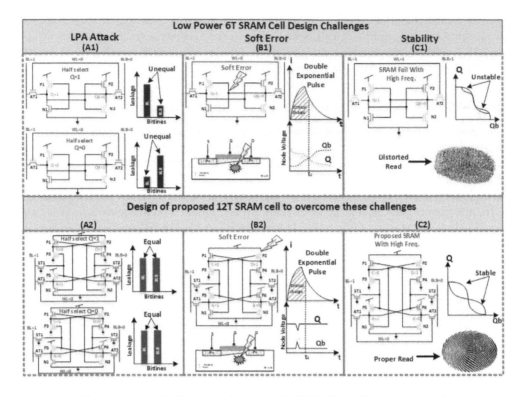

Figure 7.2 6T SRAM design challenges and proposed 12T SRAM cell to overcome these.

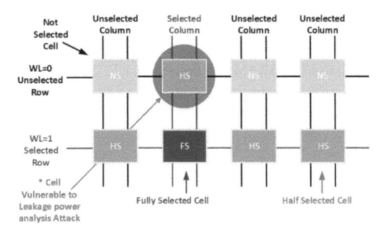

Figure 7.3 Half selection issue in memory array.

current of proposed cell is 98% Correlation observed with mean leakage current 3.57pA and 3.55pA when cell stores 0 and 1 respectively. It shows high correlation and LPA attack resistance of proposed cell in Figure 7.4(d). This indicates effectiveness of sleep cells S1 and S2 in Figure 7.2(d) providing similar BL leakage current for both stored values.

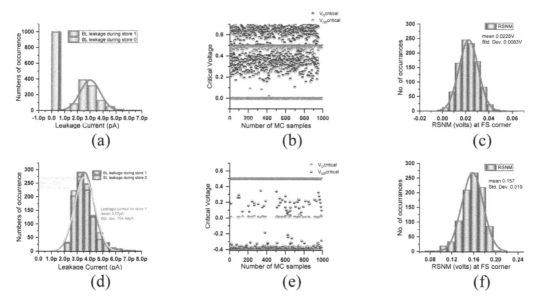

Figure 7.4 MC simulation results (1000 Samples) (a) LPA for 6T, (b) 6T critical voltage MC, (c) 6T RSNM MC, (d) LPA for proposed cell, (e) proposed cell critical voltage MC, (f) proposed cell RSNM MC.

(ii) Soft Error Analysis

Soft error analysis is an important aspect of SRAM reliability. As shown in Figure 7.3 during the hold state, when the row is selected by WL=1 and the column is deselected, the cell is in a another half-select condition and vulnerable to soft errors (Figure 7.2(b & e). This type of error occurs when WL is high, and both bit-lines are at same voltage level, causing the storage nodes "Q and QB" lose initial stored value & flip due to positive feedback between them [13, 14]. Once this error occurs, the SRAM cell cannot recover its stored value. Figure 7.4 (b and e) shows MC simulation results of 6T cell and proposed cell under soft error condition on injecting current noise at the storage node and observe that critical voltages of the storage node in 6T has the large variations in the threshold voltage than the proposed 12T cell.

(iii) Read Stability Analysis

Performance of the static random-access memory is notably influenced by its stability, making it a crucial factor to consider. Analysing previous research, it is found that conventional 6T cell has poor read stability especially at low power supply voltages, due to which cell fails to perform read operation as shown in Figure 7.2(c). MC simulation results of the RSNM analysis for 1000 MC samples at the worst FS (fast-slow) corner. It is observed in Figure 7.4(c) that RSNM is less than 26mV that indicates failure of 6T cell [6, 7] with failure probability of 50% which is defined as:

$$failure\,Probability = \frac{Numner\,of\,samples\,fail}{total\,numbers\,of\,sample\,taken}$$

Read operation of proposed 12T cell shown in Figure 7.2(f) the failure probability was found to be less than 0.01% which is significantly less, as shown in Figure 7.4(f).

As discussed above, it is clear that the conventional 6T SRAM is vulnerable to LPA attacks, susceptible to soft errors, and has poor stability during read operations. These drawbacks make the 6T cell unsuitable for use in fingerprint authentication systems, where LPA attacks and the probability of image data loss are major concerns.

Simulation Results

In this section, an analysis of simulation results will be presented, comparing the performance of the 12T and 6T SRAM cells. The focus will be on key metrics such static noise margins (SNM), read access time (RAT) and write delay (WD). These matrices will provide insight into the respective behaviors of the 12T and 6T cells during different operation scenarios, shedding light on their relative strengths and weaknesses. To compare the 6T and 12T cells fairly, all transistor sizes were set equal (W/L=120n/45n) and simulations were done at TT corner, 27°C, and 500mV supply voltage. Figure 7.5(a) illustrates the read butterfly curve for both 12T and 6T cells. The RSNM was calculated to be 226mV for the 12T cell, whereas it measured 125mV for the 6T cell. This demonstrates that the 12T cell boasts a RSNM 1.8x better than that of the 6T cell. For both cell designs, write and hold margins simulation shows in Figure 7.5 (b and c) illustrated that as anticipated the 6T cell

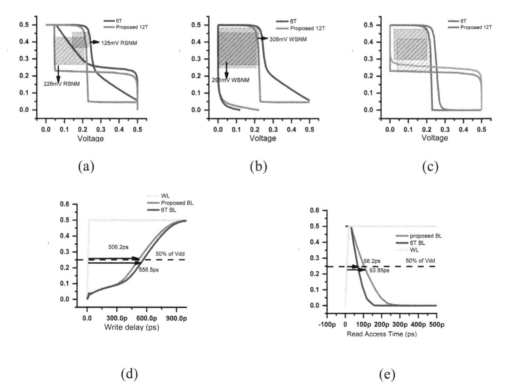

Figure 7.5 Simulation results of 6T and proposed 12T cell (a) RSNM (b) WSNM (c) HSNM (d) WD (e) Read Access Time (RAT).

Table 7.1: Comparison simulations results of 6T and proposed 12T.

Cell	RSNM	HSNM	WSNM	RAT	WD	PSNR
6T	125mV	296mV	308mV	58.2ps	556ps	6.1dB
12T	226mV	266mV	291mV	98.8ps	506.2ps	34.71dB

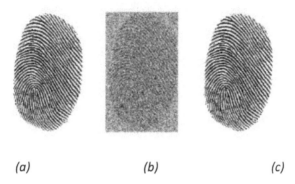

(a)   (b)   (c)

Figure 7.6 Finger print image (a) original (b) stored in 6T cell (c) stored in proposed 12Tcell.

exhibits a superior write SNM. The proposed 12T cell 1.1x time more hold SNM than 6T cell. A longer RAT and WD than the 6T cell as show in Figure 7.5 (d and e). By above simulation observed that 6T failure due to soft-error, and less stability during Read operation. Table 7.1 exhibits 6T and suggested 12T simulation results. calculates read failure probability of 6T is 5% and for proposed 12T cell is 0.01% in section III. With the help of this failure probability noise is injected in stored sensitive fingerprint image & simulate using MATLAB. And observed image quality in terms of PSNR. The 6T cell had 6.3dB PSNR and the 12T cell 34.71dB as shows in Figure 7.6 (b and c) respectively.

Conclusion

In conclusion, this research paper highlights the vulnerabilities and limitations of conventional 6T SRAM cells in biometric information processing systems, specifically in the context of storing sensitive biometric data like fingerprint templates. LPA poses significant risks to data security by compromising information confidentiality. Furthermore, the study demonstrates the shortcomings of 6T SRAM cells regarding soft error susceptibility and read stability, underscoring the need for improved memory solutions in biometric systems. The research proposes a novel 12T SRAM cell design to address these challenges. The 12T SRAM cell incorporates features aimed at enhancing security and reliability, including resistance against LPA attacks, prevention of soft errors, and improved read margin. MC Simulation results indicate that the proposed 12T cell 98% correlation in leakage current during hold 1 and 0 respectively which make its higher resistance to LPA attacks, reduced soft error susceptibility, and improved read margin 1.6x times compared to the standard 6T Static RAM cell during 1000 Monte Carlo simulation. The proposed design offers a more high-quality image, secure and reliable memory solution for biometric information processing systems.

References

1. Liu, S., Kanniwadi, S., Schwarzl, M., Kogler, A., Gruss, D., and Khan, S. (2023). Side-channel attacks on optane persistent memory. In 32th USENIX Security Symposium (USENIX Security 23).

2. Jain, A., Hong, L., and Bolle, R. (1997). On-line fingerprint verification. *IEEE Transactions on Pattern Analysis and Machine Intelligence*, 19(4), 302–314.

3. Alioto, M., Giancane, L., Scotti, G., and Trifiletti, A. (2009). Leakage power analysis attacks: a novel class of attacks to nanometer cryptographic circuits. *IEEE Transactions on Circuits and Systems I: Regular Papers*, 57(2), 355–367.

4. Maltoni, D., Maio, D., Jain, A. K., and Prabhakar, S. (2009). Handbook of Fingerprint Recognition. (Vol. 2). London: Springer.

5. Giterman, R., Vicentowski, M., Levi, I., Weizman, Y., Keren, O., and Fish, A. (2018). Leakage power attack-resilient symmetrical 8T SRAM cell. *IEEE Transactions on Very Large Scale Integration (VLSI) Systems*, 26(10), 2180–2184.

6. Mondal, D., Naz, S. F., and Shah, A. P. (2023). Radiation hardened and leakage power attack resilient 12T SRAM Cell for secure nuclear environments. In Proceedings of the Great Lakes Symposium on VLSI 2023, (pp. 227–228).

7. Giterman, R., Vicentowski, M., Levi, I., Weizman, Y., Keren, O., and Fish, A. (2018). Leakage power attack-resilient symmetrical 8T SRAM cell. *IEEE Transactions on Very Large Scale Integration (VLSI) Systems*, 26(10), 2180–2184.

8. Sharma, P., Neema, V., and Panchal, A. (2023). Proposed approximate hybrid memory architecture for handheld multimedia devices. *Materials Today: Proceedings*, 2023, https://doi.org/10.1016/j.matpr.2023.03.377.

9. Sharma, P., Neema, V., Vishvakarma, S. K., and Chouhan, S. S. (2023). MPEG/H256 video encoder with 6T/8T hybrid memory architecture for high quality output at lower supply. *Memories-Materials, Devices, Circuits and Systems*, 4, 100028.

10. Hore, A., and Ziou, D. (2010). Image quality metrics: PSNR vs. SSIM. In 2010 20th International Conference on Pattern Recognition, (pp. 2366–2369). IEEE.

11. Pasandi, G., and Pedram, M. (2018). Internal write-back and read-before-write schemes to eliminate the disturbance to the half-selected cells in SRAMs. *IET Circuits, Devices and Systems*, 12(4), 460–466.

12. Abbasian, E., Birla, S., and Gholipour, M. (2022). A 9T high-stable and low-energy half-select-free SRAM cell design using TMDFETs. *Analog Integrated Circuits and Signal Processing*, 112(1), 141–149.

13. Kavitha, S., Reniwal, B. S., and Vishvakarma, S. K. (2023). Design of radiation hardened 12T SRAM with enhanced reliability and read/write latency for space application. In 2023 36th International Conference on VLSI Design and 2023, 22nd International Conference on Embedded Systems (VLSID), (pp. 104–108). IEEE.

14. Ibe, E., Taniguchi, H., Yahagi, Y., Shimbo, K. I., and Toba, T. (2010). Impact of scaling on neutron-induced soft error in SRAMs from a 250 nm to a 22 nm design rule. *IEEE Transactions on Electron Devices*, 57(7), 1527–1538.

8 BlockCure—An Anonymized Patient Data Sharing Platform Using Consortium Blockchain

Shivesh Krishna Mukherjee[1,a], Maheak Dave[1,b], Rituparna Bhattacharya, and Jayanta Poray[3,d]

[1]BTech Student, Techno India University, West Bengal, Kolkata, India

[2]Assistant Professor, Techno India University, West Bengal, Kolkata, India

[3]Associate Professor, Techno India University, West Bengal, Kolkata, India

Abstract

The contemporary healthcare data management systems do not sanction divulging patient data to external healthcare organizations or even medical practitioners within the same healthcare organization as it is imperative to protect the patient's privacy. But apposite dissemination of any patient's diagnostic and treatment history may lead to knowledge building among researchers and medical practitioners. They can gain valuable insights referring to such healthcare records and use the knowledge in patient treatment and further research. However, to protect the patient privacy and establish data security, the personally identifiable information of the patients must be anonymized before purveying the records to third parties in a secure and transparent manner. In this paper, we address this challenge and propose BlockCure system as a solution. We designed BlockCure leveraging consortium blockchain to enable decentralized data sharing and machine learning for anonymisation incumbent for effective and efficient publishing of electronic healthcare records without compromising the patient's privacy. Ergo, we aim to delineate a novel patient data management system through BlockCure that facilitates knowledge sharing among the medical practitioners and researchers utilizing anonymised records of patients admitted to the healthcare organizations without negotiating the patient privacy.

Keywords: Anonymization, BERT, consortium blockchain, encryption, named entity recognition, pattern matching, transformers.

Introduction

Medical research necessitates sharing of healthcare patient records; the demand for such information has always existed for advanced study and development of new remedies for diseases. In the present-day healthcare systems, patient data stored and managed by the hospitals cannot be distributed directly amongst other organizations or individuals since these data contain sensitive details such as patients' personally identifiable information (PII), not suitable for public disclosure. While such data has the ability to benefit medical research, the patient data is usually discarded after certain period of time, especially if

[a]shivkrish501@gmail.com, [b]maheakdave@gmail.com, [c]rituparna.b@technoindiaeducation.com, [d]jayanta.p@technoindiaeducation.com

the data is no longer required by the hospital and/or the patient. Dispersing patient data containing PII calls for that particular patient's consent; otherwise, it would compromise the patient's privacy and may lead to legal arbitrations between the patient and the health-care organization. The second issue concerns the distribution of patients data containing PII even with their consent posing a potential risk to the patient's anonymity and patient identity leaks may damage the reputation of the associated hospital and contravene patient privacy.

In the research work we present in this paper, we envisage and build a prototype of a system that will share anonymised patient data through a decentralized platform thus resolving the challenge. We have devised the novel BlockCure platform in such a way that the patient's privacy and anonymity are not compromised and the dissemination of data is done in a transparent and secure manner with potential researchers and medical practitioners. The patient details will be filtered and/or anonymised before being displayed to the user based on his authorized role, preventing him from accessing unnecessary information. It will not only retain but also increase the value of the data which would otherwise have been dispensed with. Consortium blockchain will enable decentralized functioning of the BlockCure platform such that all hospitals participating in the system will share equal responsibilities in terms of data sharing, block addition and validation.

Thus, our contributions in this paper are the followings:

- A novel decentralized patient data management platform that manages patient records and shares it with apposite users to enrich their knowledge. Such platform defies autonomy of a single hospital and fosters decentralization through consortium blockchain.
- An anonymisation technique applied to the data sharing platform to anonymize patients data and protect their privacy such that when any patient data will be shared, the patient cannot be identified from such data.

This paper is organized as follows: In Section 2, we briefly introduce the technologies we utilized in developing BlockCure such as blockchain and transformers. Section 3 includes related research work done in the arena of patient data sharing between/by hospitals employing cutting-edge technologies such as blockchain. In Section 4 we discuss the system description of BlockCure including its design and architecture. Section 5 delineates the implementation details of BlockCure. Finally, we conclude and outline future work in Section 6.

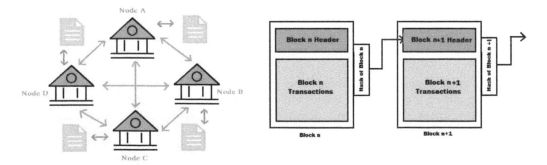

Figure 8.1 Diagram depicting the architecture of a blockchain.

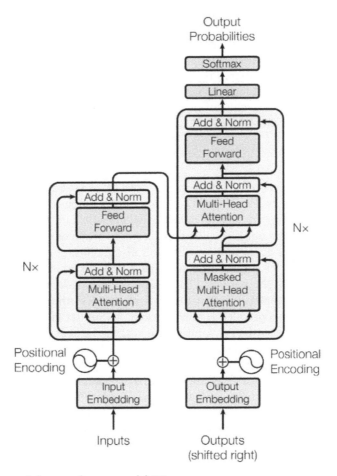

Figure 8.2 Diagram of the transformer model [3].

Preliminaries

BlockCure utilizes two key technologies to keep the data private, secure, anonymized and easily accessible, namely blockchain, see Figure 8.1 and transformers, see Figure 8.2. It uses blockchain to keep a secure and transparent decentralized transaction environment whereas the transformer architecture is used to identify and anonymize PII.

Blockchain

Blockchain is a decentralized, distributed digital ledger which stores data in blocks. Other than the transactions, in each block there are various details pertaining to the block stored, such as block hash, timestamp and previous block hash. Since, each block contains the previous block hash, blocks are being linked together to form an endless chain or the blockchain. The system of calculating hash, and thereby using it to link to the next block makes it impossible for any change to go unnoticed; if anyone attempts to tamper data in a block, the block's hash code will change thus breaking the link with the next block.

Transformers

To anonymize the PII, the Transformer model can be used. In practice, the Transformers are a state-of-the-art machine learning model which relies on a self-attention mechanism [3] to achieve faster results compared to Recurrent Neural Network (RNN). It consists of a pair of encoder and decoder blocks based on attention layers and feed forward network. We have used this particular architecture for the purpose of Named Entity Recognition (NER) [4, 5, 6] to identify PII.

Related Research

A plethora of research work is ongoing in the area of using blockchain for managing electronic healthcare records (EHR). In [7], the issues pertaining to sharing health records across hospitals and institutions such as deficit of security or data misuse have been highlighted. Further, the study mentions a blockchain-based framework that prioritizes data privacy thus enhancing the accessibility of EHR. It also facilitates making EHR accessible over a broader network. Such information sharing is patient-mediated where information-exchange is driven by the patients. The hospital collects data from the patient and stores it in the existing databases when the patient decides who can access their data. The access to such data is controlled by the blockchain. Similar efforts can be seen in [8] that presents a smart contract-based solution for managing patient consent to achieve a secure health record exchange and its access in automated way and in [12] where a blockchain-based medical data sharing application between hospitals is delineated.

Additionally, in [1], Tidke et al. has implemented a blockchain-based consultation system in which the patient's data being uploaded is secured by encryption, and can be viewed in raw form by healthcare experts. In [2], a similar effort can be found where a key change is that the user can also act as recipient. Unlike [1], the system designed by Lee et al. as described in [2] allows patients to observe other patient's data without anonymization. In this paper, a dynamic consent system has been delineated such that the application asks for patient's consent before disseminating his data. However in the implementation of both the systems, few important issues have not been focused, which is the un-anonymity of the patient's data and the need of consent of patient to prevent any exploitation of patient data which is cumbersome for non-tech-savvy users. In these two papers, the only security feature implemented is encryption, but the PII has been kept as it is. This poses a problem as a patient can be easily identified and their personal information can be exploited even if the patient has given consent for the transaction of the records.

While EHR sharing between hospitals has many advantages, such system fails to address a few challenges:

1. Hospitals may share the data generated at their end instead of collecting it from the patient, an issue with the work presented in [1].
2. Instead of having a patient-driven system which depends on the patients who may or may not be competent enough to use such systems, hospitals themselves can anonymize and share patient data using blockchain.
3. An incentive is necessary to motivate the healthcare organizations in furnishing useful data to interested third-parties.

Such problem associated with patients controlling access to the data owned by them and shared between hospitals is also conspicuous in the work of Jaramillo-Alvarado et al. [9]

who propose a universal EHR, Zhuang et al. [11], Enescu et al. [13] and Harshini et al. [15] who propose EHR sharing using blockchain. An interesting application in this context can be found in the work of Zhao et al. [10]. They propose an EHR matching system that matches patients to be diagnosed with existing records of corresponding diseases finding required doctors, hospitals, prescriptions, etc. thus sharing of EHR resources. Again, we find an application of consortium blockchain that supports data sharing between multiple healthcare providers operating in a single platform allowing data access among authorized members [14]. However, the disadvantage of the system mainly involves sharing un-anonymized data between organizations thus posing a threat to patient's privacy. We have addressed this issue by introducing an anonymization layer in the system. Patient consent is not required since their PII has been anonymized.

BlockCure System Description

We rely on consortium blockchain to build BlockCure underpinned by a consortium of healthcare organizations where each organization manages a node of the blockchain network. They agree to share the anonymized patient data within the network of healthcare organizations. Anonymization is essential as all the organizations should be able to store the same copy of the blockchain and data will eventually be furnished to external entities. An employee of a participating healthcare organization belonging to the consortium should be able to visualize data anonymized, based on his role without any charges but such anonymized data will be accessible to external entities not employed with the participating healthcare organizations only after making a payment. Such system will mutually benefit the organizations within the consortium in general and medical researchers and practitioners and outside the organizations in particular.

Each hospital in addition to its contribution to the blockchain, will maintain a central database where all details related to patients treated there will be stored. This includes PII as well as financial details and treatment history. The admin of the hospital will be able to access such data but other employees will be able to access filtered/anonymized data based on his role. The patient himself should be able to access all his details.

Next, we include some important aspects of the BlockCure system.

Users

We discuss here the different actors involved in BlockCure. There can be doctors employed by the participating organizations whom we call internal doctors and medical practitioners outside the consortium of the organizations whom we call external doctors. These external doctors not associated with the hospitals but accessing anonymized healthcare data make payment for the BlockCure service. There will be administrator and accounts department in each hospital. The patients and insurance companies are other users of the system. There may be more consumers of such platform, for instance, any government or research organization or individual requesting anonymized patient records for research purpose.

Functionality

The blockchain under consideration will be permissioned and only organization with known identities should be able to participate in block addition process. It may or may not include cryptocurrency generation; sharing of data between different organizations will mutually benefit each other and money will be earned from payments made by external entities accessing the anonymized data. The healthcare organization submitting the greatest number of authentic patient records or adding the maximum number of valid blocks in a certain time period will be able to claim the maximum amount from the payments made by external entities or earn maximum rewards if the system supports cryptocurrency generation.

When any new patient record is added by the healthcare organization or additional details are appended to an existing patient's records, the same is broadcasted to the blockchain network after anonymization. The patient's treatment history will be stored in the blockchain as well as the healthcare's internal database. While blockchain will store the anonymized version, the local database will store the un-anonymized version of the patient's records. His financial details and PII will be stored only within the internal database of the organization as the purpose of the BlockCure system is medical knowledge building and sharing financial details is not the necessity here. The block addition process will follow a Proof of Reputation consensus protocol. After every interval of a few minutes or hours depending on the frequency of data addition by the healthcare organizations, the patients' treatment histories are collected in a block. One organization among the top few most recently active in valid data submission will be chosen randomly to mine the block. If any hospital submits invalid data repeatedly, it will be banned from the block addition process. Once the block is broadcasted to the remaining nodes, there is a validator in every node who will validate the block. If the block is valid, it will be added to the blockchain.

The result of every appointment with a doctor for a particular patient or his admission in the hospital will be appended to the blockchain. Once the anonymized data is stored in

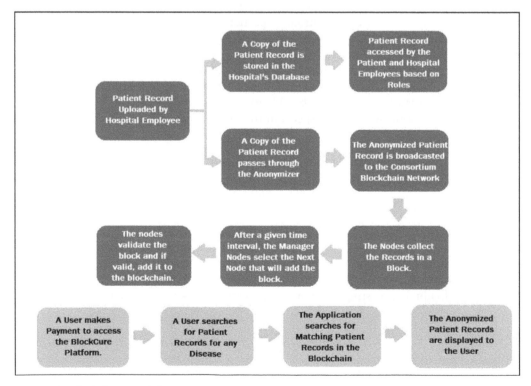

Figure 8.3 BlockCure workflow.

the form of blocks in the blockchain, it will be available to all the participating healthcare organizations and based on the agreement between such organizations, the data will be read from the blockchain. Post validation of the data in the blocks and the block getting appended to the blockchain, it will also be available to external entities such as external doctors or researchers. The workflow is depicted in Figure 8.3.

Functional Requirements

Here, we include a list of the functional requirements. The related use case is included in Figure 8.4.

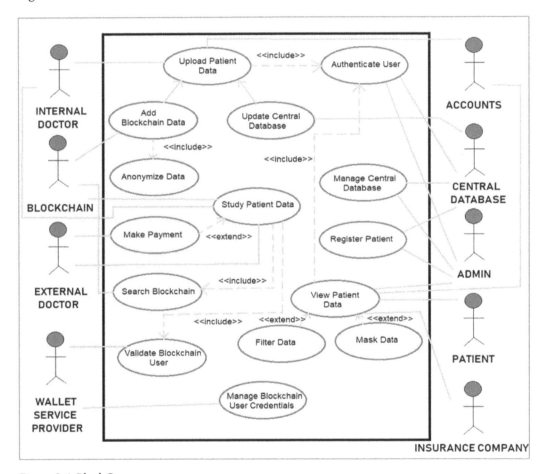

Figure 8.4 BlockCure use cases.

a. *Upload Patient Data*—The doctor should be able to upload new and existing patient's data in the system. This excludes the financial information. The Accounts division should be able to upload the financial and insurance details of the same patient. Such details will be stored in the organization's internal database. In addition, the data uploaded by the doctor will be added to the blockchain after anonymization.

b. *Authenticate* User—Any user whether doctor, administrator, patient, representative of insurance company or an employee within the accounts department should be granted access to the application only after authentication. Any hospital will centrally store the credentials of such users associated with that particular hospital and will authenticate the user before allowing access to the data stored in its centralized repository.

c. *Study Patient Data*—Any doctor whether internal or external should be able to study patient data stored in the blockchain. A doctor will view the anonymized version of the patient record when the patient is not treated by him and un-anonymized but filtered/masked (excluding financial details) version of the data if the patient is treated by him.

d. *Filter Data*—Not all data is required to be anonymized, some data may be filtered from being displayed to the user. Say for example, the employee in the Accounts team does not need to see the patient's prescription and the same must be filtered out before displaying patient's data instead of anonymization based on user's role. Again, the doctor does not need to see his patient's financial details and the same is filtered out before displaying patient's data to him. In both these cases, PII of the patient will be shown to the user and hence anonymization is not required, instead we apply filtering. But when the patient's treatment history is made available to a different doctor, his PII such as name, address, phone number will be anonymized and his financial details will again be filtered out.

e. *View Data*—The patient, accounts, admin and the insurance company will be able to see the un-anonymized/filtered version of the patient's records based on the roles after authentication.

Architecture

The BlockCure has multiple User Interfaces (UI), one using which the users such as doctors, patients and insurance companies related to a hospital view required patient data from its central data repository and another using which any user can read blockchain data. The second one, we call the wallet user interface. An external doctor when requests data through the second UI, it interacts with the payment getway to process the payment received from the external doctor for the service. The Accounts division, patients, internal doctors, insurance companies view patient data through the UI of the portal. The centralized database stores the authentication details of all users of the portal and all details of the patients. The Admin communicates with centralized database to add/update/delete the authentication details of the different users of the system.

In the image included here, we portray the components such as the portal, centralized database, anonymizer, etc functioning at a single node. The same components will be present at all the peer nodes of the blockchain network building the decentralized platform. The centralized database managed by each healthcare organization will store the financial details of only the patients treated at that healthcare organization and the authentication details of all the parties associated with that organization such as the patients, accounts, admin, insurance companies and internal doctors. The credentials pertaining to external doctors will be maintained by the wallet service provider.

The system design is divided into two key components—the consortium blockchain application and the anonymization layer.

Consortium Blockchain Application

To build BlockCure, we rely on consortium blockchain, which is backed by a consortium of healthcare organisations, each of which manages a node of the blockchain network, see Figure 8.5. They agree to provide the anonymized patient data across the healthcare network. Among the network of the organizations, few manager nodes will also be present to implement the consortium consensus.

When a user wants to upload a particular patient's data, he will be able to do so via a user interface designed for that task, each organization/hospital will have their own local database. When the patient data has been uploaded, another copy of the data is made which is kept unfiltered and non-anonymized, that particular copy is stored in their own local databases. The original copy of the data is passed through an anonymization layer which anonymizes the PII which is usually found in the patient medical records and which cannot be filtered due to the risk of loss of valuable information.

After the data has been anonymized and filtered, it is then encrypted and broadcasted throughout the network of organisations and the manager nodes, a manager node's purpose is to track each transaction and from which node it was broadcasted. It will also implement the consensus protocol for the blockchain. Each organisation node will then store that data in its own transaction pool. Our implementation of consensus protocol utilizes the manager node to keep track of the number of records received from each node and also the total number of records received, in a given time interval, the node with the maximum number of valid uploads will be selected to mine a block in the blockchain. The mined block will also be broadcasted to the other nodes of the network including the manager nodes as well.

The retrieval of the patient data is done by searching the blockchain maintained by the consortium with the search parameters being the disease, then the patient id which is assigned by the system before it was uploaded and then the patient records. The patient id is hashed to conceal patient identity. For the retrieval application, each user will be categorized into either of two categories which is external and internal, external being those user who are not part of the consortium blockchain but are still interested in the data available, and the internal users who are part of the consortium and contribute to it. External users will have to pay a fee to access a particular patient record if they wish to do so but for the internal users the data is available for free. When the user selects a particular patient record to be accessed, the data is first decrypted and then made available for the user to see.

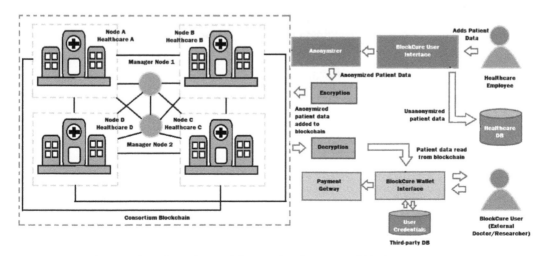

Figure 8.5 A diagram describing the workflow and architecture of blockcure.

Anonymization

Flaws in security measures may lead to data breaches where patient data possesses the risk of getting leaked. Anonymization of such data is crucial in order to protect the identity of the individual. Patient data consists of several fields which can lead to the identification of the individual. Some of the common PII includes name, email address, contact number and address while other quasi-identifiers include admission number and other hospital related unique identifiers. It is essential to appropriately mask them in order to eliminate re-identification risk.

Another factor that must be considered during anonymization is the usefulness of the data. Anonymization should be done, ensuring that the value of the data is retained. In case of patient related data, we must ensure no such fields must be masked which are needed by healthcare professionals for their analysis. Hence it is crucial to ensure diagnostic fields like medications and treatment procedures are not completely obfuscated. However, it is possible that these fields may themselves contain PII which need to be properly masked. Thus, we describe a multi-layer anonymizer.

In this paper, we have trained a BERT model to perform Named Entity Recognition (NER) in order to identify the PII's, see Figure 8.6. The model is trained on the "CoNLL 2003" dataset to identify name, location and organisation. This dataset is primarily used for language independent NER. It consists of sentences and has four columns. Each word of the sentence is in a different line, and each sentence is separated by an empty line. The first column consists of words, second consists of a parts-of-speech (POS) tag, then a syntactic chunk tag. The fourth column contains the named entity tag. There are four kinds of NER tags:—person, location, organization and miscellaneous tags i.e. those which do not belong to the other three groups.

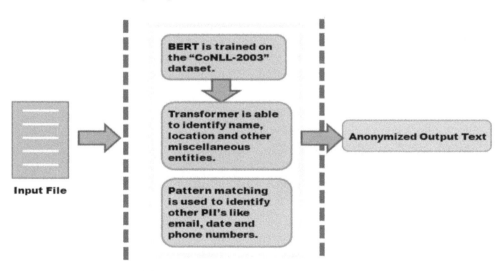

Figure 8.6 Anonymizer module.

Implementation

The consortium blockchain application was developed using Java. The framework used for developing the User Interfaces (UI) for the application was JavaFX and the networking part of the system is done via sockets. The consortium blockchain was developed from scratch.

As for the anonymization layer, we have used the Hugging Face API to use BERT for token classification. For prediction we have used a classification pipeline consisting of three parts mainly:—preprocessing, forwarding and post processing. BERT can only handle a token of size 512 IDs. Therefore, in order to handle text of any length small modifications are required. During the pre-processing, we make chunks of size 512 each. For chunks less than 512 IDs, we apply padding to the tokens to adjust the token to the corresponding input size. We then forward each of these tokens to the model, concatenate the outputs and fit it according to the size of the post process inputs. Since we have already reshaped our outputs, no modifications are needed in the post-processing method. We can now use the pipeline directly and fetch the named entities. These entities can be masked in order to anonymize these fields.

The remaining fields are masked using pattern matching as the remaining PIIs are always found in a given format in any text. The 'Matcher' and 'PhraseMatcher' libraries are used from the 'spacy' package in order to accomplish this task. The required formats are stored in an object and the text input is passed. The 'spacy' library recognizes these entities based on its pattern. Once identified, they can be easily masked. Ergo, our proposed anonymizer anonymizes the data in two steps—using transformers and pattern matching and the output obtained has eliminated any re-identification risks.

The BERT architecture being trained on the CoNLL-2003 boasts the following performance metrics [16]:

Precision	0.9438
Recall	0.9525
F1–Score	0.9482
Accuracy	0.9911
Loss	0.0355

Hence, with such a performance record this transformer is the ideal candidate for the de-identification using NER task.

We achieved the following results with BlockCure.

Time taken by the system to upload 1000 data records by 2 nodes.	1562s–26 mins
Time taken for each record to be uploaded.	3.124 s
Time taken to anonymize each records	2.34 s

We include screenshots of the BlockCure prototype application, see Figures 8.7 and 8.8.

Conclusion and Future Work

We are shifting to a digital era in the medical sector where all data will be eventually stored in electronic form. Sharing of such data between hospitals and also with other relevant parties without compromising the privacy of the patients will lead to knowledge building of individuals such as doctor and may also endow medical researchers with valuable cognizance. Systems like BlockCure will help to realize such objectives. Here, we have described such system only briefly as it is a complex platform and we have built a prototype of the

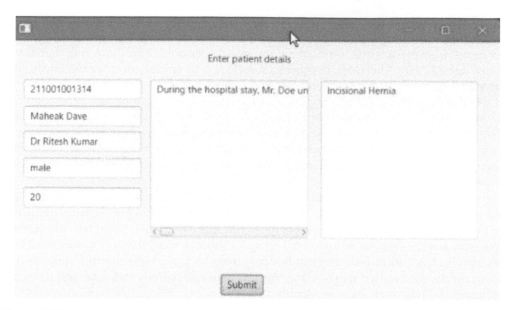

Figure 8.7 Data entry UI.

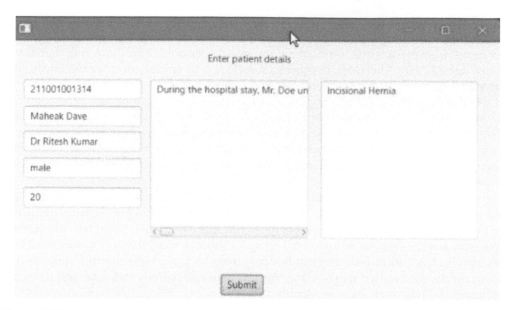

Figure 8.8 Data search and view UI.

system to prove the idea presented here. Our future efforts will encompass anonymizing QR codes and medical reports and images such as X-rays. In addition, novel data search and retrieval techniques may be devised to search efficiently the patient data stored in the blockchain. Also, further research will be directed towards enhancing the security of the application and improve the user experience. To conclude, the BlockCure application can greatly improve the knowledge base for medical researchers and help develop new remedies for diseases at a fast rate.

References

1. Tidke, S. K., Khedkar, V., Banerjee, A., Mulik, A., Goyal, A., and Chhabaria, Y. (2022). An interactive and secure blockchain web portal for online healthcare services. In 2022 International Conference on Decision Aid Sciences and Applications (DASA), Chiangrai, Thailand, (pp. 454–459). doi: 10.1109/DASA54658.2022.9764973.
2. Lee, A. R., Kim, M. G., Won, K. J., Kim, I. K., and Lee, E. (2020). Coded dynamic consent framework using blockchain for healthcare information exchange. In 2020 IEEE International Conference on Bioinformatics and Biomedicine (BIBM), Seoul, Korea (South), 2020, (pp. 1047–1050). doi: 10.1109/BIBM49941.2020.9313330.
3. Vaswani, A., Shazeer, N., Parmar, N., Uszkoreit, J., Jones, L., Gomez, A. N., Kaiser, Ł., and Polosukhin, I. (2017). Attention is all you need. *Advances in Neural Information Processing Systems*, 30.
4. Liu, X., Chen, H., and Xia, W. (2022). Overview of named entity recognition. *Journal of Contemporary Educational Research*, 6(5), 65–68.
5. Lothritz, C., Allix, K., Veiber, L., Klein, J., and Bissyande, T. F. D. A. (2020). Evaluating pretrained transformer-based models on the task of fine-grained named entity recognition. In Proceedings of the 28th International Conference on Computational Linguistics (pp. 3750–3760).
6. Labusch, K., Kulturbesitz, P., Neudecker, C., and Zellhöfer, D.(2019). BERT for named entity recognition in contemporary and historical German. In Proceedings of the 15th conference on natural language processing, 2019, 9–11.
7. Vardhini, B., Dass, S. N., Sahana, R., and Chinnaiyan, R. (2021). A blockchain based electronic medical health records framework using smart contracts. In 2021 International Conference on Computer Communication and Informatics (ICCCI-2021). Coimbatore: IEEE.
8. Naik K. V., Murari, T. V., and Manoj, T. (2022). A blockchain based patient consent management technique for electronic health record sharing. In 2022 IEEE 7th International Conference on Recent Advances and Innovations in Engineering (ICRAIE). IEEE.
9. Jaramillo-Alvarado, A. F., Diaz-Arango, G., Garcia-Baez, J. R., Gamino-Aparicio. C., Hernandez-Capistran, J., Velandia-Caballero, O. J., Huerta-Chua. J., and Vazquez-Leal, H. (2022). A Blockchain Solution for Universal Electronic Health Record: Mexican Healthcare System Case. In 2022 IEEE International Conference on Engineering Veracruz (ICEV), pp. 1–6. IEEE.
10. Zhao, Y., and Du, K. (2022). A matching scheme from supply and demand sides of electronic health records based on blockchain. In 2022 7th International Conference on Intelligent Computing and Signal Processing (ICSP). IEEE.
11. Zhuang, Y., Sheets, L. R., Chen, Y. W., Shae, Z. Y., Tsai, J. J., and Shyu, C. R. (2020). A patient-centric health information exchange framework using blockchain technology. *IEEE Journal of Biomedical and Health Informatics*, 24(8), 2169–2176.
12. Maghraby, A., Numan, A., Mashi, A. A., Aljuhani, A., Almehdar, R., and Abdu, N. (2021). Applied blockchain technology in Saudi Arabia electronic health records. In 2021 International Conference on Computational Science and Computational Intelligence (CSCI). IEEE.
13. Enescu, F. M., Bizon, N., Cirstea, A., and Stirbu, C. (2018). Blockchain technology applied in health. In ECAI 2018—International Conference—10th ed. Iasi, Romania: IEEE.
14. Gautama, A., Rochim, A. F., and Bayuaji, L. (2022). Privacy preserving electronic health record with consortium blockchain. In 2022 6th International Conference on Information Technology, Information Systems and Electrical Engineering (ICITISEE). IEEE.
15. Harshini, V. M., Danai, S., Usha, H. R., and Kounte, M. R. (2019). Health record management through blockchain technology. In Third International Conference on Trends in Electronics and Informatics (ICOEI 2019), IEEE.
16. Kanakarajan, K. R. (n.d.). kamalkraj/bert-base-cased-ner-conll2003. Retrieved from huggingface: https://huggingface.co/kamalkraj/bert-base-cased-ner-conll2003

9 Secure Embedded SRAM from Side Channel and Data Imprinting Attacks

Aastha Gupta[1,a], Ravi Sindal[2,b], Vaibhav Neema[2,c], and Ashish Panchal[2,d]

[1]Ph.D scholar, Institute of Engineering and Technology, Devi Ahilya University, Indore

[2]Institute of Engineering and Technology, Devi Ahilya University, Indore

Abstract

The increased usage of handheld embedded devices and the rising complexity of cyber-attacks have made the security of SRAM a critical issue. SRAM is widely used in embedded systems due to its advantages of fast access times and low power consumption. However, it is susceptible to physical attacks, including side-channel attacks (SCA) and data imprinting effect which can compromise the confidentiality and integrity of sensitive data stored in these devices. This paper presents a SRAM cell which can mitigate both the attacks effectively. Usage of loop cut technique prevents power analysis SCA by consuming equal power of 193.8nW during write 1/0 operation which is 1.15x less than conventional 6T cell. Data toggling circuit does not allow static noise margin (SNM) to change with time as a result data imprinting effect can be prevented. The proposed SRAM cell require least stand-by power of 101.1nW and toggling power of 13.55μW. Also, the proposed SRAM cell consists of only 10 transistors making it area efficient as well.

Keywords: data imprinting effect, embedded devices, side channel attack, SRAM.

Introduction

SRAM memory in handheld embedded devices offers fast access, efficient storage, low power consumption, and supports key functionalities [1, 2]. In the context of the Digital India initiative, securing SRAM becomes crucial due to increased usage in digital payments, government services, and sensitive data processing. Protecting SRAM ensures privacy, integrity, and national security in critical applications, preventing unauthorized access and data tampering [3, 4]. SRAM devices can be vulnerable to both online and offline attacks, which can compromise the security of data stored in the memory. Online attacks occur when the SRAM device is actively connected and powered on. They include Side-Channel Attacks (SCA) that exploit the information leaked during the execution of cryptographic operations, such as power consumption, electromagnetic radiation, or timing variations. By analyzing these side channels, an attacker can extract sensitive data, such as encryption keys. However, the power dissipation/consumption of a chip is the most commonly used

---

[a]aasthagupta72@gmail.com, [b]rsindal@ietdavv.edu.in, [c]vneema@ietdavv.edu.in, [d]apanchl@ietdavv.edu.in

property in SCA to determine secret keys [5, 6]. Various SRAM design solutions have been proposed in the literature to mitigate Side-Channel Attacks (SCAs). These include techniques such as power cut [7], shared PMOS [2], and the loop cutting technique [8]. The underlying principle behind these techniques is to precondition the storage nodes to logic '0' before the write operation, which helps reduce vulnerability to side-channel attacks.

Offline attacks on SRAM devices occur when they are powered off or removed from the system, allowing attackers to access and manipulate memory contents. These attacks exploit the data imprinting effect caused by Positive Bias Temperature Instability (PBTI) [10] and Negative Bias Temperature Instability (NBTI) [9, 10]. Over time, PBTI and NBTI gradually alter the relative strengths of pull-up and pull-down transistors, leading to imprints of previously stored data in memory cells. These imprints persist even when power is turned off, enabling potential retrieval of stored data by attackers. The shifts in transistor threshold voltage caused by NBTI and PBTI compromise the security of the SRAM cell [11].

To overcome these drawbacks, hardware-designed solutions can offer more direct control, lower performance overhead, and improved reliability. Hardware-based periodically toggling data within storage nodes can help reduce the imprinting effect in SRAM cells. The idea behind this technique is to distribute the stress on the SRAM cells more evenly, mitigating the impact of long-term data retention and reducing the risk of imprinting. The evenly distributed stress on SRAM cells minimizes the impact of bias temperature instability (BTI) and other aging effects [3, 12]. This can help maintain more stable threshold voltages and improve the overall reliability of the SRAM cells. To mitigate data imprinting effect, a data-toggling SRAM with 14 transistors is proposed in [3], consisting of a master-slave design. This design was a modification to the previously designed master-slave SRAM cell in [12], which contains 22 transistors for the same purpose.

However, the SRAM designs proposed in the literature so far have only considered one type of attack involving large number of transistors that results in high power dissipation. The paper introduces a memory cell design that protects SRAM from both online side-channel attacks (SCA) and offline bias temperature instability (BTI) attacks using a minimum of 10 transistors. The side channel attack and data imprinting effect, along with their countermeasures, are analyzed with a flow diagram as shown in Figure 9.1. The contributions presented in this paper are as follows:

- The proposed cell incorporates a loop cutting technique to protect SRAM from power analysis side-channel attacks during the write operation.
- A data toggling circuit has been proposed in this paper, which is integrated into the proposed SRAM cell. This circuit periodically toggles the stored data to prevent the effects of bias temperature instability (BTI).

Proposed SRAM Cell

The SRAM cell design proposed in this paper aimed at safeguarding the security of confidential data temporarily stored within it by employing data toggling and loop cut mechanism. Figure 9.2 illustrates circuit diagram of this cell which can be divided into four main blocks: main memory with loop cut transistors, toggle circuit, delay circuit and toggle enable circuit.

The main memory circuit comprises two access transistors controlled by a word line (WL), that connects the storage nodes Q and QB to Bit-line (BL) and Bit Line Bar (BLB)

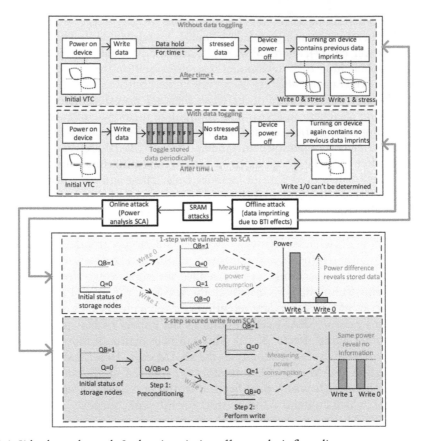

Figure 9.1 Side channel attack & data imprinting effect analysis flow diagram.

respectively. This circuit enables read and write operations within the memory. It also includes two loop cut transistors placed between the two inverters controlled by signal W. The purpose of using loop cut transistors is to disable the regenerative feedback loop of the SRAM cell, preparing the storage nodes for logic '0' before performing write operation.

The Toggle circuit consists of two NMOS transistors with drain terminal connected to storage nodes Q and QB respectively and gate being driven by the clock (CLK) signal This arrangement toggles the stored data periodically within the SRAM cell. Toggle enable circuit contains AND gate with inputs Toggle enable Signal (TEN) and CLK. TEN is a control signal that determine the toggle mode of SRAM cell. TEN is kept low for read, write and hold operation whereas, for toggle mode it is provided high input. Delay circuit contains two delay elements which provide substantial delay of half clock time to the toggle circuit and allow data to toggle periodically.

The block diagram of proposed SRAM cell as depicted in Figure 9.3 comprises of Proposed SRAM memory array, counter to count the clock pulses as even and odd, and state machine that accepts the counter output and generates the T-state signal. For odd clock cycle, T-state become high whereas it becomes low for every even cycle as depicted through timing diagram in Figure 9.4. The status of T-state determines whether the data stored in storage nodes is 'true' or 'flipped' as tabulated in Table 9.1. So, the state machine ensures actual data to be read from memory.

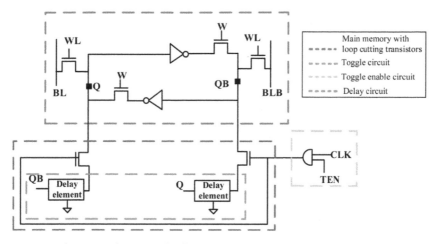

Figure 9.2 Circuit diagram of proposed cell.

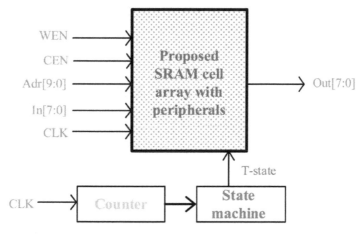

Figure 9.3 Block diagram of proposed cell.

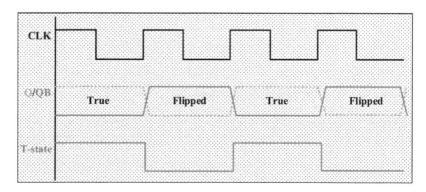

Figure 9.4 Timing diagram of proposed SRAM cell during toggle mode.

Write Enable (WEN) and Chip Enable (CEN) global signals are provided for enabling chip and write operation. SRAM accepts 10-bit address and 8-bit input data to produce 8-bit data output. The timing diagram of proposed cell is depicted in Figure 9.4.

Circuit Operation

The circuit operates in four different modes—Read, Write, Hold and Toggle.

Read mode: During read operation, BL and BLB are pre-charged to V_{DD} while WL is set to a high voltage level. To enable the loop-cut transistor and establish regenerative feedback between the inverters, the signal W must be high. The read operation occurs when the T-state is logic 1, with TEN kept low to deactivate the toggle circuit. BL discharges for read 0, while BLB discharges for read 1. The sense amplifier is used to detect the difference between discharge levels. By ensuring equal power consumption for read 0 and read 1 operation, the vulnerability to power analysis Side-Channel Attacks (SCA) is significantly reduced.

Write mode: To enhance security against SCA, the proposed cell implements a two-step write operation. Firstly, storage nodes Q and QB are preconditioned to logic 0 by setting WL to high voltage and W/TEN to low voltage, discharging BL/BLB. Then, the write operation proceeds by re-enabling the loop-cut transistor and providing the desired values to BL and BLB. This approach ensures equal power dissipation for write 1 and write 0 operations, safeguarding against SCA during write process.

Hold/Toggle mode: The hold/ toggle mode of the proposed SRAM cell is determined by TEN signal. When TEN is kept at logic 0 and T-state is high, the toggle circuit is deactivated, and the SRAM cell retains the current data. Whereas, when TEN is set to logic 1, the toggle circuit get activated, causing the data to periodically toggle.

The state diagram for various operating modes of proposed SRAM cell is illustrated in Figure 9.5.

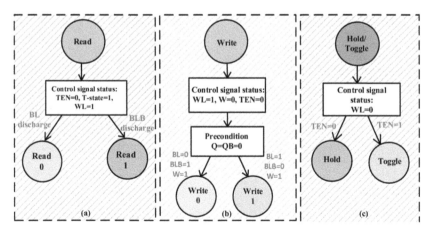

Figure 9.5 State diagram for read, write and hold/toggle mode.

Table 9.1: Data status when initially Q=0/1 & QB=1/0 during toggle mode.

Clock	Q=0 (Initial)	QB=1 (Initial)	Data Status	T-state	Q=1 (Initial)	QB=0 (Initial)	Data status	T-state
Odd	0	1	True	1	1	0	True	1
Even	1	0	Flip	0	0	1	Flip	0

Results and Discussion

The proposed cell is implemented using cadence virtuoso in 90nm technology node. Various parameters associated with proposed cell are measured and compared with conventional 6T [1], 22-T [12] and 14-T [3] cells. Figure 9.6 illustrates the amount of power consumption during read and write by conventional 6T & proposed cell. The power consumed of proposed cell during both write 1 & 0 is 193.8nW which is 1.15x less than maximum write power in conventional 6T cell which is 224.4 nW. Pre-conditioning of storage nodes to logic 0 prior to write operation makes write power dissipation equal irrespective of data stored and ensures SRAM cell security during data write. In conventional 6T cell power dissipation during write 1 & 0 is 224.4nW & 25.18nW respectively (Q=0 &QB=1). This unequal write power makes it vulnerable to SCA. Read 1/0 power dissipation in the proposed cell is 32.08nW which is 1.14x greater than conventional 6T cell (28.03nW). This increase in read power is due to the addition of two loop cut transistors in main memory circuit. However, equal power dissipation of the proposed cell during read 0 and 1 ensures secured read operation.

The toggling power dissipation also dominant power in total power dissipation. In the proposed cell toggling power is 13.55µW which is equal to 14-T SRAM cell and is 2.34x lesser than 22-T cell. 22-T cell used replica of memory cell as a slave circuit which results in large toggling power dissipation. The stand-by power of proposed SRAM cell is 101.1nW which is least as compared to 22-T and 14-T cell as illustrated in Figure 9.7. It is because the proposed cell requires a smaller number of transistors for toggling and retaining data that results in minimum power dissipation. High stand-by power of 22-T cell is due to involvement of large number of transistors in its master-slave configuration. On the other hand, the 14-T cell exhibits the highest standby power consumption among all three considered cells, primarily due to the presence of 2 transmission gates within it. The comparative analysis of power dissipation among various cells in stand-by and toggle mode is tabulated in Table 9.2.

The proposed cell consists of only 10 transistors to consider both online as well as offline attacks. Whereas, 22-T & 14-T SRAM cells require 22 and 14 number of transistors respectively to prevent only data imprinting effect. Hence, proposed SRAM cell is designed in such a way that it involves minimum number of transistors to provide security from online as well as offline attacks and dissipates less power. The proposed cell performance in terms of leakage power at various clock frequencies is measured at 1V supply voltage and is observed that the leakage power dissipation increases exponentially as increases in clock frequency as shown in Figure 9.8.

Conclusion

In this paper, introduce an SRAM cell design that prioritizes the security of the SRAM SCA and data imprinting effects. The loop cutting technique secure write operation of proposed

Table 9.2: Power dissipation during hold and toggle mode.

Cells	No. of transistors	Stand-by power	Toggling power
22-T	22	176.8 nW	31.78 µW
14-T	14	2.58 µW	13.55 µW
Proposed	10	101.1 nW	13.55 µW

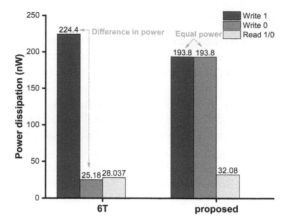

Figure 9.6 Variation of leakage power w.r.t. frequency for proposed cell.

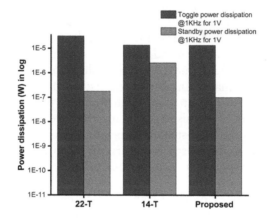

Figure 9.7 Power dissipation analysis of various cells.

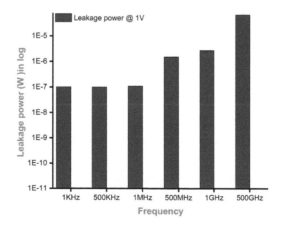

Figure 9.8 Variation of leakage power w.r.t. frequency for proposed cell.

memory with respect to power analysis. It requires 193.8 nW power to perform both write 0 & 1 operation whereas, conventional 6T cell require unequal power i.e., 224.4nW & 25.18nW for write 1 & 0 respectively. The data toggling circuit enables data to toggle periodically and therefore, prevents imprinting effect. The proposed SRAM cell has least stand-by and dynamic power dissipation i.e., 101.1nW & 13.55μW respectively as compared to other cells considered. These results conclude that proposed SRAM cell is secured from SCA along with data imprinting effect and give better results in terms of power consumption using less transistors as compared to other considered cells.

References

1. Gupta, A., Sindal, R., Sharma, P., Panchal, A., and Neema, V. (2023). Methods for noise margin analysis of conventional 6 T and 8 T SRAM cell. *Materials Today: Proceedings*, 4, 1–10.
2. Giterman, R., Keren, O., and Fish, A. (2018). A 7T security oriented SRAM bitcell. *IEEE Transactions on Circuits and Systems II: Express Briefs*, 66(8), 1396–1400.
3. Ho, W. G., Chong, K. S., Kim, T. T. H., and Gwee, B. H. (2019). A secure data-toggling SRAM for confidential data protection. *IEEE Transactions on Circuits and Systems I: Regular Papers*, 66(11), 4186–4199.
4. Skorobogatov, S. (2018). Hardware security implications of reliability, remanence, and recovery in embedded memory. *Journal of Hardware and Systems Security*, 2(4), 314–321.
5. Parameswaran, S., and Wolf, T. (2008). Embedded systems security—an overview. *Design Automation for Embedded Systems*, 12, 173–183.
6. Talaki, E. B., Savry, O., Bouvier Des Noes, M., and Hely, D. (2022). A memory hierarchy protected against side-channel attacks. *Cryptography*, 6(2), 19.
7. Konur, E., Ozelci, Y., Arikan, E., and Eksi, U. (2006). Power analysis resistant SRAM. In 2006 World Automation Congress, (pp. 1–6). IEEE.
8. Rožić, V., Dehaene, W., and Verbauwhede, I. (2012). Design solutions for securing SRAM cell against power analysis. In 2012 IEEE International Symposium on Hardware-Oriented Security and Trust, (pp. 122–127). IEEE.
9. Khan, S., and Hamdioui, S. (2010). Trends and challenges of SRAM reliability in the nanoscale era. In 5th International Conference on Design and Technology of Integrated Systems in Nanoscale Era, (pp. 1–6). IEEE.
10. Ahmad, S., Iqbal, B., Alam, N., and Hasan, M. (2018). Low leakage fully half-select-free robust SRAM cells with BTI reliability analysis. *IEEE Transactions on Device and Materials Reliability*, 18(3), 337–349.
11. La Rosa, G., Ng, W. L., Rauch, S., Wong, R., and Sudijono, J. (2006). Impact of NBTI induced statistical variation to SRAM cell stability. In 2006 IEEE International Reliability Physics Symposium Proceedings, (pp. 274–282). IEEE.
12. Backus, R. M., Duffey, C. F., Weil, A. C., and Joshi, S. V. (2009). Non-Imprinting Memory with High Speed Erase. U.S. Patent No. 7,525,836. Washington, DC: U.S. Patent and Trademark Office.

10 A Brief Exploration on the Evolutions of Consensus Protocols in the Blockchain Architecture

Suvrojyoti Biswas[1,a] and Rituparna Bhattacharya[2,b]

[1]M.Tech. Student, Techno India University, Kolkata, India

[2]Assistant Professor, Techno India University, Kolkata, India

Abstract

Blockchain technology has emerged as the fundamental pillar of Bitcoin—the primordial cryptocurrency invented by Satoshi Nakamoto in 2008. Consensus protocols in blockchain are fundamental algorithms that facilitate agreement among distributed nodes. They ensure transaction validity, data consistency, immutability and enable a single version of truth in a decentralised manner. By establishing agreement on valid transactions, these protocols enable fair interactions, eliminate the need for central authorities, and foster censorship resistance. Now there are several means by which consensus is regulated and in the context of this paper it is highlighted by differentiating generation wise evolution of blockchain architecture by its diverse variation of preferences. We are focusing here only on some renowned existing consensus regimes which have real time applications.

Keywords: Blockchain, Consensus protocols, Cryptocurrency, Decentralised peer-to-peer network, Distributed-ledger technology.

Introductory Notions

Blockchain technology enables transaction of currency and assets including tangible belongings such as property, vehicle, cash, etc and intangible belongings such as intellectual properties (patents, copyrights, etc). Precisely, blockchain offers a low-risk, less-expensive, secured mechanism of trading anything of value [15]. Such a platform ensures averting unauthorised transactions and assures consistency in the context of common view of the transactions [31]. Transactions are recorded in a decentralised, protected system—the blockchain.

It is imperative that all parties associated with monetary transactions and business deals bolstered by a blockchain platform accord with each other. This is where the consensus protocol is leveraged, discovering a block to store the authentic transactions, thus building trust among the participants [8]. Consensus, is therefore, a distributed agreement about the status of the ledger acceded by the majority of the participants [43]. In this paper, we discuss different variants of consensus protocols needed for block creation, grouped on the basis of blockchain generations. This is significant for securing information through encryption

[a]suvro5495@gmail.com, [b]rituparna.b@technoindiaeducation.com

and utilising automated group verification. Consensus mechanism is an important arena in the domain of distributed ledgers utilised by enterprises. Various applications buttressed by blockchain platforms are conceived for business and government uses, and consensus mechanisms play an important role there.

This paper is organised as follows: section II presents the background research; section III delineates history of blockchain implementation timelines and their types; section IV includes the list of some appreciable consensus model; finally, we include discussion in section V and portray conclusion notes in section VI.

Background of the Consensus Protocols

One of the primary benefits of consensus algorithms is to achieve reliability in a blockchain network that includes several unreliable and in some cases, unavailable nodes. To cope up with such a situation, consensus algorithms will have to be fault tolerant. Efficient deployment of consensus protocols may lead to more warranted and credible solutions [3]. To reiterate, the consensus in the blockchain is needed for verification of transactions including their order and inclusion in the ledger. The key point is to understand that this consensus is from almost all of the participating nodes. It is responsible for decentralisation of the network by doing these—(i) assurance of the next block in a block chain is one and only version of the truth and (ii) keeps strong adversities from derailing the system by effectively ramifying the chain [14].

There are many ways to divide the blockchain consensus protocols based on their nature, deployment and system application. In the case of categorization by nature we can divide blockchains into three kinds—permissioned, permissionless and a combination of these two. But on the basis of deployment we have four types [12, 26, 27] of model—private, public, federated or consortium, hybrid or public-private [34]. Permissionless blockchain permits anyone to participate pseudonymously like—public deployment model whereas, permissioned blockchain restricts access and limits rights of a subset of nodes on the network like—private and federated deployment model [10, 35], hybrid deployment model [31] is that type which is combination of permissioned and permissionless both. And finally on the basis of system application view [22] it can be categorised into three variations: (i) C2C—only cryptocurrency blockchain (eg. Bitcoin transactional payments), (ii) B2C—business to cryptocurrency blockchain (eg. Ethereum based DApps and smart contracts), (iii) B2B—only business blockchain (eg. Hyperledger Fabric).

There are several research endeavours pursued related to the blockchain consensus protocols, a few of them includes the categorization of the consensus algorithms. In [38], Verma et al. mapped out a categorization of consensus protocols based on the blockchain types such as permissioned, consortium or permissionless. In [35], a similar classification is presented by Kaur et al. Another work that summarises some notable protocols used only in private blockchain is outlined in [20]. An interesting work in this context can be observed in [46] where consensus algorithms are classified into categories involving Synchrony assumption and Non-determinism. In [12], Hameed et al. categorised blockchains based on their generations from 1.0 to 5.0 and also enlisted a few consensus protocols against each of the generation. However, we have not found any study that delineates a comprehensive list of the important algorithms including the most recent ones, categorised on the basis of blockchain generations with relevant descriptions and analysis.

In this paper, our contributions include a novel separation of models by blockchain's generation wise evolutionary phase, which are Blockchain 1.0, 2.0, 3.0, 4.0 and 5.0

respectively. It is totally based on their implementation as per execution for demand and supply in the five respective time periods. Therefore, we have to go through the historical timelines of blockchain implementation to get the motif of their implementation in those five different periods, and the principal thing is to notice the drawbacks of each generation and analyse the progress and transition of the algorithms throughout the evolutions of blockchain architecture.

Historical Timeline of Blockchain Implementation

In this section, we concisely discuss the history of blockchain implementation and how it has evolved over the last decade. The key contribution of this paper is based on this evolution of blockchain, also see Figure 10.1.

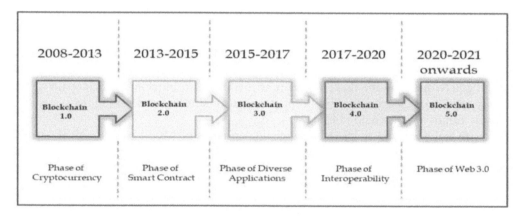

Figure 10.1 Blockchain evolutions.

Generations of blockchain architecture: Blockchain technology has experienced five important developments. Phases 1.0–4.0 [27, 11] are modelled to cater to different lookouts such as functionality, features, strengths, challenges, and security issues. Version-5.0 [12, 41] is currently developing and has many raising issues, and research communities are working on it to improve its functionality for distinct demanding models. Each of these versions are described shortly in the following.

a. **Blockchain 1.0:** From the idea of *Distributed-ledger technology (DLT)* [15] the first type of decentralised cryptocurrency '**Bitcoin**' has emerged [18]. The concepts of digital currency, '**Merkle tree**' [13], blockchain data, small value payments etc. also originated. Basically, it was the era of cryptocurrencies [28, 29].

b. **Blockchain 2.0:** It was the era of smart contracts which is an agreement between two people in the form of computer code. For this, an *Ethereum* blockchain [21] platform has been developed and the required cryptocurrency is '**Ether**'. Also a high level language '**Solidity**' was devised to execute the smart contracts along with some opensource hosting platforms like *Codius* [26], *Hyperledger* etc [28, 29].

c. **Blockchain 3.0:** Here, the decentralised peer-to-peer network concept is utilised rigorously at high speed to store a huge amount of data and to legally support a wide variety of communication mediums. Increment of *Decentralised Applications (DApps)* in umpteen fields like—Electronic voting, Health care sector, Notary system, Identity and access control system etc. are supported by open source blockchain platforms such as *EOS.IO* based on cryptocurrency *'EOS'* [28, 29].

d. **Blockchain 4.0:** This generation of blockchain is devised to resolve the industrial challenges and constraints of real-world applications [11, 27]. The aim is to produce business applications achieving controlled and balanced execution. Industry 4.0 revolution is identified with application of cybersecurity and emerging technologies such as *Industrial Internet of Things (IIoT)* and Artificial Intelligence (AI) [7]. New technology blockchain with AI emerged [23] for enterprise resource planning and asset management.

e. **Blockchain 5.0:** Combination of DLT and AI is happening, plus IoT [5] based data collection is added to it. Here, blockchain integration in *Metaverse* [17] is also an advancement of proprietary blockchain platforms to manage digital assets. Developers of few substantial projects like—*Everscale, Propersix, and Relictum Pro* have independently claimed that they have brought the fully AI-enabled blockchain technology into existence. Peak industrial performance can be noticed in this era with the 5G and 6G network technique [41]. Examples include Big data analytics for efficient supply chain management systems, infrastructures of higher educational institutions with *Internet of Everythings (IoE)* [7], smart city blueprint planning, smart healthcare projects, smart grid system for distribution of power, prevention of various cyberattacks by using suitable consensus protocols.

Numerous Notable Consensus Algorithms

From above, it is understood that consensus regimes are also developed throughout the evolutions of the blockchain network. That means, generation wise our practical preferences gradually switched. It is continuously happening by several urges and proposed designs to fulfil requirements as per demand of business networks along with their execution platforms. Then, we can segregate the protocols by generation wise updated applicability of blockchain network architecture, see the Figure 10.2.

A. *Epoch-1 of Consensus regime (Blockchain 1.0):*

  Here, the consensus protocols executed during the first generation of blockchain are discussed.

  1. *Proof-of-Work(PoW)*—This was the first implemented consensus protocol used by the Bitcoin blockchain [18]. PoW requires miners to solve complex mathematical problems to validate transactions and create new blocks. The first miner to solve the problem and validate the block is rewarded with new coins. Besides Bitcoin, several other blockchains such as Litecoin, Dogecoin, and Monero also use PoW [2, 4, 10, 32].

  2. *Proof-of-Stake(PoS)*—This consensus protocol was first proposed in 2011 and is based on the concept of **"staking"** coins or tokens as a way to secure the network and validate transactions. In a PoS blockchain network, validators are selected based on the amount of coins or tokens they have staked, and they are incentivized to act in the best interests of the network. Examples of blockchains that use PoS include Ethereum, Cardano etc [2, 4, 10, 15, 32, 36].

  3. *Proof-of-Memory(PoM)*—It combines elements of PoW and PoS mechanisms. It requires participants to allocate a significant amount of memory resources to

participate in the consensus process. The computational puzzle-solving aspect of PoW is replaced with memory-intensive operations. Participants are required to allocate and store a certain amount of memory, typically in the form of RAM, to prove their commitment to the network. This commitment is then used as a measure of their stake or influence in the consensus process. It provides a more energy-efficient alternative to traditional PoW consensus mechanisms, as it reduces the reliance on computational power and encourages the use of memory resources instead, where individuals with larger memory resources have a higher chance of being selected to validate blocks [14].

4. *Federated Byzantine Agreement (FBA)*—It is used by Stellar, Ripple, Dispatch blockchain platforms. In FBA, a network is divided into **"quorum slices"**, which are subsets of nodes that have mutually trusted relationships. Each node decides on its own set of quorum slices, and consensus is reached through a process of propagating and evaluating messages among the nodes. Nodes choose which other nodes they trust, forming a decentralised web of trust relationships. In Ripple the validators are pre-selected by the Ripple foundation for eg. XRP ledger is based on this regime. In Stellar (SCP) [25], anyone can be a validator so you choose which validators to trust [9, 14].

5. *Proof-of-Retrievability (PoRet)*—It was used by Microsoft to ensure that data can be retrieved from storage nodes and that the retrieved data matches the original version. This is done by generating a cryptographic proof that can be verified by other nodes. In decentralised storage, it is used to ensure that nodes are storing the data that they are supposed to store. This is done by having each node generate this consensus for the data that it is storing. The proofs are then stored on the blockchain, and other nodes can verify them to ensure that the data is present [15, 19].

Generation 1.0 consensus flaws:
Commonly used PoW and PoS discussed here are for mining and validation of blocks. Blockchain 1.0 version consensus algorithms' speed is on an average 6–7 TPS (transactions per second), entails high energy consumption, so cost is also high, intercommunication is not allowed, and are associated with high adversarial vulnerability like–51% attack, Sybil attack, Selfish mining, eclipse attack for PoW; long-range attack, nothing-at-stake problem, grinding attack [39], cartel formation for PoS etc. In addition to the memory exhaustion attacks on PoM that can lead to denial-of-service (DoS), sybil attacks and federated failure in FBA were happening; availability attacks on PoRet was another challenge in this context. These regimes are not scalable enough to handle a vast process. It was built on one key idea—DLT and the transaction cost is the value addition to reward for block discovery. It was expensive as well as not interoperable. Most of the protocols were bound by payment and transaction purposes i.e., the financial sector often used these regimes [16, 26, 27, 42, 43].

B. *Epoch-2 of Consensus regime (Blockchain 2.0):*
   Here, we discuss the consensus protocols that were executed during the second generation of blockchain.

1. *Practical Byzantine fault tolerance (PBFT)*—Byzantine faults refer to arbitrary or malicious behaviour exhibited by nodes in a network. In PBFT, a designated leader node is responsible for proposing the next block, and the other nodes in the network act as validators. Through a multi-round voting process and digital

signatures, the nodes collectively agree on the validity and ordering of transactions. Once a block is accepted by a two-thirds majority, it is considered finalised and added to the blockchain. PBFT is known for its fast transaction processing times and ability to handle a certain number of faulty or malicious nodes. However, it typically requires a predefined set of known and trusted nodes, *Hyperledger Fabric* uses this solution [9, 23, 26, 33].

2. *Delegated-Proof-of-Stake (DPoS)*—This was introduced by BitShares in 2014 and is based on a system of elected delegates that are responsible for validating transactions and maintaining the network. In a DPoS blockchain network, token holders vote for delegates who are responsible for validating transactions and creating new blocks. EOS and Tron are examples of blockchains that use DPoS [10, 34, 36].

3. *Delegated Byzantine fault tolerance (DBFT)*—This was introduced by NEO in 2014 and is a variation of the DPoS protocol. In a DBFT blockchain network, nodes are chosen to validate transactions based on their stake in the network, and a certain percentage of nodes must agree on a transaction for it to be considered valid [4, 19, 29].

4. *Proof-of-Capacity (PoC)/Proof-of Space*—This uses hard drive storage space instead of computational power to validate transactions. In a PoC blockchain network, validators compete to find solutions to cryptographic challenges, and the solutions are stored on their hard drives. *Burstcoin*, *Chia* and *SpaceMint* are examples of blockchains that use PoC [1, 2, 8, 9, 13, 20].

5. *Proof-of-Personhood (PoP)*—It works by having nodes prove their humanity through a variety of means, such as providing a government-issued ID, taking a selfie, or answering a series of questions about their identity. By ensuring that all nodes in the network are unique human beings, PoP can help to prevent attacks such as *Sybil attacks and double-spend attacks*. There are a variety of ways to prove humanity and uniqueness. Some common methods include: Biometric identification, Human verification and Randomised challenges [15].

6. *Proof-of-Existence (PoE)*—It is used to prove the existence of a certain piece of data or digital asset at a specific point in time. It involves hashing the data and storing a cryptographic proof (commonly used *Merkle tree root hash*) of that hash on the blockchain. This proof can then be used to verify the integrity and time-stamp of the data without revealing the actual content itself. It enables individuals or organisations to establish the existence and ownership of intellectual property, certificates, contracts, or other types of digital assets [8, 37].

7. *Proof-of-Burn (PoBr)*—It requires users to burn or destroy existing coins in order to mine new coins. In a PoBr blockchain network, users burn coins and are then rewarded with new coins based on the amount they burned. *Slimcoin* and *Counterparty* are examples of blockchains that use PoBr [14, 19].

Comparative study of the flaws of generation 2.0 and 1.0:
Now if we compare blockchain 2.0 consensus protocols with blockchain 1.0 consensus protocols, then we see a few advantages. Blockchain 2.0 consensus models are poorly scalable with an average speed of 15–16 TPS. They could be enhanced in terms of interoperability and intercommunication. They were characterised by moderate energy consumption, cheaper cost, emergence of virtual machines and flaws of previous version blockchain 1.0 were partially removed. But they still have some

adversarial vulnerabilities like—bribing or vote buying in DPoS, Distributed-denial of service (DDoS) attacks for PBFT and DPoS both, sybil attack on DBFT, time stamping attack on PoE, pre-computed attack on PoC etc. PoP overcame some of the attacks on PoW and PoS. Here, implementation is almost bound to the non-financial sector [3, 6, 16, 26].

C. *Epoch-3 of Consensus regime (Blockchain 3.0):*
   Here, we discuss the consensus protocols that were executed during the third generation of blockchain.
   1. *Leased Proof-of-Stake (LPoS)*—Here, token holders can lease or delegate their stake to a trusted node or validator, known as a **"forging node."** Also called a *"delegated node"* or *"leased node,"* is responsible for validating transactions and creating new blocks on behalf of the token holders who have leased their stake. The rewards generated through block creation and transaction validation are distributed among the token holders based on the amount of stake they have leased. It was implemented in the cryptocurrency network Waves [15, 29].
   2. *Proof-of-Importance (PoI)*—It was first proposed by the New Economic Movement (NEM) in 2015. It takes into account not only the amount of coins staked by a validator, but also their overall contribution to the network, including transaction history and network activity. Validators with higher PoI scores are more likely to be chosen to validate transactions [2, 13, 19, 23].
   3. *Proof-of-Elapsed Time (PoET)*—This was introduced in early 2016 by the Intel corporation, uses trusted execution environments to randomly select nodes to validate transactions. In a PoET blockchain network, validators are chosen randomly, and the first validator to complete the required task is rewarded with new coins. Sawtooth and Hyperledger Burrow blockchains use PoET [1, 19, 29, 37].
   4. *Proof-of-Activity (PoA)*—It combines PoW and PoS, as proposed by OpenMined Network. Here, the mining process begins like a PoW system, but after a new block has been successfully mined, the system switches to resemble a PoS system. The first miner to solve the PoW puzzle is rewarded with a block reward, but they are also given the right to nominate a group of validators to sign the block. The validators are chosen randomly from the pool of stakeholders, and they are required to stake a certain amount of tokens in order to participate. If a validator votes to approve a fraudulent block, they will lose their stake. The most well-known cryptocurrency that uses the PoA consensus mechanism is *Decred (DCR)* [9, 10, 23, 33].

Comparative study of the flaws of generation 3.0 and previous version:
Compared to blockchain 2.0, blockchain 3.0 consensus models have lower level of decentralisation therefore, some adversarial vulnerabilities have risen, but they are more scalable than previous versions of algorithms and of course the average speed of 3.0 consensus regime is 1K TPS. These models are quite interoperable, and here intercommunication has occurred, they are cheaper than 2.0 generation, energy efficiency is enough, and involve a lesser level of computational complexity than previous versions. Here, verification is based on the key concept of DApps, storage with computing and the main purpose of these consensus models is to solve the problem of business platforms only and that means enterprise level blockchain, hence values are added by organisation boundaries to reward for block discovery. Lastly some noticeable drawbacks of the 3.0 regimes are included one by one—LPoS also has nothing-at-stake

problem and collusion issue, PoET suffers from time-jacking and stake grinding problems, PoA faces sleepy slot attacks and time drifts issue, PoI has problems with stake accumulation and sybil attacks etc [1, 2, 3, 6, 27, 42, 43].

D. *Epoch-4 of Consensus regime (Blockchain 4.0):*
  Here, we discuss the consensus protocols that were executed during the fourth generation of blockchain.
  1. *Proof-of-Authority (PoAu)*—It relies on trustworthy validators instead of a decentralised network of nodes. A group of validators is chosen to validate transactions, and they are typically trustworthy organisations or individuals. The POA Network and Kovan are examples of blockchains that use PoAu [1, 2, 8, 9, 15, 23, 29]. It has been implemented in specific blockchain networks like *Ethereum's Clique consensus* algorithm, which was introduced in 2017.
  2. *Delayed Proof-of-Work (dPoW)*—It was introduced by the Komodo platform. It combines the security of a PoW blockchain with the hashing power of an already established PoW blockchain, such as Bitcoin. Komodo introduced dPoW as a mechanism to provide enhanced security to its blockchain network by leveraging the hashing power and immutability of the Bitcoin blockchain [2, 15, 19, 29].
  3. *Proof-of-Believability (PoB)*—It combines PoW and PoS algorithms. Nodes are selected to create new blocks based on their *"believability"* score, which is derived from their past network contributions and user behaviours. This approach aims to reward nodes that actively participate in the network and have a reputation for behaving honestly and reliably. The protocol also introduces a **"self-selection"** process where nodes have the opportunity to be selected to create blocks. Eg. It is used by IOST to develop IOSChain [2, 9, 15, 19].
  4. *IoT-Proof-of-Work(IoT-PoW)*—It was proposed by the IOTA network, uses a modified version of PoW that is optimised for IoT devices. It is designed to be energy-efficient and scalable, making it ideal for *IIoT DApps*.
  5. *DPoS with AI*—This consensus protocol, proposed by the DeepBrain Chain network, uses a modified version of DPoS that incorporates AI. Nodes are incentivised to contribute computing power to the network, and AI is used to optimise the allocation of computing resources.

Comparative study of the flaws of generation 4.0 and previous version:
Blockchain 4.0 consensus models' level of decentralisation is somewhat coarser than blockchain 3.0 consensus protocols. In 4.0 generation the key concept is based on blockchain with AI or IoT and here, block verification is done through an automation process called sharding [40, 42], which is for distributing a single dataset across multiple databases, which can then be stored on multiple machines [44]. Participants use expensive vector pipelining processors to get reward by discovering and adding the new block. Now compared to 3.0 models, 4.0 models are highly scalable, highly interoperable, cost effective, optimised computational complexity, highly efficient, it came off for the revolutions of industry 4.0 as well as healthcare 4.0, that means they are totally focused on industrial R&D applications. The models' average execution speed is 1M TPS which is very much higher than 3.0 models speed. The notable protocols of 4.0 like—dPoW, IoT-PoW, DPoS with AI have already partially overcome their adversarial vulnerabilities like traditional PoW and PoS algorithms. Noticeable matter that dPoW gave such outstanding results by preventing 51% attack and double-spending attack in

case of Komodo blockchain platform. Another important difference is that 3.0 model block verification is probabilistic but, in case of 4.0 models it is immediate. Since, most of them follow decentralised machine learning algorithms which help an autonomous decision making that adds the values as block discovery reward. 4.0 models also have some drawbacks such as—sybil attacks for all the 4.0 protocols are mentioned here, collusion attack for all except dPoW here, key-compromise and DDoS attack on PoAu, malicious validators issue is faced by PoB [6, 27, 42, 43].

E. *Epoch-5 of Consensus regime (Blockchain 5.0):*
   Here, we discuss the consensus protocols that were executed during the fifth generation of blockchain.

1. *Time-locked Proof-of-Stake(TPoS)*—Here, users **lock up** *(i.e., stake)* some of their cryptocurrency tokens for a set amount of time to show their long-term commitment to the network and earn rewards for helping to validate transactions. The longer the tokens are locked up, the higher the rewards that can be earned. By combining PoB and TPoS consensus mechanisms, Everscale platform aims to provide a secure and decentralised hybrid network with a stable and appreciating token value.

2. *Proof-of-History (PoH)*—It was introduced in 2021 by **Solana** and uses a *verifiable delay function (VDF)* to create a time-based consensus protocol. In a PoH blockchain network, validators are able to reach consensus based on the ordering of transactions, and the VDF [45] provides a way to ensure that transactions are ordered chronologically. Solana is the only blockchain network currently using PoH [2, 9, 15, 19].

3. *Nominated Proof-of-Stake (NPoS)*—Introduced in **Polkadot** blockchain networks in 2020, it is an updated version of DPoS. Here, token holders in the network can nominate a set of validators they trust to secure the network and validate transactions. The validators with the highest nominations become active validators and participate in block production and consensus.

4. *Proof-of-Tsar*—It combines PoS and BFT to provide high network security and scalability. It also claims to have low energy consumption compared to other consensus mechanisms like PoW. The developers of Relictum Pro claim that it provides a high level of security and performance for their blockchain platform.

5. *Proof-of-Six (PoSIX)*—Propersix platform is built on this custom regime, it is designed to provide high-speed transactions with low transaction fees while maintaining a high level of security. It combines elements of both proof PoS and DPoS consensus mechanisms, and it uses a combination of master nodes and regular nodes to validate transactions and create new blocks.

6. *Proof-of-Space-Time(PoST)*—This is proposed by Chia Network that uses unused hard drive space to secure the network. It requires nodes to prove that they have stored a certain amount of data for a specific amount of time. It was implemented in the Filecoin blockchain network. Here, participants earn rewards for storing and providing access to data. PoST ensures that the participant has allocated and maintained storage space over a designated period [15].

7. *Proof-of-Replication (PoRep)*—It guarantees to decentralised ecosystems that a miner can store and present a replica of data on a distributed storage space. It

combines PoC and PoRet both, which lets users store data on a server, check whether it is stored, and eventually support data retrieval. On the flip side, PoC gives the permission for a user to outsource storage space to a network server. PoRep makes assurance that—(i) The prover must be using as much space to produce the proof as replicas it claims to store, (ii) The prover can retrieve a committed data file, (iii) The prover can use the capacity to store this file without any overhead. Eg. *Filecoin mainnet and Storj* is the most popular cloud storage blockchain that uses the PoRep consensus algorithm [15].

8. *Proof-of-Authentication(PoAh)*—It is mainly designed to be lightweight and efficient for resource-constrained devices, such as those used in the IoT. It works by having nodes compete to be the next block producer. The node that is chosen is the one that can provide the most convincing proof of authentication. It is based on the *ElGamal* cryptographic system and uses digital signatures to verify the authenticity of transactions. It can be provided in a variety of ways, such as by using a *physical unclonable function (PUF)* or by using the mentioned cryptographic signature. It has been implemented in a number of blockchain projects, such as *IoTeX* and *Syscoin* and on a number of different IoT platforms, including *Raspberry Pi* and *Arduino*. It has also been used in a number of real-world applications, such as smart cities [24] and environmental monitoring [15].

Comparative study of the flaws of generation 5.0 and previous version:

If we notice the level of decentralisation between 5.0 models and 4.0 models then, we can see that 4.0 is to some extent finer than 5.0 but, comparing execution speed 5.0 models are far better than 4.0 as 5.0 models have nearly 1G TPS speed (one can get this on the Relictum Pro platform where proof-of-tsar is used). Key concept of this 5.0 generation is based on integration of DLT, AI and IoT based data collection, storage, processing, analysis and management. Basically here, block verification occurs via an automation process for distributing multiple dataset across multiple databases, which can then be stored on multiple machines [44, 45]; it is done by database sharding [40] and database partitioning both. Mainly these three—Recursive decomposition, data decomposition and exploratory (dynamic) decomposition methods are involved here to carry the process on. Very sophisticated types of machinery are used for the next block discovery method where multicore parallel processing is involved. Decentralised AI algorithms are utilised here for automated quick decision making processes. Compared to 4.0 generation, 5.0 models are obviously more scalable, cost effective, energy efficient, with greater optimisation level of model complexities etc. 5.0 models came of by the influence of decentralised version of web, that is web 3.0 which has affect in all of the rising technologies like—deep learning models, quantum computing, quantum cryptography, cloud computing architecture, IoE, Metaverse etc. Version 5.0 actually came off as an onset of the era of Digital society [41]. Envisage now the probable adversarial impacts on this 5.0 generation consensus protocols—instances of attacks in Solana caused by PoH are not rare. The protocol lacks thorough testing and it cannot be claimed that it will execute without any errors. There are possibilities for other attacks like—sybil and eclipse attacks on PoAh, pre-computation and time-jacking attacks on PoST, long-range and bribery attacks on NPoS [6], interconnection of nodes here can bring DDoS attacks [7].

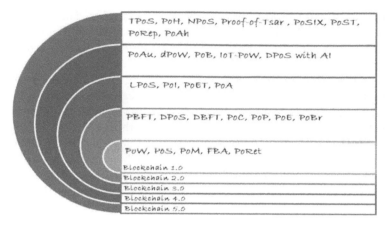

Figure 10.2 Generation wise segregation of the consensus protocols.

Discussion

We can see that every consensus regime has a different perspective of its own working procedure. For example—PoW is best suited for mining, PoS is well-suited for validation or forging with a good level of security. So the combination of these two in PoM, PoA and PoB all gave strong results as, in case of fraudulent prevention PoA works well, in case of memory intensive operation through an agreement PoM securely does etc. Same goes in case of voting purposes-FBA is for quorum based system, PBFT is for special majority system etc. In case of any summit or conference where the presence of delegates or representatives is playing a critical role, DPoS and LPoS work well.

In case of a confidential agreement of meeting, verifiers can use PoP protocol to check everyone's background, again one can use PoB protocol to maintain privacy with integrity and also can use PoAu regime to carry on the trustworthiness, sincerity among the members. While executing a bigger level of intergovernmental treaty on a far-flung flourishing project where high expenditure is involved from every organisation in that case, DBFT may be utilised where transaction finality is crucial, PoI may be used to hinder money hoarding, PoBr may be suitable to wane money circulation, and PoET to be ensured equality with fairness among participants. For digital asset management, IPRs, ownership transferring, e-Bonds selling and buying, PoE and PoI together can help a lot. Again, PoM, PoAh, PoRet and PoC can be used together in IoT layers protocol on cloud based storage, analysis and authentication where huge data processing is involved, it can be implemented in the health care sector where patients' data privacy is the key thing.

Conclusion Notes

This paper has accumulated some of the substantial consensus protocols, used in several generations of blockchain evolution. Selecting a correct consensus model is quite difficult as it is totally dependent on what the problems are. The decentralised peer-to-peer nature is very much compatible with deployment models of Cloud computing architecture. Moreover, IoT plus AI enabled models will be needed to assemble together to get more powerful results for managing large scale supply chains such as—IoT-PoW, DPoS with AI, PoRep, PoSIX, TPoS, Proof-of-Tsar etc. Therefore, to get it done further research and rigorous experiments will be needed to ready many tools for advanced model making as well as fulfil several demands by enhancing performance. It is observed that integration of consensus models is more efficient than their performance individually. We aim to study more on various mechanisms for making a broader vision of their applicability.

References

1. Azbeg, K., Ouchetto, O., Jai Andaloussi, S., and Fetjah, L. (2021). An Overview of Blockchain Consensus Algorithms: Comparison, Challenges and Future Directions. In: Saeed, F., Al-Hadhrami, T., Mohammed, F., Mohammed, E. (eds) Advances on Smart and Soft Computing. Advances in Intelligent Systems and Computing, 1188. Springer, Singapore. https://doi.org/10.1007/978-981-15-6048-4_31

2. Bhattacharjya, A., Wisniewski, R., and Nidumolu, V. (2022). Holistic research on blockchain's consensus protocol mechanisms with security and concurrency analysis aspects of CPS. *Electronics,* 11(17), 2760. https://doi.org/10.3390/electronics11172760.

3. Bosamia, M., and Patel, D. (2020). Comparisons of blockchain based consensus algorithms for security aspects. *International Journal on Emerging Technologies,* 11(3), 427–434.

4. Chaudhry, N., and Yousaf, M. M. (2018). Consensus algorithms in blockchain: comparative analysis, challenges and opportunities. In 12th International Conference on Open Source Systems and Technologies (ICOSST), December, 10. http://dx.doi.org/10.1109/ICOSST.2018.8632190.

5. Ferrag, M. A., Shu, L., Yang, X., Derhab, A., and Maglaras, L. (2020). Security and privacy for green IoT-based agriculture: review, blockchain solutions, and challenges. *IEEE Access,* PP(99) 8, 23. http://dx.doi.org/10.1109/ACCESS.2020.2973178.

6. Guru, A., Mohanta, B. K., Mohapatra, H., Al-Turjman, F., Altrjman, C., and Yadav, A. (2023). A survey on consensus protocols and attacks on blockchain technology. *Applied Sciences,* 13(4 February), 21. http://dx.doi.org/10.3390/app13042604.

7. Hewa, T., Gür, G., Kalla, A., Ylianttila, M., Braeken, A., and Liyanage, M. (2020). The role of blockchain in 6G: challenges, opportunities and research directions. In 6G Wireless Summit, March, (p. 5). http://dx.doi.org/10.1109/6GSUMMIT49458.2020.9083784.

8. Islam, S., Islam, M. J., Hossain, M., Noor, S., Kwak, K. S., and Islam, S. M. R. (2023). A survey on consensus algorithms in blockchain-based applications: architecture, taxonomy, and operational Issues. *IEEE Access,* 4(April), 17. http://dx.doi.org/10.1109/ACCESS.2023.3267047.

9. Ismail, L., and Materwala, H. (2019). A review of blockchain architecture and consensus protocols: use cases, challenges, and solutions. *Symmetry,* 11(10), 1198. https://doi.org/10.3390/sym11101198.

10. Kaur, M., Khan, M. Z., Gupta, S., Noorwali, A., Chakraborty, C., and Pani, S. K. (2021). MBCP: performance analysis of large scale mainstream blockchain consensus protocols. *IEEE Access,* 9(May), 80931–80944. https://doi.org/10.1109/ACCESS.2021.3085187.

11. Khan, M., Den Hartog, F., and Hu, J. (2022). A survey and ontology of blockchain consensus algorithms for resource-constrained IoT systems. *Sensors,* 22, 8188. http://dx.doi.org/10.3390/s22218188.

12. Hameed, K., Barika, M., Garg, S., Amin, M. B., and Kang, B. (2022). A taxonomy study on securing Blockchain-based Industrial applications: An overview, application perspectives, requirements, attacks, countermeasures, and open issues. *Journal of Industrial Information Integration,* 26, 100312, ISSN 2452-414X, https://doi.org/10.1016/j.jii.2021.100312. (https://www.sciencedirect.com/science/article/pii/S2452414X21001060)

13. Krishnamurthi, R., and Shree, T. (2021). A brief analysis of blockchain algorithms and its challenges. In N.p.: Research Anthology on Blockchain Technology in Business, Healthcare, Education, and Government, (pp. 23–39). http://dx.doi.org/10.4018/978-1-7998-5351-0.ch002.

14. Kumar, S., and Saxena, A. (2020). Blockchain Technology: Concepts and Applications. 1st ed. New Delhi: Wiley India Pvt. Ltd.

15. Lashkari, B., and Musilek, P. (2021). A Comprehensive Review of Blockchain Consensus Mechanisms. In IEEE Access, 9, 43620–43652, doi: 10.1109/ACCESS.2021.3065880.

16. Makhdoom, I., Abolhasan, M., and Ni, W. (2018). Blockchain for IoT: the challenges and a way forward. In 15th International Joint Conference on e-Business and Telecommunications, 1 July, (p. 12). http://dx.doi.org/10.5220/0006905605940605.

17. Mourtzis, D., Angelopoulos, J., and Panopoulos, N. (2023). Blockchain integration in the era of industrial metaverse. *Applied Sciences,* 13(3), 29. http://dx.doi.org/10.3390/app13031353.

18. Nakamoto, S. (2008) Bitcoin: A Peer-to-Peer Electronic Cash System. https://www.ussc.gov/sites/default/files/pdf/training/annual-national-training-seminar/2018/Emerging_Tech_Bitcoin_Crypto.pdf

19. Oyinloye, D. P., Teh, J. S., Jamil, N., and Alawida, M. (2021). Blockchain consensus: an overview of alternative protocols. *Symmetry,* 1363(13), 35. https://doi.org/10.3390/sym13081363.

20. Pahlajani, S., Kshirsagar, A., and Pachghare, V. (2019). Survey on private blockchain consensus algorithms. In 1st International Conference on Innovations in Information and Communication Technology (ICIICT), (p. 6). http://dx.doi.org/10.1109/ICIICT1.2019.8741353.

21. Paulavičius, R., Grigaitis, S., Igumenov, A., and Filatovas, E. (2019). A decade of blockchain: review of the current status, challenges, and future directions. *Informatica,* 30(4), 729–748. http://dx.doi.org/10.15388/Informatica.2019.227.

22. Sabry, S. S., Kaittan, N. M., and Majeed, I. (2019). The road to the blockchain technology: concept and types. *Periodicals of Engineering and Natural Sciences (PEN),* 7(4), 1821–1832. http://dx.doi.org/10.21533/pen.v7i4.935.

23. Salah, K., Rehman, M. H., Nizamuddin, N., and Al-Fuqaha, A. (2019). Blockchain for AI: review and open research challenges. *IEEE Access,* 7(January), 10127–10149. https://doi.org/10.1109/ACCESS.2018.2890507.

24. Salha, R. A., El-Hallaq, M. A., and Alastal, A. (2019). Blockchain in smart cities: exploring possibilities in terms of opportunities and challenges. *Journal of Data Analysis and Information Processing,* 7(3), 118–139. https://doi.org/10.4236/jdaip.2019.73008.

25. Salimitari, M., and Chatterjee, M. (2019). A survey on consensus protocols in blockchain for IoT networks. *IEEE Internet of Things Journal,* Volume 4, 15. https://doi.org/10.48550/arXiv.1809.05613.

26. Sayadi, S., Rejeb, S. B., and Choukair, Z. (2018). Blockchain challenges and security schemes: a survey. In Seventh International Conference on Communications and Networking (ComNet), November, (p. 7). http://dx.doi.org/10.1109/COMNET.2018.8621944.

27. Singh, S. K., and Vadi, V. R. (2022). Evolutionary transformation of blockchain technology. *International Journal of Engineering Research and Technology (IJERT),* 10(1), 26–30.

28. Tanriverdi, M., and Tekerek, A. (2019). Implementation of blockchain based distributed web attack detection application. In 1st International Informatics and Software Engineering Conference (UBMYK), November, (p. 6). http://dx.doi.org/10.1109/UBMYK48245.2019.8965446.

29. Thanujan, T., Rajapakse, R. A. C., and Wickramaarachchi, D. (2020). A review of blockchain consensus mechanisms: state of the art and performance measures. In 13th International Research Conference, Computing Sessions, no. Paper ID: 211, (pp. 315–326).

30. Wang, H., Zheng, Z., Xie, S., Dai, H. N., and Chen, X. (2018). Blockchain challenges and opportunities: a survey. *International Journal of Web and Grid Services,* 14(4), 352–375. http://dx.doi.org/10.1504/IJWGS.2018.10016848.

31. Wang, W., Hoang, D. T., Hu, P., Xiong, Z., Niyato, D., Wang, P., Wen, Y., and Kim, D. I. (2019). A survey on consensus mechanisms and mining strategy management in blockchain networks. *IEEE Access,* PP(99), 1-1(January), 43. http://dx.doi.org/10.1109/ACCESS.2019.2896108.

32. Yadav, A. S., Singh, N., and Kushwaha, D. S. (2021). A scalable trust based consensus mechanism for secure and tamper free property transaction mechanism using DLT. *International Journal*

of *Systems Assurance Engineering and Management*, 13(3), 17. http://dx.doi.org/10.1007/ s13198-021-01335-0.

33. Yadav, A. K., and Singh, K. (2020). Comparative analysis of consensus algorithms of blockchain technology. In Ambient Communications and Computer Systems, Advances in Intelligent Systems and Computing 1097, March, (pp. 205–218). https://doi.org/10.1007/978-981-15-1518-7_17.

34. Zhang, S., and Lee, J. H. (2020). Analysis of the main consensus protocols of blockchain. *ICT Express*, 6(2), 93–97. http://dx.doi.org/10.1016/j.icte.2019.08.001.

35. Kaur, M., and Gupta, S. (2021). Blockchain consensus protocols: state-of-the-art and future directions. In 2021 International Conference on Technological Advancements and Innovations (ICTAI), (pp. 1–8). DOI: 10.1109/ICTAI53825.2021.9673260.

36. Pan, J., Song, Z., and Hao, W. (2021). Development in consensus protocols from PoW to PoS to DPoS. In 2021 2nd International Conference on Computer Communication and Network Security (CCNS), (pp. 1–6). DOI: 10.1109/CCNS53852.2021.00020.

37. Sankar, L. S., Sindhu, M., and Sethumadhavan, M. (2017). Survey of consensus protocols on blockchain applications. In 2017 International Conference on Advanced Computing and Communication Systems, (ICACCS-2017).

38. Verma, N., Jain, S., and Doriya, R. (2017). Review on consensus protocols for blockchain. In 2021 International Conference on Computing, Communication, and Intelligent Systems (ICCCIS), (pp. 281–286). DOI: 10.1109/ICCCIS51004.2021.9397089.

39. Xiao, Y., Zhang, N., Lou, W., and Hou, Y. T. (2020). A survey of distributed consensus protocols for blockchain networks. *IEEE Communications Surveys and Tutorials*, 22(2), 1–34. doi: 10.1109/COMST.2020.2969706.

40. Chen, R., Wang, L., Peng, C., and Zhu, R. (2022). An effective sharding consensus algorithm for blockchain systems. *Electronics*, 11, 2597. https://doi.org/10.3390/ electronics11162597.

41. Wenhua, Z., Qamar, F., Abdali, T. A. N., Hassan, R., Jafri, S. T. A., and Nguyen, Q. N. (2023). Blockchain technology: security issues, healthcare applications, challenges and future trends. *Electronics*, 12, 546. https://doi.org/ 10.3390/electronics12030546.

42. Mukherjee, P., and Pradhan, C. (2021). Blockchain 1.0 to blockchain 4.0—the evolutionary transformation of blockchain technology. In Blockchain Technology: Applications and Challenges, (pp. 29–49). http://dx.doi.org/10.1007/978-3-030-69395-4_3.

43. Sayeed, S., and Gisbert, H. M. (2019). Assessing blockchain consensus and security mechanisms against the 51% attack. *Applied Sciences*, 9, 1788, (April), 17. http://dx.doi.org/10.3390/ app9091788.

44. Halgamuge, M. N., and Mapatunage, S. P. (2021). Fair rewarding mechanism for sharding-based blockchain networks with low-powered devices in the internet of things. In IEEE 16th Conference on Industrial Electronics and Applications (ICIEA), August, (p. 2). http://dx. doi.org/10.1109/ICIEA51954.2021.9516281.

45. Xi, J., Zou, S., Xu, G., Guo, Y., Lu, Y., Xu, J., and Zhang, X. (2021). A comprehensive survey on sharding in blockchains. *Hindawi, Mobile Information Systems*, 2021, 22. Article ID: 5483243. https://doi.org/10.1155/2021/5483243.

46. Jayabalan, J., and Jeyanthi, N. (2021). A study on distributed consensus protocols and algorithms: the backbone of blockchain networks. In 2021 International Conference on Computer Communication and Informatics (ICCCI), Coimbatore, India, 2021, (pp. 1–10). doi: 10.1109/ ICCCI50826.2021.9402318.

11 Distributed Denial-of-Service Attacks Detection using Machine Learning Approaches

Arpita Shome[1,a] and Munmun Bhattacharya[2,b]

[1]Assistant Professor, Abacus Institute of Engineering & Management, Magra, Hoogly

[2]Assistant Professor, Jadavpur University, Department of Information Technology, Kolkata

Abstract

With the massive proliferation of internet systems and technologies, the security systems of the networks must also be strengthened. A distributed denial of service (DDoS) attack is the most threatening type of network security attack. The purpose of this paper is to give a comparative analysis of machine learning techniques for detecting DDoS attacks. It provides a machine learning-based DDoS assault detection system that consists of four steps: The first one is data preprocessing. In this paper, Principle Component Analysis (PCA) is utilised in the preprocessing of a dataset for feature selection and dimension reduction. Following the preprocessing of the data, the retrieved attributes are sent into machine learning classifiers for classification. Numerous machine learning algorithms such as Logistic Regression, Decision Tree, Random Forest, SVM, KNN, and Naïve Bayes employ an Intrusion Detection System-based Corrected KDD9 dataset for analysis. While classifying the kinds of input features in the training step. The results were tested in the third step by using the previously learned datasets. The suggested system performance assessment prosody, such as appropriateness, detection, precision, F1-score and were gathered in fourth stage. Suggested model is trained and tested using the corrected KDD99 dataset. When all of the above mentioned classifiers are compared to their classification results, the highest accuracy of 99.98%, 100% detection rate, and 99.9% F1-score was attained when the PCA = 13.

Keywords: DDoS, KDD99, KNN, machine learning, PCA, random forest, SVM.

Introduction

With the continuous development of technology and the ceaseless expansion of network business needs with the fast expansion of the digital economy in the online world, network services with significant business and technological information have expanded to meet society's production and daily lives. Today, the vast majority of individuals rely on the internet to carry out their daily tasks. One of the most important security systems is an intrusion detection system [8]. As a result, network security is critical, especially whenever highly confidential data is involved. To ensure network security, a variety of tools are

[a]pujashome19@gmail.com, [b]munmun.bhattacharya@jadavpuruniversity.in

used. When identifying attacks with an anomaly-based intrusion detection system, many features of internet traffic must be evaluated. This system must perform an analysis on a massive amount of data, and the output must avoid a high rate of false positives.

DDoS attacks, or Distributed Denial-of-Service attacks, express a serious threat to online services. Introduction the DDoS attacks can cause anomalies in connected network services, resulting in significant financial losses and potentially disastrous outcomes. Detecting DDoS assaults effectively and rapidly is an important research area in the security industry. A DDoS attack, the attacker exploits zombie devices and takes control, flooding a network with excessive traffic. IDS use packet analysis to identify DDoS attacks using signature-based and anomaly-based methods [10]. Machine learning techniques separate DDoS and regular traffic, reducing noise and improving supervised part accuracy. DDoS attacks like HTTP flood, Smurf, and UDP flood are detected and analyzed to determine network traffic type.

Machine learning techniques are fully data-driven. The Corrected KDD99 dataset has been used in this paper. Here, in this paper, we use PCA (Principle Component Analysis) for dimension reduction and feature selection. By processing data, the set of data has been separated into testing and training sets. Following that, SVM, KNN, Decision Tree model, Gaussian Naive Bayes and logistic Regression classifiers are applied the training data. Then evaluate the model utilising test data and obtain the result.

The structure of this thesis is as follows: Section II conveys a review of the literature on various DDoS detection approaches using various classifiers, Section III describes the DDoS detection approach, Section IV represents various classifiers used in this study and Section V describes the experimental setup used in this paper. Section VI analyse the obtained result and Section VII concludes by summarising the paper and discusses the conclusions and future work.

Literature Review

Several detection systems have been developed over time to safeguard networks from various forms of assaults (both internal and external). For DDoS detection, machine learning approaches are considered as an effective technology that can detect attacks effectively. Hema & Shyni in their research [12], discussed a traffic classification technique using Naïve Bayes predictions for aggregating traffic flows with the minimal availability of training data. Using statistical features extraction, their scheme can classify packets more effectively, with an accuracy of 92.34 percent. But it will works only within a Local Area Network. According to Barati et al. [4], boosting the detection efficiency of DDoS attacks is possible. by utilizing a machine learning technique that combines a genetic algorithm and an artificial neural network. But they have tested their method only with very small dataset.

L. Schehlmann and H. Baier using the OpenFlow protocol proposed COFFEE [20] for detection and elimination of botnet activity. Machine learning techniques were used to detect the attack, and an OpenFlow protocol was used to react to the attack by installing greater priority rules. But their approach will not work in real-time environment. Idhammad et al. [13] presented a semi-supervised method to reduce normal traffic in order to improve DDoS detection by removing the noisy proportion of data. Alexander et al. [19] used SDN's centralized control feature to identify attack traffic, using genetic algorithm and kernel principal component analysis (KPCA) for SVM models. GA optimizes parameters, and N-RBF reduces noise caused by feature discrepancies. Most of the researches are focused on datasets. To increase detection accuracy, the study in [22] presented a multiple ANN-based assault detection technique to increase the false-positive rate by reducing the false-negative rate. The dataset implemented in the research, NSL-KDD, is acknowledged for the standard cyber-attack detection technique. In 2016, Alkasassbeh et al. [3] created a dataset with 27 features and 5 groups for detecting DDOS attacks targeting application and network layers. The MLP classifier achieved the highest accuracy rate at 98.63 percent, followed by RF at 98.02 percent and NB at 96.91 percent.

Abdalelah et al. [1] proposed a scheme by combining two datasets, that was trained and tested using the MIX dataset PORTMAP and LDAP, taken from the CICDDOS2019 datasets. The PCA approach is used eight times for various characteristics to minimize the dataset's dimensions. Hongmei and Jiaqing [17], proposed a novel classification system based on principal component analysis and attribute correlation coefficient. The technique eliminating redundant characteristics, the resultant attribute subset satisfies the assumption of the Naive Bayesian classifier.

As a conclusion, it can be said that for detecting or preventing DDoS attacks various machine learning (ML) approaches—artificial neural networks, genetic algorithm, and fuzzy clustering etc. have been used by the researchers, but as feature selection is limited, an approach to detect DDoS attack with low false alarm and high accuracy is still an open research issue.

Detection Approach

The difficulty with a DDoS attack is simulating the flow of traffic on the intended network because all network traffic has two characteristics: linear traffic and burst traffic. Machine learning techniques classify observed objects using training data. A suggested generic incursion is offered as a solution for detection model. There are some stages to the raised DDoS attack detection architecture, in Figure 11.1,

- Data Preprocessing
- Detection classifiers
- Classification
- Performance Evaluation

Preprocessing

A dataset, which is an assortment of data arranged for a particular issue, is necessary for a predictive model. The Corrected-KDD data set, which can be noisy, missing, or inappropriate, is used in the suggested model architecture. Preparing unstructured data for application in a model for classification is a process known as data preprocessing. Building a classification model starts with cleaning the data, which improves accuracy and productivity. This procedure involves completing blanks in the data, reducing noise, and resolving discrepancies. Getting a preprocessed dataset, employing a feature selection strategy, and examining the data are all part of the preprocessing stage. There are mainly three steps in the preprocessing stage.

Encoding: We frequently work with feature sets that involve several labels and one or even more columns in learning algorithms. Constantly readable or comprehensible phrases are used to label the training data. In learning algorithms for transforming classified variables into numeric format, label encoding is a critical preprocessing step. Better predictive

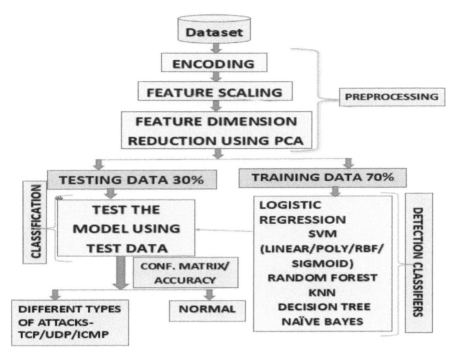

Figure 11.1 DDoS detection architecture.

models are produced as a result of machine learning algorithms being able to understand and apply labels to training data more effectively.

Feature Scaling: Data scaling is a suggested preprocessing step. The algorithm for feature scaling is: Rescaling the range of data to find the distribution mean so that the mean of observed data is 0. Find the standard deviation for each attribute so that it should be 1. The formula to calculate the new data point:

$$z = \frac{x - u}{\sigma} \tag{1}$$

Where u is the training datasets' mean or zero, and σ is the standard deviation of the training datasets. Centering is the process of subtracting the mean from the data, whereas scaling is the process of dividing by the standard deviation. As a result, the procedure is also referred to as "center scaling."

Dimension Reduction: The intrusion detection dataset in Corrected KDD contains higher-dimensional data for analysis. Statistical analyses like feature selection and dimensionality reduction are necessary. Principal Component Analysis (PCA) is a popular machine learning technique for reducing dimensionality while preserving critical information. The dataset has 42 columns, where each row corresponds to one independent variable and one dependent variable:

After standardization, we get matrix of attributes Z. Multiply it by its transpose to get the covariance matrix of Z. Mathematically, ZTZ. To discover the principal components, Eigen decomposition is required, which is the decomposition of ZTZ into AKA−1, where A is the eigenvector matrix and K is a diagonal matrix with zero eigenvalues outside of the diagonal and eigenvalues there. The Eigen values on K's diagonal will be connected with the appropriate column in A—for example. K's first element is λ_1, and the respective Eigen vector is A's first column. Arrange the eigenvalues $\lambda_1, \lambda_2 \dots \lambda_n$ in ascending order and then the eigenvectors in A according to this. After using Principal Component Analysis, a feature selection [21] strategy is used on the preprocessed data. Here is the algorithm of feature selection:

After calculating all the eigenvalues, λ_i, where i = 1,2,..., n. Divide the preserved eigenvalues by the total eigenvalues to get the proportion of variance stated for each feature. Choose the major components, q. Determine the accumulative contribution rate, Q of the main components. The q value is mostly determined by the size of the rate of cumulative contribution. That is, the cumulative contribution rate must be more than 85 percent in order for the new features to convey the most information of the original characteristics.

$$Q = \frac{\sum_{i=1}^{q} \lambda i}{\sum_{i=1}^{n} \lambda i} \tag{2}$$

Finally, we may choose all of the required features equivalent to 1, ..., q.

Out of dataset make a training set and a testing set. For example, 69.99% of the dataset's rows may be randomly selected for training, while the remaining 30% could be used for testing

Detection Classifiers

A detection classifier is used in machine learning to learn patterns in data and then to categorize data into distinct groups based on those patterns. To determine how a given set of input variables is related to the class, a classifier uses some training data. Machine learning programmes use a variety of methods to categorize future datasets into appropriate and relevant categories. We employed six classifiers in this work. Logistic regression, Naïve Bayes, KNN, Decision Tree, SVM-Kernels, and Random Forest classifier.

Logistic Regression

The algorithm of logistic regression is based on the principle of probability and is used to do predictive analysis. Logistic regression is a classification model that, in its most basic form, to represent of employs a logistic function a variable of binary dependent, requiring the output to be discrete. True or false might be 0 or 1. Rather than supplying specific numbers like 0 and 1, give a range of values

Naïve Bayes classifier: The Naïve Bayes classifier illustrates applying and learning probabilistic information and provides features-learning that is independent of the provided class [17]. It, like decision tree algorithms, is developed to be used in a supervised manner. It is founded on the Bayesian theorem and is especially useful when the input has a large dimensionality

K-nearest neighbours (KNN): The classifier is employed for classification and regression issues. It calculates the separation between training data points and test data points, predicting the relevant class. The algorithm returns the anticipated class with the highest probability after sorting distances and discovering common rows. The value in classification is the average of the 'K' training points that were selected.

Decision tree classifier: Decision trees are computational methods that divide a dataset into segments based on criteria and produce workable solutions based on the properties of the dataset. They begin at the root node, compare root property values to dataset attributes, and then branch to the next node, checking attribute values with sub-nodes. The algorithm repeats until reaching the leaf node.

SVM Classifier: An SVM model is a supervised machine learning algorithm based on the training data to identify the optimal hyperplane. Support vectors are the training samples that are closest to the hyperplane. According to statistical learning theory, best classifier is found by increasing the margin [18]. This can be stated as a problem of minimization. Maximizing, on the other hand, is the same as minimizing.

The nonlinear classification is supported by SVM, which is a linear classifier with kernel functions. The types of commonly utilized functions of kernel are:

- Sigmoid
- Polynomial
- linear Radial basis function (RBF).

SVM classifies data points by mapping them to a high dimensional feature space. The kernel function is the mathematical function that is utilized to alter the data in SVM. The nonlinear classification is supported by the Radial Basis Function (RBF) kernel.

Random Forest Classifier: Random Forest is an ensemble-based instruction classifier that integrates multifarious classifiers to enhance execution and solve challenging issues. It collects forecasts from each tree, predicting output based on majority votes. The classification stage tests the model against trained pattern data, categorising normal or multi-type attacks.

Performance Evaluation

The confusion matrix is used in the performance evaluation stage. The proposed method uses the Corrected-KDD dataset. Computer running Windows 10 Professional, we set up our experimental configuration with Intel® CoreTM i3-1115GRE processor running at 3.90 GHz, a 2.19 GHz processor, and RAM of 8 GB. A DDoS detection program built in Python was created using various libraries in our environment. During start up, we may need to install these libraries, such as Scikit-learn, Matplotlib, Pandas and NumPy. Scikit-Learn [11] is the key library for machine learning in our study. It has a large number of algorithms, performance measures, and optimization approaches.

DATASET DESCRIPTION: KDD-99 assessed IDS system employing the DARPA KDD99 Intrusion Detection Evaluation dataset. The 30 H. Altwaijry Competition, which was hosted in association with the Lincoln Laboratory at MIT, developed this dataset. It was the third international competition, the 5th International public Conference on Knowledge Exploration & Data Mining. This dataset [7, 6] is the most one authentic openly obtainable sets of genuine assault. The KDD99 data set [2] comprises three distinct groups: "full KDD," "10 percent KDD," & "corrected KDD."

The experiments in this paper use the Corrected KDD dataset as the input dataset. The features of the dataset have been presented in Table 11.1. This dataset contains 311029 single connection records with 41 features (plus the column 42, class type of the attack this is), gradually. The data in the Corrected KDD dataset has a different statistical distribution than the data in the "10 percent KDD" or "full KDD" datasets.

Basic features are listed from 1–9 followed by content features from 10–22, traffic features from 23–31, and host-based features from 32–41. A detailed list of the features [9] was established for connection records and forms of attacks in the Corrected KDD dataset.

Table 11.1: Basic features of corrected-KDD dataset.

Serial No.	Feature Name	Serial No.	Feature Name	Serial No.	Feature Name	Serial No.	Feature Name
1	Tenn	12	Loggedin	23	Amount	34	D sthostsamesrvrate
2	Protocol type	13	Numcompromised	24	Srvcount	35	Dstliostdiffsrvrate
3	Service	14	Rootsliell	25	S errorrate	36	Dsthostsamesrcportrate
4	Flag	15	Suattempted	26	Srvserrorrate	37	Dstliost srvdifihostrate
5	Srcbytes	16	Numroot	27	Rerrorrate	38	Dsthostserrorrate
6	Dstbytes	17	Numfilecreations	28	Srvrerrorrate	39	Dsthostsrvsenorrate
7	Land	18	Numshells	29	Samesrvrate	40	Dsthostrerrorrate
8	Wrongf ragmen!	19	Numaccessfiles	30	Diffsrvrate	41	Dsthostsrvrerronate
9	Urgent	20	Numoutboundcmds	31	Srvdifihostrate		
10	Hot	21	Ishotlogin	32	Dsthostcount		
11	Numfailedlogins	22	Isguestlogin	33	Dsthostsrvcount		

Table 11.2: Most relevant features of our approach.

No	Feature Name	Description	DDoS Attack
1	duration	Length of time duration of the connection	Smurf
2	Protocol type	Protocol used in the connection	Smurf
3	service	Destination network service used	Smurf
4	flag	Status of the connection—Normal or Error	Neptune
5	src bytes	Number of bytes transferred from source to destination	Smurf
6	dst bytes	Number of data bytes transferred from destination to source in single connection	Smurf
7	land	1 if source and destination IP addresses and port numbers are equal and 0 otherwise	Land
8	Wrong fragment	Total number of wrong fragments in this connection	Teadrop
23	count	Number of connections to the same destination host as the current connection in the past two seconds	Smurf
30	diff srv rate	Percentage of the connections which were going to different services, amongst the connections aggregated in (count)	Neptune
33	dst host srv count	Number of connections having the same port number	Neptune
36	dst host same srv port rate	The percentage of connections that were to the same source port, among the connections aggregated in dst host srv count	Smurf
42	Result	Attack results	

Principal component analysis was used for feature selection of the forty-two dimensions. The outcome of the PCA approach is a collection of features that is a linear combination of the original set. So, after feature selection, 13 features out of 42 are more relevant for prediction. These 13 selected features are shown in Table 11.2.

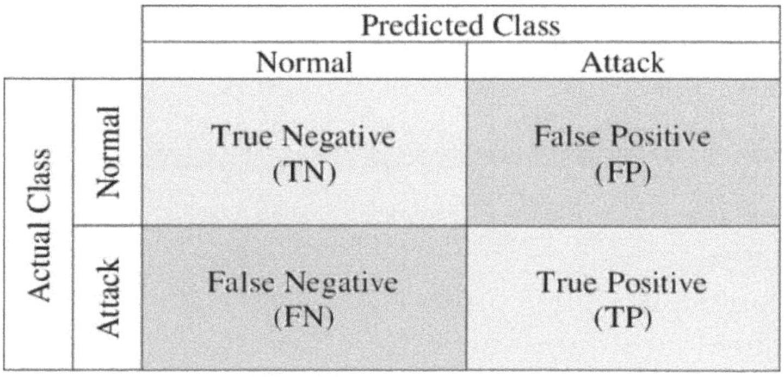

		Predicted Class	
		Normal	Attack
Actual Class	Normal	True Negative (TN)	False Positive (FP)
	Attack	False Negative (FN)	True Positive (TP)

Figure 11.2 Confusion matrix.

Result Analysis

If a model developed with machine learning properly anticipates the type of attack at the time of testing, then it is accurate. The confusion matrix estimated the model's effectiveness. Figure 11.4 highlights the confusion matrix-based accuracy assessment for each classification approach. The test result can be classified as either positive or negative, as shown by the Figure 11.2 [16, 14]. For testing, the four metrics (TP, TN, FP, and FN) are calculated, as well as the Recall, Accuracy, F1-Score, and Precision.

$$Accuracy = \frac{TP + FN}{TP + FN + FP + TN} \qquad Recall = \frac{TP}{TP + FN}$$

$$Precision\ (in\ \%) = \frac{TP}{FP + TP} \qquad F1\_Score = \frac{2(Precision\,value\ *\ Recall\,value)}{Precision\,value + Recall\,value}$$

Where,

- *True +ve Rate (TP):* TP refers to a positive sample that is intended to be positive.
- *True -ve Rate (TN):* TN refers to a predicted negative sample that is intended to be negative.
- *False +ve Rate (FP):* FP stands for a negative sample that is intended to be positive.
- *False -ve Rate (FN):* FN stands for a positive sample that is intended to be negative.

Table 11.3 displays the classification report for the model, showing variations in F1 score, accuracy, recall, precision for each classifier. The features of final numbers was experimentally tweaked, with Variance Explained accounting for feature variation, where PCA chose bars and a line representing the cumulative variance explained Figure 11.3. The negotiation of true +ve rate and the false +ve rate is shown by ROC curve. The AUC statistic measures a model's classification accuracy, with higher AUC values indicating better detection of DDoS and normal attacks. The DET Curve shows the negotiation between false +ve and false -ve rates in logistic regression, SVM kernels, and Gaussian Naïve Bayes classifiers [15].

Using AUC analyses to derive Receiver Operating Characteristic (ROC), as illustrated in Figure 11.5. The accuracy value offered isn't accurate for different operational points or changes in the model's performance.

Table 11.3: The obtained result.

No	Machine Learning Approaches	Accuracy	Precision	Recall	F1-score
1	Logistic Regression	99.98	99.98	99.99	0.99
2	Random Forest Classifier	99.98	99.98	99.99	0.99
3	SVM—Linear Model	99.98	99.98	100	0.99
4	SVM—Poly Model	99.98	99.98	99.99	0.99
5	SVM—RBF Model	99.98	99.98	100	0.99
6	SVM—Sigmoid Model	99.96	99.98	99.98	0.99
7	Decision Tree	99.98	99.99	99.99	0.99
8	Gaussian Naive Bayes	83.13	100	83.13	0.83
9	K-Nearest Neighbor	99.98	99.98	99.99	0.99

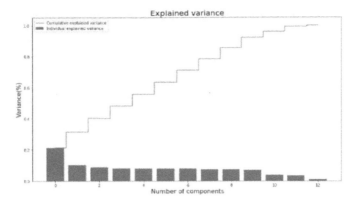

Figure 11.3 Explained variance plot.

		Predicted				Predicted	
		0	**1**			**0**	**1**
LR	0	93290	1	**POLY SVM**	0	93290	1
	1	14	4		1	12	6
RF	0	93287	4	**RBF SVM**	0	93291	0
	1	10	8		1	16	2
Linear SVM	0	93291	0	**SIGMOID SVM**	0	93277	14
	1	16	2		1	18	0
NB	0	77555	15736	**DT**	0	93283	8
	1	0	18		1	6	12
KNN	0	93285	6				
	1	10	8				

Figure 11.4 Confusion matrix.

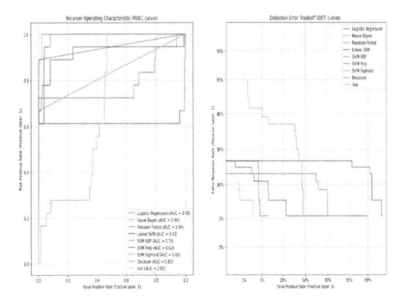

Figure 11.5 ROC and DET Curve.

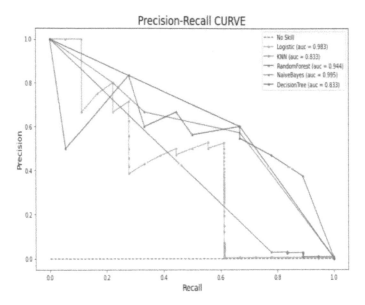

Figure 11.6 PR curve.

   The precision-recall curve in Figure 11.6 represents the comprises between recall and precision value with elevated scores indicating accurate results and positive outcomes.

Conclusion

Using the Corrected KDD dataset, this work proposed a DDoS detection algorithm. SVM-kernels, Random Forest, Naïve Bayes, KNN, and Decision trees, as well as Logistic regression, are employed as classifiers for detection. Feature scaling has lowered computing complexity; it has demon started its effectiveness by standardizing the data, which reduces all characteristics to a similar scale without distorting disparities in the range of values. The proposed model achieved the highest accuracy of 99.98% when there were 13 principal components. All the classification models, Gaussian Naïve Bayes has the lowest appropriateness of 83.1%. Almost all of the classifiers have a precision and recall value of 99.9% with a maximum precision and recall of 100 % with a 0.99 F1-score.

   In the future, there will be plenty more attacks, and the network will always be vulnerable. This study's future focus will be to incorporate a deep learning approach for example categorization, As well as handling more attack detection, it creates numerous other intriguing algorithms by combining kernel operations with different segmentation strategies. More efficient ways of dealing with high-dimensional text classification. In addition, several complicated datasets will be used. Future intrusion detection research should highlight the need to develop techniques that can identify novel threats with low false alarm rates.

References

1. Abbas, S. A., and Almhanna, M. S. (2021). Distributed denial of service attacks detection system by machine learning based on dimensionality reduction. *Journal of Physics: Conference Series*, 1804(1), 012136. IOP Publishing.

2. Aggarwal, P., and Sharma, S. K. (2015). Analysis of KDD dataset attributes-class wise for intrusion detection. *Procedia Computer Science*, 57, 842–851.

3. Alkasassbeh, M., Al-Naymat, G., Hassanat, A. B., and Almseidin, M. (2016). Detecting distributed denial of service attacks using data mining techniques. *International Journal of Advanced Computer Science and Applications*, 7(1), 436–445.

4. Barati, M., Abdullah, A., Udzir, N. I., Mahmod, R., and Mustapha, N. (2014). Distributed denial of service detection using hybrid machine learning technique. In 2014 International Symposium on Biometrics and Security Technologies (ISBAST), (pp. 268–273). IEEE.

5. Bolon-Canedo, V., Sanchez-Marono, N., and Alonso-Betanzos, A. (2011). Feature selection and classification in multiple class datasets: an application to KDD Cup 99 dataset. *Expert Systems with Applications*, 38(5), 5947–5957.

6. Salvatore, S., Wei, F., Wenke, L., Andreas, P., and Philip, C. (1999). KDD Cup 1999 Data. UCI Machine Learning Repository. https://doi.org/10.24432/C51C7N.

7. Cup, K. D. D. (1999). http://kdd. ics. uci. edu/databases/kddcup99/kddcup99. html. The UCI KDD Archive.

8. Das, S., Mahfouz, A. M., Venugopal, D., and Shiva, S. (2019). DDoS intrusion detection through machine learning ensemble. In 2019 IEEE 19th International Conference on Software Quality, Reliability and Security Companion (QRS-C), (pp. 471–477). IEEE.

9. Alsumaidaie, M. S. I., Alheeti, K. M. A., and Alaloosy, A. K. (2023). Intelligent detection of distributed denial of service attacks: a supervised machine learning and ensemble approach. *Iraqi Journal for Computer Science and Mathematics*, 4(3), 12–24.

10. Fouladi, R., Ermis, O., and Anarim, E. (2019). Anomaly-based DDoS attack detection by using sparse coding and frequency domain. In 2019 IEEE 30th Annual International Symposium on Personal, Indoor and Mobile Radio Communications (PIMRC).

11. Hao, J., and Ho, T. K. (2019). Machine learning made easy: a review of scikit-learn package in python programming language. *Journal of Educational and Behavioral Statistics*, 44(3), 348–361.

12. Hema, V., and Shyni, E. C. (2015). DoS attack detection based on naive bayes classifier. *Middle-East Journal of Scientific Research*, 23, 398–405.

13. Mohamed, I., Karim, A., and Mustapha, B. L. (2017). Dos detection method based on artificial neural networks. *International Journal of Advanced Computer Science and Applications,* 8(4), 465–471.

14. Joshi, M., and Hadi, T. H. (2015). School of Computer Sciences, North Maharashtra University, Jalgaon (M.S.) India. A review of network traffic analysis and prediction techniques. arXiv preprint arXiv:1507.05722.

15. Greenberg, C. S., Mason, L. P., Sadjadi, S. O., and Reynolds, D. A. (2020). Two decades of speaker recognition evaluation at the national institute of standards and technology. *Computer Speech and Language*, 60, 101032.

16. Moustafa, N., and Slay, J. (2016). The evaluation of network anomaly detection systems: statistical analysis of the UNSW-NB15 data set and the comparison with the KDD99 data set. *Information Security Journal: A Global Perspective*, 25(1–3), 18–31.

17. Hongmei, N., and Jiaqing, Z., (2015). A new bayesian classification algorithm based on attribute reduction. 4th International Conference on Mechatronics, Materials, Chemistry and Computer Engineering, (pp. 1–5). Atlantis Press.

18. Han, X., Grubenmann, T., Cheng, R., Wong, S. C., Li, X., and Sun, W. (2020). Traffic incident detection: a trajectory-based approach. In 2020 IEEE 36th International Conference on Data Engineering (ICDE), (pp. 1866–1869). IEEE.

19. Sboev, A., Selivanov, A., Rybka, R., Moloshnikov, I., and Rylkov, G. (2021). Evaluation of Machine Learning Methods for Relation Extraction Between Drug Adverse Effects and Medications in Russian Texts of Internet User Reviews. 5th International Workshop on Deep Learning in Computational Physics 28–29 June, 2021 Moscow, Russia.

20. Schehlmann, L., and Baier, H. (2013). COFFEE: a concept based on OpenFlow to filter and erase events of botnet activity at high-speed nodes. In INFORMATIK 2013—Informatik Angepasst

an Mensch, Organisation und Umwelt. Bonn: Gesellschaft für Informatik e.V.. PISSN: 1617–5468. ISBN: 978-3-88579-614-5. (pp. 2225–2239). Regular Research Papers. Koblenz. 16–20, September 2013.

21. Sadioura, J. S., Singh, S., and Das, A. (2019). Selection of sub-optimal feature set of network data to implement Machine Learning models to develop an efficient NIDS. In 2019 International Conference on Data Science and Engineering (ICDSE), (pp. 120–125). IEEE.

22. Zhao, H., Feng, Y., Koide, H., and Sakurai, K. (2019). An ANN based sequential detection method for balancing performance indicators of IDS. In Proceedings—2019 7th International Symposium on Computing and Networking, CANDAR 2019, (pp. 239–244). [8958477] (Proceedings—2019 7th International Symposium on Computing and Networking, CANDAR 2019). Institute of Electrical and Electronics Engineers Inc. https://doi.org/10.1109/CANDAR.2019.00039.

12 Efficient Attribute-based Unbounded Inner Product Functional Encryption for Many-authority

Prashnatita Pal[1,a], Kalyan Adhikary[2,b],
Rituparna Bhattacharya[2,c], and Jayanta Poray[2,d]

[1]St Thomas College of Engineering & Technology Kolkata, India

[2]Computer Science & Engineering, Techno India University, Kolkata, India

Abstract

The study introduces a novel approach called Functional Inner Product Encryption based on Many Authorities (FIPE-MA). This method focuses on decentralization and supports vectors with an infinite length. The core idea revolves around attribute-based type encryption process (ABE) and the capability of using encrypted data to assess linear functions. Different stakeholders may autonomously generate several keys with suitable parameters within the scope of a multi-authority environment. FIPE-MA extends the ABE component to this setting, enabling decentralization. In the presented FIPE-MA scheme for unbounded vectors, encryptors have the freedom to encrypt vectors of any length, based on predefined access policies. Administrators can provide users with secret keys that encompass attributes and vectors of any length. Like FIPE-MA, decryption is successful when the secret key and ciphertext have matching vector lengths.

Keywords: attribute-based encryption, inner product functional encryption, linear secret sharing schemes, prime-order bilinear groups, random oracle model.

Overview

A sophisticated kind of public key encryption called Functional Encryption (FE) allows calculating encrypted data while maintaining the secrecy of both the data and the calculated results. A particular family of FE techniques called Inner Product Functional Encryption (IPFE) enables computation on linear functions while requiring the decryption algorithm for estimation of the features vector for the input variables, which is provided inside a chipper text of a single vector and a private pair associated with other features vectors. Access control, secure computing outsourcing, statistical analysis on encrypted data, and other areas have seen extensive use of IPFE. Attribute-Based IPFE (AB-IPFE) was created through merging of associated encryption attributes. This is knowns as Attribute-Based Encryption (ABE). The roll of ABE is to alleviate the inherent data loss in IPFE. In order to execute calculations on the encrypted data, AB-IPFE links access policies/attributes with

[a]prashnatitap@gmail.com, [b]kalyan.babai@gmail.com, [c]rituparna.b@technoindiaeducation.com, [d]jayanta.p@technoindiaeducation.com

ciphertexts/secret keys and necessitates attribute validation. Existing AB-IPFE systems, however, presuppose a single authority controlling all characteristics, limiting their usefulness in situations when many authorities oversee certain attributes. Agrawal et al. [1] started the research on Many-Authority Functional Inner Product Encryption (FIPE-MA) after being inspired by the idea of Multi-Authority Attribute-Based Encryption (MA-ABE). This is one of the decentralizes platform for management of the attribute-based encryption. Without communicating with other authorities, FIPE-MA enables individual authorities to create individual mapping of the master key among two pairs. And generate associated private key, aiming characteristics with their own preview. Because diverse traits may be managed by appropriate authorities, this decentralization increases confidence and practicality.

However, the inclusion of computationally costly instigated orders groups (bilinear groups). Additionally, the scheme's set vector lengths and constrained authority/attribute numbers limit its capacity to scale and adapt. Moreover, the method restricts the expressive power and effectiveness of attribute appearances inside access rules to a single usage.

These restrictions are addressed in this study, which also makes the following contributions:

1. Decentralization of multi authority-based attributes for Unbounded Inner Product Functional Encryption (MA-ABUIPFE) is created, and it permits arbitrary authority additions, numerous independent authorities, and unbounded vector lengths for processing messages and keys [2].
2. Building MA-ABUIPFE schemes enable prime-order bilinear groups to access structures for Linear Secret Sharing Schemes (LSSS) [3]. Two different systems are offered, each with a different trade-off between adaptability and underlying presumptions.
3. The large-universe MA-ABUIPFE method improves the small-universe system by allowing exponentially more characteristics per authority to be supported without enumeration during setup. The parameterized variant of the DBDH assumption (L-DBDH) provides the foundation for security.

The suggested systems are more deployable since efficiency gains are made as compared to the current FIPE-MA scheme. The static model, in which the attacker discloses all requests in advance, is the focus of the security study of the schemes. No vulnerabilities were found, although security against picky attackers was not shown. The random oracle concept is used to create the schemes.

The FIPE-MA strategy for small universes is developed using methods from MA-ABE and unbounded IPFE constructs based on DBDH. In order to handle hash value evaluation, encryption/key generation/decryption, and adversary queries differently and increase security, a unique hash-decomposition approach is presented.

An addition to ABE that emphasizes decentralizing attribute management is called multi-authority attribute-based inner product functional encryption (FIPE-MA). Without engaging other authorities, it enables individual authorities to create their own master key pairs and give secret keys for characteristics under their control. FIPE-MA offers a wide range of possible uses, including calculating average employee pay and researching university mental health data, among others. The current FIPE-MA system, however, has drawbacks and inefficiencies, including the usage of pricey composite-order bilinear groups, fixed vector lengths, and constrained authority and characteristics. We address these issues and provide effective MA-ABUIPFE techniques based on prime-order bilinear groups in this study.

Enhancements to achieve Efficiency and Practicality:

1. Employing Bilinear Groups of Prime Order: [4] Composite-order bilinear groups are computational and resource-intensive, and they are used in the present MA-ABIPFE technique. Our proposed systems make use of prime-order bilinear groups, which offer substantial advantages in processing, communication, and storage.
2. Unbounded Length and Arbitrary Number of Authorities: Our MA-ABUIPFE techniques enable unbounded vector lengths and an unlimited number of authorities, in contrast to the prior system. This guarantees scalability and supports changing system needs, enabling the addition of additional authorities and characteristics at any moment.
3. Overcoming the "One-Use" Restriction: A restriction existed in the older MA-ABIPFE system that limited the number of times an attribute could appear in an access policy. Our schemes get around this limitation by allowing characteristics to occur arbitrarily many times inside access rules, broadening the range of supported policies, and improving effectiveness.

This MA-ABUIPFE (Multi-authority Attribute-Based Unbounded-Size Inner-Product Functional Encryption) methods presented in the excerpt have many advantages over the MA-ABIPFE (Multi-authority Attribute-Based Inner-Product Functional Encryption) scheme suggested by Agrawal et al. [9].

The following are the key benefits:

1. Efficiency: The FIPE-MA methods with the group of prime-order bilinear groups are substantially quicker compere are to the Attribute-Based General Transform (AGT-FE) scheme described by Agrawal et al., which consisted with compounded order groups. Due to their lower group element sizes and quicker pairing procedures, prime-order architectures have an inherent advantage in efficiency.
2. Public Key Size: The suggested methods, as compared to AGT-FE, lower the public key size by over 99% while keeping the ciphertext size same. This large key size decrease aids in the scheme's practical implementation.
3. Security Assumptions: The source-group assumptions that underpin AGT-FE's security, particularly its ranging the order of small groups. These are fundamentally new addition compare to the targeted groups, which use the group-based schemes. The transitions from the system of group of the prime order to the compounded order system of groups. Thus, the proposed methods yield notable gains in performance at the same time as being safe with fewer assumptions.
4. Versatility: The FIPE-MA methods that have been presented may be applied to the big universe scenario, enabling the system to include an infinite number of characteristics under various authorities. Compared to AGT-FE, this flexibility has advantages.
5. Static Security: The recommended methods primarily emphasize performance and adaptability, even if static security may not be the best security model for FIPE-MA. The techniques have not found any weaknesses against specific adversaries as discussed in [9], even though they do not have demonstrable security against stronger adversaries. Schemes with better performance but less robust verifiable security have often been found to be appropriate for real-world deployments and may operate as a first step towards more sophisticated security concepts.

In comparison to the current AGT-FE method by Agrawal et al., the proposed FIPE-MA schemes provide increased efficiency, reduced key sizes, flexibility in the large universe scenario, and weaker but reasonable security assumptions. The remaining sections of the essay are structured as follows.

The background is presented in Section 2. Section 3 describes the pertinent related work on the integration of Efficient encryption function for reveal the unbounded attribute based inner product that has been published in the literature. The Efficient Attribute-Based Unbounded Inner Product Functional Encryption setup and procedure used to create the proposed algorithm testing platform are covered in Section 4. The report is concluded and future research is discussed in Section 7 after Section 6 provides and discusses our preliminary findings.

Attributes of the Proposed Model

Under decisional assumptions based on target groups, two FIPE-MA systems enabling access policies supported by LSSS in much faster prime-order bilinear groups are described. These hypotheses are acknowledged to be less strong than those based on source groups. The suggested plans show various trade-offs between adaptability and underlying presumptions. The first method, which is safe according to the well-researched DBDH assumption, gives each authority control over a finite number of characteristics. The second technique supports a huge attribute universe by allowing authority to govern exponentially many attributes without necessitating attribute enumeration at startup. The suggested approaches also permit the occurrence of the same characteristics inside an access policy an unlimited number of times, in contrast to Agrawal et al. The emphasis of this study is on effectiveness and practicality, and as a result, it follows previous studies on MA-ABE with similar goals and presumptions that the security of the structures is shown in the random oracle model against static attackers. Technically speaking, the work introduces a unique hash-decomposition approach to expand the unbounded inner product functional encryption techniques of Dufour-Sans and Point cheval [8] to the setting of MA-ABUIPFE.

A pair of FIPE-MA schemes are included next.

1. The Small-Universe MA-ABUIPFE Scheme: It enables a single attribute or a finite set of attributes to be under the jurisdiction of a single authority. The system's capacity for adding authority continues to be arbitrary. The DBDH assumption, a well-researched computational assumption in the target group, serves as the foundation for the security of this technique. When compared to the earlier MA-ABIPFE system, the efficiency gains accomplished by this plan are substantial.
2. Large-Universe MA-ABUIPFE Scheme: We provide an improved version of the small-universe scheme that accommodates a large attribute universe. There is no need that all characteristics be listed at setup, and each authority may manage an exponential number of attributes. This scheme's security is based on the L-DBDH assumption, a parameterized variation of the DBDH assumption. We provide evidence for the accuracy of this supposition using the general bilinear group model. When it comes to practicalities, the large-universe design is just as effective as the small-universe approach.

Related Work

The contribution done by Abdalla et al. [5], we find two new schemes for a type of encryption that allows controlled access and computation on encrypted data Abdalla et al. [6].

The schemes can handle data and computation of any length and multiple groups that can issue permissions Agrawal et al. [7]. The schemes are consist with the groups of prime-order and the assumptions are not strong enough and also faster compared to the existing technique proposed by Agrawal et al. The paper [8] also develops a new technique to adapt the methods of Dufour-Sans and Point cheval [8] for this type of encryption. The paper gives formal definitions, constructions, proofs, and comparisons for both schemes.

Within the larger subject of functional encryption, the attribute base functional encryption Multi-authority Attribute-Based Unbounded-Size Inner-Product Function, this functional encryption for the inner product of unbounded size attribute based multi aureity technique is somewhat narrow and specialized issue. As a result, there is little research on MA-ABUIPFE. However, we can provide a quick literature review that outlines the most important pieces of work in this field. The idea of ABIPFE was established in [9] the publication "Attribute-Based Inner-Product Encryption with Hidden Structures and Improved Efficiency" by Fang et al. [9]. The study also suggested an effective method based on composite-order bilinear groups. It served as the basis for further field study performed later. The study [10] presented the idea of MA-ABE and suggested an effective method for broadcast messages, although not being exclusively focused on MA-ABUIPFE. The concepts and methods used in MA-ABE schemes are often built upon in MA-ABUIPFE systems. The idea of MA-ABE was expanded to include unbounded-size attributes in "Multi-authority Attribute-Based Encryption" by Ling et al. (2021). Although unrelated to MA-ABUIPFE, the ideas and methods presented in this work serve as the foundation for further study into unbounded-size attribute-based functional encryption. An effective MA-ABUIPFE method having fixed size encrypted data was suggested in the article "Fully Secure Unbounded Inner Product Functional Encryption with Constant Size Ciphertexts" by Chen and Wee [12]. In contrast to the limited or static security models, it was fully secure under the selective model, which has a better security concept.

The paper of Kwon et al. (2022) "Efficient Attribute-Based Unbounded Inner Product Functional Encryption" presented an effective MA-ABUIPFE technique based on prime-order bilinear groups [13]. The technique outperformed earlier designs in terms of efficiency and was able to produce ciphertexts of constant size. A completely secure MA-ABUIPFE method with constant-size ciphertexts consist with the bilinear groups of prime order was reported in this work "Multi-Authority Unbounded Attribute-Based Inner-Product Functional Encryption with Constant-Size Ciphertexts" by Kim et al. [14] (2023). Unbounded-size attribute-based functional encryption was effectively solved by the approach, which also achieved adaptive security. It's crucial to remember that there isn't a lot of literature on MA-ABUIPFE, thus the exact references listed here may not be all-inclusive.

Model Assessing

A FIPE-MA scheme is created technically via the execution of many crucial procedures. An executive summary of the technical procedure is provided below:

1. Defining the security and functionality requirements for the MA-ABUIPFE scheme accurately is the first stage in problem formulation. The intended security qualities, the attributes, the encryption and decryption methods, and the access control rules must all be specified.

2. Establish the essential mathematical and cryptographic underpinnings as a first step. Assumptions used in cryptography, such as the DBDH assumption, and the underlying mathematical structures, such as bilinear groups, are often defined in this process.
3. Attribute-Based Inner-Product Functional Encryption (ABIPFE) construction: MA-ABUIPFE schemes often rely on ABIPFE as a foundation. Designing algorithms for key generation, encryption, and decryption is a necessary step in building ABIPFE. The objective is to make it possible for approved parties to encrypt data with characteristics and provide useful decryption keys linked to certain inner products of attributes.
4. Extending to numerous Authorities: In MA-ABUIPFE, there are numerous authorities, each of which oversees a certain set of characteristics. Multiple authorities must be supported by the scheme for them to issue attribute keys and take part in the encryption and decryption procedures.
5. Unbounded-size attributes: Handling unbounded-size attributes, which allow for variable lengths rather than being restricted to fixed sizes, is one of the main issues faced by MA-ABUIPFE. To achieve this, it is necessary to create data structures and cryptographic methods that can effectively manage characteristics of any length.
6. Conduct a thorough investigation of the MA-ABUIPFE scheme's security characteristics. Typically, this entails demonstrating the security of the scheme under relevant cryptographic assumptions, such as the DBDH assumption. The scheme should satisfy the required security features, such as confidentiality, attribute privacy, and unforgeability, according to the security analysis.
7. Efficiency considerations: Reduce computational and communication overhead to make the system as efficient as possible. This entails ciphertext size reduction, optimization of the encryption and decryption algorithms, and consideration of methods like pairing-based algorithms and bilinear map compression.
8. Implementing the FIPE-MA scheme and assessing its performance regarding scalability, communication overhead, and computing efficiency. To determine the scheme's viability, tests are conducted, and the results are compared to those of other strategies.

Establish of Model

The technical procedure may differ based on the precise building and design decisions made by the researchers. The stages listed above provide a broad foundation for comprehending how FIPE-MA schemes are created.

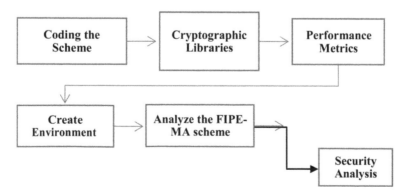

Figure 12.1 Block diagram representation of the proposed efficient attribute-based unbounded inner product functional encryption.

FIPE-MA systems need practical implementation as well as performance tests to determine the effectiveness and scalability of the scheme shown in Figure 12.1.

The following are the main components of developing and accessing FIPE-MA schemes:

1. Coding the Scheme: The FIPE-MA scheme must first be put into practice in accordance with the technical guidelines given in the research paper or design proposal. Writing the necessary code for the key generation, encryption, and decryption algorithms is essential, as well as any auxiliary operations or data structures needed by the scheme.
2. Using cryptographic libraries or frameworks that already exist to offer the necessary mathematical primitives, such as bilinear pairings, elliptic curve operations, and cryptographic hash functions, is a good idea. These libraries may aid in ensuring the implementation's accuracy and security.
3. Performance Metrics: Choose the performance indicators that will be used to assess the effectiveness of the plan. Common measurements include ciphertext and key sizes, communication overhead, key generation time, and encryption and decryption times. These measures aid in evaluating the scheme's effectiveness and expandability.
4. Create a proper test environment for carrying out the performance assessment. This can include deploying certain hardware resources, such servers or cloud instances, to model real-world situations and gauge the performance of the plan under various workloads.
5. The FIPE-MA scheme should be tested using a collection of test cases that cover each different feature of the system. To evaluate the functionality and performance of the scheme under various circumstances, test cases should contain various attribute combinations, attribute lengths, and inner-product searches.
6. Analyse the FIPE-MA scheme's capacity to grow if system size, such as the number of attributes, authorities, or users, is increased, and note how this affects the scheme's performance. This helps in figuring out if the system can handle bigger and more complicated situations without suffering significantly from decreased efficiency.
7. Comparison with Other current Schemes: Examine how the FIPE-MA scheme performs in comparison to other current schemes or comparable benchmarks. This makes it possible to evaluate how competitive the plan is in terms of efficiency and scalability.
8. Security Analysis: Although the FIPE-MA scheme offers security assurances, installation and assessment largely concentrate on performance factors. Make sure the implemented scheme complies with the intended security characteristics listed in the research paper and do a comprehensive security evaluation to find and fix any possible flaws.

Software Implementation of FIPE-MA Scheme

Programming is required to implement the FIPE-MA scheme. It cannot provide you a detailed implementation since it would go beyond the boundaries of a text-based dialogue but can give you a high-level rundown of the essential stages and methods used to code the system.

1. Generating a Key:
    a. Produce the system parameters, including the prime order and the description of the bilinear group.
    b. Create secret keys for every system authority or attribute authority (AA).
    c. Give the relevant authorities access to the public parameters and secret keys.

2. Encryption:
    a. Obtain a message and a list of user-provided characteristics.
    b. Create a key for symmetric encryption at random.
    c. To acquire a decryption key, compute the inner product of the characteristics and the AA keys.
    d. Create the ciphertext by encrypting the message with the symmetric key.
3. Decryption:
    a. Obtain from the user the ciphertext and the appropriate set of characteristics.
    b. Ask the user's authorization for the AA keys related to the characteristics.
    c. Calculate the inner product of the obtained AA keys and the characteristics.
    d. To get the final decryption key, combine the decryption key you got from the inner product with the AA's secret key.
    e. To retrieve the original communication, decode the ciphertext using the final decryption key.
4. Internal Product Calculation
    a. Calculate the inner product of the two vectors that stand for the AA keys and attributes.
    b. Carry out the required calculations, such as multiplying the respective components and adding the outcomes.
5. Group operations with bilinear pairings:
    a. Compute pairing operations between group components by using the characteristics of bilinear pairings.
    b. Exponentiate and multiply the components of the bilinear group using group operations.
6. It's crucial to remember that the precise implementation details may change based on the cryptography libraries, programming language, and design decisions. Additionally, a thorough comprehension of the underlying cryptographic principles and associated mathematical operations is necessary to execute the whole FIPE-MA method.

Implementing the complete FIPE-MA scheme in Python would involve a significant amount of code and require familiarity with cryptographic libraries. However, it can provide that demonstrates the basic steps involved in the encryption and decryption processes. Additionally, the example assumes the availability of a suitable cryptographic library that provides the necessary primitives like bilinear pairings and encryption algorithms. Please note that this code snippet shown in Figure 12.2 provides a high-level representation and does not include the specific details of generating keys, performing bilinear pairings, or applying encryption algorithms. Those details would depend on the specific cryptographic library you choose to use.

To implement the complete FIPE-MA scheme, you would need to further expand these functions and handle additional components such as key management, attribute management, and security considerations. Additionally, you should consult cryptographic literature and reliable implementations to ensure the correctness and security of your implementation.

It's important to emphasize that building a secure and production-ready cryptographic scheme implementation requires expert knowledge and thorough testing. Therefore, it is strongly recommended to consult with cryptography experts and rely on established and audited libraries when dealing with real-world cryptographic systems.

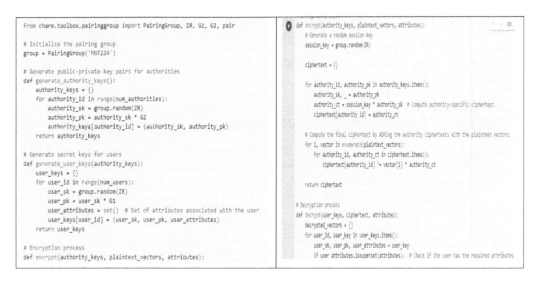

Figure 12.2 Here's a Python code snippet that illustrates a basic encryption and decryption process for the FIPE-MA scheme.

Conclusion

Future research is required to strengthen security against adaptive attackers, even if our methods' security has been shown in the static model, where the opponent must reveal all requests up front. Additionally, the techniques provide more assurance that FIPE-MA exists based on computational presumptions. Our work enhances the groundwork for this practical primitive and adds to the set of computational assumptions for internal product encryption with function. This research work is introducing two effective FIPE-MA methods based in order to get past prime-order bilinear groups the shortcomings and restrictions of previous MA-ABIPFE technique. These methods get beyond the "one-use" constraint while supporting unlimited vector lengths and an infinite number of authorities. Our designs greatly increase practicality and efficacy, allowing the deployment of MA.

References

1. Agrawal, S., Goyal, R., and Tomida, J. (2021). Multi-party functional encryption. In Nissim, K., and Waters, B. eds. Theory of Cryptography. TCC 2021. Lecture Notes in Computer Science, (Vol. 13043). Springer, Cham. https://doi.org/10.1007/978-3-030-90453-1_8.
2. Boneh, D. (2007). Bilinear groups of composite order. In Takagi, T., Okamoto, T., Okamoto, E., and Okamoto, T. eds. pairing-based cryptography—pairing 2007. Lecture Notes in Computer Science, (Vol. 4575). Berlin, Heidelberg: Springer..https://doi.org/10.1007/978-3-540-73489-5_1.
3. Mehta, S., and Saraswat, V. (2021). Linear secret sharing schemes with finer access structure. In Giri, D., Buyya, R., Ponnusamy, S., De, D., Adamatzky, A., and Abawajy, J. H. eds. Proceedings of the Sixth International Conference on Mathematics and Computing. Advances in Intelligent Systems and Computing, (Vol. 1262). Singapore: Springer. https://doi.org/10.1007/978-981-15-8061-1_15.

4. Takabatake, Y., I, T., Sakamoto, H. (2022). Information Processing on Compressed Data. In: Katoh, N., et al. Sublinear Computation Paradigm. Springer, Singapore. https://doi.org/10.1007/978-981-16-4095-7_6

5. Abdalla, M., Gay, R., Raykova, M., and Wee, H. (2017). Multi-input inner-product functional encryption from pairings. In Coron, J.-S., and Nielsen, J. B. eds. Eurocrypt 2017, Part I. LNCS, (Vol. 10210, pp. 601–626). Springer, Cham. https://doi.org/ 10.1007/978-3-319-5662.

6. Abdalla, M., Chevalier-Mames, B., Gay, R., and Pointcheval, D. (2020). Attribute-based inner-product functional encryption. In Advances in Cryptology–ASIACRYPT 2020, (pp. 3–33).

7. Agrawal, S., Garg, S., and Tripathi, A. (2021). Decentralized multi-authority attribute-based inner-product functional encryption. In Theory of Cryptography Conference, (pp. 1–32).

8. Dufour-Sans, E., and Pointcheval, D. (2019). Unbounded inner-product functional encryption with succinct keys. In: Deng, R., Gauthier-Umaña, V., Ochoa, M., Yung, M. (eds) *Applied cryptography and network security. ACNS 2019. Lecture Notes in Computer Science*, 11464. Springer, Cham. https://doi.org/10.1007/978-3-030-21568-2_21

9. Fang, S., Yang, S., and Zhang, Y. (2020). Inner product encryption from ring learning with errors. *Cybersecur*, 3, 22. https://doi.org/10.1186/s42400-020-00062-6.

10. Hei, Y., Liu, J., Feng, H., Li, D., Liu, Y., and Wu, Q. (2021). Making MA-ABE fully accountable: a blockchain-based approach for secure digital right management. *Computer Networks*, 191, 108029. ISSN 1389–1286. https://doi.org/10.1016/j.comnet.2021.108029.

11. Castagnos, G., Laguillaumie, F., and Tucker, I. (2018). Practical fully secure unrestricted inner product functional encryption modulo p. In Peyrin, T., and Galbraith, S. eds. Advances in Cryptology—ASIACRYPT 2018. ASIACRYPT 2018. Lecture Notes in Computer Science, (Vol. 11273). Springer, Cham. https://doi.org/10.1007/978-3-030-03329-3_25.

12. Gorasia, N., Srikanth, R. R., Doshi, N., and Rupareliya, J. (2016). Improving security in multi authority attribute based encryption with fast decryption. *Procedia Computer Science*, 79, 632–639. ISSN 1877–0509. https://doi.org/10.1016/j.procs.2016.03.080.

13. Xiong, H., Yang, M., Yao, T., Chen, J., and Kumari, S. (2022). Efficient unbounded fully attribute hiding inner product encryption in cloud-aided WBANs. *IEEE Systems Journal*, 16(4), 5424–5432. doi: 10.1109/JSYST.2021.3125455.

14. Fouque, P. A., Georgescu, A., Qian, C., Roux-Langlois, A., and Wen, W. (2023). A generic transform from multi-round interactive proof to NIZK. In Boldyreva, A., and Kolesnikov, V. eds. Public-Key Cryptography—PKC 2023. PKC 2023. Lecture Notes in Computer Science, (Vol. 13941). Springer, Cham. https://doi.org/10.1007/978-3-031-31371-4_16.

15. Ling, J., Chen, J., Chen, J., and Gan, W. Multiauthority attribute-based encryption with traceable and dynamic policy updating, *Security and Communication Networks*, vol. 2021, Article ID 6661450, 13 pages, 2021. https://doi.org/10.1155/2021/6661450.

16. Okamoto , Datta and K. Takashima (2022) Efficient Attribute-Based Signatures for Unbounded Arithmetic Branching Programs, *IEICE Transactions on Fundamentals of Electronics, Communications and Computer Sciences*, Volume 104-A, No.1, pp. 25–57 .

Part 2

Survelience & Real Life Applications

13 FPGA Based Accelerator for Human Activity Recognition using LSTM

Pradipta Roy

Scientist, Integrated Test Range, Chandipur, DRDO

Abstract

Human Activity Recognition (HAR) has gained a lot of research importance after the development of wearable mobile sensing devices like smartwatches. With the advent of deep learning approaches the recognition and prediction of human activities like sleep, walk, run swim etc. are performed more accurately which has tremendous use in healthcare, human robot interaction etc. Many deep learning methods like Convolutional Neural Network (CNN) based methods, Recurrent Neural Network (RNN) and Long Short Term Memory (LSTM) are used for solving the problem of data augmentation and getting desired accuracy. But most of the methods are too computationally complex to be implemented in real time hardware. In this paper, I used a Hybrid CNN and LSTM based method which produces satisfactory performance for UCI-HAR database. The proposed method is targeted to Xilinx System on Chip (SoC) Zynq 7000 (FPGA and ARM Processor combination). The hardware modules of pipelined CNN and LSTM is carefully designed in conjunction with the ARM processor inside Zynq SoC carrying out data normalization and other network tuning operation. The proposed architecture produces encouraging result for real time application consuming very low power which is suitable for wearable devices.

Keywords: FPGA, HAR, LSTM.

Introduction

Wearable devices like smart watches with sensors like accelerometer, gyros and magnetometer is widely used for Human Activity Recognition (HAR). With combination of smartphones with Bluetooth connection they find application in health monitoring Wang et al. [1], Tang et al. [2], activity logging Faurholt-Jepsen et al. [4] and localization and tracking [3]. Typical human activity that are recognized as walking, sitting, jogging, and climbing stairs etc. [5]. Jin et al. [6] reviewed usage of wearable sensors in the mines for detecting hazardous gas and to monitor brain functions of workers and for fatigue monitoring. Bahadir et al. [7] used a wearable system to detect obstacles for visually impaired people.

Nowadays, with the progress of machine learning and deep learning techniques, there are numerous methods proposed to infer sensible information from collected data. These powerful processes like Convolutional Neural Network (1D or 2D CNN) [9], Long Short

---

pradiptar@yahoo.com

Term Memory (LSTM) Tufek et al. [10], Recurrent Neural Network (RNN) Bai et al. [12] etc. can extract meaningful high level features from the sensor data.

Though these supervised deep learning methods are quite efficient, they have obvious drawback of high computational cost, slow training speed and large memory overhead. This limits their direct use for real time application in the portable devices having embedded processor or CPU. The obvious choice for porting these algorithms for real time application is GPU, ASIC or FPGA. Multicore GPUs exploit parallel processing, but they consume a lot of power and has memory bottleneck when multiple cores try to access same global memory. ASICs are too costly and not reconfigurable. In order to meet the requirements of a flexible deep learning accelerator, a reconfigurable and scalable design is required. Reconfigurability, low cost and low power make FPGA the best choice for implementation [8].

In this work, I have adopted a hybrid deep learning model of CNN and LSTM to recognize human activity. The basic functional block like convolution layers, activation function and LSTM are designed for hardware implementation using basic building blocks. LSTM is used for its ability to learn long term dependencies and find patterns in sequential data more efficiently. To the best of my knowledge, FPGA implementation of LSTM for time series classification is the first attempt reported so far. I have used Zynq SoC of Xilinx to utilize both software and hardware co-processing. The ARM processor of the SoC is used for data normalization and network tuning. I have evaluated the model on UCI—HAR dataset. A fixed point approximation was taken for implementing the data types in FPGA. This sacrifices the accuracy for less than 2% but improves the processing speed of 70 times of that of CPU and 20 times of the GPU implementation. Also the power consumption is very less (5% of that of GPU) compared to GPU implementation which makes it perfect choice for deployment in portable devices.

Rest of the paper is constructed as follows. In Section-2, I briefly review about the methodologies for HAR. Section-3 describes the proposed method with functional blocks. Section-4 discusses the hardware architecture and implementation issues. Section-5 presents the results with analysis. In Section-6 the paper is concluded.

Literature Review

Initially, traditional handcrafted feature extraction and classical machine learning methods like support vector machines, decision tree and Bayesian methods were used for Human Activity Recognition [11]. There are some works which classifies activities of human using kNN algorithm. Real time implementation is the key challenge for this algorithm [13]. The classical machine learning algorithms achieves roughly 85% accuracy while deep learning algorithms like LSTM and CNN produces more than 90% accuracy [14]. Jain and Kanhangad [15] used Fourier descriptor and histogram of gradient for feature extraction while kNN and SVM is used for recognition. Sikder et al. [16] uses 2 Channel CNN to achieve 95.25% accuracy on UCI—HAR dataset. Many works used accelerometer and gyroscope for human activity recognition. UCI HAR is the first big dataset which includes daily activities of people measured via inertial sensors in mobile phone Anguita et al. [19]. CNN and LSTM combined model was used by Xia et al. [20] to achieve an accuracy of more than 95%.

Due to high computational demand of deep learning networks, high performance accelerators like FPGA, ASIC and GPU are gaining a lot of interest as implementation platform for Deep Neural Network (DNN). Parallel processing to achieve higher throughput and low power are the key features of FPGA, which make it a suitable candidate for DNN implementation. Mittal [18] presented a study on how FPGA can be used for deep learning applications. Nurvitadhi et al. [17] compared the performance of GPU and FPGA and showed that FPGA has a better speed/ power ratio compared to GPU.

Proposed Method

The proposed method builds a hybrid CNN and LSTM classifier to recognize 6 different human actions perceivable from wearable sensors namely "sitting", "walking", "lying", "standing", "down-stairs" and "up-stairs". The dataset used to train and test the model is very well known UCI-HAR. The model is first built in Python and them implemented on Xilinx Zynq SoC. The CNN is used for extraction of features and LSTM is used for classification. I have used 1D CNN for feeding the time series data. The overall block level system architecture is given in Figure 13.1.

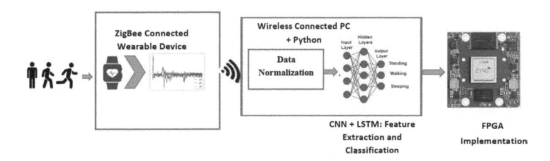

Figure 13.1 Block diagram of proposed HAR implementation in FPGA.

The human activity data is collected through the sensor (accelerometer) fitted in wearable device. This data in time series format is transmitted via ZigBee to the initial processing station (a CPU with Python and wireless connection). The important parameters that is generated from the sensor is the time series of three axes accelerometer and gyroscope data. The data is normalized and sent to a hybrid model of CNN and LSTM. I have used 1D CNN which is most suitable for this type of time series data. The CNN is used for extraction of features and LSTM is used for classification. The CNN Model consists of 3 convolution block as shown in Figure 13.2. Each convolutional block has 1 Dimensional convolution followed by ReLU (non-linear activation function) and 1 Dimensional max-pooling with stride 1. The detailed hardware structure is discussed in next section. Due to multiple convolution block, vanishing gradient may arise. LSTM is very efficient RNN for combatting vanishing gradient and long term dependencies. Two LSTM layers were used for classification with every hidden layer consist of 128 neurons. The network is trained and tested with UCI-HAR dataset. The model is trained with a dropout of 0.5.

CNN Feature Extractor

The three layer CNN is shown in Figure 13.2. Inertial sensor data from 3 channels (x, y and z) is sampled in 64 × 3 blocks as the network input. The three Convolution block

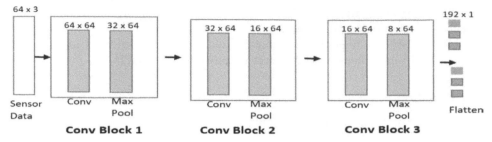

Figure 13.2 CNN architecture for feature extraction.

follows each containing one CNN filter (3 × 1 kernel size) and one Max Pool (2 × 1 size). The three convolution blocks are tapered progressively and finally flattened to 192 featured output which is fed to LSTM block.

LSTM Classifier

A LSTM provides an extension of RNN (Recurrent Neural Network). It consists of 3 gates namely an input gate (i), an output gate (o), and a forget gate (f), which does an overwrite and retrieval of memory from memory cell (c) at time instant t.

The expression for input gate (i_t) and forget gate (f_t) are given by:

$$i_t = sigm\left(W_i * x_t + H_i * h_{t-1} + C_i * c_{t-1} + b_i\right) \tag{1}$$

$$f_t = sigm\left(W_f * x_t + H_f * h_{t-1} + C_f * c_{t-1} + b_f\right) \tag{2}$$

Where, W, H, and C are weight matrices between input and gates, recurrent connections, and between cell and gates and *sigm* is sigmoid activation function, x is the cell state, h represents hidden state. b is the bias.

These steps are followed by using the previous contents (c_{t-1}), the current memory cell (c_t) content to produce the new memory (c-bar). They can be calculated by following equations:

$$c - bar_t = tanh\left(W_{c-bar} * x_t + H_{ic-bar} * h_{t-1} + b_{c-bar}\right) \tag{3}$$

$$c_t = f_t * c_{t-1} + i_t * c - bar_t \tag{4}$$

Finally, the final activation at the present instance (h_t) is calculated with the output gate (o_t)

$$o_t = sigm\left(W_0 * x_t + H_0 * h_{t-1} + C_o * c_{t-1} + b_o\right) \tag{5}$$

$$h_t = o_t * tanh(c_t) \tag{6}$$

The basic block wise structure of an LSTM cell is given in Figure 13.3.

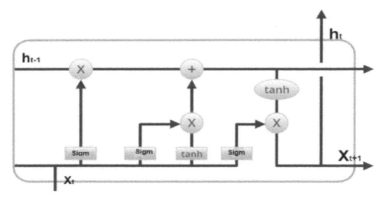

Figure 13.3 LSTM cell architecture.

Hardware Architecture

System Overview

The system is targeted for the Xilinx SoC Platform of Zynq 7000 Zed Board which contains the device XC7Z020-CLG484-1. The CNN and LSTM are implemented inside FPGA fabric as shown in Figure 13.4. A hardware software co-processing strategy is adopted for the implementation.

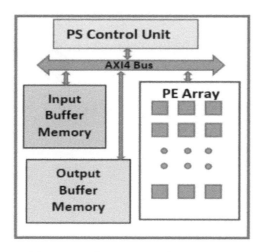

Figure 13.4 FPGA system architecture with processing element (PE).

There is mainly two functional units inside the SoC namely Processing System (PS: ARM Processor) and Programmable Logic (PL: FPGA). Both these units with memory controller is connected with AXI4 bus for seamless data communication. The hardware architecture for parallel replicated units were implemented in FPGA and data normalization, network tuning and overall control is done by a program written in C running inside ARM processor. The data from PC is transferred through Ethernet to the System configuring the weights of the network. The CNN and LSTM are implemented inside parallel Processing Element (PE) array which is connected to the AXI4 bus for taking control command from ARM processor (PS) and taking data from input buffer and output buffer memory. The memory modules are also connected to the common Bus which is controlled by PS. Input and Output buffers are accessible to the PC through Ethernet.

Convolution Block

Convolution Block is the core of processing for the feature extraction function. They are implemented inside each processing element which are connected in parallel to form a stack or array. The arrays are connected in pipeline to produce output in different layers (input layers, hidden layers and output layers). The basic functional blocks of the Convolution Block are 1. Conv. Filter 2. ReLU Activation Block and 3. Max Pooling. The basic hardware architectures of these blocks are described below.

1. Conv. Filter: This is the major block of extracting features using 1D CNN. Basic multiplication (with weight vectors) and addition of all weighted inputs are the primary operation of this block. The basic architecture of this block is shown in Figure 13.5. The multiplier is taken as Wallace Tree based multiplier and adder is carry look ahead adder for their performance superiority.

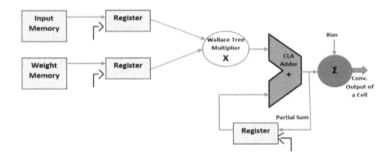

Figure 13.5 CNN hardware architecture.

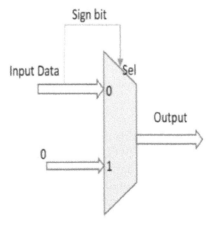

Figure 13.6 ReLu architecture.

2. ReLU Block: Rectified Linear Unit (ReLu) is a popular activation function which has been used here. *ReLu* suppresses the negative value and passes the positive value. It has been implemented using a multiplexer only and using the sign bit of the input to control the multiplexer as shown in Figure 13.6.

3. Max Pooling Block: The function of Max Pooling Block is to find out maximum value within a patch of a data to down-sample it. In this work a 4 value Max Pool Unit is implemented as shown in Figure 13.7. A sliding window is designed to select Max Values from 4x1 window in a sequence. 3 dual rank comparators are used to determine the max value of the block.

LSTM Block

The functional block diagram of LSTM is shown in Figure 13.3. The multiplier and adder are implemented using Wallace Tree and CLA Adder respectively. The *tanh* and *sigmoid* functions are implemented using Look Up Table (LUT) based decoder, where the values corresponding to linear and non-linear sections of the functions are stored in LUTs.

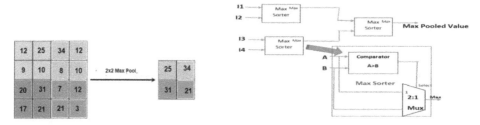

Figure 13.7 Max pool operation and its hardware architecture.

Results and Discussion

The CNN and LSTM is first trained in Python using Keras and Tensorflow. UCI-HAR dataset was used to both train and evaluate the model. The coefficient of the neurons and other parameters are transferred to Zxnq SoC using Ethernet connection. The hardware implementation is targeted to Zedboard containing Zynq 7000 SoC with 100 MHz Clock. The EDA tool used was Vivado from Xilinx. The model accuracy with epochs is plotted in Figure 13.8. The Test accuracy was settled around 98.3% after 20 epochs. The learning rate was 0.0001.

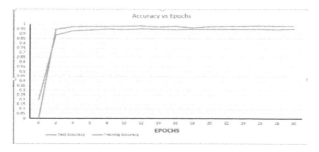

Figure 13.8 Model accuracy plot.

In Table 13.1, hardware resource utilization is summarized. Table 13.2 provides the performance comparison with CPU and GPU implementation. The CPU in the experiment was Intel Xeon Gold 6148 CPU (40 cores) and GPU was chosen as NVIDIA Titan RTX.

From Table 13.2 it is seen that the power consumption is very less in CPU and it has almost 70x faster implementation of that of CPU and 20x of that of GPU. The accuracy is down little bit due to fixed point implementation of FPGA weight registers. Table 13.3 Compares the performance with other related work implemented in software. Proposed method performs better because the combination of 1D CNN and LSTM networks are layer wise optimal and tuned to extract the features (though Xia et al. [20] method was applied on different dataset).

Table 13.1: FPGA design summary (XC7Z020).

LUT	Max Frequency	BRAM	Power
36100 out of 53200	190 MHz	27Kb (including Processor. Memory)	12 W

Table 13.2: Performance comparison.

Device	Throughput	Accuracy	Power
CPU	30 ksamples/s	98.3	250W
GPU	100 ksamples/s	98.3	10 W
FPGA	2100 ksamples/s	96.2	0.5 W

Table 13.3: Performance comparison (CPU) with other methods in literature.

Work By	Method	Accuracy Reported	Dataset
[16]	CNN	95.6%	UCI-HAR
[20]	CNN + LSTM	95.85%	WISDM
Proposed (CPU)	CNN + LSTM	98.3%	UCI-HAR

Conclusion

In this paper a novel learning based method and its hardware implementation is proposed for Human activity recognition. The low power and high throughput makes this method a suitable candidate for porting in wearable devices. In future, the tital tuning and full implementation will be targeted to FPGA only (Here in Zynq SoC, ARM Processor is used with FPGA).

References

1. Wang, Y., Cang, S., and Yu, H. (2019). A survey on wearable sensor modality centred human activity recognition in health care. *Expert Systems with Applications*, 137, 167–190.
2. Tang, C. I., Perez-Pozuelo, I., Spathis, D., and Mascolo, C. (2020). Exploring contrastive learning in human activity recognition for healthcare. ArXiv preprint arXiv:2011.11542.

3. Zou, H., Chen, Z., Jiang, H., Xie, L., and Spanos, C. (2017). Accurate indoor localization and tracking using mobile phone inertial sensors, wifi and ibeacon. In IEEE International Symposium on Inertial Sensors and Systems (INERTIAL), IEEE, (pp. 1–4).

4. Faurholt-Jepsen, M., Vinberg, M., Frost, M., Debel, S., Margrethe Christensen, E., Bardram, J. E., and Kessing, L. V. (2016). Behavioral activities collected through smartphones and the association with illness activity in bipolar disorder. *International Journal of Methods in Psychiatric Research*, 25(4), 309–323.

5. Kwapisz, J. R., Weiss, G. M., and Moore, S. A. (2011). Activity recognition using cell phone accelerometers. *ACM SigKDD Explorations Newsletter*, 12, 74–82.

6. Jin, H., Abu-Raya, Y. S., and Haick, H. (2017). Advanced materials for health monitoring with skin-based wearable devices. *Advanced healthcare materials*, 6(11), (Art. no. 1700024).

7. Bahadir, S. K., Koncar, V., and Kalaoglu, F. (2012). Wearable obstacle detection system fully integrated to textile structures for visually impaired people. *Sensors and Actuators, A: Physical*, 179, 297–311.

8. Farrukh, F. U. D., Xie, T., Zhang, C., and Wang, Z. (2018). Optimization for efficient hardware implementation of CNN on FPGA. In IEEE International Conference on Integrated Circuits, Technologies and Applications (ICTA). IEEE, (pp. 88–89).

9. Ahmad, Z., and Khan, N. (2020). CNN-based multistage gated average fusion (mgaf) for human action recognition using depth and inertial sensors. *IEEE Sensors Journal*, 21(3), 3623–3634.

10. Tufek, N., Yalcin, M., Altintas, M., Kalaoglu, F., Li, Y., and Bahadir, S. K. (2019). Human action recognition using deep learning methods on limited sensory data. *IEEE Sensors Journal*, 20(6), 3101–3112.

11. Lara, O. D., and Labrador, M. A. (2012). A survey on human activity recognition using wearable sensors. *IEEE Communications Surveys and Tutorials*, 15(3), 1192–1209.

12. Bai, S., Yan, M., Wan, Q., He, L., Wang, X., and Li, J. (2019). DL-RNN: an accurate indoor localization method via double RNNS. *IEEE Sensors Journal*, 20(1), 286–295.

13. Wang, A., Chen, G., Yang, J., Zhao, S., and Chang, C. Y. (2016). A comparative study on human activity recognition using inertial sensors in a smartphone. *IEEE Sensors Journal*, 16(11), 4566–4578.

14. Ronao, C. A., and Cho, S. B. (2016). Human activity recognition with smartphone sensors using deep learning neural networks. *Expert Systems with Applications*, 59, 235–244.

15. Jain, A., and Kanhangad, V. (2018). Human activity classification in smartphones using accelerometer and gyroscope sensors. *IEEE Sensors Journal*, 18(3), 1169–1177.

16. Sikder, N., Chowdhury, M. S., Arif, A. S. M., and Nahid, A. A. (2019). Human activity recognition using multichannel convolutional neural network. In 5th International Conference on Advances in Electrical Engineering (ICAEE). IEEE, (pp. 560–565).

17. Nurvitadhi, E., Venkatesh, G., Sim, J., Marr, D., Huang, R., Jason Ong Gee Hock, Liew, Y. T., Srivatsan, K., Moss, D., Subhaschandra, S., and Boudoukh, G. (2017). Can FPGAs beat GPUs in accelerating next—generation deep neural networks. In Proceedings of the ACM/SIGDA International Symposium on Field Programmable Gate Arrays.

18. Mittal, S. (2018). A survey of FPGA-based accelerators for convolutional neural networks. *Neural Computing and Applications*, 32(4), 1109–1139.

19. Anguita, D., Ghio, A., Oneto, L., Parra, X., and Reyes-Ortiz, J. L. (2013). A public domain dataset for human activity recognition using smartphones. In Proceedings ESANN, (pp. 1–6).

20. Xia, K., Huang, J., and Wang, H. (2020). Lstm-cnn architecture for human activity recognition. *IEEE Access*, 8, 56 855–56 866.

14 Multi-sensor Image Fusion using an Interconnected Auto-encoder like Network with Visible Edge Emphasis

A. K. Singh[1,a], D. Chaudhuri[1,b], M. P. Singh[2,c], S. Mitra[1,d], and B. B. Chaudhuri[1,e]

[1]CSE Department, Techno India University, Kolkata, India

[2]DRDO Young Scientist Laboratory—CT, IIT-M Research Park, Chennai, India

Abstract

Multi-sensor image fusion has emerged as a crucial initial step in target detection. Currently, various approaches such as spatial and transform domain methods have been utilized alongside several deep-learning-based techniques for image fusion. However, the results obtained are not consistently visually appealing. Implementing edge enhancement can be an effective strategy to enhance spatial resolution since edges play a critical role in image perception. Additionally, most deep learning-based algorithms primarily focus on feature extraction from source images before fusion but tend to overlook the specific properties inherent within these source images. In this study, we propose a novel approach for image fusion utilizing both thermal and visible images along with the more detailed edge map of the visible image compared to that of the thermal image due to differences in overall imaging performance. To accomplish this, we employ an interconnected network designed after a CNN auto-encoder structure for seamless integration of these features during the merging process.

Keywords: Auto-encoder, deep-learning, image fusion, interconnected network, multi-sensor, target detection.

Introduction

Image fusion is a vital aspect of image processing, combining multimodal or multi-focus images to extract more information than the original images. By integrating diverse modalities into a single fused image, we are able to enhance the richness of available information for a specific region of interest. This technique has seen increasing applications in various domains such as computer-aided medical diagnosis Wang et al. [1], automated target identification Ma et al. [2], security and surveillance Paramanandham and Rajendiran [3], remote sensing Jinju et al. [4], among others. In a recent study Singh et al. [21], a method using fast discrete curvelet transform for feature extraction and a bounded PCA methodology for fusing the precise coefficients of thermal and visible images has been proposed.

Thermal images can capture the heat radiation of a scene in low-light or dark settings, making the target visible. However, these images often have poor imaging capabilities, low contrast, and lower resolution. In contrast, visible images may render the target invisible

[a]aniket.kumarsingh.ece22@heritageit.edu.in, [b]debasis.chaudhury@technoindiaeducation.com, [c]director.dysl-ct@gov.in, [d]sntnmtr@gmail.com, [e]bbcisical@gmail.com

and distorted due to factors such as smoke and bad weather. Nonetheless, they can still provide valuable information about the rest of the scene. Combining thermal and visible images enhances target recognition accuracy and overcomes limitations posed by thermal and visible sensors' restricted imaging capabilities.

Numerous image fusion techniques using deep learning have emerged for effective real-time fusion, employing CNNs to extract features from input images and fuse them using various fusion rules such as Select Maximum, Averaging, Select Minimum, and Principal Component Analysis. The fused visuals are then restored to human-comprehensible form.

In this article, a novel approach has been proposed for image fusion. Instead of utilizing a conventional CNN-based autoencoder Hou and Yan [5], our method introduces an interconnected network with dicephalic architecture. Unlike typical autoencoders, our network incorporates two independent encoder networks dedicated to thermal and visible images respectively, while the decoder section is designed to be dependent on the outputs from both encoders' layers. Notably, we enhance the visible image encoder structure by incorporating the edge map information along with the original image data. This additional step aims to exploit the higher level of detail present in the visible image's edge map to optimize feature extraction performance. The key novelty in our network lies within its utilization of both edge maps and interconnections between encoder and decoder layers, allowing for enhanced collaboration among different stages of processing. We have coined the term "ConvFuse" for our proposed image fusion model. The aim of this paper is to create a fused image by extracting and smoothly integrating corresponding features from both thermal and visible images using the provided network. This involves analyzing the grayscale thermal image as well as the RGB+E visible image, where RGB represents red, green, and blue color channels respectively, while E denotes the edge map derived from morphological processing applied to the visible image. Moreover, these input images are used in subsequent layers of the network's encoders in order to facilitate feature learning. The proposed network architecture can be seen in Figure 14.1.

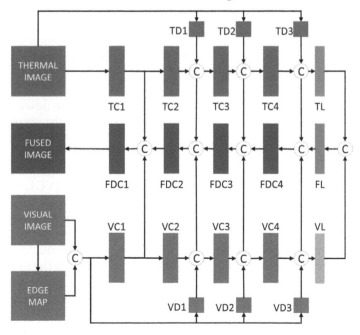

Figure 14.1 ConvFuse architecture.

The paper is organized in the following manner. Initially, we provide an overview of the pre-processing step, our suggested network architecture, and the specifications for training. Subsequently, we present the results and analysis of our experiments along with a comparison to other existing methods. Lastly, we conclude by summarizing our findings and discussing potential future avenues of research.

Proposed Network

The model utilizes two encoder networks to extract distinctive features from both the thermal and visible images. Additionally, each encoder network incorporates three residual connections in subsequent layers. In contrast to traditional autoencoders, the decoder stage is not independent; it is connected with the outputs of the encoder layers.

Pre-Processing

The visible image's edges provide crucial information about the scene and are far more apparent and detailed than the thermal image's edges, which is critical to the fusion process. Before giving the visible image as input to the neural network, the suggested model retains the edges of the visible images. The visible image and its edge map are fed into the visible image encoder network. To detect edges in visible images, the typical morphological erosion and dilation procedures are used. The operation of erosion is written as: $I_{erode} = I \ominus S$ and the dilation operation is expressed as: $I_{dilate} = I \oplus S$. The difference between dilated image I_{dilate} and I_{erode} eroded image, define the morphological gradient that serves as the visible image's edge map. This edge map is referred to as I_{edge}. ($I_{edge} = I_{dilate} - I_{erode}$). Here, I is the visible image and S is the structuring element for the morphological operations.

Network Architecture

From Figure 14.1, it is evident that the image fusion model consists of three interconnected sub-networks: the thermal image encoder network, the visible image encoder network, and the decoder network. Tables 14.1, 14.2, and 14.3 show the specifications of the blocks in the encoder and decoder networks, respectively.

The first network (Table 14.1) is the thermal image encoder network, which consists of four base convolutional blocks (TC1, TC2, TC3, and TC4) and three down-sample blocks (TD1, TD2, and TD3), the outputs of which are concatenated with the outputs of the base convolutional blocks. The output of the last block is fed into a linear block, TL, which flattens and projects the input to a 1-dimensional space.

The visible image encoder network comes next (Table 14.2). This network is the same as the thermal image encoder, with four base convolutional blocks (VC1, VC2, VC3, and VC4), three down-sampled blocks (VD1, VD2, and VD3), and a final linear block (VL). However, this network accepts both the visible image and its edge map as input. Instead of a single-channel input, this network accepts a 4-channel input, with the RGB channels acting as the first three channels and the edge map acting as the fourth.

Finally, we arrive at the decoder network, which provides us with the final fused image (Table 14.3). This decoder network receives the outputs of both encoder networks' linear blocks as input. It is made up of four deconvolutional blocks, FDC4, FDC3, FDC2, and FDC1, which transform the input in the 1-dimensional feature space back to a 3-dimensional feature space. The encoder blocks' outputs, together with the preceding block's

Table 14.1: Thermal image encoder block structure.

Block Type	Sub-Block	Input Size	Output Size
TC1 (Convolution Block)	Conv+BN+GELU Conv+BN+GELU	24,1,512,512	24,8,512,512
TD1 (Downsample Block)	Conv+BN+GELU	24,1,512,512	24,32,128,128
TC2 (Convolution Block)	Conv+BN+GELU Conv+BN+GELU	24,8,512,512	24,32,128,128
TD2 (Downsample Block)	Conv+BN+GELU	24,1,512,512	24,128,32,32
TC3 (Convolution Block)	Conv+BN+GELU Conv+BN+GELU	24,64,128,128	24,128,32,32
TD3 (Downsample Block)	Conv+BN+GELU	24,1,512,512	24,512,8,8
TC4 (Convolution Block)	Conv+BN+GELU Conv+BN+GELU	24,256,32,32	24,512,8,8
TL (Linear Block)	Flatten+Linear+GELU Linear+GELU Linear+GELU	24,1024,8,8	24,256

Table 14.2: Visible image encoder block structure.

Block Type	Sub-Block	Input Size	Output Size
VC1 (Convolution Block)	Conv+BN+GELU Conv+BN+GELU	24,4,512,512	24,8,512,512
VD1 (Downsample Block)	Conv+BN+GELU	24,4,512,512	24,32,128,128
VC2 (Convolution Block)	Conv+BN+GELU Conv+BN+GELU	24,8,512,512	24,32,128,128
VD2 (Downsample Block)	Conv+BN+GELU	24,4,512,512	24,128,32,32
VC3 (Convolution Block)	Conv+BN+GELU Conv+BN+GELU	24,64,128,128	24,128,32,32
VD3 (Downsample Block)	Conv+BN+GELU	24,4,512,512	24,512,8,8
VC4 (Convolution Block)	Conv+BN+GELU Conv+BN+GELU	24,256,32,32	24,512,8,8
VL (Linear Block)	Flatten+Linear+GELU Linear+GELU Linear+GELU	24,1024,8,8	24,256

output, are provided as inputs to the decoder deconvolutional blocks in the network. Thus, the architecture becomes interconnected.

Training Dataset and Specifications

The suggested model was trained on sets of recorded thermal and visual images. These image pairings are derived from several different sources, including the OSU Color-Thermal Database Davis and Sharma [6], the Pedestrian Infrared/Visible Stereo Video Dataset Bilodeau et al. [7], the M3FD Dataset Liu et al. [8], and the Thermal Visual Paired Dataset Goswami et al. [9]. Because the collection only comprised a few registered images, augmentation techniques were used on the images, resulting in a dataset of 13592 thermal-visible image pairings. The model was trained using mini-batches of 24 image pairings for 20 epochs. For training, the Adam optimizer Kingma et al. [10] was used, with a learning rate of 0.001 and a weight decay of 10^{-6}. The network was tuned using mean-squared error (*MSE*) as the loss function. The network was trained in a *PyTorch* environment on a workstation equipped with *Intel(R) Xeon(R) Silver* 4210R 2.40GHz CPU, 96 GB RAM, and an 8 *GB NVIDIA Quadro RTX* 4000 *GPU*.

Table 14.3: Decoder block structure.

Block Type	Sub-Block	Input Size	Output Size
FL (Linear Block)	Linear+GELU	24,512	24,512,8,8
	Linear+GELU		
	Linear+GELU+Unflatten		
FDC4 (Deconvolution Block)	Deconv+BN+GELU	24,2560,8,8	24,128,32,32
	Deconv+BN+GELU		
FDC3 (Deconvolution Block)	Deconv+BN+GELU	24,640,32,32	24,32,128,128
	Deconv+BN+GELU		
FDC2 (Deconvolution Block)	Deconv+BN+GELU	24,160,128,128	24,8,512,512
	Deconv+BN+GELU		
FDC1 (Deconvolution Block)	Deconv+BN+GELU	24,24,512,512	24,3,512,512
	Deconv+BN+GELU		

Experimental Results

Our tests were carried out on a set of 25 thermal-visible image pairs. 21 image pairs were taken from the VIFB dataset [11]. These images depict a range of locations and working conditions, including indoor and outdoor situations, low lighting, and overexposure. The image resolutions in the VIFB collection range from 320 × 240 to 630 × 460, or 512 × 184 to 452 × 332. The remaining 4 image pairs (images in the 3rd and 4th row 3rd column in Figure 14.2(a), 1st and 2nd row 1st column in Figure 14.2(b), and 2nd and 4th column in Figure 14.2(c)) were captured using an *AXIS Q8741-LE Bi-spectral PTZ Network Camera*. Each pair of visible and thermal images has been registered to guarantee that the image fusion can be performed successfully. The resolution of these four registered image pairings is 800 × 600. Figure 14.2 displays a few illustrations of the images in the dataset. This section presents the results of the experiments. The image fusion results are displayed and compared to previously available image fusion algorithms. Various quantitative analytical measures are also compared along with the runtime speed comparison. Aside from the suggested method, all other algorithms' source codes were borrowed from the VIFB code library.

(a)   (b)

(c)

Figure 14.2 Testing dataset.

Qualitative Analysis

The final images after fusion are compared with the results of standard fusion methods that are cross bilateral filter (CBF) Shreyamsha Kumar [12], convolutional neural network (CNN) Liu et al. [13], guided filter-based context enhancement (GFCE) Zhou et al. [14], gradient transfer fusion (GTF) Ma et al. [15], latent low-rank representation (LatLRR) Li and Wu [16], non-subsampled contourlet transform and sparse representation (NSCT SR) Liu et al. [17], and ration-of-low-pass pyramid and sparse representation (RP SR) [17]. The results of these methods, as well as the proposed "ConvFuse" method, are shown in Figure 14.3(c–j).

(a) Thermal image   (b) Visible image   (c) CBF output image

(d) CNN output image   (e) GFCE output image   (f) GTF output image

(g) LatLRR output image   (h) NSCT SR output image   (i) RP SR output image

(j) Proposed "ConvFuse" method output image
Figure 3: Image fusion results

Figure 14.3 Image fusion results.

Quantitative Analysis

The objective analysis approach of quantitative evaluation is well recognized and is based on mathematical modeling. It compares the spectral and spatial similarities between the fused image and the raw input images using a set of predetermined quality parameters to determine the quality of the fused image. Many assessment strategies for thermal-visible image fusion have been developed. The assessment of image quality based on just one of the metrics is inappropriate. We have used four assessment measures to compare performance comprehensively and objectively: peak signal-to-noise ratio (PSNR) Jagalingam and Hegde [18], root mean squared error (RMSE) Jagalingam and Hegde [18], Chen-Varshney metric (Q_{CV}) Chen and Varshney [19], and structural similarity index measure (SSIM) [20]. Their average values for the set of 25 test image pairs are computed and are shown in Table 14.4.

Runtime Comparison

The performance time is governed by how long the image fusion algorithm takes to run. When assessing an algorithm for real-time use cases, performance time becomes an important issue, and it is impacted by both the method and the platform on which it is executed. Table 14.5 displays the runtime analysis for the collection of 25 test image pairs computed for each algorithm. For better comparison, the runtime is calculated using only the core fusion functionality, while ignoring the time taken to load the images. However, these results are obtained while running on a CPU for lack of valid GPU executable codes for the VIFB algorithms. Our proposed method can utilize GPU for faster performance. Table 14.6 shows the performance of our proposed method when processed using a GPU.

Table 14.4: On 25 image pairs, the average assessment metrics for all procedures. (The top three values for each statistic are indicated by the colors Green, blue, and red, respectively).

Metric	CBF	CNN	ConvFuse	GFCE	GTF	LatLRR	NSCT SR	RP SR
PSNR	57.809	58.097	58.32	56.322	58.08	56.412	57.668	57.976
Q_{CV}	1426.7	471.4	704.9	814.2	1966.2	672.6	1299.4	809.5
RMSE	0.119	0.113	0.105	0.161	0.112	0.16	0.124	0.116
SSIM	1.163	1.343	1.37	1.117	1.331	1.152	1.256	1.295

Table 14.5: Execution tme for all algorithms (Best result is marked by **bold**).

Method	Avg. Time (s)	Total Time (s)	FPS	Method	Avg. Time (s)	Total Time (s)	FPS
CBF	15.979	399.482	0.063	LatLRR	237.553	5938.837	0.004
CNN	30.556	763.906	0.033	NSCT SR	95.603	2390.076	0.01
GFCE	1.905	47.615	0.525	RP SR	0.7672	19.181	1.303
GTF	5.179	129.483	0.193	Proposed	**0.148**	**3.699**	**6.757**

Table 14.6: Results of our proposed method on GPU.

Avg. Time (s)	0.0524
Total Time (s)	1.3123
FPS	19.084

Discussion

When comparing each image in Figure 14.3(c-j), it is clear that CBF and NSCT SR has introduced a lot of noise in the background, making the images unsuitable for human vision. GTF has produced a hazy, extremely unsaturated image. Color noise has been introduced to the CNN and RP SR outputs, making them unsuitable for visual interpretation. The GFCE, LatLRR, and proposed method, all produced satisfactory fusion results.

However, the proposed approach appears to be preferable because it has performed well with both key image fusion objectives, namely merging hidden information from thermal images and retaining color information from visible images.

Table 14.4 shows the quantitative analysis of the above fusion methods. For the evaluation metrics used, the higher the PSNR and SSIM values, the better the quality of the fused image. On the other hand, a lower and RMSE value indicates that the image is superior. From the table, it is evident that the proposed method performs better in three out of the four given evaluation metrics while giving the 3rd best result in the other one.

In Table 14.5, we compare the runtime speeds of each method. All the algorithms are run on a CPU. In comparison to all other methods, the suggested method produces the best results, proving a frame rate (FPS) of 6.757. We have also processed our proposed "ConvFuse" method on a GPU and from Table 14.6 we can see that we get an FPS of 19.084, which is almost real-time.

Conclusion

This paper proposes an interconnected network for the real-time merging of thermal and visible images. Unsupervised learning is applied in the deep learning-based model. Thermal and visible image feature maps are extracted using two independent encoder networks. The edges of the visible images represent essential characteristics that are recovered using morphological operations and utilized along with the visible image in the deep learning network. The encoder and decoder networks are linked, and the decoder network makes use of the feature maps from each encoder block. The novelty of this paper lies in the network itself. The network leverages the edge map of a visible image in conjunction with deep feature maps to enhance feature learning. The decoder's reliance on the output from encoder layers also enhances the relationship between features at both stages.

The main objective behind our work was to create a good image fusion method, which can be used in real-time video processing. From our experiments, we have found that our proposed method gives good results and works in real time. In the future, we will try to optimize this method further by utilizing cross-stage partial networks for encoder blocks, which might give us faster inference speeds and decrease memory consumption.

Acknowledgment

We would like to express our sincere gratitude to DYSL-CT, DRDO for funding the project.

References

1. Wang, Z., Cui, Z., and Zhu, Y. (2020). Multi-modal medical image fusion by laplacian pyramid and adaptive sparse representation. *Computers in Biology and Medicine*, 123, 103823.
2. Ma, J., Ma, Y., and Li, C. (2019). Infrared and visible image fusion methods and applications: a survey. *Information Fusion*, 45, 153–178.
3. Paramanandham, N., and Rajendiran, K. (2018). Infrared and visible image fusion using discrete cosine transform and swarm intelligence for surveillance applications. *Infrared Physics and Technology*, 88, 13–22.
4. Jinju, J., Santhi, N., Ramar, K., and Bama, B. S. (2019). Spatial frequency discrete wavelet transform image fusion technique for remote sensing applications. *Engineering Science and Technology, an International Journal*, 22(3), 715–726.

5. Hou, B., and Yan, R. (2018). Convolutional auto-encoder based deep feature learning for finger-vein verification. In 2018 IEEE International Symposium on Medical Measurements and Applications (MeMeA), (pp. 1–5). IEEE.
6. Davis, J. W., and Sharma, V. (2007). Background-subtraction using contour-based fusion of thermal and visible imagery. *Computer Vision and Image Understanding,* 106(2–3), 162–182.
7. Bilodeau, G. A., Torabi, A., St-Charles, P. L., and Riahi, D. (2014). Thermal–visible registration of human silhouettes: a similarity measure performance evaluation. *Infrared Physics and Technology,* 64, 79–86.
8. Liu, J., Fan, X., Huang, Z., Wu, G., Liu, R., Zhong, W., and Luo, Z. (2022). Target-aware dual adversarial learning and a multi-scenario multi-modality benchmark to fuse infrared and visible for object detection. In Proceedings of the IEEE/CVF Conference on Computer Vision and Pattern Recognition, (pp. 5802–5811).
9. Goswami, S., and Singh, S. K. (2022). A simple mutual information based registration method for thermal-optical image pairs applied on a novel dataset. In 2022 3rd International Conference for Emerging Technology (INCET), (pp. 1–5). IEEE.
10. Kingma, D. P., and Ba, J. (2014). Adam: a method for stochastic optimization. arXiv preprint arXiv:1412.6980.
11. Zhang, X., Ye, P., and Xiao, G. (2020). VIFB: a visible and infrared image fusion benchmark. In Proceedings of the IEEE/CVF Conference on Computer Vision and Pattern Recognition Workshops, (pp. 104–105).
12. Shreyamsha Kumar, B. K. (2015). Image fusion based on pixel significance using cross bilateral filter. *Signal, Image and Video Processing,* 9, 1193–1204.
13. Liu, Y., Chen, X., Cheng, J., Peng, H., and Wang, Z. (2018). Infrared and visible image fusion with convolutional neural networks. *International Journal of Wavelets, Multiresolution and Information Processing,* 16(03), 1850018.
14. Zhou, Z., Dong, M., Xie, X., and Gao, Z. (2016). Fusion of infrared and visible images for night-vision context enhancement. *Applied Optics,* 55(23), 6480–6490.
15. Ma, J., Chen, C., Li, C., and Huang, J. (2016). Infrared and visible image fusion via gradient transfer and total variation minimization. *Information Fusion,* 31, 100–109.
16. Li, H., and Wu, X. J. (2018). Infrared and visible image fusion using latent low-rank representation. arXiv preprint arXiv,1804.08992.
17. Liu, Y., Liu, S., and Wang, Z. (2015). A general framework for image fusion based on multi-scale transform and sparse representation. *Information Fusion,* 24, 147–164.
18. Jagalingam, P., and Hegde, A. V. (2015). A review of quality metrics for fused image. *Aquatic Procedia,* 4, 133–142.
19. Chen, H., and Varshney, P. K. (2007). A human perception inspired quality metric for image fusion based on regional information. *Information Fusion,* 8(2), 193–207.
20. Wang, Z., Bovik, A. C., Sheikh, H. R., and Simoncelli, E. P. (2004). Image quality assessment: from error visibility to structural similarity. *IEEE Transactions on Image Processing,* 13(4), 600–612.
21. Singh, A., Chaudhuri, D., Mitra, S., Singh, M., and Chaudhuri, B. (2023). Bounded PCA based Multi Sensor Image Fusion Employing Curvelet Transform Coefficients. *Defence Science Journal,* 73(06), 675–687.

15 Emotion Detection from EEG by Gated Recurrent Unit Along with Particle Swarm Optimization

Madhuchhanda Basak[a], Diptadip Maiti[b], and Debashis Das[c]

Dept. of CSE, Techno India University, Kolkata, West Bengal, India

Abstract

Emotion recognition derive from electroencephalography (EEG) signals pick up Significant recognition in recent years. This paper hand out a novel approach that combines EEG-based emotion recognition with particle swarm optimization (PSO) for feature reduction. PSO-based feature reduction technique is employed to choose the best discriminative features among the initial feature set, which achieve a feature reduction of 48%. Gated Recurrent Unit (GRU) is utilized as the classifier to recognize emotional states with the reduced feature set. The validation accuracy achieved using reduced feature set is 97%, which is comparable to the accuracy obtained using the full feature set. Additionally, the reduced feature set is tested with other classifiers, and similar accuracy levels are achieved, further validating the effectiveness of the proposed feature reduction technique.

Keywords: EEG, emotion detection, gated recurrent unit, particle swarm optimization.

Introduction

Emotion is a psychological experience characterized by high mental activity, encompassing attributes such as knowledge, expression, reaction, and action inclinations. While traditional methods rely on outwardly reported emotions, electroencephalogram (EEG) impulses provide insights into hidden information and reveal emotional patterns. EEG signals are break down into 5 frequency bands. The different frequency of EEG signals depending on the activity is shown in Figure 15.1(a). It's possible to capture EEG signals using monopolar or bipolar methods, with monopolar recording involving the voltage differential between the scalp electrode and reference electrode at the earlobe, and bipolar recording measuring the voltage differential between two scalp electrodes [17]. Participants wear an electrode cap as stated by the 10/20 international electrode positioning scheme and view stimuli while their EEG is recorded using specialized software [5]. Ekman identified happiness, sorrow, surprise, fear, and anger as the six primary emotions experienced universally across cultures. Emotions can be expressed and articulated using the orthogonal dimensions of valence (pleasant or unpleasant) and arousal (calm or excited). Long-term

---

[a]madhuchhanda.basak@gmail.com, [b]diptadipmaiti@gmail.com, [c]debashis.d@technoindiaeducation.com

visual examination of EEG data is laborious, leading to the proposal of feature extraction-based techniques to quantify the underlying complexity and non-linearity of EEG signals [6]. These techniques enhance the analysis and interpretation of emotional states. Figure 15.1(b) shows the 10/20 electrode arrangement scheme and Figure 15.1(c) shows the emotion classification. EEG signals present challenges in terms of noise contamination, limited spatial resolution, and restricted frequency range [14]. Artefacts caused by patient movement or electrode-related factors further affect the quality of EEG recordings.

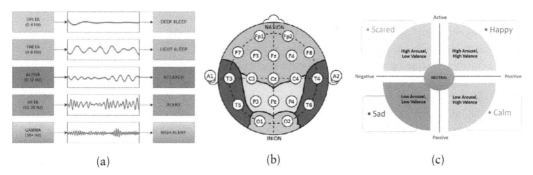

(a)   (b)   (c)

Figure 15.1 (a) Frequency of EEG signal based on activity. (b)10/20 Electrode placement scheme (c) Classification of emotion.

The scalp and skull also distort and attenuate EEG signals, while volume conduction leads to overlapping electrical activity from different brain regions [19]. Overcoming these limitations requires advanced signal processing techniques and the integration of EEG with other imaging modalities to enhance accuracy and reliability [12]. Machine learning approaches have gained attention for emotion detection from EEG data, utilizing models such as SVM, random forests, or deep learning architectures trained on labelled EEG data obtained during emotional induction tasks or stimulus-based elicitation [9]. This work proposed a method to assess the emotion from EEG signal by extraction of different features and identify the necessary features by implementing particle swarm optimization-based dimensionality reduction and finally deployment of a gated recurrent unit classifier to identify the emotion. This work is divided into the following sections: Section 1: Introduction; Section 2: Literature Review; Section 3: Proposed Method; Section 4: Results and Discussion; and Section 5: Concluding Remarks.

Literature Review

A feature vector was then obtained by backward sequential search. For identifying the categories of Low/High Arousal and Low/High Valence, a linear kernel LS-SVM containing an ANN produced cross-validated accuracies of 64.84% and 61.17%, respectively. ASFM (Adaptive Subspace Feature Matching), a quick domain adjustment technique, is suggested by Chai et al. [3]. They provide a linear translation function that, without the use of regularization, matches the borderline distributions of the origin and destination subpopulations. Their mean accuracy with logistic regression (LR) is 80.46%. In order to create a focused time-frequency rendering of EEG signals and record their spectrum fluctuations over time, Alazrai et al. [1] used a QTFD (Quadratic Time-Frequency Distribution). The obtained average categorization accuracy lies between 73.8% and 86.2%. Three DNN (Deep Neural Network) layers and three convolutional layers make up the SincNet-R classifier that Zeng et al. [16] propose. The classifier achieves an average accuracy of 94.5%. One of the best strategies for emotion recognition using EEG have been proposed by Sorinasa et al. [11] and include QDA and KNN classifiers as well as population rating criterion for stimulus labelling. The suggested model has a 98% average accuracy. For EEG-based emotion identification, Zhong et al. [18] suggest a RGNN (Regularized Graph Neural Network). They use an adjacency matrix inside a GNN (Graph Neural Network) to reproduce the inter-channel relationships among EEG data. They achieve an accuracy of 85.30% on average. To increase the precision of emotion identification and draw out additional differential features out of EEG data, Tao et al. [13] propose an ACRNN (Attention-based Convolutional Recurrent Neural Network). A CNN have been used to pull out the spatial particulars from encoded EEG waveforms, and ACRNN utilizes a channel-wise observation technique to adaptively allocate the weights on various channels. ACRNN has a 93.38% accuracy rate. Using DCNN (Deep Convolutional Neural Network), Ozdemir et al. [10] suggest a technique for classifying the emotional spirits of valence, arousal, dominance, along with liking. In terms of test accuracy, they obtained 90.62% for negative as well as positive valence, 86.13% for high together with low arousal, 88.48% for high together with low dominance, and finally 86.23% on like along with unlike. He et al. [7] analysed the EEG waves over days to determine the brain patterns of emotions. The appropriate dimensionality in various temporal domains is adaptively determined using the Transfer Component Analysis (TCA) technique. Cross-day emotion identification accuracy may be successfully increased by 3.55% using a TCA-based domain adaptation technique.

Proposed Method

For this study we have used EEG Brainwave Dataset by Bird et al. [2]. Data was accumulated from two human being—a male along with a female—with duration of three minutes in all of the 3 states—positive, neutral, as well as negative. The experiment employed a Muse EEG headgear to record the TP9, AF7 with AF8 along with TP10 EEG locations utilizing dry electrodes. Class distribution of the dataset is shown in Figure 15.2. Sample of positive, negative together with neutral emotion signal is exhibited Figure 15.3.

Work Flow

We have proposed an amalgamation of dimension reduction using particle swarm optimization and a state-of-art classifier called Gated Recurrent Unit to classify emotion of

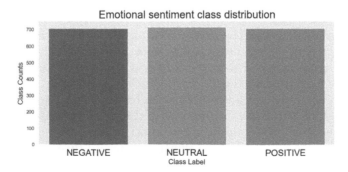

Figure 15.2 Dataset class distribution.

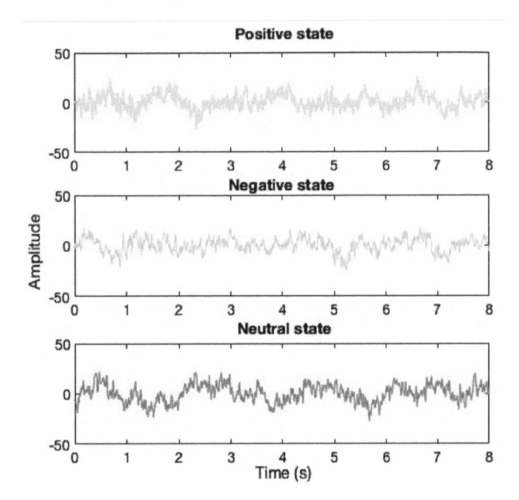

Figure 15.3 Sample emotion signal.

human from the obtained EEG signals. The feature extracted from the EEG signal is first optimized and reduced by using particle swarm optimization and fed to the classifier to acquire the final recommendation. The work flow with regard to suggested approach is illustrated in Figure 15.4.

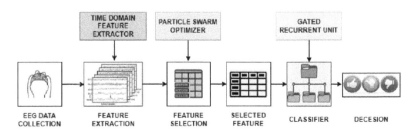

Figure 15.4 Workflow of the proposed method.

EEG Acquisition and Feature Extraction

The Muse headband, which houses the EEG sensor used for data collection, interfaces accompanied by computer through BLE (Bluetooth Low Energy). Following statistical attributes are draw out from the acquired EEG signals in company of a sliding time windows of total length 1s.

1. **Mean:** For a temporal window of S1, S2, S3, ... Sm

$$\text{mean} = \frac{1}{M} \sum_{M}^{i} S_i$$

2. **Standard Deviation:**

$$std = \sqrt{\frac{1}{M} \sum_{M}^{i} (x_i - \text{mean})^2}$$

3. **Skewness:**

$$sk = \frac{\text{mean}^k}{std^k}$$

    where k is the moment order.

4. **Kurtosis:**

$$kt = \frac{1}{M} \sum_{M}^{i} (s_i - \text{mean})^k$$

5. **Max Value:** Max value of the time window.
6. **Min Value:** Min value of the time window.
7. **Derivative 1:** Minimum and maximum value derivatives obtained by splitting the time frame in half.
8. **Derivative 2:** Minimum and maximum value derivatives obtained by splitting the time frame in fourth.
9. **Euclidean Distance:** The Euclidean distance between the four time periods is calculated for each min, max, and mean value.
10. **Covariance Matrix:** Upper triangular value of covariance matrix of derivative 1 and derivative2.

$$CM_{ij} = \frac{1}{M \sum_{M}^{k} (s_{ik} - \text{mean}_i)(s_{kj} - \text{mean}_j)}$$

11. **Shannon Entropy:**

$$SE = -\sum_{j} ES_j \times \log(ES_j)$$

    where ES is the normalized time window value.

12. **Log-Energy Entropy:**

$$LE = \sum_{i} \log(SE_i^2) + \sum_{j} \log(SE_i^2)$$

13. **Fast Fourier Transform:**

$$FFT_k = \sum_{m=0}^{M-1} SE_m^t e^{-i2\pi k(m/M)}$$

where k = 0, . . ., M − 1

With the above feature extraction techniques 120 no's mean, 20 no's standard deviation, 40 no's moments, 120 no's max, 120 no's min, 288 no's covariance matrix values, 24 no's eigen values, 156 no's logarithm value, 10 no's entropy values, 150 correlated values and finally 1500 no's FFT, a total of 2548 feature values are calculated from the obtained EEG Signals.

Dimension Reduction Using Particle Swarm Optimization (PSO)

To direct the particles in their quest for global optimum solutions, the PSO employs a straightforward technique that replicates swarm way of behaving in birds flocking along with fish schooling. It deviates from alternative optimization methods in particular that is, it requires just the objective function along with it is not affected by the gradient nor by any distinctive form of the target. Additionally, it has a few hyper-parameters. The mathematical derivation of PSO are as follows:

$$V_i^{t+1} = W \cdot V_i^t + c_1 U_1^t \left(P_{b_1}^t - P_i^t \right) + c_2 U_2^t \left(g_b^t - P_i^t \right)$$

Where $W \cdot V_i^t$ is inertia or diversification, $c_1 U_1^t \left(P_{b_1}^t - P_i^t \right)$ is personal influence and $c_2 U_2^t \left(g_b^t - P_i^t \right)$ is social influence. The personal and social influence in total called intensification of the optimizer Kennedy and Eberhart [8], Yang and Du [15]. We have employed PSO on the feature set obtained from the previous step with a population size of 10. After optimization the number of features selected for further classification is 1238 which is almost 48% reduction in the feature space dimensionality. In the reduce feature set we have 58 no's of mean, 11 no's of standard deviation, 18 no's of moments, 53 no's max, 62 no's min, 139 no's of covariance matrix value, 11 no's of eigen value, 80 no's logarithm value, 4 no entropy value, 73 of correlation value and finally 729 no's of FFT values.

Gated Recurrent Unit (GRU) Classifier

A step up from the standard RNN, or recurrent neural network, is the GRU, or gated recurrent unit. It was first used by Cho et al. [4]. Figure 15.5 depicts one basic unit of a GRU cell. We have employed a simple one-unit GRU classifier on the reduced feature set to classify the emotion from EEG data. The GRU network has 512 unit and dense layer accompanied by SoftMax acting as activation function and 3 outcomes which will predict the emotion as positive, negative or neutral. The proposed GRU classifier is exhibited in the Figure 15.6. The details of the proposed model are presented in the Table 15.1.

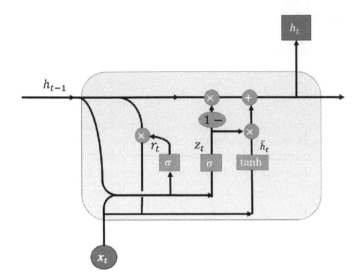

Figure 15.5 GRU cell.

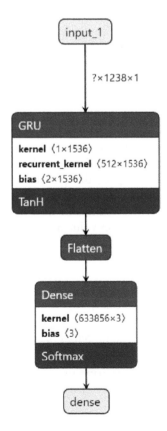

Figure 15.6 Proposed GRU classifier.

Table 15.1: Proposed GRU classifier's parameter details.

Model: "GRU model"		
Layer (type)	Output Shape	Param #
input 1 (Input Layer)	[None, 1238, 1]	0
gru (GRU)	(None, 1238, 512)	791040
flatten (Flatten)	(None, 633856)	0
dense (Dense)	(None, 3)	1901571
Total parameters: 2,692,611 Trainable parameters: 2,692,611 Non-trainable parameters: 0		

Results and Discussion

The GRU classifier is trained together by Adam optimizer together with loss function categorical cross entropy. Learning rate chosen for Adam optimizer is 0.001. The classifier is trained for 10 epochs. The training and validation accuracy with loss values is presented in the Table 15.2. Table 15.3 with Table 15.4 represents the confusion matrix and classification report of the GRU classifier.

Table 15.2: Training & validation accuracy and loss for emotion classification.

Emotion Classification	Acc.	Loss
Training	99.65%	0.0136
Validation	97.19%	0.1045

Table 15.3: Confusion matrix of proposed GRU model.

		Predicted		
	Neutral	149	4	0
Actual	Negative	3	136	3
	Positive	0	4	128
		Neutral	Negative	Positive

Table 15.4: Classification Report of proposed GRU model.

	Precision	Recall	F1-Score	Support
Neutral Emotion	0.98	0.97	0.98	153
Negative Emotion	0.94	0.96	0.95	142
Positive Emotion	0.98	0.97	0.97	132
Accuracy			0.97	427
Macro Average	0.97	0.97	0.97	427
Weighted Average	0.97	0.97	0.97	427

Figures 15.7–15.11 represents the plot of accuracy, loss, receiver operation characteristics and precession-recall of the GRU classifier.

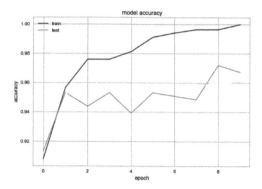

Figure 15.7 Proposed GRU model's accuracy.

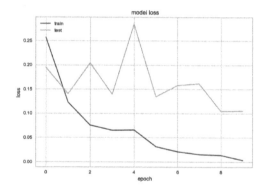

Figure 15.8 Proposed GRU model's loss.

We have tested the effectiveness of the feature reduction technique by applying different classifier shown in Figures 15.11 and 15.12, found that the reduce feature set is almost as good as the full feature set for the classification of the emotion. Figures 15.11 and 15.12 shows the accuracy of classification on reduce feature dataset and full feature dataset.

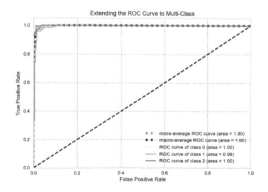

Figure 15.9 Proposed GRU model's ROC curve.

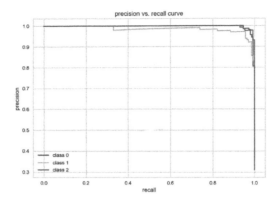

Figure 15.10 Proposed GRU model's precession & recall curve.

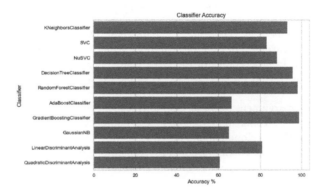

Figure 15.11 Accuracy on reduce feature.

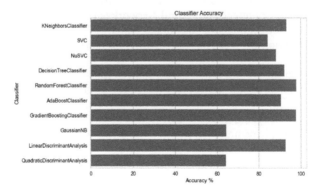

Figure 15.12 Accuracy on full feature.

Conclusion

EEG signals. By applying PSO, a significant 48% reduction in feature dimensionality is achieved, while maintaining high accuracy in emotion recognition. The GRU classifier effectively models temporal dependencies in the reduced feature set, resulting in an impressive classification accuracy of 97%. Advancements in machine learning algorithms, preprocessing techniques, and the availability of large EEG datasets hold promise for improving emotion detection accuracy. Further research is required to enhance the robustness, generalization, and interpretability of the models and to explore their applications in areas like affective computing, mental health estimation, and human-computer interactivity.

References

1. Alazrai, R., Homoud, R., Alwanni, H., and Daoud, M. I. (2018). Eeg-based emotion recognition using quadratic time-frequency distribution. *Sensors*, 18(8), 2739.
2. Bird, J. J., Faria, D. R., Manso, L. J., Ekárt, A., and Buckingham, C. D. (2019). A deep evolutionary approach to bioinspired classifier optimisation for brain-machine interaction. In complexity (Vol. 2019, pp. 1–14). Hindawi Limited. https://doi.org/10.1155/2019/4316548
3. Chai, X., Wang, Q., Zhao, Y., Li, Y., Liu, D., Liu, X., and Bai, O. (2017). A fast, efficient domain adaptation technique for cross-domain electroencephalography (eeg)-based emotion recognition. *Sensors*, 17 (5), 1014.
4. Cho, K., Van Merriënboer, B., Gulcehre, C., Bahdanau, D., Bougares, F., Schwenk, H., and Bengio, Y. (2014). Learning Phrase representations using rnn encoder decoder for statistical machine translation. In proceedings of the 2014 conference on empirical methods in natural language processing (EMNLP), p. 1724–1734, Doha, Qatar. Association for Computational Linguistics.
5. Ekman, P. (1992). An argument for basic emotions. *Cognition and Emotion*, 6(3–4), 169–200.
6. Feldman, G., Hayes, A., Kumar, S., Greeson, J., and Laurenceau, J.-P. (2006). Mindfulness and emotion regulation: the development and initial validation of the cognitive and affective mindfulness scale-revised (CAMS-R). In journal of psychopathology and behavioral assessment (Vol. 29, Issue 3, pp. 177–190). Springer Science and Business Media LLC. https://doi.org/10.1007/s10862-006-9035-8
7. He, Z., Zhuang, N., Bao, G., Zeng, Y., and Yan, B. (2022). Cross-day eeg-based emotion recognition using transfer component analysis. *Electronics*, 11(4), 651.
8. Kennedy, J., and Eberhart, R. (1995). Particle swarm optimization. In Proceedings of icnn'95—International Conference on Neural Networks (Vol. 4, pp. 1942–1948).
9. Liu, Y. J., Yu, M., Zhao, G., Song, J., Ge, Y., and Shi, Y. (2017). Real-time movie-induced discrete emotion recognition from eeg signals. *IEEE Transactions on Affective Computing*, 9(4), 550–562.
10. Ozdemir, M. A., Degirmenci, M., Izci, E., and Akan, A. (2021). Eeg-based emotion recognition with deep convolutional neural networks. *Biomedical Engineering/Biomedizinische Technik*, 66(1), 43–57.
11. Sorinas, J., Fernandez-Troyano, J. C., Val-Calvo, M., Ferrández, J. M., and Fernandez, E. (2019). A new model for the implementation of positive and negative emotion recognition. Cold Spring Harbor Laboratory. https://doi.org/10.1101/638205
12. Soroush, Z., Maghooli, K., Setarehdan, S. K., and Nasrabadi, A. M. (2018). Emotion classification through nonlinear eeg analysis using machine learning methods. *International Clinical Neuroscience Journal*, 5(4), 135–149.
13. Tao, W., Li, C., Song, R., Cheng, J., Liu, Y., Wan, F., and Chen, X. (2023). EEG-based emotion recognition via channel-wise attention and self attention. In IEEE transactions on affective computing (Vol. 14, Issue 1, pp. 382–393). *Institute of electrical and electronics engineers (IEEE)*. https://doi.org/10.1109/taffc.2020.3025777

14. Wirawan, I. M. A., Wardoyo, R., and Lelono, D. (2022). The challenges of emotion recognition methods based on electroencephalogram signals: A literature review. *International Journal of Electrical and Computer Engineering*, 12(2), 1508.

15. Yang, H., and Du, Q. (2011). Particle swarm optimization-based dimensionality reduction for hyperspectral image classification. In 2011 IEEE International Geoscience and Remote Sensing Symposium. IEEE.

16. Zeng, H., Wu, Z., Zhang, J., Yang, C., Zhang, H., Dai, G., and Kong, W. (2019). EEG emotion classification using an improved sincnet-based deep learning model. In *Brain sciences* (Vol. 9, Issue 11, p. 326). MDPI AG. https://doi.org/10.3390/brainsci9110326

17. Zhang, J., Chen, M., Zhao, S., Hu, S., Shi, Z., and Cao, Y. (2016). Relieff-based eeg sensor selection methods for emotion recognition. *Sensors*, 16(10), 1558.

18. Zhong, P., Wang, D., and Miao, C. (2020). Eeg-based emotion recognition using regularized graph neural networks. *IEEE Transactions on Affective Computing*, 13(3), 1290–1301.

19. Zhuang, N., Zeng, Y., Yang, K., Zhang, C., Tong, L., and Yan, B. (2018). Investigating patterns for self-induced emotion recognition from EEG signals. *Sensors*, 18(3), 841.

16 A Technique for Aligning Thermal-visible Images by Utilizing a Bounding-box Matching Approach

A. K. Singh[1,a], D. Chaudhuri[1,b], M. P. Singh[2,c], S. Mitra[1,d], and B. B. Chaudhuri[1,e]

[1] CSE Department, Techno India University, Kolkata, India

[2] DRDO Young Scientist Laboratory—CT, IIT-M Research Park, Chennai, India

Abstract

Image registration is now essential in various fields due to practical obstacles. Registering thermal and visible images poses challenges due to differences in imaging principles. Traditional methods rely on point feature matching but face limitations such as the need for complex feature descriptors and eliminating mismatched points, resulting in reduced accuracy, and increased computational requirements. To tackle these issues, a new approach using corner points extracted from bounding boxes is proposed in this research paper. This method involves three main steps: employing an object detection algorithm to extract four corners of the bounding box, which eliminates the need for intricate feature descriptors while enhancing precision through higher-level semantic information. A new method for matching bounding boxes is used to determine connections between points in the images. Then, an affine transformation matrix is calculated using these matched point pairs.

Keywords: bounding-box matching, registration coefficient, thermal-visible image registration.

Introduction

Multiple image modalities have been shown to be valuable in improving decision-making processes for various applications, including medical diagnosis Wang et al. [1], remote sensing Zhou and Gao [2], and surveillance tasks [3]. To achieve accurate fusion between different images acquired at different times or with different sensors, image registration is crucial. This process involves finding a reliable spatial transformation Zimmer et al. [4] that aligns the images properly. However, registering thermal and visible images poses a challenge due to their inherent differences in imaging principles and information content. Henceforth, this study aims to propose an innovative framework facilitating precise registration between thermal and visible images.

Various techniques have been proposed to improve the precision, efficiency, and reliability of image registration. These approaches can be broadly classified into intensity-based methods, deep learning-based methods, and feature-based methods [5]. However, intensity-based approaches face challenges such as image distortion, noise-induced changes

[a]aniket.kumarsingh.ece22@heritageit.edu.in, [b]debasis.chaudhury@technoindiaeducation.com, [c]director.dysl-ct@gov.in, [d]sntnmtr@gmail.com, [e]bbcisical@gmail.com

in appearance, variations in illumination conditions, and discrepancies caused by different imaging sensors. Intensity-based methods are not well-suited for thermal-visible image registration tasks due to significant differences in pixel information content between these types of images. While deep learning networks have gained popularity across various domains, their effectiveness is limited when establishing spatial relationships between multiple points during wide baseline image registration. Achieving superior performance in thermal-visible image registration remains a challenge for deep learning methods. Feature-based methods are widely favored due to their robustness and flexibility, relying on correspondence features to determine spatial transformation parameters.

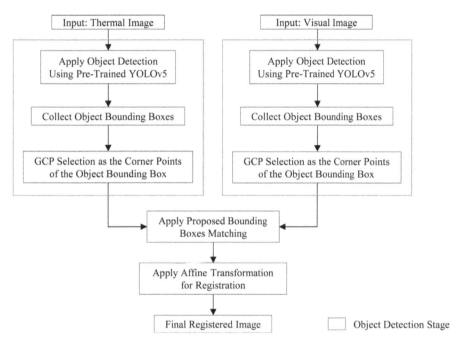

Figure 16.1 Registration process workflow.

The challenge of accurately registering thermal and visible images arises due to their diverse information. To address this, our proposed framework suggests using corner points from bounding boxes as features for precise and efficient registration. An object detection algorithm helps in identifying these corner points accurately. This study employs a distinct method for matching bounding boxes based on a registration coefficient that considers mutual information, aspect ratio, and relative position of the bounding box. Correspondences are established by utilizing this technique for the corner points. Subsequently, a transformation matrix is derived based on the matched point pairs which facilitates alignment between both thermal and visible images.

This paper begins by evaluating the viability of the proposed method while providing an overview of techniques for extracting and comparing corner points. Following this, we present our experimental results and analysis, including observations from feature point extraction and matching experiments as well as image registration experiments. Additionally, we engage in a comprehensive discussion on our research findings derived from these experiments. In conclusion, we summarize the importance of our approach.

Methodology

The paper presents a novel approach for accurate and efficient thermal-visible image registration using the four corner points of bounding boxes as point features. The proposed method aligns the images by utilizing constrained points obtained through object detection. The framework, depicted in Figure 16.1, consists of three main components: GCP selection, bounding box matching, and affine transformation. This method is referred to as 'BoxReg' throughout the paper.

Ground Control Points (GCP) Selection

This study uses a method for accurately extracting ground control points using object detection techniques [8]. Rather than relying on complex feature descriptors, this approach utilizes and the corresponding corner points of the bounding boxes as GCPs. To accurately define these bounding boxes, a powerful one-shot detector, YOLOv5 Glenn et al. [7], known for its speed and accurate box production was chosen instead of two-shot detectors like Faster RCNN [6]. In this work, the focus has been placed on prioritizing detection speed over accuracy in obtaining the required object bounding boxes.

For our experiments, we utilize a pre-trained version of the detector model for visible images, and we re-train the model for better detection of objects in thermal images, as the pre-trained model does not perform too well with thermal images due to the big difference in modality. To conduct our experiments, we employ a pre-trained detector model designed for visible images. To improve object detection in thermal images, which present significant differences in modality compared to visible images, we re-train the model. This is necessary as the performance of the pre-trained model on thermal images is not satisfactory due to these modal variations.

Bounding Box Matching

In this paper, we utilize a novel approach for matching GCP pairs by matching respective object bounding. The LV rule method Li et al. [8] has a major drawback. It is limited in real world application as it relies on having number of bounding boxes being the same in both images and assumes that all pairs are matched one to one after sorting. In this paper we plan to overcome these conditions by introducing specific values to each bounding box pair, and the matching depends on this specific parameter. We sort the bounding boxes by sorting all left values of an image. The object bounding boxes of one image are then checked with bounding boxes, with the same class labels, of the corresponding image of the other modality in that order. We introduce some parameters for each bounding box pair and a registration coefficient for the bounding box pair is calculated using those parameters. The core idea of matching bounding boxes is that an object bounding box of one image corresponds to the object bounding box with the minimum registration coefficient of the respective image with the other modality. We apply some bounds on the bounding box matching for optimization. The process of matching bounding boxes is outlined in the following manner.

The representation of object bounding boxes from an image can be depicted as

$$\Phi^{2 \times N} = \left(\phi_1, \phi_2, \phi_3, \ldots \phi_i, \ldots \phi_n \right) \tag{1}$$

$$\phi_i = \begin{pmatrix} x_{i1}\ x_{i2}\ x_{i3}\ x_{i4} \\ y_{i1}\ y_{i2}\ y_{i3}\ y_{i4} \end{pmatrix} \tag{2}$$

where $i = 1, 2, 3, \ldots$ n, represents the order of sorted bounding box ($i = 1$ denoting the leftmost box and so on), ϕ_i represents the coordinate information for ith bounding box (sorted), n denotes the number of bounding boxes in an image, while N = 4n represents the total number of ground control points in that same image. To summarize, Φ can be thought of as a matrix with dimensions $2 \times$ N.

Since we have a pair of thermal-visible image, we represent the bounding boxes of visible image as $\Phi_{vis}^{2 \times N} = \left(\phi_1^v, \phi_2^v, \phi_3^v, \ldots \phi_i^v, \ldots \phi_n^v \right)$ and the bounding boxes of thermal image as $\Phi_{therm}^{2 \times M} = \left(\phi_1^t, \phi_2^t, \phi_3^t, \ldots \phi_i^t, \ldots \phi_m^t \right)$.

For linking corresponding points, we employ an affine transformation Min et al. [9] matrix consisting of six variables. This implies that for the system to provide a unique solution, it is necessary for at least three feature points to not lie on a straight line. It is evident that our proposed registration method fulfills this requirement as each object bounding box possesses four vertices with three of them being non-collinear.

To achieve box matching, we employ a registration coefficient ϱ. This coefficient is computed by considering three parameters associated with a pair of bounding boxes from corresponding thermal-visible images. A mutual information-based measure acts as the parameter for structural similarity, while the ratio of aspect ratios measures spatial similarity. Additionally, a centroid based distance parameter serves as an indicator of relative distance. We select these three parameters to establish similarities between the given bounding boxes.

Spatial Ratio Calculation

The thermal and visible bounding boxes have aspect ratios called A_{therm} and A_{vis}, respectively. Then the spatial ratio ζ is the ratio of both aspect ratios.

$$\zeta = \frac{A_{therm}}{A_{vis}} \tag{3}$$

If ζ is greater than 1, we take its inverse as the new spatial ratio. If we ensure precise object detection and non-linear distortion-free images, it is ideal to have a spatial ratio close to or equal to 1. The further away from this value we are, the more dissimilar the bounding boxes become.

Mutual Information Based Measure Calculation

To process each pair of bounding boxes, we cropped the corresponding images to match the dimensions of these boxes. Subsequently, we applied the Canny edge detector. We also employed adaptive histogram equalization on the cropped thermal images as a preprocessing step. To ensure uniformity in size, one of the generated edge maps was adjusted to match the dimensions of its counterpart. A mutual information-based distance measure between this pair of edge maps was computed as µ. By using this mutual information-based distance calculation on edge maps instead of directly calculating mutual information on original images with diverse modalities, potential inaccuracies due to differences in modality were avoided. Let the edge point coordinates of the edge maps of thermal and visible bounding boxes be ϕ_e^t and ϕ_e^v, respectively. The centroids of these edge points are P_e^t and P_e^v, respectively. The distances of each point in ϕ_e^t and ϕ_e^v are calculated with respect to their corresponding centroids P_e^t and P_e^v and are denoted as 1-dimensional matrices D_e^t and D_e^v, respectively. The K-bin histograms of D_e^t and D_e^v are λ_k^t and λ_k^v. These histograms are the *shape context* Belongie et al. [10] relative to the respective centroids. $SC(X, Y) = 0.5 * \Sigma(X - Y)^2/(X + Y)$ is a mutual information-based measure between X and Y. Therefore,

$$\mu = SC\left(\lambda_k^t, \lambda_k^v\right) \tag{4}$$

Relative Distance Calculation

The image coordinate system is standardized due to the varying resolutions of the images. To determine the relative position of the bounding box centroid, we calculate the distance (δ) from both top-left and bottom-left corners of the image to its centroid. This approach allows for a simplified representation of the bounding box centroid by condensing it into a single parameter. If the centroid of a bounding box is (x_0, y_0), then,

$$\delta = \frac{\sqrt{x_0^2 + y_0^2} + \sqrt{x_0^2 + (y_0 - 1)^2}}{2} \tag{5}$$

Let the centroid-based distance (δ) of bounding box of thermal image be δ_t and that of the bounding box of the corresponding visible image be δ_v. Then the relative distance parameter becomes

$$\Delta = \text{abs}(\delta_t - \delta_v) \tag{6}$$

Registration Coefficient Calculation

The matching process can sometimes lead to errors when relying solely on these three parameters, particularly if the parameter of the corresponding box does not possess the optimal value. However, by incorporating and analyzing all three parameters together, we are able to calculate our registration coefficient ϱ. When both ζ and μ have higher values, it indicates a greater degree of resemblance between the bounding boxes. Conversely, when Δ has a lower value, it signifies better alignment among the bounding boxes. The registration coefficient is then expressed as $\varrho = \zeta + \mu - \Delta$. Combining equations 3, 4, and 6, the registration coefficient becomes

$$\varrho = \frac{A_{therm}}{A_{vis}} + SC\left(\lambda_e^t, \lambda_e^v\right) - abs\left(\delta_t - \delta_v\right) \tag{7}$$

Let ϕ_i^t is the thermal image bounding box and ϕ_j^v be one of the visible image's bounding boxes. If $\varrho_{ij} = \max(\varrho_{i1}, \varrho_{i2}, \varrho_{i3}, \dots \varrho_{ij}, \dots \varrho_{in})$, then ϕ_j^v is the corresponding bounding box of ϕ_i^t. If ϕ_j^v has already been matched with another bounding box before, we discard the thermal bounding box, to avoid one-to-many matching.

GCP Matching Optimization

The GCP matching method can be quite time-consuming, especially when dealing with many boxes. To match each bounding box from the thermal image with all the boxes in the visible image, we need to go through $m \times n$ loops. To address this issue and improve efficiency, we have implemented a strategy to reduce the total number of boxes in the thermal image. We do this by comparing each current box's top-left point (x_i, y_i) with that of the previous box (x_{i-1}, y_{i-1}). If $x_i < x_{i-1} + \dfrac{W}{8}$ and $y_i < y_{i-1} + \dfrac{H}{8}$ hold true, we discard that particular current box. This approach helps minimize computational overheads while still maintaining accuracy. Here W and H refer respectively to width and height measurements pertaining to our thermal image. Furthermore, if number of boxes in the thermal image is less than the ones detected in visible image, we start our matching with thermal image, else we start with the visible image, to reduce unnecessary computations.

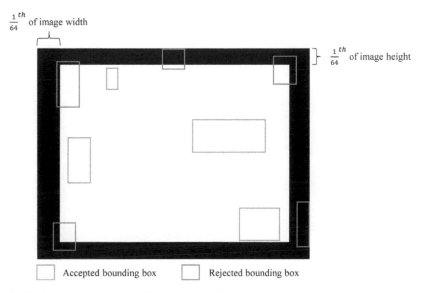

Figure 16.2 Mask for the image for filtering bounding boxes.

Since the bounding boxes were sorted initially, we compare the index of each thermal bounding box with a subset of visible image's bounding boxes using a $\min(n, 7)$ nearest indices. By doing this, we effectively decrease the number of bounding boxes to be analyzed for both images.

To improve the accuracy of our algorithm, we have implemented an additional step that involves eliminating bounding boxes whose corner points fall outside a specific boundary within the image. To achieve this, we generate a white mask of our image and add black padding on the inside on all four sides. The size of this padding is set to be $\dfrac{1}{64}$ th of the width and height of the original image for x and y directions respectively. Any bounding box that has at least one corner point lying outside this inner white area, as illustrated in Figure 16.2, is excluded from further consideration. This approach serves two purposes: first, it prevents us from including bounding boxes that might contain only partially visible objects; secondly, by removing these unnecessary boxes altogether, it helps reduce computational overheads.

Affine Transformation

The affine transformation is a commonly utilized technique in image processing that includes four main types: translation, rotation, scaling, and shearing. This method enables spatial transformations of both thermal and visible images using an affine transformation matrix denoted as P. To achieve this, corresponding points are identified in both images to compute the matrix P. Once derived, the resulting matrix can be implemented to generate an aligned image. Let the final GCP list of the thermal image be $\tilde{\Phi}_{therm} = \begin{pmatrix} x_1 \, x_2 \, x_3 & x_{\acute{n}} \\ y_1 \, y_2 \, y_3 & y_{\acute{n}} \end{pmatrix}$,

and the matched point pairs from visible image be $\tilde{\Phi}_{vis} = \begin{pmatrix} u_1 \, u_2 \, u_3 & u_{\acute{n}} \\ v_1 \, v_2 \, v_3 & v_{\acute{n}} \end{pmatrix}$. Here, $\acute{n}$ is the

number of matched points. Since for affine transformation matrix, we need a minimum of 3 points, if $\acute{n} > 3$, we need to optimize the points to find the best fit for generating the affine transformation matrix. We utilize the best 2D affine transformation method [11].

Figure 16.3 Subset of test images.

(a)SIFT   (b)ORB   (c)LV   (d)Proposed method (BoxReg)

Figure 16.4 Matching results between feature points from various methods.

Experiments and Results

Dataset Used

We conducted our experiments using a sample of 37 unregistered pairs of thermal-visible images. Out of these, 32 image pairs were captured using an *AXIS Q8741-LE Bi-spectral PTZ Network Camera*. These images covered various locations and working conditions, including indoor and outdoor environments with different lighting levels such as low lighting or overexposure. Some visible images in the set were IR-illuminated. The resolution for the thermal images in these image pairs was 800 × 600, whereas the resolution for the corresponding visible images was higher at 1920 × 1080. The remaining five image pairs (images in the 3rd and 4th row, columns 6 to 10 in Figure 16.3) were sourced from the VIFB dataset [12]. These VIFB collection's image resolutions ranged from 320 × 240 to 630 × 460. Given that these five specific image sets had already undergone pre-registration process beforehand, we further applied random affine transformation on their corresponding visible imagery to render them as a new set of unregistered visible images.

Qualitative Analysis

In Figure 16.4, we compared the proposed method to SIFT Lowe, [13], ORB Rublee et al. [14], and the LV-rule-based method boxes [8]. The comparison focused on evaluating their performance in terms of feature point extraction and matching. Both SIFT and ORB can extract many feature points, but this can result in redundant information that affects accurate matching operations. Finding corresponding point pairs using either SIFT or ORB is particularly challenging as it is rare to achieve sufficient matches for successful thermal-visible image registration. Additionally, both methods require deleting mismatched and excessive point pairs to improve accuracy which further prolongs the already time-consuming matching process. In contrast, the LV method overcomes these limitations by precisely matching all constrained points without having to eliminate any redundant point pairs initially extracted from images. Nonetheless, its drawback arises when there is a disparity in the number of detected boxes between both images as this leads to inaccurate pairing. In contrast, the suggested approach can achieve enough paired points for obtaining an exclusive affine transformation matrix, and by examining registration coefficients for each pair of boxes, it effectively minimizes matching errors.

Table 16.1: On 37 image pairs, the average assessment metrics for all procedures.

Method	ERR (%) (+)	RMSER (%) (+)	ARMSE (-)	Time (s) (-)
SIFT	0.00	32.43	0.508	1.802
ORB	5.41	32.43	0.481	**0.112**
LV	75.67	72.97	**0.223**	0.493
BoxReg	**83.78**	**75.67**	0.247	0.364

The proposed framework was compared with three registration algorithms using visible float images and thermal reference images. In Figure 16.5, the W/O. R columns show the images without registration, while the 1st and 2nd rows display qualitative measures of registration accuracy. In the 1st row, image pairs represent alignment between registered visible images and thermal images. Evaluation box pairs in the 2nd row reflect overlap between registered visible and thermal images. The red, green, and white areas represent the box pixels of the thermal image, visible image, and overlapping pixel area accordingly. Figure 16.5 clearly demonstrates that our 'BoxReg' method outperforms other methods in terms of image alignment as well as box overlap. Notably, both SIFT and ORB methods were found to be ineffective. In certain cases, the LV method achieved satisfactory registration when the number and order of boxes were consistent.

Quantitative Analysis

For further comparison we used the ERR measure Li et al. [8], along with 2 root-mean-squared error-based scores. RMSER (Root Mean Squared Error Rate) is a metric like ERR, where instead of utilizing RE we utilize the root mean squared error. ARMSE (Average Mean Squared Error) is basically the RMSE average value for all image pairs. In Table 16.1, we compare the registration results obtained from four different methods. The most favorable value is highlighted in bold. It should be noted that a positive sign (+) indicates larger values are preferable, while a negative sign (-) signifies that smaller values are more desirable. The analysis is done on our abovementioned sample of 37 image pairs, and runtime is taken as an average of all experiments, excluding the object detector model loading times.

The performance of SIFT and ORB was unsatisfactory when applied to our test images. Although the LV method showed some similarities with our approach, its limitation of maintaining a consistent number of boxes hindered its performance in scenarios where the number of detected boxes varied between modalities. On the other hand, our proposed 'BoxReg' method demonstrated superior ERR and RMSER values. When comparing processing times, it is worth noting that excluding ORB from this comparison is necessary since although it has a faster speed, its accuracy is compromised due to being prone to errors. With the exclusion of ORB, our method is also the fastest amongst the rest.

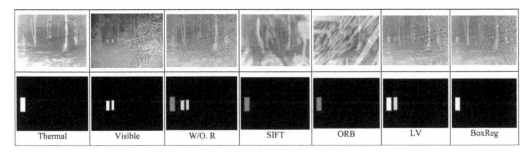

Figure 16.5 Registration results for various methods.

Conclusion

This paper presents a novel bounding box matching method aimed at improving GCP matching for thermal-visible image pairs. The objective of our work was to devise an effective image registration technique that mitigates the drawbacks associated with the LV rule method. Through extensive experimentation, we have attained promising results using our proposed approach. Our method leverages three key factors in determining similarity between bounding boxes: shape dissimilarity, relative positional change across respective images, and edge feature similarity. These parameters collectively contribute to accurate box alignment within the registration process. Furthermore, it should be noted that the overall accuracy of the registration process hinges on the quality of bounding boxes obtained from object detection algorithms utilized beforehand. Error in the detection can result in subsequent steps being compromised, hindering proper matching and registration. Our method utilizes an affine transform model and can also be adapted for a perspective transformation model. Even with just one successful bounding box match, we have a minimum of four points for working with the perspective transform model. However, our method may not work efficiently when dealing with non-linear distortions.

In future, we aim to enhance the efficiency of this method by improving inference speeds and reducing memory consumption. Additionally, we have plans to refine our registration coefficient through the implementation of a cross-correlation-based measure. Our ultimate goal is to achieve real-time performance for optimal use in various applications.

Acknowledgment

We would like to express our sincere gratitude to DYSL-CT, DRDO for funding the project.

References

1. Wang, Z., Cui, Z., and Zhu, Y. (2020). Multi-modal medical image fusion by laplacian pyramid and adaptive sparse representation. *Computers in Biology and Medicine*, 123, 103823.
2. Zhou, H., and Gao, H. (2014). Fusion method for remote sensing image based on fuzzy integral. *Journal of Electrical and Computer Engineering*, 2014, 26.
3. Paramanandham, N., and Rajendiran, K. (2018). Infrared and visible image fusion using discrete cosine transform and swarm intelligence for surveillance applications. *Infrared Physics and Technology*, 88, 13–22.
4. Zimmer, V. A., Ballester, M. Á. G., and Piella, G. (2019). Multimodal image registration using Laplacian commutators. *Information Fusion*, 49, 130–145.
5. Ma, J., Jiang, X., Fan, A., Jiang, J., and Yan, J. (2021). Image matching from handcrafted to deep features: a survey. *International Journal of Computer Vision*, 129, 23–79.
6. Ren, S., He, K., Girshick, R., and Sun, J. (2015). Faster r-cnn: towards real-time object detection with region proposal networks. *Advances in Neural Information Processing Systems*, 28, 91–99.
7. Jocher, G., Chaurasia, A., Stoken, A., Borovec, J., Kwon, Y., Michael, K., Fang, J., Yifu, Z., Wong, C., Montes, D. and Wang, Z. (2022). Ultralytics/yolov5: v7.0—YOLOv5 SOTA realtime instance segmentation (v7.0). zenodo.
8. Li, Q., Han, G., Liu, P., Yang, H., Luo, H., and Wu, J. (2021). An infrared-visible image registration method based on the constrained point feature. *Sensors*, 21(4), 1188.
9. Min, C., Gu, Y., Li, Y., and Yang, F. (2020). Non-rigid infrared and visible image registration by enhanced affine transformation. *Pattern Recognition*, 106, 107377.
10. Belongie, S., Malik, J., and Puzicha, J. (2002). Shape matching and object recognition using shape contexts. *IEEE Transactions on Pattern Analysis and Machine Intelligence*, 24(4), 509–522.
11. Shapiro, L. G., and Stockman, G. C. (2001). Computer Vision. Prentice Hall, (p. 339).
12. Zhang, X., Ye, P., and Xiao, G. (2020). VIFB: a visible and infrared image fusion benchmark. In Proceedings of the IEEE/CVF Conference on Computer Vision and Pattern Recognition Workshops, (pp. 104–105).
13. Lowe, D. G. (2004). Distinctive image features from scale-invariant keypoints. *International Journal of Computer Vision*, 60, 91–110.
14. Rublee, E., Rabaud, V., Konolige, K., and Bradski, G. (2011). ORB: an efficient alternative to SIFT or SURF. In 2011 International Conference on Computer Vision (pp. 2564–2571). IEEE.

17 Non-invasive Dietary Monitoring using Chewing Sounds

Saubhik Bandyopadhyay[1,a], Ujjwal Mandal[2,b], Pratham Majumder[3,c], and Punyasha Chatterjee[4,d]

[1]Computer Science & Engineering Department, Amity University, Kolkata, India

[2]Techno India University, Kolkata, India

[3]CMR Institute of Technology, Bengaluru, India

[4]School of Mobile Computing and Communication, Jadavpur University, India

Abstract

This study explores the potential of using chewing sounds to automatically monitor food intake. Traditional methods, such as food diaries, are prone to bias and inaccuracies, while automatic monitoring systems have the potential to provide more objective and precise data. In this study, we collect Mel-Frequency Cepstral Coefficients (MFCC) from chewing sounds and use these features to train two neural network models, an Artificial Neural Network (ANN) and a Long Short-Term Memory (LSTM) network. The study's findings demonstrate that the models are useful for categorizing various food categories based on chewing sounds, suggesting that automatic monitoring systems based on sound analysis have the potential as a non-invasive and accurate method for dietary monitoring.

Keywords: ANN, chewing sound, deep learning, eating detection, LSTM, MFCC.

Introduction

Overweight and obesity are described by the World Health Organization (WHO) as the abnormal or excessive accumulation of fat that poses a health concern. A wide range of health issues, such as diabetes, cardiovascular disease, musculoskeletal diseases, and several types of cancer, are more likely to occur in people who are overweight or obese. A person who consumes inadequate amounts of necessary nutrients may also be at risk for major health issues such sarcopenic obesity, anorexia nervosa, pancreatic cancer, rheumatoid arthritis, and organ failure. A nutritious diet reduces the risk of chronic diseases and encourages normal growth and development. Even in the midst of the COVID-19 pandemic, WHO recommended eating a nutritious diet to improve immunity Organization et al. [12]. Therefore, it is essential to adopt healthy lifestyle habits, such as exercising regularly, taking a balanced diet, etc to prevent and manage overweight, obesity and illness.

To evaluate food intake, one must first keep track of one's regular diet. Food intake monitoring Päßler et al. [15] is an important tool that may provide details of an individual's eating episodes, such as how much food and what type of food was eaten, what is the

[a]saubhikit@gmail.com, [b]ujjwal@gmail.com, [c]pratham.majumder1989@gmail.com, [d]punyasha.chatterjee@gmail.com

eating duration, etc. The traditional approach for food intake monitoring Fallaize et al. [3] allows self-reporting and manual recording of eating activities throughout time, as well as nutrition analyses. These manual methods are prone to self-bias and recollection mistakes, as well as poor patient compliance. As a result, automated dietary monitoring Selamat and Ali [16] may be useful in certain situations. These methods depend largely on wearable sensors that detect human body activity when put at various points on the body, and hence have the capacity to overcome the limitations of manual systems. Different applications of automated dietary monitoring are an objective measure of caloric intake, eating behaviour detection, hospitalization services Heighington-Wansbrough and Gemming [6] etc.

In this paper, we have concentrated on automated dietary monitoring using chewing sound analysis. The masseter, temporalis, medial, and lateral pterygoid muscles collaborate during chewing or mastication to produce a sequential sound pattern that may be further analysed. In this study, we collect Mel-Frequency Cepstral Coefficients (MFCC) Kalantarian and Sarrafzadeh [7] features from chewing sounds and analyse those by Artificial Neural Network (ANN) Mishra and Srivastava [11] and Long Short-Term Memory (LSTM) Yu et al. [17] to identify different types of foods.

This paper is organized as follows: Section 2 refers to the literature survey. Section 3 refers to our proposed data analysis approach based on chewing sounds. Section 4 refers to the simulations and results. Section 5 gives the conclusion.

Related Work

In this section, we will discuss the existing works on dietary monitoring using chewing sounds.

In Khan et al. [8], the chewing sound is acquired by iHearken, a wireless headphone-like device. In Farooq and Sazonov [4] the piezoelectric sensors are taped to the temporalis muscle to collect chewing sounds. Piezoelectric sensors have a variety of shortcomings, such as soreness because of the skin's adhesive application. In Zhang et al. [19], based on the activity of the temporalis muscle, authors utilized Electromyography (EMG) sensor to determine the chewing movement. By integrating a vibration sensor, authors in Zhang and Amft [18] enhanced the design. The EMG sensor is restricted in a number of ways, such as the requirement for direct sensor placement and the potential to distinguish a substantially larger number of foods is unlikely. Authors in Chun et al. [1] employed a VL6180X proximity sensor that was mounted on a 3D printed chain around the neck. The VL6180X sensor detects eating by capturing jaw movement. The performance of the suggested sensing design is strongly reliant on sensor orientation and is influenced by tiny movements caused by physical activity.

Several sensing methods are combined in a multi-sensor technique to identify chewing alone or in conjunction with other detection systems e.g., bite, swallow, and motion. In Fontana et al. [5], the authors used piezoelectric and accelerometer sensors to detect hand-to-mouth gestures. SPLENDID is the name of the second widely used multimodal sensing method proposed by Papapanagiotou et al. [14] to detect chewing action using a PPG sensor and a microphone integrated into an in-ear placement. SPLENDID was used in conjunction with a mandometer Papapanagiotou et al. [13] in order to better understand human eating behaviour during single meals Diou et al. [2]. Similarly, Photoplethysmography (PPG) are photo, proximity, and doppler sonar sensors, that are utilized to detect chewing actions.

In this paper, we have analysed the chewing sounds using deep learning to detect different food types. We captured the chewing sounds by an acoustic approach using microphone as it is simple, inexpensive and can be easily embedded with a Body Sensor/Area Network (BSN). For your kind reference, in one of our earlier works, we proposed a BSN prototype named *DietSN* Mamud et al. [10] for automatic dietary monitoring, where the chewing sound is captured by wearable microphone.

Chewing Sound Analysis

To analyze the chewing sounds, we use deep learning LeCun et al. [9] approaches. At first, we extract the features of the audio samples using MFCC. Then we apply two different Neural Network models, i.e., *Artificial Neural Network (ANN)* and *Long Short-Term Memory (LSTM)* for correctly identifying food types from chewing sounds. The data set, and model specifications, are illustrated in the next subsections.

Dataset Overview

We have considered a data repository Selamat et al. [16], where there are 157 chewing sounds. This chewing data set is divided into six classes i.e., Burger, Chips, Carrot, Pizza, Grapes and Soup. These sounds are in .wav format. The models are trained using 80% of the data, while the remaining 20% is used for random testing.

Feature Extraction by MFCC

The steps of feature extraction by MFCC are illustrated below:

Sampling: At first, the continuous audio signal $S(t)$ consisting of N frames is splitted into M samples where M < N. Discrete time and discrete amplitude signal is represented as $S(n)$. In literature, the most preferred pair value of (N, M) is considered to be (256,100) for error-free reconstruction of the signal $S(t)$.

Windowing: It is the process of minimizing disruption in the start and end frames in $S(n)$. The output signal.

$$Y(n) = \sum_{i=0}^{N} \left[S(i)W(i) \right],$$

where $W(i)$ is the system transfer function of window size i.

Fast Fourier Transform (FFT): It is the process that converts the spatial signal $Y(n)$ to frequency domain $Y(f)$,

i.e. $y(f) = FFT[Y(n)]$.

Mel Scale: Mel-frequency spectrum Mf is created by scaling these spectrum $Y(f)$ with the *Mel-frequency triangular window-based Filter bank*.

Discrete Cosine Transform (DCT): Mel-scale results highly correlated MFC coefficients Mf. DCT is applied to Mf to compute and extract the de-correlated Mf coefficients, known as MFCC. For each audio sample, 40 MFCCs were generated per frame with a window size of a few milliseconds.

Model Specifications

We have used two types of Neural Networks for classification: Artificial Neural Network (ANN) and Long Short-Term Memory (LSTM). Specifications of the ANN model and LSTM model are described in the Table 17.1.

Table 17.1: Specifications of ANN and LSTM models.

Model	ANN	LSTM
No. of layers	1 (input layer) + 2 (hidden layers) + 1 (Output layer)	3 (LSTM layers) + 1 (Dense layer)
No. of nodes/layer	128 (input layer) + 64 (first hidden layer) + 32 (second hidden layer) + 6 (Output layer)	128 (first LSTM layer) + 64 (second LSTM layer) + 32 (third LSTM layer) + 6 (dense layer).
Activation function	Rectified Linear Unit (ReLU) (first 3 layers) + Soft max activation function (output layer)	Rectified Linear Unit (ReLU) (first 3 layers) + Soft max activation function (output dense layer)

In ANN the number of nodes in our output layer is 6 which corresponds to the number of labels. Because the soft max function makes the total of outputs equal to one, we utilize it as the activation function for the output layer, such that the output may be treated as probability. The prediction is then determined by selecting the label with the highest probability. The dropout is considered as 50%.

In LSTM, in the output dense layer, we employed a soft max activation function to transform the inputs into a discrete probability distribution. The Adam optimizer is used to reduce the categorical cross-entropy loss function during training. Because a thorough search of hyper parameters would take too much time due to the extended training time, the most promising combination was discovered via single fold evaluation.

Simulation and Results

Software Used

The code is done in Python 3.9.7. The following libraries are used:

- Ipython.display.Audio: This enables us to play audio directly.
- LibROSA: librosa is a Python tool for music and audio processing that allows to load audio as a NumPy array for analysis and editing.

Performance Metric

We have used the following metrics for performance analysis:

- **Accuracy:** It measures the proportion of correctly classified samples in a given data set. It is expressed as:

$$Accuracy = \frac{Number\ of\ correctly\ classified\ samples}{Total\ number\ of\ samples}$$

- **Loss Function:** Mathematically, the loss function is defined as a function of the model's parameters θ and the training data D. It takes as input a predicted output $\hat{y}$ and the corresponding true output y, and outputs a scalar value that represents the difference between the predicted and true outputs. It is expressed as:

$$L(\theta, D) = \frac{1}{N} \sum_{i=1}^{N} l(\hat{y}_i, y_i)$$

where N is the number of training samples.

Results

We obtained the following training and testing accuracy from two different Neural Network models, as depicted in Table 17.2. Based on these accuracy values, the ANN model outperforms the LSTM model in terms of testing accuracy, indicating that it generalizes better to unseen data. The higher training accuracy of the LSTM model suggests that it might have overfit the training data, resulting in a lower testing accuracy. The corresponding accuracy graphs with respect to epochs are shown in Figure 17.1. We can see that as the number of epochs rises, Data used for training and testing become more accurate. The curve changes from underfitting to optimal as the number of epochs increases and the weights in the neural network are modified more often.

The loss functions analysis is depicted in Figure 17.2. The Loss of the model decreases exponentially When the quantity of epochs for both the training and testing data grows, as seen in the graphs. Comparing the two models based on the loss analysis, we can observe that the ANN model shows a decreasing loss trend over epochs, starting at a higher value

Table 17.2: Simulation results.

Model	Training Accuracy	Testing Accuracy
ANN	98%	94%
LSTM	99%	81%

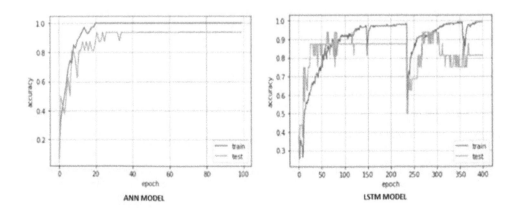

Figure 17.1 ANN and LSTM model accuracy graphs.

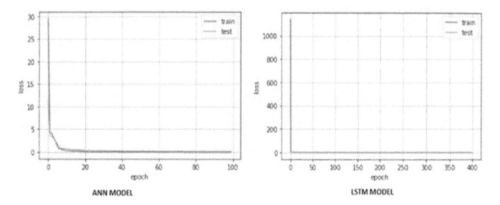

Figure 17.2 ANN and LSTM model loss graph.

and gradually decreasing to a minimum value of 0. It indicates a continuous improvement in the model's performance as the training progresses. On the other hand, the LSTM model achieves a loss of 0 after the initial epoch and maintains this minimum loss throughout the remaining epochs. This suggests that the LSTM model quickly converges and achieves a good fit to the training data. In terms of the loss vs epoch performance, the LSTM model appears to have a better performance compared to the ANN model.

Conclusion

Our study demonstrates the successful categorization of chewing sounds through the utilization of machine learning approaches. We extract relevant features from audio files of the chewing sound dataset using the MFCC method. Then, based on specific sound characteristics, we apply two neural network models, ANN and LSTM, to categorize the chewing sounds of different food items. We have compared the performance of two models with respect to accuracy and loss function. This study showcases the effectiveness of multiple neural network models in accurately classifying chewing sounds.

References

1. Chun, K. S., Bhattacharya, S., and Thomaz, E. (2018). Detecting eating episodes by tracking jawbone movements with a non-contact wearable sensor. *Proceedings of the ACM on Interactive, Mobile, Wearable and Ubiquitous Technologies,* 2(1), 1–21.
2. Diou, C., Sarafis, I., Ioakimidis, I., and Delopoulos, A. (2017). Data-driven assessments for sensor measurements of eating behavior. In 2017 IEEE EMBS International Conference on Biomedical and Health Informatics (BHI), (pp. 129–132). IEEE.
3. Fallaize, R., Forster, H., Macready, A. L., Walsh, M. C., Mathers, J. C., Brennan, L., Gibney, E. R., Gibney, M. J., and Lovegrove, J. A. (2014). Online dietary intake estimation: reproducibility and validity of the food4me food frequency questionnaire against a 4-day weighed food record. *Journal of Medical Internet Research*, 16(8), e190.
4. Farooq, M., and Sazonov, E. (2016). A novel wearable device for food intake and physical activity recognition. *Sensors*, 16(7), 1067.
5. Fontana, J. M., Farooq, M., and Sazonov, E. (2014). Automatic ingestion monitor: a novel wearable device for monitoring ingestive behavior. *IEEE Transactions on Biomedical Engineering*, 61(6), 1772–1779.

6. Heighington-Wansbrough, A. J., and Gemming, L. (2022). Dietary intake in hospitals: a systematic literature review of the validity of the visual estimation method to assess food consumption and energy and protein intake. *Clinical Nutrition ESPEN*.

7. Kalantarian, H., and Sarrafzadeh, M. (2015). Audio-based detection and evaluation of eating behavior using the smartwatch platform. *Computers in Biology and Medicine, 65*, 1–9.

8. Khan, M. I., Acharya, B., and Chaurasiya, R. K. (2022). ihearken: chewing sound signal analysis based food intake recognition system using bi-lstm softmax network. *Computer Methods and Programs in Biomedicine*, 221, 106843.

9. LeCun, Y., Bengio, Y., and Hinton, G. (2015). Deep learning. *Nature*, 521(7553), 436–444.

10. Mamud, S., Bandyopadhyay, S., Chatterjee, P., Bhandari, S., and Chakraborty, N. (2021). Dietsn: a body sensor network for automatic dietary monitoring system. In Data Management, Analytics and Innovation: Proceedings of ICDMAI, 2021, (Vol. 1, pp. 367–381). Springer.

11. Mishra, M., and Srivastava, M. (2014). A view of artificial neural network. In 2014 International Conference on Advances in Engineering and Technology Research (ICAETR-2014), (pp. 1–3). IEEE.

12. World Health Organization. Connecting the World to Combat Coronavirus. *Healthy at Home, Healthy Parenting*, 2020.

13. Papapanagiotou, V., Diou, C., Langlet, B., Ioakimidis, I., and Delopoulos, A. (2015). A parametric probabilistic context-free grammar for food intake analysis based on continuous meal weight measurements. In 2015 37th annual International Conference of the IEEE Engineering in Medicine and Biology Society (EMBC), (pp. 7853–7856). IEEE.

14. Papapanagiotou, V., Diou, C., Zhou, L., van den Boer, J., Mars, M., and Delopoulos, A. (2016). A novel approach for chewing detection based on a wearable ppg sensor. In 2016 38th Annual International Conference of the IEEE Engineering in Medicine and Biology Society (EMBC), (pp. 6485–6488). IEEE.

15. P"aßler, S., Wolff, M., and Fischer, W. J. (2012). Food intake monitoring: an acoustical approach to automated food intake activity detection and classification of consumed food. *Physiological Measurement*, 33(6), 1073.

16. Selamat, N. A., and Ali, S. H. M. (2020). Automatic food intake monitoring based on chewing activity: a survey. *IEEE Access*, 8, 48846–48869.

17. Yu, Y., Si, X., Hu, C., and Zhang, J. (2019). A review of recurrent neural networks: Lstm cells and network architectures. *Neural Computation*, 31(7), 1235–1270.

18. Zhang, R., and Amft, O. (2016). Bite glasses: measuring chewing using emg and bone vibration in smart eyeglasses. In Proceedings of the 2016 ACM International Symposium on Wearable Computers, (pp. 50–52).

19. Zhang, R., Bernhart, S., and Amft, O. (2016). Diet eyeglasses: recognising food chewing using emg and smart eyeglasses. In 2016 IEEE 13th International Conference on Wearable and Implantable Body Sensor Networks (BSN), (pp. 7–12). IEEE.

18 *AttOnTrack*: Estimating Sustained Attention using Visual and Facial Cues with Unmodified Webcams

Pragma Kar[1,a], Souvik Halder[2,b], and Saugat Bhattacharyya[3,c]

[1]Computer Application Techno India University Kolkata, India

[2]Information Technology, Jadavpur University, Kolkata, India

[3]School of Computing, Eng & Intel. Sys, Ulster University, Londonderry, UK

Abstract

Recent years have witnessed the emergence of diverse and novel approaches for studying human attention with the help of advanced technology. In the field of education, tracking students' attention over a prolonged period of time is crucial to understand their cognitive state and degree of boredom, along with the comprehensibility of the course. While existing works have utilised commercial trackers, custom devices and physiological signals to understand the characteristics of human attention, there still exists a gap between sophisticated systems and non-intrusive methodologies for tracking learner's attention. In this paper, we propose *AttOnTrack*, a novel and lightweight vision-based approach that combines learners' visual and facial features to trace their sustained attention during online courses, using an unmodified commercial webcam. In an exhaustive lab-scaled evaluation, *AttOnTrack* has proved its significance with an overall accuracy of 80.1%, thus revealing the potential of a pervasive and low-cost system in tracking human attention.

Keywords: Expressions, isolation forest, sustained attention, visual features.

Introduction

Human attention has been defined and studied in numerous forms. These studies range from the Biological definition of attention that involves the selection of competing information to gain access to the working memory, through top-down or bottom-up controls to its Psychological definition involving contextual alertness, information selection and cognitive processing [16]. In an attempt to apply William James' classic definition [8] of attention along with Posner's theory in the context of online education, we perceive attention as the mental process allowing a learner to be aware of, orient to, and process the relevant (educational) information. In online learning, distractions can be more prevalent due to the lack of active communication, manual supervision and motivation. In such scenarios, it becomes easier for a learner to lose attention. This in turn, leads to further boredom and mind wandering, often leading to high dropout rates in online courses. This very

---

[a]pragyakar11@gmail.com, [b]souvikhalder32@gmail.com, [c]s.bhattacharyya@ulster.ac.uk

observation necessitates the estimation of sustained attention [3], through an automated technique, for online courses.

Traditionally, sustained attention has been measured through manual inspection, vigilance tasks, electroencephalogram (EEG) signals [2], and so on. However, one of the primary challenges arise from the fact that automated attention estimation should be pervasive, and non-intrusive in nature. This is to ensure that the system does not violate its very purpose by creating distractions for the user. Consequently, the second challenge is to minimise the requirement of custom or commercial hardware as they are not always available to the common mass, and might impose a severe restriction on the utility of the system. Thirdly, the availability of sufficient attention-related data is difficult to collect. Thus, the final challenge is to develop a system that can effectively learn from fewer unlabelled data points, that can be collected seamlessly from the user.

In this paper, we propose *AttOnTrack,* a novel and lightweight model for tracking sustained attention of learners by jointly considering five visual and facial features. The system uses a commodity webcam or an integrated laptop camera for extracting these features in real time. These features are then learnt by an anomaly detector in an unsupervised manner for detecting any deviation from normal gaze or facial patterns, caused by possible distractions. In doing so, we make the following contributions in the paper: (a) We propose and develop a non-intrusive, automated model for estimating sustained attention of online learners. The system requires no additional sensors, and can learn the attentional patterns in an unsupervised manner. (b) We explore the joint contribution of visual and facial features in understanding attentional patterns, and (c) prove *AttOnTrack*'s efficacy through a rigorous lab-scaled evaluation. Evaluation results show that *AttOnTrack* performs with an overall accuracy of 80.1% when tested with 5 real-world users.

Rest of the paper is structured as follows: In Section 2, an extensive discussion on *state-of-the-art* literature has been presented. Section 3 describes the proposed model whose performance has been discussed in Section 4. The paper has been concluded in Section 5.

Literature Review

In this section, we provide a background for *AttOnTrack* by discussing some of the existing works that considers different forms of attention and their estimation.

Attention has been studied in multitudinous forms, in different contexts and domains [11]. In a temporal spectrum, attention can range from being divided, alternating, sustained [1] and so on. In [18], the authors have aimed at estimating the learner's divided attention using photoplethysmography through a smartphone's camera. While the model requires no external hardware, it imposes some restrictions on the users to place their fingers on the back camera of the device, for continued attention monitoring. In the context of online education, we consider sustained attention to be relevant [7] as the learner needs to maintain their focus throughout the course, with occasional multitasking. Such relevance has been supported by previous studies [4] that have shown a positive correspondence of sustained attention with a learner's cognition, and hence, their academic performance and procurement. Previous works [5, 6] have also focused on analysing visual attention based on gaze patterns. However, these methods are mostly suitable for particular video categories like videos including lecturers, animations, etc. Utilising visual features for estimating attention, without depending on any visual stimuli (video objects) still requires attention. In the dimension of visibility, attention can be characterised by its visibility or covertness. Observable indicators of attention can be tracked through audio-visual processing [9], while covert indicators often require physiological sensing [2] to be accurately tracked. In the case of virtual presence, physiological sensing through dedicated sensors can be difficult to achieve due to deployment issues, availability, or remote data sharing-related security concerns. Thus, in this work, we analyse the sustained attention of the learners using commodity webcams on a desktop or laptop.

AttOnTrack: The Proposed Model

In this section, we present a detailed description of the proposed model through its architectural analysis, feature relevance and other functional aspects.

Overview of the Proposed Model

The overall idea of *AttOnTrack* is to continuously track a learner's state of attention as either *'attentive'* or *'distracted',* to analyse the characteristic of their sustained attention, while they watch Massive Open Online Course (MOOC) videos. The objective of *AttOnTrack* is to understand whether it is possible to track the attention of a learner **without depending on** the (1) characteristics of the course video like movement of video objects (e.g. location of texts, movement of lecturer, etc.), (2) auditory information (e.g. spoken statements, speech intents, etc.) of the lecturer or learner, (3) necessity of conversational aspects (e.g. active communication between the attendees), or (4) necessity of remote user-related data transfer. Based on empirical evidence, *AttOnTrack* assumes that, for the initial *n (=4)* minutes of watching online videos, the occurrence of distractions will be negligible. To eliminate the requirement of collecting data from the users separately for training the core model, *AttOnTrack* ubiquitously collects the visual and facial features of the users for these first *n* minutes of the video. This set of data is used to train an unsupervised anomaly detector so that the model can learn the visual and facial patterns of the learner, indicative of their normal (attentive) behaviour. This also ensures that the model can identify the anomalies (distractions) during the prediction phase. The prediction phase starts from (*n+1*)th minute of the video. This method also eliminates the requirement of explicit calibrations for individual users. Other potential approaches [12] of tracking attention could involve acoustic sensing, WiFi-based attention monitoring, etc. However, these systems would either require additional hardware or would impose some restriction on the learner's behaviour (e.g. motion, position, etc.). Moreover, using supervised models for detecting attention levels would require ample labelled data to train the model. The intuition behind using the vision-based features with an unsupervised learning was

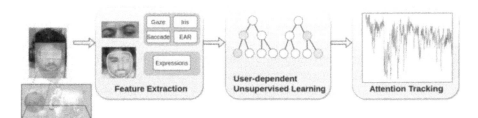

Figure 18.1 Overview of *AttOnTrack*.

to overcome these limitations. Figure 18.1 presents the overview of the proposed model. Next, we discuss the details of the individual modules of *AttOnTrack*.

Feature Extraction

One of the major issues of vision-based face detection involves missed detection due to poor lighting conditions and varied facial angles. To minimise the effects of these factors, *AttOnTrack* integrates two face detection methods: (a) the Google MediaPipe framework [14] and (b) the approach proposed in [10]. The first method, that returns a total of 478 facial landmarks, is used as the primary approach. For those video frames, where MediaPipe fails to detect any face, the second approach is used to check for the presence of a face and its 68 landmarks. From these landmarks, the (X,Y) coordinates of the pupils are estimated as *pupil_left_x, pupil_left_y, pupil_right_x*, and *pupil_right_y,* respectively. Using the pinhole camera model, *AttOnTrack* projects these pupil locations on the device's screen that correspond to the gaze location of the user on the screen. The system estimates the gaze location for both the eyes distinctly. These individual gaze points are considered to maximise the inclusion of their peripheral vision points, instead of considering one central point of fixation. The (X,Y) coordinates of these gaze points are identified by *gaze_left_x, gaze_left_y, gaze_right_x*, and *gaze_right_y.* For each frame *f*, the system then evaluates the learners eye movements (saccades) as *(saccade_left_x, saccade_left_x) = (abs (pupil_left_x$_f$ -pupil_left_x$_{f-1}$), abs (pupil_left_y$_f$- pupil_left_y$_{f-1}$))* and *(saccade_right_x, saccade_right_x) = (abs (pupil_right_x$_f$- pupil_right_x$_{f-1}$), abs(pupil_right_y$_f$- pupil_right_y$_{f-1}$))*, where, *feature$_f$* denotes the corresponding feature value in frame *f*. Here, we consider the magnitude of the saccades, instead of its direction. This is due to the fact that the previous gaze-point estimation sufficiently corresponds to the direction of the vision. The intention behind using saccades as one of the features is based on the previous finding that attentiveness is correlated with controlled saccades [19]. Next, *AttOnTrack* considers the learner's drowsiness (or blinking) by estimating the average Eye Aspect Ratio (*EAR*) [15] from 12 landmarks from both the eye regions. For detecting the facial expression (*exp*) of the learner, we first train a Convolution Neural Network (CNN) with the extended Cohn-Kanade dataset, as described in [9]. By estimating this set of features from the learner's facial preview, *AttOnTrack* creates the training dataset for the first *n* minutes of the video and then starts making predictions. The model is described in the following section.

User-Dependent Unsupervised Learning of Features

The idea behind using an unsupervised learning model for *AttOnTrack* arises from the fact that the manual annotation of frame-wise features can be both biased and erroneous. Moreover, using physiological sensing for collecting the ground truth can be expensive, intrusive and might lack robustness. Thus, we take advantage of the observation that attention, in online classes, drops with time [17]. According to this finding, we assume that for about *n* minutes of the online video, a learner will be attentive with few instances of distraction. We use the data from this time frame to train an Isolation Forest [13] anomaly detector model. The model assumes that outliers (anomalies) are less and number and can be distinguished easily. This aligns with the nature of a learner's initial sustained attention in online courses. We use an ensemble of 100 estimators in the Isolation Forest and train it with about 7200 data samples collected for the initial 4 minutes of the video (30 frames per second). During prediction, the model assigns a mean normality score (N_{score}) to the predicted sample, which is then compared to a threshold to predict it as either -1 (distraction) or $+1$ (attention) (B_{score}). Since the model trains on data collected from each of the users separately, every time they watch a video, it is a user-dependent model that can adapt to the user's behavioural patterns.

Attention Tracking

In this work, we aim at considering a simplistic binary classification of attention. Thus, the binary detection of the features as either anomalous or normal is more relevant to our model. However, the level of attention can vary in a continuous form and is not always essentially binary. For this reason, we use the normalised N_{score} to calculate the overall attention score for every 2 minutes of the video (A_{seg}) and the B_{score} as an overall attention indicator. The formula for A_{seg} is as follows:

$$A_{seg} = \left(\left(\sum_{i=1}^{n} N_{score\_i}\right) \div n\right) \times 100$$

where n= total number of frames in the segment. In the next section, we discuss the user study and evaluation methodology for analysing the performance of *AttOnTrack*.

User Study and Results

In this section, we discuss the methodology for evaluating *AttOnTrack,* along with the obtained results.

Participants and Methodology

In the user study, we consider 5 participants (2 females, 3 males) from India. Three of these participants were between 24–35 years of age and the rest were above 50 years. The participants belong to various professional backgrounds including Academia, Industry and Business. The participants were asked to watch two different online videos during which their visual and facial features were tracked and predicted by *AttOnTrack*. Each of these videos was of ~20 minutes duration and covered educational topics like advanced signal processing and neural networks. The participants were informed of the experimental setup and gave consent for their data to be recorded for the purpose of this research. However, no restriction was imposed on their video viewing patterns other than their seating position, which was about 50 cms away from the device. To avoid the scope of biased performance, no further details about the experiment were communicated to them. In total, we consider 10 videos for the evaluation of *AttOnTrack*.

Annotation and Ground Truth Collection

Although *AttOnTrack* uses an unsupervised learning model, we collect ground truth for validating the predictions made by the system. We employ two methods for collecting the ground truth. Firstly, we consider the online videos into 10 segments of 2 minutes duration and generate 3 questions from each of these segments. Thus for each video, the participant was asked to answer 30 questions, based on the video content. The answers were evaluated manually and each question was scored between 0–100. For each segment S, the average of the 3 questions was considered as the attentiveness score for S. These segment-wise attentiveness scores are compared with the generated A_{seg} scores. It is to be noted that the first 6 questions (during the training data collection phase) were asked to confirm their high attentiveness.

In the second approach, we asked two independent annotators to watch the individual facial recording of the participants (slowed version), along with the frame numbers. The annotators were asked to mark the starting and ending frames during which the corresponding participants appeared to be distracted. These frames were programmatically marked as –1 (distracted) and all other frames were marked as 1 (attentive). We compare these frame-wise classification of attention with the corresponding B_{score}.

Results

Figure 18.2 shows the distribution of average manual (M) and average *AttOnTrack*-generated (G) scores (considering A_{seg}) for participants P1-P5, while watching videos V1-V2. Although the scores do not exhibit an exact match, due to the difference between the manual estimation techniques and the score generation, the graph indeed reveals a significant correlation between the two scores, with an average difference of 11.3. Interestingly, for P3V2 and P4V2, we can see that the manual scores are distributed over a wider spectrum, while the generated scores are lower and higher for P3V2 and P4V2, respectively. Such differences arise from positive multitasking like taking notes (in P3V2) or inattentional blindness(in P4V2).

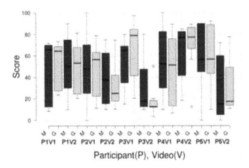

Figure 18.2 Segment-wise scores distribution.

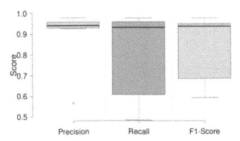

Figure 18.3 Frame-wise classification of attention.

Figure 18.3 depicts the distribution Precision, Recall and F1-score, considering the frame-wise generated classification (B_{score}) and the manual classification of attention. The system performs with a significant accuracy of 80.1% along with an average Precision of 0.9, Recall of 0.7 and F1-score of 0.83.

Conclusion

In this work, we present *AttOnTrack*, a non-intrusive system that relies on the learner's visual and facial features to track their sustained attention during online videos. In its current form, *AttOnTrack* performs with significant accuracy, as revealed through the thorough user study. However, the model can further be improved by including visual features like pupil dilation and a time-series based anomaly detector, that capture the long-term temporal relations between the consecutive feature points. Moreover, since the model is based on the generic behaviour of learners being attentive during the initial timeframe of the course, its performance can be affected by exceptional scenarios where a learner shows inattentiveness from the beginning of the session. Despite its current limitations, a system like *AttOnTrack* holds a significant potential in shaping the future of online education. Future work will involve a more extensive non-binary analysis of *AttOnTrack* with a larger group of learners by considering a wider demographic variation.

References

1. Abdelrahman, Y., Khan, A. A., Newn, J., Velloso, E., Safwat, S. A., Bailey, J., and Schmidt, A. (2019). Classifying attention types with thermal imaging and eye tracking. *Proceedings of the ACM on Interactive, Mobile, Wearable and Ubiquitous Technologies*, 3(3), 1–27.
2. Al-Nafjan, A., and Aldayel, M. (2022). Predict students attention in online learning using EEG data. *Sustainability*, 14(11), 6553.
3. Chen, C. M., and Wu, C. H. (2015). Effects of different video lecture types on sustained attention, emotion, cognitive load, and learning performance. *Computers and Education*, 80, 108–121.
4. Chen, C. M., and Wang, J. Y. (2018). Effects of online synchronous instruction with an attention monitoring and alarm mechanism on sustained attention and learning performance. *Interactive Learning Environments*, 26(4), 427–443.
5. Das, S., Chakraborty, S., and Mitra, B. (2022). I cannot see students focusing on my presentation; Are they following Me? continuous monitoring of student engagement through stungage. In Proceedings of the 30th ACM Conference on User Modeling, Adaptation and Personalization, (pp. 243–253).
6. Hutt, S., Mills, C., Bosch, N., Krasich, K., Brockmole, J., and D'mello, S. (2017). Out of the Fr-Eye-ing pan towards gaze-based models of attention during learning with technology in the classroom. In Proceedings of the 25th Conference on User Modeling, Adaptation and Personalization, (pp. 94–103).
7. Hwu, S. L. (2023). Developing SAMM: a model for measuring sustained attention in asynchronous online learning. *Sustainability,* 15(12), 9337.
8. James, W. (2007). The Principles of Psychology. (Vol. 1). Cosimo, Inc.
9. Kar, P., Chattopadhyay, S., and Chakraborty, S. (2022). Bifurcating cognitive attention from visual concentration: utilizing cooperative audiovisual sensing for demarcating inattentive online meeting participants. *Proceedings of the ACM on Human-Computer Interaction,* 6(CSCW2), 1–34.
10. Kazemi, V., and Sullivan, J. (2014). One millisecond face alignment with an ensemble of regression trees. In Proceedings of the IEEE Conference on Computer Vision and Pattern Recognition, (pp. 1867–1874).

11. Knudsen, E. I. (2007). Fundamental components of attention. *Annual Review of Neuroscience*, 30, 57–78.

12. Kosch, T., Karolus, J., Zagermann, J., Reiterer, H., Schmidt, A., and Woźniak, P. W. (2023). A survey on measuring cognitive workload in human-computer interaction. *ACM Computing Surveys*, 55, 1–39.

13. Liu, F. T., Ting, K. M., and Zhou, Z. H. (2008). Isolation forest. In 2008 Eighth IEEE International Conference on Data Mining, (pp. 413–422). IEEE.

14. Lugaresi, C., Tang, J., Nash, H., McClanahan, C., Uboweja, E., Hays, M., and Grundmann, M. (2019). Mediapipe: a framework for perceiving and processing reality. In Third Workshop on Computer Vision for AR/VR at IEEE Computer Vision and Pattern Recognition (CVPR), (Vol. 2019).

15. Maior, C. B. S., das Chagas Moura, M. J., Santana, J. M. M., and Lins, I. D. (2020). Real-time classification for autonomous drowsiness detection using eye aspect ratio. *Expert Systems with Applications*, 158, 113505.

16. Posner, M. I., and Boies, S. J. (1971). Components of attention. *Psychological Review*, 78(5), 391.

17. Tang, T., Abuhmaid, A. M., Olaimat, M., Oudat, D. M., Aldhaeebi, M., and Bamanger, E. (2023). Efficiency of flipped classroom with online-based teaching under COVID-19. *Interactive Learning Environments*, 31(2), 1077–1088.

18. Xiao, X., and Wang, J. (2017). Understanding and detecting divided attention in mobile MOOC learning. In Proceeding of the 2017 CHI Conference on Human Factors in Computing Systems, (pp. 2411–2415).

19. Zhao, M., Gersch, T. M., Schnitzer, B. S., Dosher, B. A., and Kowler, E. (2012). Eye movements and attention: the role of pre-saccadic shifts of attention in perception, memory and the control of saccades. *Vision Research*, 74, 40–60.

19 Enhancing Object Detection through Target Orientation in Multi-sensor Images

A. Pareek and D. Chaudhuri

C.S.E. Department, Techno India University, West Bengal, India

Abstract

Over the course of the last few decades, pattern recognition and image processing have both used the concept of the bounding box for object detection as well as target tracking. In most areas of research and not just defence, the bounding boxes used so far, serve as guides or fundamental steps to address the objects of interest before moving onto bigger video processing goals like object/target detection and target tracking. While the bounding box is effective, it generally has a restriction of being aligned to the horizontal axis of the frame of the image. Testing out various alternatives to this restriction and induce a tilted bounding box into the object detection via modern deep learning techniques, specifically various types of convolutional neural networks (CNN) like YOLACT (You Only Look At CoefficienTs) has been the main motivation behind this research. We have built our model to cater to both visual spectrum as well as thermal spectrum imagery from any standard multi-sensor camera.

Keywords: Bounding box, CNN, image processing, object detection, pattern recognition, target tracking, thermal imagery, visual spectrum imagery.

Introduction

Image Segmentation results in a collection of shapes that are defined by connectedness, contiguity and consistency. Every section is an individual object or a portion of the object and has a bounded region. The minimum rectangle that encompasses every single unique point in the region is what we refer to as the minimal bounding rectangle. In this case, the both the major and minor axes of the object region that is being bound are parallel to the rectangle's edges. The bounding rectangle may also be represented as a bounding box. The major axes of the blob are not always aligned with the bounding box's axes, which are instead aligned and oriented with the image axes.

Figures 19.1. and 19.2 are bounding boxes generated from the algorithm from Chaudhuri et al. [1]. Figure 19.1 shows an example of bounded boxes with centroid position, major and minor axes with different orientation. Another typical example of non-convex objects with different orientation and their corresponding bounded boxes are shown in Figure 19.2.

---

aryanp3128@gmail.com

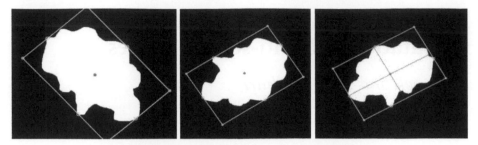

Figure 19.1 Bounding boxes of different steps with different orientation.

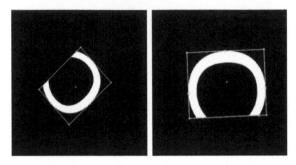

Figure 19.2 Bounding boxes of different shape objects with different orientation.

Bounding boxes help define the region of presence for the object within the image. For our research, we want to implement rotated bounding boxes for all the objects within the image as shown in these images.

Methodology

Object Detection

In order to gain a thorough comprehension of an image, it is crucial not only to classify different images but also estimate the concepts and locations of objects present in each image with good accuracy. This process is commonly known as object detection and typically involves various subtasks.

Object detection, being a fundamental problem in computer vision, has a critical aspect in improving the semantic understanding of image data from various sensors. It offers significant insights that are crucial for numerous applications like image classification, analysing human behaviour, facial recognition, and self-driving automobiles. By effectively identifying and locating objects within visual data, object detection enables machines to comprehend their surroundings at a deeper level. This capability opens up avenues for advanced tasks involving intelligent scenes interpretation and comprehensive perception in various domains.

Dataset

In this study, we utilised data that was obtained with the consent of DYSL-CT (DRDO) Lab, Chennai through the AXIS Q8741-LE Bi-spectral PTZ Network Camera. To diversify our object classes further, we also incorporated videos from YouTube for testing purposes and datasets sourced from public data centres such as Roboflow Universe. The dataset provided by DYSL-CT contains images captured by various sensors including visual cameras, thermal cameras, and infrared cameras. These images have different resolutions and are taken using different sensors. Each image contains at least one object of interest such as humans, cars, motorcycles, bicycles, airplanes and boats. The dataset is properly annotated with rotated bounding boxes and class labels for these objects. Moreover, it has been partitioned into training validation and testing sets.

Mask R-CNN and YOLACT for target mask and edge map detection

Mask R-CNN

Mask R-CNN is an extended and advanced version of Faster R-CNN, a popular object detection algorithm. While Faster R-CNN provides two outputs for every potential object—a classification label and a bounding-box offset, Mask R-CNN introduces an additional branch that predicts the mask of the detected object. The additional output from the class and bounding box outputs is autonomous and independent, enabling a more comprehensive spatial representation of objects. The main innovation in Mask R-CNN lies in its ability to align pixels accurately. This addresses one of the shortcomings found in previous variants such as Faster R-CNN. Just like Faster R-CNN's two-stage process with Region proposal network, Mask RCNN also employs this procedure but adds on another parallel branch dedicated to predicting masks for each region proposal. Notably different from recent systems where classification relies solely on mask predictions.

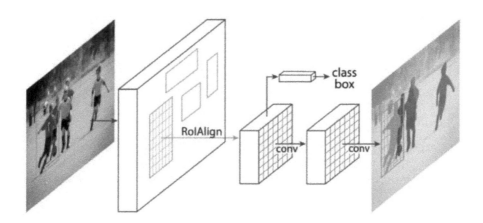

Figure 19.3 Mask RCNN architecture [7].

YOLACT

The YOLACT model architecture shown above in Figure 19.4 employs a two-part methodology to achieve instance segmentation, which involves the generation of prototype masks and the prediction of per-instance mask coefficients. This approach aims to attain accurate masks that are spatially coherent and semantic vectors that represent each individual object within an image. In the initial stage, YOLACT utilises Fully Convolutional

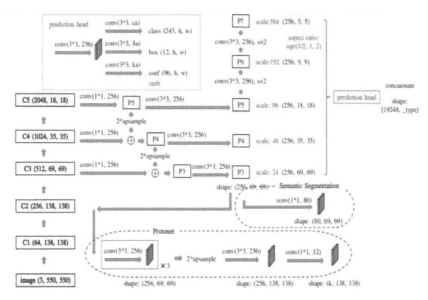

Figure 19.4 YOLACT architecture [8].

Networks to generate a collection of prototype masks. Subsequently, through fully connected layers, the model predicts per-instance mask coefficients. These coefficients serve as semantically meaningful vectors that characterise each specific object present in the scene. To seamlessly combine these components together during assembly, YOLACT leverages matrix multiplication to perform a simple linear combination. It is noteworthy that unlike traditional approaches where prototype masks are generated based on specific categories, this method adopts category-independent prototypes.

Later in the paper, we try to use both Mask R-CNN and YOLACT and give a comparative study of how these two networks generate masks and thereby impact the edge map selection.

Rotated Bounding Box

An object (target) in a video scene will change location and orientation. It is not effective to track such a target using a bounding box that is parallel to the x and y axes. The detection and tracking outcome will be better if the bounded box is aligned along the target

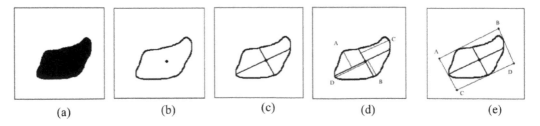

| (a) | (b) | (c) | (d) | (e) |

Figure 19.5 Various processes of forming the bounding box rectangle. (a) original image, (b) the centroid position, (c) the major and minor axes, (d) the upper and lower furthest points and (e) the oriented bounding rectangle [1].

orientation. With this in mind, the rotating bounding box is carefully implemented while targeting a specific target. The bounding box has been implemented using the Chaudhuri and Samal approach [1]. Figure 19.3 shows the stepwise different parameters extraction procedure and formation of bounding boxes.

We get the vertices, tl_x, tr_x, bl_x, br_x and tl_y, tr_y, bl_y, and br_y angle of orientation θ from equations 1, 2, 3, 4 (shown below) given in Chaudhuri and Samal's work [1].

$$\left(tl_x = \frac{x_1 tan\theta + x_3 cot\theta + y_3 - y_1}{tan\theta + cot\theta}, tl_y = \frac{y_1 cot\theta + y_3 tan\theta + x_3 - x_1}{tan\theta + cot\theta} \right) \tag{1}$$

$$\left(tr_x = \frac{x_1 tan\theta + x_4 cot\theta + y_4 - y_1}{tan\theta + cot\theta}, tr_y = \frac{y_1 cot\theta + y_4 tan\theta + x_4 - x_1}{tan\theta + cot\theta} \right) \tag{2}$$

$$\left(bl_x = \frac{x_2 tan\theta + x_3 cot\theta + y_3 - y_2}{tan\theta + cot\theta}, bl_y = \frac{y_2 cot\theta + y_3 tan\theta + x_3 - x_2}{tan\theta + cot\theta} \right) \tag{3}$$

$$\left(br_y = \frac{x_2 tan\theta + x_4 cot\theta + y_4 - y_2}{tan\theta + cot\theta}, tl_y = \frac{y_2 cot\theta + y_4 tan\theta + x_4 - x_2}{tan\theta + cot\theta} \right) \tag{4}$$

where (tl_x, tl_y), (tr_x, tr_y), (bl_x, bl_y) and (br_x, br_y) are the top left (tl), top right (tr), bottom left (bl) and bottom right (br) (x, y) coordinates for fitting the bounding rectangle or square, respectively while θ is the angle of orientation of the object of interest along the horizontal *x* axis.

Experimental Results

Object Detector Comparison

The Mask R-CNN paper [7] (Figure 19.4) is treated as the standard for mask extraction in image processing, but we soon found out that the masks generated had deviations that severely depreciated our results and implementation of the rotated bounding boxes. In certain situations, we could not find the correct masks and as a result our output image and video frames had the wrong bounding boxes. The situation shows below in Figure 19.5. Original image with multiple bounding boxes is presented in Figure 19.5(a) and then, the enlarged image of the motor bike is shown in Figure 19.5(b). We have seen that Mask R-CNN cannot find a proper bounding box and as a result it is very difficult to move with ambiguous results for the next step of bounding box formation.

In certain cases, the masks were distorted as the object changed positions over different frames. Truck moving behind a tree in a live video. In such a situation, we found that the mask broke the main object into multiple different unique segments and gave both segments an incorrect class label.

Another drawback of Mask R-CNN was that after applying the image processing bounding box method to a live video feed, the video output was severely constrained to ~1.2 frame per second. Mask R-CNN was not a practical solution for our use case even if there were previous claims of it operating at 5fps. It was not at the same level as our thermal camera sensor live stream configuration, which streams at a minimum frame rate of 8fps. This would be far lower than required real time results.

(a)   (b)

Figure 19.6 Thermal image with multiple masks which lead to multiple bounding boxes. (a) thermal image with multiple bounding boxes of an object and (b) enlarged image of the same object.

The above shown discrepancy comes in due to two main factors:

1. Mask R-CNN fails to create precise masks for thermal imagery and objects at unusual angles.
2. Mask R-CNN's output time for this still frame failed to satisfy the video fps. requirements. It operated at 1.2 fps while thermal video input was estimated at 8 fps.

There are various issues with the Mask R-CNN model:

1. Mask RCNN consists of multiple stages, each with distinct components at each level. Therefore, it cannot be trained completely.
2. To facilitate later training of the SVMs, the extracted features from the pretrained CNN are stored on disk. Numerous hundred Gigabytes of storage are needed for this.
3. R-CNN utilizes the selective search method, which can be time-intensive, in order to generate regions. In the process of feature extraction, each proposed area is sent separately to a CNN. This approach prevents real-time execution of R-CNN.

YOLACT *Detector Results*

(a)   (b)

Figure 19.7 Thermal image with multiple masks which lead to multiple bounding boxes. (a) thermal image with multiple bounding boxes of an object and (b) enlarged image of the same object.

Given below are the mask and object detector results for YOLACT, that clearly depict the mask and normal bounding box of the objects in the same thermal image as shown above.

Now we use this mask to perform the mathematical calculations given in the reference paper to get the oriented bounding boxes.

Oriented Bounding Box generation

Given below are the entire steps of oriented bounding box generation showcased via images given below. Now we use this mask to perform the mathematical calculations given in the reference paper to get the oriented bounding boxes as shown below.

Figure 19.8 (a) Original mask of the target object of interest within the image (marked in blue). (b) Edge of the mask and its centroid computed. (c) Farthest points from the major and minor axes of the target object of interest are computed. (d) Final oblique(oriented) bounding box along the orientation of the target object and original mask.

Final Results and Inference

Custom training on the dataset gives elevated results and Average Precision AP scores as shown below:

Table 19.1: Average Precision scores for the three object detector models.

	Visual images			Thermal images		
	AP small	AP@0.5	AP@0.75	AP small	AP@0.5	AP@0.75
Mask RCNN	30.2	36.7	38.9	29.2	31.0	35.6
YOLACT (resnet101)	43.7	45.2	46.6	39.7	40.2	42.6
YOLACT (transfer learning) *	59.2	61.3	67.1	48.5	56.7	61.1

*trained on i7-12700 with 64GB RAM and Nvidia RTX 3060 with 12 GB of VRAM

Table 19.2: Comparison of area coverage percentages of bounding boxes i.e., ratio of mask area/ oriented bounding box area.

Area coverage percentages	Visual images (%)	Thermal images (%)
Ordinary Bounding Box	76.26	71.87
Oriented Bounding Box	85.12	86.01

(a) (b)

(c) (d)

Figure 19.9 (a) Ordinary bounding box on visual image. (b) Oriented bounding box on visual image. (c) Ordinary bounding box on thermal image. (d) Oriented bounding box on thermal image.

For the given testing dataset, the general trend of oriented bounding box being the minimal bounding box follows suit. The testing dataset used has only one object of interest within a single image. For area coverage comparison, we have not focused on more complex oriented bounding box techniques due to feasibility of this method and also because our object detection system has practical life use case of delivering results in real-time video feeds.

In Figure 19.9 above, we have noted the comparison of the two types of bounding boxes, the left ones being the ordinary bounding box, the ones on the right showcasing the oriented bounding boxes generated from the given algorithm from Chaudhuri et al. [1]. After the implementation of these bounding boxes, the overall results between oriented bounding boxes generated by Mask RCNN and YOLACT are shown.

The initial visual image of the identical car is displayed in Figure 19.10(a). The bounded boxes using the Mask R-CNN and YOLACT method approach are shown in Figures 19.10(b) and 19.11(c), respectively. Here again, the proposed method performs oriented bounding boxes better than the Mask R-CNN method.

(a)   (b)   (c)

Figure 19.10 Visual image performance. (a) original visual image, (b) bounding box by using Mask R-CNN and (c) bounding box by proposed fine-tuned YOLACT detector.

(a)   (b)   (c)

Figure 19.11 Thermal image performance. (a) original thermal image, (b) bounding box by using Mask R-CNN and (c) bounding box by fine-tuned YOLACT detector.

The thermal image of the corresponding car image captured by the thermal sensor is shown in Figures 19.11(a). Figure 19.11(b) and 11(c) are the resultant oriented bounding boxes that are generated using the Mask R-CNN and transfer learned YOLACT approach respectively.

These results show that the suggested strategy of transfer learning onto YOLACT architecture and then generating masks and oriented bounding boxes performs more effectively in all aspects than the Mask R-CNN method and satisfies the real-time requirement of our system for live video feeds at a respectable frame rate comparable to the input stream's frame rate.

Acknowledgment

We would like to express our sincere gratitude to DYSL-CT, DRDO for funding the project and providing us with the necessary multi-sensor dataset.

References

1. Chaudhuri, D. and Samal, A. (2007). A simple method for fitting of bounding rectangle to closed regions. *Pattern Recognition*, 40(7), 1981–1989. https://doi.org/10.1016/j.patcog.2006.08.003
2. Chaudhuri, D., Kushwaha, N.K., Sharif, I. et al. (2012). Finding best-fitted rectangle for regions using a bisection method. *Machine Vision and Applications*, 23, 1263–1271. https://doi.org/10.1007/s00138-011-0348-6
3. Gonzalez, R., and Woods, R., (2018). *Digital Image Processing (4th Edition)*, England: Pearson Education.
4. Sonka M., Hlavac, V., Boyle, R. (1999). *Image Processing, Analysis and Machine Vision, (3rd Edition)*, Brooks/Cole.
5. Roboflow Universe Projects (2022). People Detection—Thermal Dataset [Open-Source Dataset]. In Roboflow Universe. https://universe.roboflow.com/roboflow-universe-projects/people-detection-thermal/
6. He, K., Gkioxari, G., Dollár, P., and Girshick, R. (2017). Mask R-CNN. *2017 IEEE International Conference on Computer Vision (ICCV)*, pp. 2980–2988. https://doi.org/10.1109/ICCV.2017.322
7. Bolya, D., Zhou, C., Xiao, F., and Lee, Y. J. (2019). YOLACT: Real-time instance segmentation. ICCV.
8. Ren, S., He, K., Girshick, R., and Sun, J. (2017). Faster R-CNN: towards real-time object detection with region proposal networks. *IEEE Transactions on Pattern Analysis and Machine Intelligence*, 39(6), 1137–1149. https://doi.org/10.1109/TPAMI.2016.2577031
9. Girshick, R., Donahue, J., Darrell, T., and Malik, J. (2014). Rich feature hierarchies for accurate object detection and semantic segmentation. *2014 IEEE Conference on Computer Vision and Pattern Recognition*, 580–587. https://doi.org/10.1109/CVPR.2014.81
10. Girshick, R. B. (2015). Fast R-CNN. CoRR, abs/1504.08083. http://arxiv.org/abs/1504.08083
11. Tumas, P., Nowosielski, A., and Serackis, A. (2020). ZUT-FIR-ADAS. IEEE Dataport. https://dx.doi.org/10.21227/7f37-hx89S.
12. Goswami, S., Yadav, N. K., and Singh, S. K. (2020). Thermal visual paired dataset. *IEEE Dataport*. https://dx.doi.org/10.21227/jjba-6220

20 A Comprehensive Study on Internet of Vehicles Intelligent Systems with Machine Learning Model Development

Metuku Pranudeep[a], Harsha Lokesh[b], Ramanathan Lakshmanan[c], and Murali Subramanian[d]

School of Computer Science & Engineering, Vellore Institute of Technology, Vellore, India

Abstract

Road accidents are the leading causes of unnatural death and injuries world over, with the recent strides in technological and intelligent applications this can not only be prevented but also avoided all together. The increase in number of smart vehicles has given us an increasing number of data points to analyse and justify reasons for fatal accidents, this data can be optimized to help in predictive analysis, further on interconnection of Vehicles and wireless communication can help pave way to driverless automation. In this paper we provide a comprehensive study and novel approach for road accident detection and accident avoidance using predictive intelligent models trained using various parameters and feature extraction. The machine learning models generated provide an accurate model for accident severity detection of a vehicle and also accident probability prediction.

Keywords: Internet of vehicles (IoV), collision detection, machine learning (ML), severity detection.

Introduction

The role of transportation in modern society is crucial. Thanks to the advancements in transport technology, the transportation of people and goods has become more efficient and faster. Nevertheless, the number of vehicles on the road has also significantly increased, leading to a rise in the number of accidents. Sadly, the statistics reveal that more than one million people die annually due to vehicle accidents. Many risk factors are associated with these accidents, such as over speeding, driving under the influence of alcohol or drugs, distracted driving, inadequate safety equipment, unsafe roads and vehicles, poor post-crash care, and insufficient traffic enforcement. These accidents occur predominantly in developing and underdeveloped countries, causing young adults to be at higher risk. Fatalities often occur due to lack of proper post-accident healthcare, particularly in backward and middle-income countries, while many non-fatal victims suffer long-term disabilities. In

[a]metuku.pranudeep2019@vitstudent.ac.in, [b]harsha.lokesh2019@vitstudent.ac.in, [c]lramanathan@vit.ac.in, [d]murali.s@vit.ac.in

2017, the World Health Organization (WHO) released a safety related technical package, known as "save LIVES," that focuses on management of speed, leadership, infrastructure design and improvement, and standards of vehicle safety. With the advancement of technologies like the Internet of Things (IoT), it is now possible to prevent accidents and provide post-accident emergency healthcare to victims, which can potentially save lives. The services that IoT provides are a boon to healthcare systems.

IoT applications have become increasingly prevalent in vehicles in recent years, with the primary aim of enhancing passenger safety. One example of this is the automatic crash response system, which is now standard in many vehicles and can utilize cloud services to contact the nearest medical center in the event of an accident [1]. In addition to this, sensors can detect over-speeding vehicles and alcohol levels, thereby helping to prevent accidents from occurring. Biomedical sensors such as pulse sensors and muscle sensors can also detect any sudden changes in vital signs, sending an emergency alert to the nearest medical center. Such systems not only enable fast post-accident healthcare, but can also notify other drivers in the vicinity of the incident through VANETs. An IoV system for detecting, analyzing, and alerting emergency services about high-speed head-on and single-vehicle accidents comprises of sensors, actuators, heterogeneous networks, rescue service centers, and a cloud-based management platform. The vehicle's sensors and actuators can detect and respond in the event of an accident, with the cloud-based platform receiving emergency alerts and relaying them to nearby medical centers and ambulances.

The purpose of this paper is to introduce an intelligent Internet of Vehicles (IoV) system that utilizes machine learning models to alert drivers of approaching accident-prone areas. Additionally, the paper aims to demonstrate the construction and functionality of an automated emergency alert system that activates when an accident occurs. By monitoring various parameters, the system can determine the severity of an accident and provide valuable information to emergency healthcare providers, potentially expediting the delivery of life-saving treatment.

Literature Review

Internet of vehicles is nothing but the interactions and interconnectivity among various vehicles in a network. As people buying new vehicles has been rising at a great pace due to growth in urban population and city development. The use of electric cars (EV), both fully electric and plug-in hybrid models, has also increased. These mobile vehicles necessitate enhanced interconnection and communications. An exclusive subset of the IoT is the IoV. With information exchange between vehicles, people, and infrastructure along the route, it has developed into a platform that is absolutely necessary.

It has garnered a lot of interest. The idea of IoV is now a national and international issue rather than just a question of IT applications in the automotive industry. IoV will eventually play a significant role in our lives and enable intelligent transportation systems in order to operate without traffic signals, accidents, and other associated issues. Millions of people will benefit from more convenient, cosy, and secure traffic service [2].

We can use dynamic prediction and real-time object detection to forecast future vehicle positions for collision detection and avoidance. Machine learning techniques can be combined with the idea of vehicle dynamics to improve the system's ability to predict future vehicle positions. Many active/advanced safety measures have been created for various scenarios, including due to the increased usage of electronics in the automotive industry such as, parking assistant system (PAS), forward collision warning (FCW), departure warning system (LDW), blind spot warning system (BSW) [3].

Given the situation of the country as it is, a computerised system that can instantly notify users of accidents and prevent other accidents in the same area is required. To overcome these difficulties, a vehicle network should be used carefully.

With smart and intelligent equipment along the roadside, the vehicular network can greatly reduce traffic while also preventing many fatalities from accidents. Vehicle to vehicle (V2V) communications, which are a type of vehicular network where a collection of cars can join, employ wireless transmission to share data amongst the vehicles. One of the main challenges with V2V communications is accident prevention [4].

Vehicular FOG

The idea of a car and the software inside it are developing towards an intelligent agent working in close proximity to other cars and sharing contents. We assert that the primary system setting that enables this evolution is a vehicular fog. The networking and computational aspects of the fog are covered in this paper. By using this concept of fog we have drawn certain conclusions with regards to collision detection [5]. The Internet of Vehicles will be able to communicate, store information, be intelligent, and learn to predict the intentions of the users.

Effective rescue efforts could save many lives. The methods include machine learning methods, vehicle ad hoc networks, GPS/GSM-based systems, and smartphone-based crash prediction, in this paper we will focus primarily on usage of Machine Learning techniques for accident detection.

There has been a corresponding increase in the number of accidents as vehicles have increased on our roads. In order to respond, a concept for an Internet of Vehicles which combines IoT and mobile internet has been developed in order to cope with this increasing concern. The IoV, especially in the automotive sector, is rapidly growing as a result of its ability to offer solutions for Intelligent Transport Systems.

IoV Architecture: A Three-Layered Approach

The IoV architecture is composed of three layers: the network layer, application layer, and perceptions layer. These layers form the foundation for IoV to overcome various challenges and promote the development of intelligent transportation systems. One crucial aspect is collision detection, which involves utilizing cameras positioned at road junctions to anticipate future vehicle positions and enable collision avoidance. Real-time object identification and dynamic prediction play a key role in analyzing upcoming vehicular positions, leading to the development of collision avoidance models [2].

Beyond Traffic Reduction: Saving Lives through Vehicular Networks

Vehicular networks have the potential to extend their applications beyond traffic reduction, playing a critical role in saving lives. A proposed system, known as Smart Accident Detection and Control System (SAD-CS) [4], focuses on automatically detecting car accidents at plus intersection road junctions. This system utilizes Raspberry Pi and onboard units equipped with microcontrollers in each vehicle to swiftly detect accidents and provide necessary assistance.

Towards a Decentralized Transportation Network

The vision of the Internet of Vehicles extends to a decentralized transportation network capable of autonomously determining the best routes for passengers. Additionally, wearable embedded smart technologies, such as smart helmets, have gained attention for their ability to minimize physical damage in accidents. Testing and simulation environments, both real and virtual, are essential for the development and evaluation of self-driving algorithms [20].

Harnessing Machine Learning for Road Safety

Machine learning techniques have proven effective in incident detection, traffic pattern analysis, and collision prediction. These techniques excel in differentiating abnormal vehicle behaviors from normal ones, making them invaluable tools in ensuring road safety. Furthermore, monocular vision technologies have had a profound impact on society and the economy, enabling the extraction of collision data from traffic monitoring videos using image processing techniques [24].

Addressing Challenges through Vehicular Fog Computing

Efficient communication and collaboration among vehicles are crucial for road safety. Vehicular fog computing facilitates the sharing of resourceful data and contents, predominantly location and positional information. However, challenges such as delayed emergency response, traffic congestion, and interruptions in communication with medical assistance teams need to be addressed [5].

Routing Protocols and Localization in IoV

Routing protocols based on software defined networking (SDN) have shown promise in IoV. These protocols offer flexibility, scalability, and enhanced intelligence for vehicular networks. Localization and mapping algorithms are pivotal in the development of self-driving vehicles and are currently an active area of research [15].

The IoV, through its various applications and technologies, presents promising solutions for enhancing road safety, detecting and preventing collisions, and providing timely medical assistance. These advancements rely on a combination of IoT, machine learning, communication protocols, and intelligent transportation systems. The continuous development of IoV holds great potential for creating safer and more efficient road networks [24].

Methodology and Model Specifications

Our platform offers two models for detecting accident severity intelligently and avoiding accidents in the IoV domain. Our accident-avoidance model leverages machine learning techniques to analyse a dataset, providing precise estimations of accident zones up to a few kilometres ahead of the accident-prone area. By utilizing this technology, we can alert the driver or slow down the vehicle automatically, reducing the likelihood of a collision. You can find a visual representation of our workflow diagram in Figure 20.1.

Our second model analyses an accident detection dataset to provide precise estimations of the victim's condition, allowing for timely critical assistance to be delivered in the event

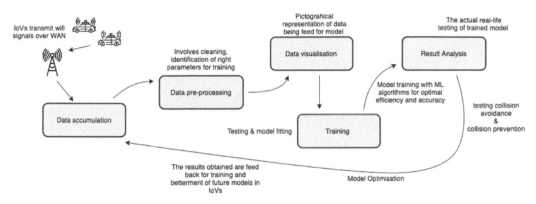

Figure 20.1 Methodology workflow diagram for model development.

of an accident, particularly in the case of multiple vehicular collisions. This data is primarily gathered using various sensors, such as accelerometers and GPS technology. The goal of our models is to deliver precise estimates with minimal error rates, reducing the likelihood of accidents while simultaneously providing the necessary assistance in the event of an accident.

Through the use of LIDAR and other IoT sensors in cars, data can be gathered and interpreted to determine the likelihood of an impending accident. By utilizing a reliable pretrained model, the vehicle can be slowed down or the driver can be alerted to slow down or stop the car in order to prevent the accident. Our model not only calculates accident detection and prevention probabilities but also estimates accident severity, providing valuable information for ambulance services to take appropriate action in the event of an accident.

Data and Visualization

The dataset contains 15 parameters for model building, i.e., TRAFFIC_CONTROL_DEVICE, WEATHER_CONDITION, LIGHTING_CONDITION, POSTED_SPEED_LIMIT, FIRST_CRASH_TYPE, MOST_SEVERE_INJURY, PRIMARY_CAUSE, ROAD_DEFECT, CRASH_TYPE, CRASH_HOUR, CRASH_DAY_OF_WEEK, ROADWAY_CONDITION, CRASH_MONTH, LATITUDE, LONGITUDE.

Due to unavailability of real-time monitor the dataset used for the model training is obtained from https://data.cityofchicago.org/Transportation/Traffic-Crashes-Crashes/85ca-t3if that is visualized and understood as seen in Figures 20.2–20.3 for the town on Chicago.

CRASH_TYPE is considered to be the target parameter.

Algorithms

Gradient Boosting Machine

Boosting is an ensemble learning technique used for regression and classification-based tasks. It involves weaker prediction models, specifically decision trees, that correct errors made by previous models, resulting in increasingly accurate predictions with each subsequent model. This sequential process is particularly effective with under-sampled or

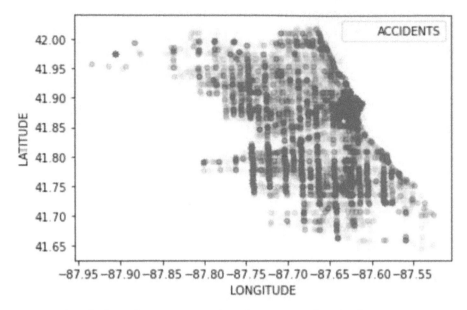

Figure 20.2 Darker shades indicate more accidents have occurred at that location.

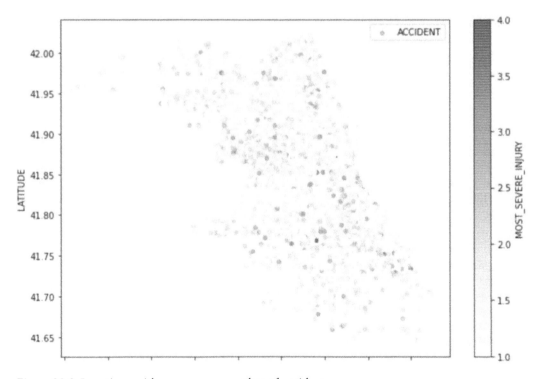

Figure 20.3 Locations with more severe number of accidents.

imbalanced data as the subsequent models focus more on the incorrect predictions made by earlier models. However, the implementation of this algorithm can be challenging due to its higher computational demand.

XGB Classification

XGBoost is a variant of gradient boosting machines that prioritizes fast computation and efficient model execution over excessive features. Despite its minimalism, the library offers a wide range of advanced functions. This technique involves developing various models that predict the errors or covariances of previous versions, and then combining them to produce the final output. XGBoost utilizes a gradient descent approach to minimize losses while building new versions, hence its name. This method is suitable for both regression and classification predictive modelling tasks, and it involves an iterative process of creating additional models to improve the accuracy of the previous ones. Models are developed in a sequential manner until no further improvements can be made. The two most notable advantages of XGB classifier are its superior model performance and speed.

Gaussian Naïve Bayes

Naïve Bayes algorithms belong to the category of machine learning algorithms which are supervised and Bayes theorem forms its basis. This simple classification technique is highly functional and is particularly useful in complex classification problems where the input data has a high dimensionality. Bayes theorem is used in probability to calculate conditional probability. This algorithm assumes that the features of the data are strongly independent of each other. This assumption works well for predicting the values of the input set of features in our accident severity dataset since the features are independent. Gaussian Naïve Bayes is one of the Naïve Bayes techniques that assumes the values associated with each feature/class are normally distributed if the data is continuous and there is no covariance between dimensions. If the features in the dataset are discrete, the optimal classification technique to use is multinomial Naïve Bayes.

Logistic Regression

Logistic regression is a vital technique for classification purposes that belongs to the family of linear classifiers and has similarities to linear and polynomial regression. This technique is typically employed for binary classification, but it can also be utilized for multi-class classification. The primary objective of logistic regression is to forecast the likelihood of a categorical dependent variable. However, it is crucial to keep in mind that this approach requires a binary representation of the dependent variable as 1 in binary regression. Besides, independent variables should not have any covariance or multi-collinearity, and the dataset should be considerably large. Logistic regression calculates the dependency of a categorical independent variable on other independent variables using a logistic function (sigmoid function) to determine their probabilities, which assists in dividing data into discrete categories. However, overfitting can be caused in logistic regression if the number of features is greater than the number of observations, making it unsuitable. Furthermore, logistic regression assumes a linear relationship between the categorical dependent variable and the independent variable, limiting its efficacy in resolving nonlinear issues.

K Nearest Neighbours (KNN)

KNN is a supervised learning algorithm whose usage comprises of both classification and regression problems. It relies on labelled input data to train the model and generate accurate predictions for unlabelled data. Unlike some other techniques that assume a specific distribution of data, KNN does not make any assumptions. KNN can be useful for assigning missing values and resampling input data. The K nearest neighbours algorithm uses instance-based learning, lazy learning, and non-parametric learning techniques. However, it's important to note that there are much faster algorithms available for producing regression and classification results. KNN is not suitable for large datasets, and the input data should not be high-dimensional. To make sure the predictions are accurate, feature scaling of data should be performed before applying KNN algorithms Figure 20.4.

Random Forest Classification
A random forest consists of numerous decision trees that operate as an ensemble, each producing a class prediction, and the class with the most votes is chosen as the final prediction. The low correlation between models is the key to the model's success, allowing for ensemble forecasts that are more accurate than any individual prediction. The decision tree is the foundation of random forest classifiers, with nodes divided based on characteristics and subsets formed using bootstrapping. The random forest is limited to a depth of seven layers, resulting in a simpler model, as deeper forests can become overly complex and prone to noise, leading to an overstatement of the number of classes in features.

Support Vector Machines
Support vector machine (SVM) is a type of supervised machine learning technique mainly used for classification problems. In SVM, each data point is represented as a point in an n-dimensional space with features as coordinates. To classify the data, a hyperplane is positioned to separate the two groups. The dimension of the hyperplane depends on the number of features, with a line for two features, a two-dimensional plane for three, and so on. Support vectors are data points that influence the hyperplane's orientation and position, and their removal can alter the position of the hyperplane. The objective of SVM is to find a hyperplane with maximum margin (MMH) that evenly separates the data into classes.

Neural Networks
Neural network-based machine learning algorithms have the advantage of not needing precise rules for input interpretation. Instead, they learn from a large number of labelled instances during training, and use this information to construct the appropriate output. As more examples and input types are processed, the neural network becomes more accurate. Neural networks are useful in solving a variety of problems and can handle different input types, such as photos, videos, and databases, without requiring explicit programming. They have a broad range of applications, including image recognition, self-driving vehicle trajectory prediction, medical diagnosis, and cancer research. The popularity of neural networks is rapidly growing due to their versatility. In deep-learning networks, each layer of nodes trains on a different set of features based on the output of the preceding layer. To improve accuracy, the following model implements a double layered approach to the neural network.

Results

The Tables 20.2 and 20.3 depicts the results for training accuracies for the mentions algorithms, Table 20.2 shows the model accuracies accident detection and Table 20.3 depicts the accident severity detection models for the models that are run. Formulas from 1–4 depict the various measuring parameters for how well the model is functioning. Our test cases depicts accuracy (1) and F-score (4) for our understanding, but other parameters are important considerations. A noteworthy point is when RFC algorithm is used, the presence of significant number of nodes or branches in the forest results in higher accuracy by solving the common problem of overfitting as compared to other algorithms. Below the table we have mathematical formulas for calculating performance metrics of each algorithm.

Table 20.1: Accident prevention accuracy of classifications.

Algorithm	Accuracy	F-Score
GUASSIAN NAÏVE BAYER CLASSIFIER (GNB)	0.857	0.909
K-NEAREST NEIGHBOURS (KNN)	0.738	0.865
SCALAR VECTOR MACHINE (SVM)	0.861	0.911
RANDOM FOREST CLASSIFICATION (RFC)	0.870	0.998
NEUTRAL NETWORK (ANN)	0.859	0.981

Table 20.2: Accident severity accuracy of classifications.

Algorithm	Accuracy	F-Score
GRADIENT BOOSTING MACHINE (GBM)	0.63	0.64
K-NEAREST NEIGHBOURS (KNN)	0.57	0.58
XGB CLASSIFICATION (XGB)	0.48	0.57
LOGISTIC REGRESSION (LR)	0.61	0.62
NAÏVE BAYES CLASSIFIER (GNB)	0.61	0.64
Random forest classification (RFC)	0.62	0.65

TP–True positives, TN–True negatives, FP–False positives, FN–False negatives

$$\text{Accuracy} = (TP + TN)/\text{Total} \tag{1}$$

$$\text{Recall} = TP/(TP+FN) \tag{2}$$

$$\text{Precision} = TP/(TP+FP) \tag{3}$$

$$\text{F-score} = 2TP/(2TP + FP + FN) \tag{4}$$

Conclusion

In this paper, we present a comprehensive conclusion that demonstrates the practical implementation of various classification and techniques. We propose using these models to train vehicles in order to enable prevention of accidents and detect accident severity for emergency services. To determine optimal data model for the vehicle, the accuracy of predictions is compared with other test case datasets. The model with the minimal error rate can then be used for further technological advancement. Cloud-based services can

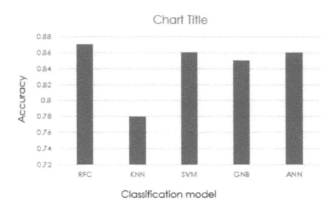

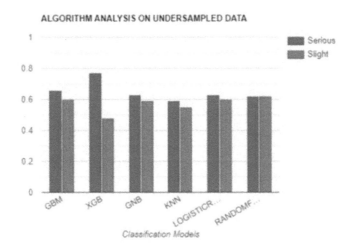

Figure 20.4 Accuracy histograms for both models.

be utilized for intelligent vehicle communication through APIs to warn drivers of potential accidents and send emergency services in case of accidents. We tested two models to obtain the most accurate results possible for numerical data. However, the effectiveness of machine learning models is significantly determined by the quality of data they are given. Therefore, data cleaning, modification, and effective parameter selection are necessary for accurate results. The paper also includes an analysis of various Internet of Vehicle (IoV) techniques such as IoV safety, multi vehicle communication, vehicle safety, networking, and V2X communication. Moving forward, we aim to explore other aspects of IoV transportation and real-world applications.

References

1. Kumar, N. Acharya, D., and Lohani, D. (2021). An IoT-based vehicle accident detection and classification system using sensor fusion. *IEEE Internet Things Journal*, 8(2). doi: 10.1109/JIOT.2020.3008896.

2. Sadiku, M. N. O., Tembely, M., and Musa, S. M. (2018). Internet of vehicles: an introduction. *International Journal of Advanced Research in Computer Science and Software Engineering*,8(1). doi: 10.23956/ijarcsse.v8i1.512.

3. Chang, C. C., Ooi, TY. M., and Sieh, B. H. (2021). IoV-based collision avoidance architecture using machine learning prediction. *IEEE Access*, 9. doi: 10.1109/ACCESS.2021.3105619.

4. Hadi, M. S. A., Saha, A., Ahmad, F., Hasan, M. S., and Milon, M. H. (2019). A smart accident detection and control system in vehicular networks.In *Proceedings of 2018 5th International Conference on Networking, Systems and Security, NSysS 2018*, 2019. doi: 10.1109/NSysS.2018.8631386.

5. Lee E. K., Gerla, M., Pau, G. Lee, U., and Lim, J. H. (2016). Internet of vehicles: from intelligent grid to autonomous cars and vehicular fogs. *International Journal of Distributed Sensor Networks*, 12(9). doi: 10.1177/1550147716665500.

6. Alvi, U., Khattak, M. A. K., Shabir, B., Malik, A. W., and Muhammad, S. R. (2020). Comprehensive study on IoT based accident detection systems for smart vehicles. *IEEE Access*, 8. doi: 10.1109/ACCESS.2020.3006887.

7. Kumar, M. B., Basit, A., Kiruba, M. B., Giridharan, R. and Keerthana, S. M. (2021). Road accident detection using machine learning. In *2021 International Conference on System, Computation, Automation and Networking, ICSCAN 2021*. doi: 10.1109/ICSCAN53069.2021.9526546.

8. Rill R. A. and Faragó, K. B. (2021). Collision avoidance using deep learning-based monocular vision. *SN Computer Science*, 2(5). doi: 10.1007/s42979-021-00759-6.

9. Qu, H. Li, W., and Zhao, W. (2020). Human-vehicle collision detection algorithm based on image processing. *International Journal of Pattern Recognition and Artificial Intelligence*, 34(8). doi: 10.1142/S0218001420550150.

10. Song, W., Yang, Y., Fu, M., Qiu, F., and Wang, M. (2018). Real-time obstacles detection and status classification for collision warning in a vehicle active safety system. *IEEE Transactions on Intelligent Transportation Systems*, 19(3). doi: 10.1109/TITS.2017.2700628.

11. Mukhtar, A., Xia, L., and Tang, T. B. (2015). Vehicle detection techniques for collision avoidance systems: a review. *IEEE Transactions on Intelligent Transportation Systems*, 16(5). doi: 10.1109/TITS.2015.2409109.

12. Bhuvaneswari, S. and Saranya, R. (2020). Internet of vehicle based accident detection and management techniques by using vanet: an empirical study. In *Proceedings of the 4th International Conference on Inventive Systems and Control, ICISC 2020*. doi: 10.1109/ICISC47916.2020.9171224.

13. Kang, Y., Yin, H., and Berger, C. (2019). Test your self-driving algorithm: an overview of publicly available driving datasets and virtual testing environments. *IEEE Transactions on Intelligent Vehicles*, 4(2). doi: 10.1109/TIV.2018.2886678.

14. Alouache, L., Nguyen, N., Aliouat, M., and Chelouah, R. (2019). Survey on IoV routing protocols: Security and network architecture. *International Journal of Communication Systems*, 32(2). doi: 10.1002/dac.3849.

15. Bresson, G., Alsayed, Z., Yu, L., and Glaser, S. (2017). Simultaneous localization and mapping: a survey of current trends in autonomous driving. *IEEE Transactions on Intelligent Vehicles*, 2(3). doi: 10.1109/TIV.2017.2749181.

16. Chang, W. J., Chen, L. B., and Su, K. Y. (2019). DeepCrash: A deep learning-based internet of vehicles system for head-on and single-vehicle accident detection with emergency notification. *IEEE Access*, 7.doi: 10.1109/ACCESS.2019.2946468.

17. Dogru, N. and Subasi, A. (2018). Traffic accident detection using random forest classifier. In *2018 15th Learning and Technology Conference, L and T 2018*. doi: 10.1109/LT.2018.8368509.

18. Zualkernan, I. A., Aloul, F., al Qasimi, S., AlShamsi, A., al Marashda, M., and Ahli, A. (2018). DigiMesh-based social internet of vehicles (SIoV) for driver safety. *In 2018 International Symposium in Sensing and Instrumentation in IoT Era, ISSI 2018*, 2018. doi: 10.1109/ISSI.2018.8538167.

19. Hasan, M., Mohan, S., Shimizu, T., and Lu, H. (2020). Securing vehicle-to-everything (V2X) communication platforms. *IEEE Transactions on Intelligent Vehicles*, 5(4). doi: 10.1109/TIV.2020.2987430.

20. Milaat, F. A. and Liu, H. (2018). Decentralized detection of GPS spoofing in vehicular ad hoc networks. *IEEE Communications Letters*, 22(6). doi: 10.1109/LCOMM.2018.2814983.

21. Singh, D. and Singh, M. (2016). Internet of vehicles for smart and safe driving. *In 2015 International Conference on Connected Vehicles and Expo, ICCVE 2015—Proceedings.* doi: 10.1109/ICCVE.2015.93.

22. Vadhwani D. N. and Buch, S. (2019). A novel approach for the ITS application to prevent accidents using wireless sensor network, IoT and VANET. (2019). *In Proceedings of 2019 3rd IEEE International Conference on Electrical, Computer and Communication Technologies, ICECCT 2019.* doi: 10.1109/ICECCT.2019.8869157.

23. Khaliq, K. A., Qayyum, A., and Pannek, J. (2018). Prototype of automatic accident detection and management in vehicular environment using VANET and IoT. *In International Conference on Software, Knowledge Information, Industrial Management and Applications, SKIMA.* doi: 10.1109/SKIMA.2017.8294107.

24. Silva, P. B., Andrade, M., and Ferreira, S. (2020). Machine learning applied to road safety modeling: A systematic literature review," *Journal of Traffic and Transportation Engineering (English Edition)*, 7(6). doi: 10.1016/j.jtte.2020.07.004.

25. Yadav, D. K., Ankita, R., and Anjum, I. (2020). Accident detection using deep learning. *In Proceedings—IEEE 2020 2nd International Conference on Advances in Computing, Communication Control and Networking, ICACCCN 2020.* doi: 10.1109/ICACCCN51052.2020.9362808.

21 IoT-based Smart Door System for Contactless Face Mask and Body Temperature Monitoring in Public Spaces: A Secure Framework for Enhanced Health Safety

Subhadip Nandi[1,a], Sunil Karforma[2,b], Rajesh Bose[3,c], and Sandip Roy[3,d]

[1]Assistant Professor, Brainware University, Kolkata, India

[2]Professor and Dean, The University of Burdwan, West Bengal, India

[3]Professor, JIS University, West Bengal, India

Abstract

The use of contactless face masks and body temperature monitoring are indeed effective measures in preventing many infectious diseases like COVID-19. Combining these measures with Internet of Things (IoT) technology and machine learning can enhance their effectiveness and provide a reliable method for maintaining a healthy environment in public places. Smart door based on IoT for monitoring face masks and body temperature without touching; which can be implemented in various settings such as shopping malls, hotels, apartment entrances, and other public spaces. The proper management and maintenance of patient data are essential for the technological improvement of hospitals and healthcare facilities. However, sending huge data, like medical data, through traditional IoT protocols like MQTT, MLLP, and CoAP can be challenging and time-consuming. Moreover, these protocols are more suitable for transferring text-based patient information rather than handling complex image data like body images, ultrasound, or CT scans. Our proposed framework includes a Face Mask Detection algorithm that can analyze video or image input to identify whether individuals are wearing face masks or not. Additionally, the system monitors the body temperature of individuals using different contactless temperature sensor. Abnormal human body temperature readings can trigger alerts or preventive actions to mitigate the potential spread of infectious disease. The implementation of the Arduino Uno R3 as a processing unit for HL7 messages received from medical devices, specifically an OV7670 camera module, is an innovative approach. By extracting the image data from the HL7 messages, the Arduino Uno R3 enables the transfer of only the necessary data to an FTP server. The Arduino Uno R3 acts as a gateway, to the IoT server from the FTP server, sending users' information and a special encrypted data link. This architecture ensures that only the essential data is transmitted which is optimizing bandwidth usage and reducing latency.

Keywords: Artificial intelligence, constrained application protocol (CoAP), Internet of Things (IoT), Minimal Low Layer Protocol (MLLP), Health level seven (HL7), message queuing telemetry transport (MQTT).

Introduction

Due to the huge population in India, many infectious diseases such as coronavirus (COVID-19) have got infected huge people in India. As reported by World Health Organization

[a]deep07mca@gmail.com, [b]sunilkarforma@yahoo.com, [c]rajesh.bose@jisuniversity.ac.in, [d]sandip.roy@jisuniversity.ac.in

(WHO), from 3rd January 2020 to 21st June 2023, there have been reported about forty four million confirmed cases and about five lakh confirmed deaths [17].The symptoms mentioned, such as fever, cough, headache, runny nose, body aches, loss of smells and taste, are indeed common symptoms of COVID-19 infection [9]. Various measures have been taken by governments, including in India, to reduce the transmission of the coronavirus (COVID-19). It is important for individuals exhibiting these symptoms to take necessary precautions, including self-isolation and seeking medical advice. These measures include the use of face masks, sanitizers, social distancing, quarantine, and restrictions on travel and large gatherings [8]. These initiatives aim to minimize direct transmission through respiratory droplets as well as indirect transmission through contaminated surfaces.

Therefore, temperature and mask checks at entrances to public places, workplaces, malls, and hospitals are implemented to identify individuals with potential infections and ensure compliance with preventive measures [15]. These checks help identify the individuals, who are considered at a higher possibility of infection, and those not wearing masks. To enhance the effectiveness of these measures, a smart entrance system is mentioned, which automatically keeps track of body temperature, spots masks, and counts the number of visitors [1]. This integrated approach aims to improve efficiency and accuracy in identifying potential risks and enforcing preventive measures.

It has been noticed that transmitting huge of data, such as medical data, through traditional IoT protocols like Message Queuing Telemetry Transport (MQTT), Minimal Lower Layer Protocol (MLLP), and Constrained Application Protocol (CoAP) can be challenging and time-consuming. MQTT cannot support video streaming [13]. It is also important facts that MQTT implementation in IoT has the lack of security. Furthermore, MQTT, MLLP, and CoAP are the message-centric wire communications [20]. These protocols enable to transmit of medical data, including medical images, from IoT devices to servers or message brokerages, even over high latency networks. In the context of IoT monitoring systems, real-time transmission of picture data can be challenging due to limitations such as low rates of data transmission rates and payload size. However, there are protocols designed specifically for IoT communications that can address these challenges.

Bottlenecks of Existing Protocol

Message Queuing Telemetry Transport
The protocol is simple and quick that excels at efficient message handling and low message overhead [21]. It is globally adopted in IoT applications for its simplicity and scalability. However, when it comes to real-time picture data monitoring systems, MQTT alone may struggle to handle the demands of IoT servers due to the large size of medical images.

MLLP (Minimal Lower Layer Protocol)
MLLP, on the other hand, is a protocol commonly used in healthcare environments for transmitting HL7 (Health Level Seven) messages [18]. It is specifically designed for reliable message transmission in healthcare systems. While MLLP can handle larger payloads, it is not inherently suitable for IoT monitoring systems, as it may lack the lightweight nature required for efficient IoT communications.

CoAP (Constrained Application Protocol)
It is another IoT protocol planned for constrained devices and networks. It offers lightweight messaging and supports RESTful principles, making it well-suited for IoT applications

[14]. CoAP is often used in resource-constrained environments, such as sensor networks, but it may also face challenges when dealing with real-time picture data due to the data size and transmission requirements.

To remedy the above problem, our contribution aims to address the challenges related to image-related data transmission in IoT-based monitoring systems. The study introduces several key contributions to address the problem:

Firstly, HL7 is a widely adopted standard for healthcare data exchange [7]. By leveraging HL7 message processing, the study suggests using machine-to-machine communication between IoT devices and an IoT server. This approach ensures interoperability and standardized communication, enabling seamless integration of IoT devices into the healthcare environment. Secondly, we proposed utilizing an FTP (File Transfer Protocol) server. The images are stored on the FTP server, and each image is assigned a unique encrypted link. This approach ensures secure and organized storage of medical image data, while the encrypted links allow authorized users, such as clinicians or public store managers, to access the images efficiently. Lastly, we suggested the implementation of a medical data display on the IoT server [6]. This display would provide clinicians with immediate access to relevant medical images, along with associated patient information [4]. By integrating medical data display into the IoT server, our aim to streamline the diagnostic process and improve the efficiency of healthcare services.

The following structure describes how the manuscript is organized; Section 2 surveys the body of literature. The research technique for our secure architecture is described in Section 3. Secttion-4 discusses the result analysis. Section-5 concludes the paper with future directions.

Literature Review

Traditionally, pulse oximeters were primarily used in hospital settings, where they are routinely employed to monitor patients' oxygen levels during surgeries, in intensive care units (ICUs), and for patients with respiratory or cardiovascular conditions. They are also used during sleep studies to assess oxygen levels during sleep.

In recent years, the availability of low-cost pulse oximeters and advancements in wearable technology has made it possible for individuals to monitor their SpO_2 levels outside of clinical settings [16]. Many smart watches and fitness trackers now incorporate pulse oximeters features, allowing users to check their oxygen saturation levels conveniently at home or on the go. This integration has made SpO_2 monitoring more accessible to a wider population [3].

Contactless devices, such as contactless payment terminals, touchless thermometers, or automatic doors, allow individuals to interact with objects without physical contact [25]. By eliminating the need for touch, these devices help reduce the potential spread of the virus through contaminated surfaces [10]. COVID-19 can survive on surfaces for extended periods, making it important to minimize contact with frequently touched objects. Contactless devices decrease the need for multiple people to touch the same surface, thereby reducing the risk of contamination. By using contactless devices, individuals can maintain better hygiene practices. Contactless devices are adaptable and can be implemented in various settings [5]. They can be integrated into existing infrastructure with relative ease, making them suitable for a wide range of industries and environments, including retail, hospitality, healthcare, transportation, and more.

Contactless devices have gained popularity since they can lower the danger of virus transmission, they were used during the COVID-19 pandemic. Remote photoplethysmography (rPPG) is indeed an emerging technology that allows for contactless monitoring of blood volume variations using a camera [24]. It leverages the principles of photoplethysmography, which is the measurement of blood volume changes using light, to assess various physiological parameters, including peripheral oxygen saturation (SpO_2) [2]. The basic concept behind rPPG is utilized when light is shone onto the skin, it gets absorbed by the underlying blood vessels and tissues. The absorption light is influenced by the pulsatile nature of blood flow, and this modulation can be detected by analyzing the changes in light intensity reflected or transmitted through the tissue. By capturing these subtle variations in light, it becomes possible to approximate parameters like, blood pressure, SpO_2 and heart rate.

Pulse transit time (PTT) and photoplethysmogram (PPG) measurements have been obtained using body-worn sensors. The time is PTT for to propagate between two specific points in the cardiovascular system in blood volume in peripheral blood vessels [23].

However, recent technological developments have resulted in the creation of wearable devices and smart watches equipped with optical sensors that can measure PPG signals. These sensors use light-emitting diodes (LEDs) to illuminate the skin and photo detectors to capture the reflected light, allowing for the measurement of PPG signals. By analyzing the PPG waveform, researchers have found that certain features of the waveform, such as the time lag between the upstrokes of the PPG signal and the ECG R-wave, can be used to estimate PTT. The advantage of using wearable devices for PTT and PPG measurement is that they offer a more convenient and continuous monitoring option compared to traditional body-worn sensors. With the ability to collect data over extended periods, these devices enable better understanding of blood pressure dynamics and provide valuable insights for managing hypertension and other cardiovascular conditions [19].

Medical image information monitoring systems using MQTT and MLLP protocols can indeed pose challenges for IoT servers [12]. While MQTT and MLLP are lightweight protocols suitable for transmitting small message sizes and managing telemetry-style data, they may not be well-suited for handling large payloads such as real-time picture data. To overcome the challenges of transmitting medical data in IoT monitoring systems, alternative approaches can be considered. Although, one possibility is to use protocols specifically designed for multimedia streaming, such as RTSP (real-time streaming protocol) or WebRTC (web real-time communication) [22]. These protocols are better suited for transmitting and handling real-time audio and video data. RTSP is not supported natively by all web browsers, which can limit its compatibility and require additional plug-in or software. Although WebRTC is supported by major web browsers, there can still be inconsistencies in implementation and support across different platforms and versions, requiring additional effort for cross-browser compatibility. RTSP introduces some latency due to the buffering and processing required in the streaming process, which may not be suitable for real-time applications that require minimal delay. WebRTC can consume significant bandwidth, especially in multi-party conferences, which may be a concern for users with limited or expensive data plans.

Our aim is to design a secure comprehensive monitoring solution by combining image display, patient data, and IoT technology. The Arduino Uno R3 is responsible for extracting the image data from the HL7 messages and transferring it to an FTP server. By extracting only the necessary image data, the Arduino Uno R3 ensures efficient data transfer. The image data is stored on the FTP server and assigned a particular encrypted link.

Additionally, the encrypted picture data link and user information are forwarded to an IoT server by the Arduino Uno R3. It offers benefits to by facilitating remote healthcare and enabling efficient monitoring of medical data.

Research Methodology

The complete process is divided into four sections by our proposed secure framework, which are as follows:

- Camera module (OV7670) to the Arduino UNO R3 HL7 messaging procedure;
- Data on output outcomes is kept in the MDB (Master Database);
- Arduino UNO R3 to image transfer of FTP server;
- Image files sent over the MQTT and MLLP protocols from an FTP server to an IoT server.

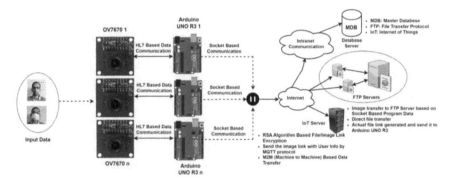

Figure 21.1 Schematic diagram.

Camera Module (OV7670) to Arduino UNO R3 HL7 Messaging process

Hardware Setup

Connect the OV7670 camera module to the Arduino Uno R3. The connections typically involve connecting the camera's power and ground pins to the corresponding pins on the Arduino. Connect the camera's data pins (such as SCCB_SDA, SCCB_SCL, VSYNC, HREF, etc.) to the appropriate digital pins on the Arduino Uno R3 [26].

Software Setup

Install the required libraries for the OV7670 camera module. Initialize the camera module and configure its settings, such as resolution, format, and frame rate.

Capture and Process the Image

Use the camera module's library functions to capture an image frame and store the captured frame in a suitable format, such as an array or a buffer in the Arduino's memory. If required, perform any image processing operations on the captured frame using suitable algorithms or libraries.

Convert the Image Data to HL7 Format

HL7 is a messaging standard primarily used in healthcare. It is typically used to transmit patient information and medical data. Our objective is to transmit image data using HL7 and need to convert the image data into a format that can be included in an HL7 message. Encoding the image using a suitable image format, such as Base64 encoding, this converts binary data into ASCII text. This encoded image data can then be included as a segment or a field within an HL7 message.

Transmit HL7 Message

Once the image data is encoded and included in the HL7 message, the MLLP and socket programming, two protocols are used for transmitting message from one end point to another end point. Socket programming is a programming technique that allows network communication between two endpoints over a TCP/IP network. It provides a set of functions and APIs (Application Programming Interfaces) that enable the creation, configuration, and management of network sockets [11]. Sockets act as endpoints for sending and receiving data over a network. In the proposed model, socket programming processes are utilized to establish communication between the sender and receiver of HL7 messages. The sender creates a socket and connects it to a listener port on the receiver's side. The receiver also creates a socket and listens on a distinct port to accept incoming connections from the sender. Once the connection is established, the HL7 messages are transmitted as data through the sockets. The MLLP framing characters are added to the HL7 messages before sending them over the network. The receiver can then identify and extract the HL7 messages by detecting the framing characters. The systems are linked via an intranet, which means the communication is limited to a local network environment.

Result Data Extraction

The HL7 result message contains MD (Message Header), UID (User with ID), SID (Sample ID), and SUD (Sample User Data) [27]. The extracted users' data are stored in the MDB (Master Database).

Arduino UNO R3 to FTP Server User Data Transfer

After receiving HL7 message, extract the user information, sample ID, and result data using Base64 image and converted to the formatted image from Base64 image. Then, the GUID (Globally unique identifier) is constructed for each formatted image. The new name has been assigned and sends it to the FTP server (Rashid et al., 2023). Following transmission to the FTP server, it returns to the Arduino UNO R3 with a special picture link. Following transmission to the FTP server, it returns to the Arduino UNO R3 with a special picture link. RSA public key uses in the Arduino UNO R3 image link that can be encrypts and delivers the link of the image which is encrypted pass to the IoT server then FTP server sends the response message.

Arduino UNO R3 to IoT Server User Data Transfer

The link to the secure image sends to the utilizing the IoT server MQTT protocol. Then, receive the response message after successfully transferred the encrypted image link. The decryption of the image connection allows the IoT server to show it.

Result Analysis

We have configured our proposed model in things board with custom widgets Rashid et al. (2023). After configuring; the store managers can easily identify the victimized person. We examine different size images (1–5 MB) on different bandwidths. Our aim is to check the response time at IoT server for displaying these images. Table 21.1 illustrates how the IoT server's picture display delay time varies with image size and bandwidth.

Table 21.1: Display 2D image in IoT server.

Total image per 30 s	Size of data	Used in network	Minimum delay (s)	Correctly display medical image (s)	Information found from patient(s)
5 > picture	picture < 1 MB	2G	20	8	4
5 < picture < 10	1 MB < picture < 3 MB	3G	15	6	3
15 < picture < 20	3 MB < picture < 5 MB	4G (LTE)	10	4	4
picture > 20	picture > 5 MB	Broadband above 5 Mbps	5	4	4

Using NS3, our framework shows the image transfer response time which can determine the effectiveness of the suggested model. Here, we used three standards: The receipt rate of packet, throughput, and received the packet size. There are two parts; (i) OV7670 camera module to MDB (Master Database), and (ii) MDB to IoT and FTP. Tables 21.2, 21.3, and 21.4 illustrate the findings and display how response and request times vary with packet size and bandwidth. Our result (Table 21.4) shows that the response time in the IoT server is less than 2 s and almost constant not dependant on different bandwidth. Here, in Table 21.5, the proposed algorithm has been compared with the traditional algorithms such as MQTT and CoAP which prove the significance of the comparison between the suggested approach and other approaches.

Table 21.2: OV7670 camera module to MDB (Intranet).

Data (MB)	BW (Mbps)	RT (s)
5	1	44.4553194
	5	13.51360727
	10	8.927368439
	15	7.478791573
	20	3.775868355
10	1	86.71886258
	5	18.74190387
	10	12.53946954
	15	9.126944264
	20	5.295862876
15	1	128.6253848
	5	26.8484959
	10	16.98238621
	15	11.33565927
	20	8.833292869

Table 21.3: MDB to FTP (Internet).

Data (MB)	BW (Mbps)	RT (s)
	1	44.75862732
	5	12.83962464
5	10	7.845329558
	15	3.885889452
	20	4.823753938
	1	84.68857585
	5	18.43182732
10	10	11.32786523
	15	8.54536534
	20	5.38335693
	1	125.4298472
	5	28.45688286
15	10	15.8648583
	15	9.95876853
	20	7.586549625

Table 21.4: MDB to IoT (Internet).

Data (MB)	BW (Mbps)	RT (s)
9.44135236	1	1.161306672
8.847268542	5	1.835431623
9.327510226	10	1.322939128
9.537125123	15	0.436060522
8.735652654	20	0.823141132

Table 21.5: Comparison between MQTT, CoAP and proposed algorithm.

Feature	MQTT	CoAP	Proposed algorithm
Application	Publish/Subscribe model	IoT and constrained environments	Healthcare data exchange
Communication	TCP/IP or MQTT over WebSocket	UDP/IP or SMS (CoAP over SMS)	TCP/IP
Message format	Publish/Subscribe messages	RESTful requests/responses	HL7 messages
QoS levels	3 levels: 2 (Exactly once), 1 (At least once), 0 (Fire and Forget)	4 levels: 0–3 (Unreliable to Reliable)	N/A
Payload format	Supports various payloads	Supports various media types	Primarily used for HL7 messages
Security	TLS/SSL encryption	DTLS for secure communication	SSL/RSA/FTP encryption
transport	Uses TCP primarily	Uses UDP primarily (Can use SMS)	Uses TCP
Use cases	IoT, home automation, messaging systems	IoT devices in constrained networks	Healthcare systems integration
Overhead	Moderate	Low	Low
efficiency	Good for large networks	Designed for constrained networks	Efficient for healthcare systems

Conclusion

The development secure framework of an IoT-enabled contactless smart door to enhancing public safety, particularly in the context of the pandemic situation. This technology can offer several benefits; including reducing manpower requirements and adding another layer of defence against infection spread. Smart door can incorporate infrared cameras to measure the body temperature, SpO_2, BP level of individuals passing through it. By detecting elevated temperatures, which may indicate fever or illness, similarly, determines the SpO_2 and BP value of the individuals from the threshold level. Then, the system can alert authorities or concerned individuals to take appropriate action, such as conducting further screening or denying entry if necessary. Moreover, it can also identify whether individuals are wearing face masks or not using camera sensor module. It can quickly scan faces and determine if masks are properly worn, covering the nose and mouth. This feature helps reinforce the adherence to mask-wearing protocols, which are crucial in reducing the transmission of respiratory diseases. This integration allows real-time monitoring and data collection, facilitating efficient management and decision-making processes. The authors express their hope that this technology, along with further research and improvements, will contribute to progress in the healthcare industry. It is worth emphasizing that the continuous advancement of IoT technology holds promise for enhancing the overall service and efficiency of the healthcare sector worldwide. As technology continues to improve, it is likely that new modifications and improvements to the system will be made, leading to further advancements in the healthcare industry and benefiting patients and healthcare providers alike.

References

1. Abirami, G., Helan, R. R. H., Vivekanandan, S. J., Haritha, D., Nancy, S., and Nalini, R. (2022). An Intelligent Approach for Monitoring Social Distancing, Temperature and Detecting Face Mask for Reducing COVID Spread. *International Journal of Modern Developments in Engineering and Science*, 1(6), 12–17.
2. Al-Naji, A., Khalid, G. A., Mahdi, J. F., and Chahl, J. (2021). Non-contact SpO_2 prediction system based on a digital camera. *Applied Sciences*, 11(9), 4255.
3. Ali, M. M., Haxha, S., Alam, M. M., Nwibor, C. and Sakel, M. (2020). Design of Internet of Things (IoT) and Android Based Low Cost Health Monitoring Embedded System Wearable Sensor for Measuring SpO 2, Heart Rate and Body Temperature Simultaneously. Wireless Personal Communications, 111(2020): 2449-2463.
4. Bose, R., Roy, S., and Mondal, H. (2022). A novel algorithmic electric power saver strategies for real-time smart poultry farming. *e-Prime-Advances in Electrical Engineering, Electronics and Energy*, 2(2022), 100053.
5. Casalino, G., Castellano, G. and Zaza, G. (2022). Evaluating the robustness of a contact-less mHealth solution for personal and remote monitoring of blood oxygen saturation. *Journal of Ambient Intelligence and Humanized Computing*, 1–10.
6. Debnath, B., Dey, R., and Roy, S. (2019). Smart switching system using bluetooth technology. *In 2019 Amity International Conference on Artificial Intelligence (AICAI), IEEE*. 760–763.
7. Dolin, R. H., Alschuler, L., Beebe, C., Biron, P. V., Boyer, S. L., Essin, D. and Mattison, J. E. (2001). The HL7 clinical document architecture. *Journal of the American Medical Informatics Association*, 8(6), 552–569.
8. Dzisi, E. K. J. and Dei, O. A. (2020). Adherence to social distancing and wearing of masks within public transportation during the COVID 19 pandemic. *Transportation Research Interdisciplinary Perspectives*, 7, 100191.

9. Ghosh, A., Roy, S., Mondal, H., Biswas, S., and Bose, R. (2022). Mathematical modelling for decision making of lockdown during COVID-19. *Applied Intelligence*, 52(1), 699–715.

10. Han, D., Khan, Y., Ting, J., King, S. M., Yaacobi-Gross, N., Humphries, M. J., and Arias, A. C. (2017). Flexible blade-coated multicolor polymer light-emitting diodes for optoelectronic sensors. *Advanced Materials*, 29(22), 1606206.

11. Rashid, M, H., Chowdhury, O., Hossain, M. M., Rahman, M. M., Muhammad, G., AlQahtani, S. A., and Hossain, M. S. (2023). IoT-based medical image monitoring system using HL7 in a hospital database. *In Healthcare*, MDPI. 11(1), 139.

12. Joshi, L. M., Bharti, R. K., and Singh, R. (2022). I nternet of things and machine learning-based approaches in the urban solid waste management: Trends, challenges, and future directions. *Expert Systems*, 39(5), e12865.

13. Mahmood, M. S., and Al Dabagh, N. B. (2023). Improving IoT security using lightweight based deep learning protection model. *Tikrit Journal of Engineering Sciences*, 30(1), 119–129.

14. Majumder, S., Ray, S., Sadhukhan, D., Khan, M. K., and Dasgupta, M. (2021). ECC-CoAP: Elliptic curve cryptography based constraint application protocol for internet of things. *Wireless Personal Communications*, 116(2021), 1867–1896.

15. Manikandan, N. (2020). Are social distancing, hand washing and wearing masks appropriate measures to mitigate transmission of COVID-19?. *Vacunas*, 21(2), 136.

16. Nemomssa, H. D. and Raj, H. (2022). Development of low-cost and portable pulse oximeter device with improved accuracy and accessibility. *Medical Devices: Evidence and Research*. 121–129.

17. Ogunsalu, C., Arunaye, F., Barton, E., Fagbola, O., Ajike, S., and Ogunsalu, D. (2023). Statistical analysis and mathematical modelling for the utilization and efficacy of menthol crystals vapor inhalation for the control of the Covid-19 pandemic. *Journal of Dentistry and Oral Sciences*, 5(1), 1–17.

18. Park, H. D. (2018). Development of web-based telemedicine system with HL7 converter. *International Information Institute (Tokyo) Information*, 21(3), 1121–1128.

19. Premkumar, S. and Hemanth, D. J. (2022). Intelligent remote photoplethysmography-based methods for heart rate estimation from face videos: A survey. *In Informatics*, 9(3), 57.

20. Radhika, R., Bhuvaneswari, A., and Kalpana, G. (2022). An intelligent semanticification rules enabled user-specific healthcare framework using IoT and deep learning techniques. *Wireless Personal Communications*, 1–25.

21. Sanjuan, E. B., Cardiel, I. A., Cerrada, J. A., and Cerrada, C. (2020). Message queuing telemetry transport (MQTT) security: A cryptographic smart card approach. *IEEE Access*, 8, 115051–115062.

22. Santos-González, I., Rivero-García, A., Molina-Gil, J., and Caballero-Gil, P. (2017). Implementation and analysis of real-time streaming protocols. *Sensors,* 17(4), 846.

23. Sinhal, R., Singh, K., and Raghuwanshi, M. M. (2020). An overview of remote photoplethysmography methods for vital sign monitoring. In Computer Vision and Machine Intelligence in Medical Image Analysis: International Symposium, ISCMM 2019, *Springer Singapore*, 21–31.

24. van Es, V. A., Lopata, R. G., Scilingo, E. P., and Nardelli, M. (2023). Contactless cardiovascular assessment by imaging photoplethysmography: A comparison with wearable monitoring. *Sensors*, 23(3), 1505.

25. Varshini, B., Yogesh, H. R., Pasha, S. D., Suhail, M., Madhumitha, V. and Sasi, A. (2021). IoT-Enabled smart doors for monitoring body temperature and face mask detection. *Global Transitions Proceedings*, 2(2), 246–254.

26. Yang, X. (2021). An approach of project-based learning: Bridging the gap between academia and industry needs in teaching integrated circuit design course. *IEEE Transactions on Education*, 64(4), 337–344.

27. Zhu, Q., Chen, M., Wong, C. W., and Wu, M. (2020). Adaptive multi-trace carving for robust frequency tracking in forensic applications. *IEEE Transactions on Information Forensics and Security*, 16, 1174–1189.

22 Enhancing Road Safety: Pothole Detection and Traffic Efficiency using YOLOv8 Deep Learning Model

A. Naveen Kumar[a], J. Akilandeswari[b], P R Mathangi[c], S Dhanush Prabhu[d], Kavya P[e], and Ashwin Kumar V[f]

Department of IT, Sona College of Technology, Salem, India

Abstract

Road's act as a means of transportation between various locations and are obviously necessary to daily life. Roads are kept safe and functional by routine maintenance. Potholes, on the other hand, are a result of road surface problems and may lead to more accidents. Roads can be prevented from getting worse by spotting potholes and notifying their presence to the relevant departments. In most areas of the road network, artificial intelligence and machine learning can reduce traffic congestion and increase traffic efficiency. To identify potholes, this research implemented and tested various deep learning architectures. For real-time object recognition, this model was created via the YOLO (You Only Look Once) algorithm. The YOLO v8 is being used by a pretrained algorithm to identify the pothole. Sequential CNN (convolutional neural network) algorithms YOLO v4, v5 and v7 were previously employed, but we later discovered through comparison testing that YOLO v8 produced better results in real time. This research provides a detailed analysis of pothole detection using YOLOv8 sub-versions such as small, nano, and medium. This is further compared with the existing YOLOv7 model. The experimental finding shows that the YOLOv8 nano version gives higher accuracy.

Keywords: YOLO algorithm, pothole detection, deep learning, object recognition.

Introduction

Potholes are a common road problem, leading to vehicle damage, safety hazards, and costly repairs. They are caused by factors like weather, poor maintenance, heavy traffic, and aging infrastructure. Potholes pose a serious threat to road safety, especially at high speeds, resulting in accidents. Timely detection and repair of potholes are crucial to prevent such incidents. Pothole detection systems have the potential to revolutionize maintenance and enhance motorist safety as technology advances. These systems, utilizing cameras installed on vehicles or fixed infrastructure, can automatically identify, and locate potholes. Machine learning algorithms analyse images of road surfaces to detect and pinpoint potholes, improving their effectiveness over time.

[a]naveenkumar@sonatech.ac.in, [b]akilandeswari@sonatech.ac.in, [c]mathangislm@gmail.com, [d]dhanushprabhusenthil@gmail.com, [e]kavya.palanivelu@gmail.com, [f]ashwinrockstar555@gmail.com

Implementing a deep learning-based pothole detection system improves detection efficiency and accuracy, reducing accident risks and vehicle damage. It enables prioritization of maintenance and repairs by road authorities to address potholes promptly. Potholes pose detection challenges due to their diverse geometrical characteristics. Convolutional neural networks (CNN) are effective recognition techniques for pothole detection. This research utilizes newly constructed road image databases and employs You Only Look Once (YOLO), a specialized CNN algorithm, to identify potholes.

This study aims to enhance the detection of potholes in roadways using the YOLO v8 (You Only Look Once, Version 8) object identification algorithm. The proposed model outperforms existing ones, reducing prediction errors. The research contributions include:

- Proposing the YOLO v8 pothole detection model for improved precision.
- Evaluating the performance of the proposed YOLO v8 models compared to other pothole detection techniques.
- Demonstrating the superior accuracy and robustness of the suggested YOLO v8 model.
- Comparing the YOLO v8 model with YOLO v7.

Figure 22.1 The proposed technique.

Literature Review

Dhiman et al. [1] summarize and classify several strategies for automatically identifying potholes on streets. They propose two approaches based on stereo-vision examination of roadway environments and deep-learning techniques for pothole identification. Experimental assessments were conducted on all four methods, and the results are discussed.

Hiremath et al. [2] have developed an application that utilizes the YOLO object detection approach to locate and identify potholes on roads. This app allows users to upload images of potholes in their neighbourhood, and the YOLO algorithm analyses the photos to pinpoint the position of the potholes on a map. This enables local civic authorities to promptly address the identified potholes. By implementing this cutting-edge and environmentally friendly solution, the researchers demonstrate the effective assessment of road conditions using a mobile phone.

Mednis et al. [3] propose a mobile sensing system based on Android smartphones for identifying potholes. The study explains selected data processing techniques and evaluates their effectiveness using real-world data, achieving high true positive rates of up to 90%. The researchers also determine the optimal parameters for the algorithms and provide recommendations on their practical application.

Prakash et al. [4] investigate the YOLOX techniques for pothole identification. They train the YOLOX algorithm using datasets containing potholes and analyse its accuracy, size, and recall in comparison to other YOLO methods. With a model size of 7.22 MB, the models achieve an average precision of 85.6%. This research contributes to cost reduction and faster pothole detection.

Pachaivannan et. al. [5], Statistics indicate that there are several factors that increase the probability of traffic accidents. According to the author, there are many causes of traffic accidents, hence strong methodologies and computer-aided support are needed to address these issues. Hence, the author creates a programme that uses deep learning to study traffic incidents.

Using object detection algorithms to find road damage is the most effective technique. CNN may be used to extract features from videos or pictures based on the pothole data-set, and it is cost-effective. In this work, Omar et. al. [6] annotates the datasets of pothole images and videos, and YOLOv4 is used to train the model. The model is assessed using the mAP, recall, and precision. The authors have also verified this model using various pothole videos and photos.

Goswami et. al. [7] examine various object detection techniques before choosing one. By utilising the YOLOv7 object identification technique, they suggest a solution for the rec-ognition of potholes as well as road craters. The accuracy and recall values for the trained model are 0.94 and 0.98, respectively. The mAP for the suggested approaches is 94.76%. Real-time identification of carters as well as potholes is possible using this technology in both low-light and daylight conditions.

Mohanraj et. al. [8] has proposed hybrid deep learning model, the rank-based multi-layer perceptron (R-MLP), to predict and target vaccination rates in under-immunized regions.

Akilandeswari et. al. [9] has used the CNN to detect pulmonary embolism in lung CT scan images. Four popular CNN architectures, namely Inception, VGG-16, ResNet50, and Mobilenet, are evaluated for their performance using the RSNA STR Pulmonary Embolism Chest CT scan image dataset.

Naveenkumar et. Al. [10] has compared the performance of three popular models, namely SSD, YOLOv3, and YOLOv4, for object detection in indoor environments.

Methodology

In this study, a pothole detecting learning model employing YOLO is built. For effective image categorization, a pothole photo collection is gathered and annotated. Groups for training, testing, and validation are created from the dataset. The YOLO version 8 method is used, which provides great object identification accuracy, particularly for potholes. To increase the model's realism, epoch values and weights are used during training. Testing is done to make sure the model is accurate and consistent. The classification process entails several processes, including dataset collecting, image labelling, dataset splitting, model training, and model evaluation.

Custom Dataset

For training the YOLOv8 model, numerous hyperparameters are considered to optimize performance without overfitting. To ensure an extensive dataset, over 7500 pothole images were collected from various sources, including Roboflow's dataset, scholarly works, YouTube videos, and RDD2022 dataset. After processing and annotation corrections, the dataset was refined to include 6962 training images and 272 validation images. This dataset is then used for the custom training of the YOLOv8 model.

YOLO

YOLO (You Only Look Once) is a fast and effective real-time object detection system based on deep neural networks. In this research, we train the YOLO v8 model using custom pothole datasets, which include numerous images with small and challenging-to-detect potholes. To achieve an efficient pothole detection model, the study conducts experimental training on three different sub-versions of YOLO v8 models: YOLO v8-Nano, YOLO v8-Small, and YOLO v8-Medium. The well-performing model is then compared with the YOLO v7 model to assess its superiority.

Experimentation and Results

Preliminary Setup

The experimentations in this research were conducted using the NVIDIA GeForce RTX 30 series 4GB GPU, an Intel Core i7 processor, and 16 GB RAM. The training run would take around 4 to 5 hours to complete all the epoch runs. The proposed methods were implemented using Python, Ultralytics YOLOv8, and the Ultralytics Python API.

Custom Data Processing

The following hyperparameters options are considered for custom data processing:

- Each model is trained for 50 epochs, taking a considerable amount of time due to the large dataset of around 7500 images and the batch size is set for fair comparison with other models.
- To improve detection of tiny and difficult-to-detect potholes, the image size is increased to 1280 resolution during training, surpassing the default value of 640.
- The custom images of different varying weather conditions of the roads were also included for example potholes with water.

YOLO v8 Nano Model

The YOLO v8 nano model is the smallest model in the YOLO v8 family. This model can reach 3.2 million parameters and it could run in real-time CPU. Training is held with the epoch size 50, batch size of 8 and with image resolution of 1280. The final plot of the training is shown in Figure 22.2.

This provides mean average precision (mAP) at 0.50 Intersection over Union (IoU) is 40.4 and at 0.95 mAP the IoU is 18.9.

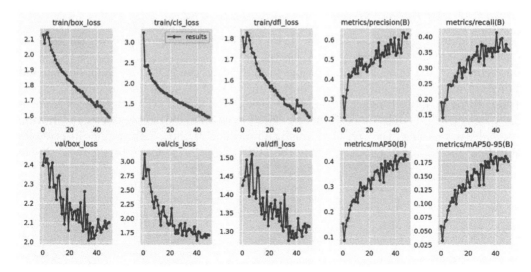

Figure 22.2 mAP vs loss plots of YOLO v8n model.

YOLO v8 Small Model

The training performance of the YOLO v8 small model on the pothole dataset is shown in Figure 22.3. This reaches utmost an mAP of 50 at 0.50 IoU. This model attains an mAP of 49 in the evaluation.

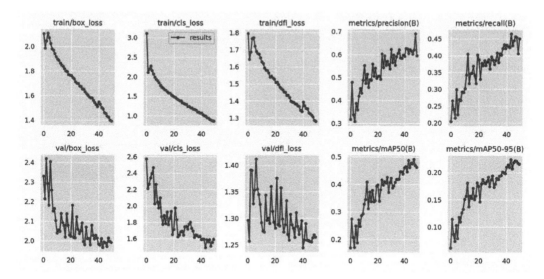

Figure 22.3 mAP vs loss plots of YOLO v8s model.

YOLO v8 Medium Model

The final training of the YOLO v8m model also reaches an mAP of nearly 50. The evaluation result states that YOLO v8m model attains an mAP of 48 with the comparison of 49 mAP of YOLO v8s model.

YOLO v8 Models Training Results

A detailed comparison between all the three YOLO v8 training models on the pothole dataset with 0.50 IoU is shown in the ClearML graph.

In Figure 22.5 the green, blue, and orange lines indicates the YOLO medium, normal and small method. From the ClearML graph both the YOLO v8 small and medium keeps on learning throughout the training. These two models can be trained even longer to get eminent results, but the nano model outperforms the others in terms of efficient and instant detection of potholes in real time. The YOLO v8 small and medium methods fails to find the potholes instantly in the real time. The accuracy of the YOLO model is analysed by the mAP (Mean

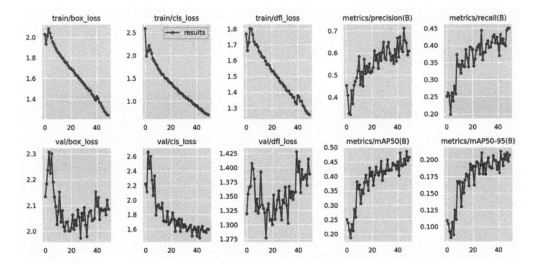

Figure 22.4 mAP vs loss plots of YOLO v8m model.

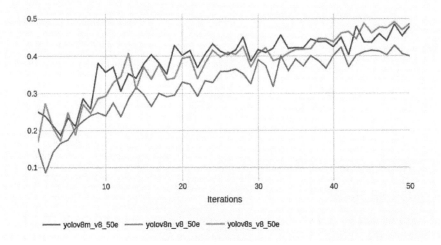

Figure 22.5 YOLO V8-medium vs YOLO V8-nano vs YOLO V8 small.

Average Precision) score. The higher mAP score indicates that the accuracy of the model is higher. The mAP scores of these models are shown the Table 22.1. The comparison between the mAP value of these models are made at 0.50 IoU. The mAP value of the YOLO v8 Nano model is higher than the other two models and it shows to be more accurate.

Table 22.1: Mean average precision analysis.

Model	mAP val
YOLOv8 medium	39.2
YOLOv8 nano	49.4
YOLOv8 small	35.8

Interface

An inference run is conducted on all the trained models. For this purpose, a sample video with numerous potholes is used. The results show that the medium mode detects more potholes that are farther away from the source. This is shown in Figure 22.6. The YOLO v8s runs at 35 FPS and YOLO v8m with 14 FPS.

Figure 22.6 YOLO v8m detects more distant potholes.

The results almost look identical because of the closer mAP. But in some frames of the clip, the YOLO v8m detects smaller potholes. The performance analysis between these three sub versions is shown in Table 22.2. From the real time inference test, we could confirm that YOLO v8 nano model to be more accurate compared with other sub-versions.

Table 22.2: Performance analysis between YOLO v8 small, medium, nano.

Pothole constraints	YOLO v8 medium	YOLO v8 small	YOLO v8 nano	Total pothole count
Small pothole counts: 14	3 (Not mostly detected)	14 (All detected)	14 (All detected)	14
Medium pothole counts: 13	13 (All detected)	1 (Not detected)	13 (All detected)	13
Tiny pothole counts: 6	0 (Not detected)	1 (Not detected)	6 (All detected)	6

YOLO v7 Analysis with YOLO v8 Nano Model

The custom dataset developed in this research is again trained with the previous version of the current YOLO v8 model which is YOLO v7. This model is more accurate and applied for major computer vision tasks. This trained model is further compared with the YOLO v8 nano model. By this experimentation analysis we found that YOLO version 7 fails to find many potholes which is accurately detected by YOLO v8 nano version. The sample image comparison between YOLO v7 and YOLO v8 nano version is shown in the Figure 22.7.

Figure 22.7 YOLO v7 VS YOLO v8 nano.

The Figure 22.7 shows the real time pothole detection of the YOLO v7 model on the left side and the YOLO v8 nano model on the right side. The experimentation findings of YOLO v8 nano and YOLO v7 are shown in Table 22.3.

Table 22.3: Comparison of YOLO v7 and YOLO v8 model.

Constraint	YOLO v7	YOLO v8 nano
Number of potholes found	Found 13 (less)	Found 20 (Reached maximum level)
Distorted potholes	Failed to detect	Detected
Accuracy	89.67%	94.48%

Conclusion and Future Scope

The research concludes that the YOLO v8 nano model is the most suitable for real-time implementation, surpassing other sub-versions such as YOLO v8 small, medium, and YOLO v7. The YOLO v8 nano model exhibits superior performance in detecting various types of potholes, including noisy ones affected by vehicle and pedestrian activity. With a training dataset of custom images, this model achieves an impressive accuracy of nearly 95% in real-time detection. This model will detect the potholes with varying weather conditions since the custom dataset is developed and trained with different weather condition road images. There are potential areas for further improvement, such as training the model to estimate the depth and width of detected potholes, allowing for automatic classification based on different ranges. Additionally, expanding the custom dataset to include low-light and nighttime pothole images would enable detection during darker conditions. The proposed solution holds promise for implementation in automated vehicles and by government highway agencies to streamline pothole detection and repair processes, reducing reliance on human labour. Further a reinforcement learning approach could be applied for the YOLOv8 models. This method could also be extended to the detection of other road defects, like cracks and bumps.

References

1. Dhiman, A. and Klette, R. (2020). Pothole detection using computer vision and learning. *In IEEE Transactions on Intelligent Transportation Systems*, 21(8), 3536–3550. doi: 10.1109/TITS.2019.2931297.
2. Hiremath, R., Malshikare, K., Mahajan, M., and Kulkarni, R.V. (2021). A Smart App for Pothole Detection Using Yolo Model. In: Fong, S., Dey, N., Joshi, A. (Eds.) ICT Analysis and Applications. *Lecture Notes in Networks and Systems*, 154. Springer, Singapore.
3. Mednis, G. Strazdins, R. Zviedris, G. Kanonirs, and L. Selavo.(2011). Real time pothole detection using Android smartphones with accelerometers. *International Conference on Distributed Computing in Sensor Systems and Workshops (DCOSS)*, pp. 1–6. Barcelona, Spain. doi: 10.1109/DCOSS.2011.5982206.
4. Mohan Prakash, B. and Sriharipriya K. C. (2022). Enhanced pothole detection system using YOLOX algorithm. *Autonomous Intelligent Systems*, 2, 22.
5. Pachaivannan, P., Hemamalini Ranganathan, R., Navin Elamparithi, P., and Dhanagopal, R. (2020). Indian road conditions and accident risk predictions using deep learning approach—a review. *3rd International Conference on Intelligent Sustainable Systems (ICISS)*, Thoothukudi, India, pp. 199–202. doi: 10.1109/ICISS49785.2020.9316128.
6. Omar, M. and Kumar, P. (2020). Detection of roads potholes using YOLOv4. *2020 International Conference on Information Science and Communications Technologies (ICISCT)*, Tashkent, Uzbekistan, pp. 1–6. doi: 10.1109/ICISCT50599.2020.9351373.
7. Goswami, K., Chattopadhyay, S., and Kundu, K. (2023). Deep learning based approach to detect potholes using YOLO version 7. *International Journal of Creative Research Thoughts*.
8. Mohanraj, G., Mohanraj, V., Senthilkumar, J., and Suresh, Y. (2020). A hybrid deep learning model for predicting and targeting the less immunized area to improve children's vaccination rate. *Intelligent Data Analysis*, 24(6), 1385–1402.
9. Akilandeswari, J., Jothi, G., Naveenkumar, A., Iyyanar, P., and Paramasivam, M. E. (2021). Detecting pulmonary embolism using deep neural networks. *International Journal of Performability Engineering*, 17(3), 322–332.
10. Naveenkumar, A. and Akilandeswari, J. (2022). Deep learning algorithms for object detection—a study. *In Evolution in Computational Intelligence: Proceedings of the 9th International Conference on Frontiers in Intelligent Computing: Theory and Applications*, pp. 65–75. Singapore: Springer Nature Singapore.

23 Through Internet of Thing and UAV Technology, Automation Enables Smart Agricultural Analysis Based on a Range of Methodologies and Applications

Monojit Manna[1,a], Subrata Dutta[2,b], Arindam Giri[3,c], Anwesa Naskar[1,d], Soumya Paul[5,e], and Animesh Samanta[4,f]

[1]Assistant Professor, HIT Haldia, Haldia Institute of Technology, West Bengal, India

[2]Assistant Professor, National Institute of Technology Jamshedpur, Jamshedpur, India

[3]Associate Professor, HIT Haldia, Haldia Institute of Technology, West Bengal, India

[4]Assistant Professor, BCET Durgapur, Bengal College of Engineering and Technology, West Bengal, India

[5]Assistant Professor, MCKV Howrah, MCKV Institute of Engineering, Howrah, West Bengal, India

Abstract

In the current world, the expansion of smart devices with help of the internet can be expanded by integrating them into a network of communicating actuators, a physical empire network. When using Internet of Thing (IoT), the actuator and sensor work together flawlessly based on the surroundings. When IoT and UAV occurred in smart agriculture have emerged to offer innovative solutions, utilization of these technologies will bring blessing in the smart agriculture platform. To achieve a specific task both are collaborating vice versa through the internet. To overcome the specific challenges between anything WSN can be segregated into IoT. This platform tested real time scenarios and deployed automated and real time monitoring sensors. In addition, other environmental parameter for collecting huge amount of surroundings data, in this piece of work focuses on giving a framework to optimizing resources (like human effort, water, fertilizers and insecticides) and uses unmanned aerial vehicles for monitoring and fetching the horticulture data, sending the information and utilizing it through IoT technology. During implementation, the arising issues due to the integration of IoT sensors and UAV technology in the application are also explored and by the use of these approaches that boost productivity and reduce use of natural resources explored in this paperwork. The name of the framework can be considered as smart IoT agriculture.

Keywords: Wireless sensor network, agricultural automation, IoT, UAV.

Introduction

In today's world information and communication technology is a prototype for Internet of Things (IoT) [1]. IoT can be considered as the different composite for physical objects globally. Physical objects i.e. devices, sensors, actuators, mobile phones and RFID [2] and various more items have network connectivity built in, e-devices, sensors which enable establishing and exchanging the data collection in smart surrounding. With the quick

---

[a]monojit.rcciit@gmail.com, [b]sdutta.cse@nitjsr.ac.in, [c]ari_giri111@rediffmail.com, [d]anwesanaskar@gmail.com, [e]soumyapaul5a1@gmail.com, [f]animesh.cseeng@yahoo.com

development of technology, modern technology is used to provide a smart environment and ease daily life. IoT though it is well known globally, there are four layers to the IoT reference model, which will bring out a secure and overall management module that delivers a secure and efficient model [3]. Given below figure the overall idea about the various IoT communication layer. In the IoT domain, there are various application areas such as smart health care model, smart home model, smart agriculture system etc. Although there has been no global validation of the usage of IoT in agriculture in developing nation. Therefore, there is a need for a proper structure for these domains [4]. This paper serves as a structure. The outcome of merging current agriculture systems with IoT will be a smart agricultural system. With the continue with this paperwork will be designed as follows: Section 2 we illustrate about some literature review In Section 3 we illustrate about the Research challenges to realize the IoT. In Section 4 a framework is proposed for UAV and IoT where the agricultural industry may employ IoT Section 5 of this article covered several message formats used in IoT agriculture. In Section 6 we looked into UAV and IoT based applications in agriculture domain. In Section 7 we illustrate about flawless connectivity and communication technology used for IoT domain. In Section 8 we discuss about related research issues, finally the conclusion is drawn in Section 9.

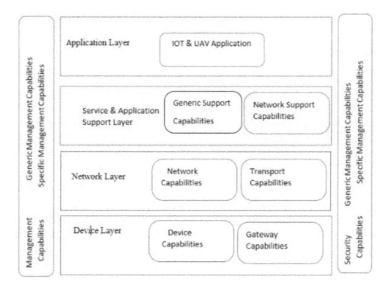

Figure 23.1 IOT architecture reference model.

Literature Review

We discuss recent IoT literature in this section. UAV technology used for smart farming and agriculture domain. Since there is no specific literature in this field, modern UAV technology, in our opinion, is the best suitable for methodical basement using two-literature article review by which we understood about the development aspect of agriculture environment.

- As per a comprehensive review paper by Boursianis et al. [2].
- As per a comprehensive review paper by Dhanaraju et al. [4]equipment, and sensors in network-based hi-tech farm supervision cycles. Innovative technologies, the Internet of Things (IoT).

IoT and Agriculture UAV in Smart Farming

In recent days, the use of UAVs and IoT technology in agriculture has revolutionized traditional farming methods and ushered in a new era of precision agriculture. According to Boursianis and Papadopoulou's research on the fundamentals of IoT, covering IoT applications and solutions for smart farming, IoT applications and solutions, types, networks, and protocols of IoT sensors used in agriculture. Additionally, they analyze the usage of UAVs in a variety of scenarios, such as irrigation, fertilization, the application of pesticides, observation of plant development, and control of crop diseases, and field-level phenotyping, in order to illustrate the role of UAV technology in smart agriculture.

IoT Based Sustainable Agriculture for Bright Agriculture

Current scenario of smart farming environment technologies are utilized IOT and UAV as blessing for development the field of farming as per Dhanaraju et al [4] review article work they contribute to the advancement of hi-tech farm monitoring cycles that emphasizes the use of machinery, equipment, and sensors in conjunction with information and communication technologies. It is projected that cutting-edge technology, the IoT, and cloud computing would spur development and user in the usage of robotics and artificial intelligence in agriculture.

Research Challenges of IoT

IoT's major goal is to give everyday things the ability to interact and use computing. Although IoT is still in its infancy with wireless technologies like sensors and actuators, among others, there are still substantial hurdles that must be overcome from both a technological and a business perspective. Analyzing the potential of IoT's rise in a firm at a strategic and operational stage is necessary. It is necessary from a technology standpoint, but researchers are also having trouble deploying IoT applications as depicted in Figure 23.1.

Standardization

To successfully install smart devices in a plug-and-play environment across all platforms. Standardization of methods and frequency bands is necessary to do that. The IETF, EPCglobal, ISO, and ITU are some of the major participants in the standardization of the IoT paradigm. Table 23.1 can be used to create an entire set of standards. The ISO is concentrating on protocols for collision resolution, modulation techniques, and frequency used. Electronic tags and other industrial standards are goals of EPCglobal. Several Auto-ID labs, governmental and non-governmental organizations share the EPCglobal Architecture Framework [7]. It makes sense that the 6LoWPAN neighbor discovery, use cases, and routing requirements shown in Table 23.1 have been merged into a comprehensive architecture. [6].

Table 23.1: IOT standardization project.

Standard	Focus area	Status
6L0WPAN	Low-power Ipv6 protocol devices	ongoing
ROLL	Protocols for Routing in Heterogeneous Network	ongoing
EPC global	Promoting EPC for RFID Tag	Advanced

Wireless Sensor Network Energy-Efficient Data Collection From the Object

When it comes to wireless sensor networks, one of the crucial and major resources is energy. It is quite difficult to recharge the sensor node battery at a glance. If the battery charge runs out quickly then it will decrease the performance and quality of IOTagriculture technology. To optimize the energy conservation we must choose an efficient routing protocol.

Networking with Addressing

Nevertheless, about the location IoT must require interaction with objects or models. So the object or model has the unique identification of the individual. For low power wireless sensor nodes, a 128-bit IPv6 addressing node is utilized. All devices connected to the network may be given IPV6 addresses since 32-bit IPv4 addressing is insufficient. RFID tags have 64–96 bit addressing [7].

Quality of Service

In the case of IoT, various QoS requirements for various applications throughput, noise, latency, loss, and security are a few examples. The development of QoS parameters for IoT applications heavily depends on resource utilization optimization. for all contrasted devices. By the rapid change of topology of object mobility, existing QoS parameters are not applicable in the WSN environment.

Security and Privacy

Wireless sensor network security is an important concern for a specific IoT application. In addition, with the physical attack (major time devices are unattempt) IoT is vulnerable to internet attacks including eavesdropping, virus, jamming, capture, or creating a hostile node from any location in the world. While there have already been numerous attempts to guarantee the security of WSN and RFID, IoT privacy and security have received a lot of research [8].

Framework OF UAV and IoT agriculture

In this segment of this article, using the IoT paradigm, we propose the framework for agricultural automation. It is known as the suggested strategy IoT agriculture as per

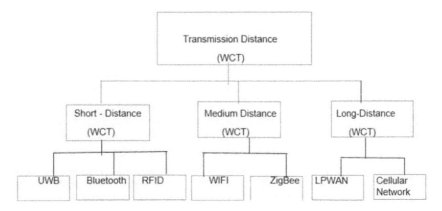

Figure 23.2 Classification for communication technology in IoT agriculture environment.

Figure 23.2. There are mainly various four steps such as Things, Local Gateway, Internet, and Cloud.

Items or Things Require for Surveillance

According to the First layer of proposed IoT agriculture framework, item or objects can be determined. Numerous physical items, including RFIDs, cell phones, sensors, actuators, and other essential items, are used in the application domains for data collection and surveillance. Most often, certain sensor nodes can align with several sensors, including humidity detectors, bug detectors, temperature detectors, etc. In single hope or multiple hop dissimilar, the object must communicate with the neighbor base station (LBS) [9] and by using internet user have to utilize the capability to control or access the objects. With the help of various efficient algorithm objects or sensor nodes need to be localized other than using the expensive global positioning system (GPS).

Local Corridor for Data Transmission

The local gateway receives the data collected by things or objects in either a single hop or multiple hops. Local gateway will compile the data after data collection to eliminate redundant or irrelevant data. The data will be transmitted to the cloud through the internet following the data aggregation process. Between the item and the user, the local gateway will serve as a communication intermediary.

Information Highway

The outcome of data by the deployed application will be transferred through an information highway to cloud storage. Users will get notified about the phase via the internet. For IoT, the internet is the high-level common communication platform. Through information highway, our model will get a weather report in advance. By the help of the weather report, Our IOTagriculture model synchronizes the water spraying of pesticides and other insecticides in the affected area with ease.

Cloud data and Mobile Connectivity

As is common knowledge, cloud computing is on-demand computing that does not need the purchase of effective hardware, essential system software, or application-base software. The fourth layer of the AgriTech framework is where the cloud is located. SaaS, PaaS, and (IaaS), which are the three essential components of the cloud, will also be included in the AgriTech cloud [10]. IOTagriculture The user must purchase the necessary infrastructure to set up layers 1 and 2. We need to design various applications for the IOTagriculture [11] user to use their smartphone to control the agricultural operation. In this work, we focused on the layer of the IoT network where a gateway is required to connect the sensor and network to their neighboring node in order to create a network infrastructure. Layer different terminal devices, sensors (WSN), RFID tags. Through an embedded device, the data is transferred for extra processing and analysis to a higher tier [14].

UAV and IOT Platform Application in Agriculture Environment

In the field of agriculture environment UAV Technology is growing rapidly, replacing satellite and other aircraft technologies now a days [23] during the initial time when UAVs were originally created, they were frequently used in military and surveillance settings. At low cost UAV, technology can capture high-resolution images. Whereas satellite surveillance requires many perspectives compared to UAV technology [13].

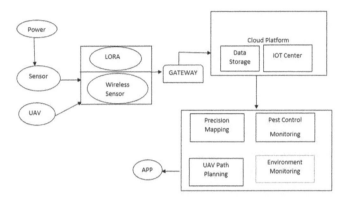

Figure 23.3 UAV and IOT platform working application in agriculture environment.

Field Monitoring

Monitoring indicates a used UAV photos to identify weeds early in an Australian field of capsicum. Additionally, the crop height of banana plants in a field of agriculture was measured using UAV photos. A smart UAV platform was linked with a wireless sensor network (WSN) to measure the influence of grape production quantity and quality in real time [27]. UAVs have been successfully used to manage crop yield in agriculture. Sensors are more deeply analyzed and propagate crop parameters like leaf index factor, color shape size of the leaves.

Agricultural Mapping

UAV Robots are useful to create 2 dimensional and 3-dimensional maps for agriculture. [16]As an instances the particular area of a land can be examine like status of the agricultural soil conditions allow for the detection of fungus within the crop. by examine those maps. These UAV photos and maps also make use of profitable precision agriculture jobs, like the control of homogeneity zones, separating fruit-quality zones, and the identification of areas of deforestation.

Seed Plantation in Field Through Modern Technology

UAV technology made more efficient and easy the process of plantation of seeds, as an example according to an efficient application of UAV technologies on a big paddy field [17]. This work shows an effective on time seed distribution fertilization along with plant nutrition and plant growth. Seed planting using UAV robots is still in is still in its infancy. Additionally, they are creating an improved method of seed planting utilizing UAV outfitted with image recognition technology.

Various Techniques Towards Fertilization and Pesticides Spraying

According to various surveys compared to a wide-range UAV drone's speed sprayer robots are more effective and quick spray pesticide and fertilizers in the same plot. The measurement of spreading pesticides in farmland correlated with the risk of health issues for workers and caused environmental pollution. Utilizing UAV robots to replace labor is one of the key goals. Because these UAV technologies spread pesticides and covered large-scale farmland within a few minutes as per Figure 23.3.

Controlling Field Through UAV Technology

In IOTagriculture Active monitoring leads to controlling agriculture using an automated method, as an example irrigation due to drought conditions hence yield losses can be decreased. Hence, the soil temperature and nutrient rate can also be controlled by sensors put in tractors and UAV drones. from time to time as depicted in Figure 23.4.

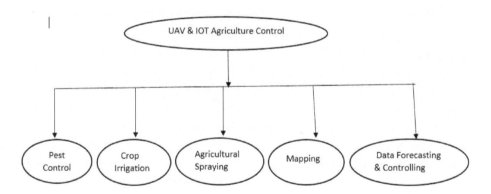

Figure 23.4 Framework for controlling in UAV and IOT agriculture platform.

Flawless connectivity and Communication Technologies used for IoTArgitech

In IoT based agriculture system, flawless connectivity is a major and important task. Connectivity and diverse value-added services have a significant impact on the system's whole supply chain. According to the connectivity, services [32] for doing some a variety of services should be offered by cellular operations. where requirement of i. connectivity and network functioning, ii. Availability and communication challenges using technology, iii. various IoT factor and iv. Connectivity limitation in remote areas.

Connectivity and Network Functioning

In Agriculture IoTArgitech require high bandwidth connectivity to process large amounts of data. Such huge amounts of data can be collected and stored in an agriculture server and downloaded and updated as per the requirement. In the case of green houses and horticulture land, different scenarios can be visualized as they are connected with an electrical grid operated through a gateway.

Availability and Communication Challenges Using Technology

In agriculture UAV technology are utilized in a smart manner collect information from sensors and nodes dispersed around the agricultural farm. This UAV technology enables drones to overcome any obstacle like a tree or big obstacle. It can easily fly and overcome those obstacles and monitor the agriculture land over a wide outdoor area. In the case of a cellular module, people can pay charge to utilize 5G technology for monitoring and fetching data by using cellular data along with IOTargitech and UAV Technology.

Table 23.2: Comparison of various wireless technology in IOT and UAV.

Parameter	LoRa	Bluetooth	LPWAN	WiMAX	Wi-Fi
Standard	LoRa r1.0	IEEE802.15.1 (ZigBee)	IEEE802.15. 4 (zigbee)	IEEE 802.16	IEEE 802.11 a to n
Energy Consumption	Very Low	Very low	Low	Medium	High
Data Rate	0.3–50kbs	1–24Mb/s	40–250kbs	1Mbs -1Gb/s	1Mb - 6.75Gb/s
Transmission Rate Cost	<30KM High	8–10 m Low	10–20 m Low	<30KM High	20–100m High

Various IoT Factors

Including various parameters like device latency, battery lifetime, coverage, scalability should be taken into account for IoT enable architecture application. The agricultural farm area can be covered by a single base station. at the highest coverage range of 40 km. It covered by using LTE infrastructure, LPWAN technology and LoraWAN etc.

Table 23.3: Various platform of agriculture UAV.

Platform (Payload)	Sensors	Communications
Helicopter (22 kg)	IMU and RGB Camera	WLAN
	IMU	Bluetooth
Quadcopter(1.25 kg)	Thermal camera	Wi-Fi Wireless Radio
	GPS Receiver	Wireless Radio
	RGB Camera	Wi-Fi
	GPS	–

Related Research Issues

UAV and IOT-based agriculture has gained enormous popularity, but there are still a number of problems that need to be solved, including those listed below.

Automated administration and control of irrigation in remote areas is necessary to preserve the quality of agricultural crops. To limit the amount of nutrients extracted from the stream by too much water in the grounded layer of soil as depicted in Figure 23.4. Cannabis detection and management. In horticulture crop management cannabis detection and control management is one of the important aspects. Failing to adequate control of cannabis leads to adverse environmental impact if not taken care of properly. Pest control and multi spectral representation for disease. For detecting pests and diseases in crop multi spectral image based remote sensing technique used high potentially as depicted in Figure 23.4.

Conclusion

In this work, we proposed one framework called Internet of Thing (IoT) Argitech that involved smart devices, wireless sensor networks with Internet access will be used to computerize agricultural activities. By intelligent devices in hand, users can monitor the farmland and crops from any location without reaching the exact location. Using smart devices, the farmer is in control of the equipment used in agriculture, like a water sprayer that runs on autopilot. Because of this, this technology can lessen the effort that humans put into the agriculture sector. Because of this, this technology can lessen the effort that humans put into the agriculture sector. Finally, we can say this framework is new of its kind, and easily adaptable along with the extent of any already used comparable model.

References

1. Anagnostopoulos, T., Xanthopoulos, T., and Psaromiligkos, Y. (2020). A smartphone crowd-sensing system enabling environmental crowdsourcing for municipality resource allocation with lstm stochastic prediction. *Sensors,* 20(14). https://doi.org/10.3390/s20143966
2. Boursianis, A. D., Papadopoulou, M. S., Diamantoulakis, P., Liopa-Tsakalidi, A., Barouchas, P., Salahas, G., Karagiannidis, G., Wan, S., and Goudos, S. K. (2022). Internet of Things (IoT) and agricultural unmanned aerial vehicles (UAVs) in smart farming: A comprehensive review. *Internet of Things*, 18, 100187. https://doi.org/https://doi.org/10.1016/j.iot.2020.100187
3. Chen, W., An, J., Li, R., Fu, L., Xie, G., Bhuiyan, M. Z. A., and Li, K. (2018). A novel fuzzy deep-learning approach to traffic flow prediction with uncertain spatial–temporal data features. *Future Generation Computer Systems*, 89, 78–88.
4. Dadashzadeh, M., Abbaspour-Gilandeh, Y., Mesri-Gundoshmian, T., Sabzi, S., Hernández-Hernández, J. L., Hernández-Hernández, M., and Arribas, J. I. (2020). Weed classification for site-specific weed management using an automated stereo computer-vision machine-learning system in rice fields. *Plants*, 9(5). https://doi.org/10.3390/plants9050559
5. Lin, Z., Lai, Y., Gao, X., Li, G., Wang, T., and Huang, G. (2016). Data gathering in urban vehicular network based on daily movement patterns. In *Proceedings 11th International Conference Computer Science Education,* pp. 641–646. doi: 10.1109/ICCSE.2016.7581655.
6. Ren, Y., Zeng, Z. Wang, T. Zhang, S., and Zhi, G. (2020). A trust-based minimum cost and quality aware data collection scheme in P2P network. *Peer-to-Peer Networking and Applications,* 1–24. doi: 10.1007/s12083-020-00898-2

7. Luo, T., Huang, Z., Kanhere, S., Zhang, J., and Das, S. K. (2019). Improving IoT data quality in mobile crowd sensing: A cross validation approach. *IEEE Internet Things Journal*, 6(3), 5651–5664. doi: 10. 1109/JIOT.2019.2904704.

8. Medaglia, C. M. and Serbanati, A. (2010). An overview of privacy and security issues in the internet of things. *The Internet of Things*, 389–395.

9. Yick, J., Mukherjee, B., and Ghosal, D. (2008). Wireless sensor network survey," Comput. Networks, vol. 52, no. 12, pp. 2292–2330, 2008.

10. Lenk, A., Klems, M., Nimis, J., Tai, S., and Sandholm, T. (2009). What's inside the cloud? An architectural map of the Cloud landscape. *In Proceedings of the 2009 ICSE Workshop on Software Engineering Challenges of Cloud Computing*, 2009, pp. 23–31.

11. Giri, A., Dutta, S., Neogy, S. (2016). Enabling agricultural automation to optimize utilization of water, fertilizer and insecticides by implementing Internet of Things (IoT). *2016 International Conference on Information Technology*. doi:10.1109/INCITE.2016.7857603

12. Hunter, M. C., Smith, R. G., Schipanski, M. E., Atwood, L. W., and Mortensen, D. A. (2017). Agriculture in 2050: Recalibrating targets for sustainable intensification. *Bioscience*, 67, 386–391.

13. Tzounis, A., Katsoulas, N., Bartzanas, T., Kittas, C. (2017). Internet of Things in agriculture, recent advances and future challenges. *Journal of Biosystems Engineering*, 164, 31–48.

14. Gubbi, J., Buyya, R., Marusic, S., and Palaniswami, M. (2013). Internet of Things (IoT): A vision, architectural elements, and future directions. *Future Generation Computer Systems*, 29, 1645–1660.

15. Orlando, F., Movedi, E., Coduto, D., Parisi, S., Brancadoro, L., Pagani, V., Guarneri, T., and Confalonieri, R. (2016). Estimating leaf area index (LAI) in vineyards using the PocketLAI smart-app. *Sensors*. 16, 2004.

16. De Castro, A. I., Torres-Sánchez, J., Peña, J.M., Jiménez-Brenes, F.M., Csillik, O., and López-Granados, F. (2018). An automatic random forest-OBIA algorithm for early weed mapping between and within crop rows using UAV imagery. *Remote Sens*. 10, 285.

17. Salaan, C. J., Tadakuma, K., Okada, Y., Sakai, Y., Ohno, K., and Tadokoro, S. (2019). Development and experimental validation of aerial vehicle with passive rotating shell on each rotor. *IEEE Robotics and Automation Letters*. 4, 2568–2575.

18. Wolfert, S., Ge, L., Verdouw, C., and Bogaardt, M. J. (2017). Big data in smart farming—A review. *Agricultural Systems*, 153, 69–80.

19. Dadashzadeh, M., Abbaspour-Gilandeh, Y., Mesri-Gundoshmian, T., Sabzi, S., Hernández-Hernández, J. L., Hernández-Hernández, M., and Arribas, J. I. (2020). Weed classification for site-specific weed management using an automated stereo computer-vision machine-learning system in rice fields. *Plants*, 9, 559.

24 Finding Optimum Number of Pollution Measurement Stations Using the Multi-Layer Perceptron Classifier

Agnish Mallick[a], Sukumar Chakrabarty[b], and Jayanta Poray[c]
Techno India University, Kolkata, West Bengal, India

Abstract

Environmental pollution is now 'the' word being discussed globally, and the authorities are spending lots of money to keep it under control, to make the world habitable. The first step towards control is measurement—accurate, real-time and at low cost. The authorities all over the world are now trying to measure density of different air pollutants round the clock, by installing some outdoor pollution measuring stations. Naturally this incurs cost. This cost, not only includes hardware and networking, but also the running cost, which is sometimes more than the one-time installation cost. Along with this, the maintenance of the infrastructure is also very high, because the sensors need to be re-calibrated at regular intervals. The challenges remain for maintaining the uninterrupted power supply and network connectivity, throughout the year, round the clock. In this paper, we have proposed an AI driven approach to optimize the number of stations, in a city. More precisely, here we have shown that with a lesser number of stations, we can estimate the pollution levels with sufficient accuracy. To do this, first we collected historical data of pollution of Kolkata. Next, we randomly split the stations into training and test stations. With the training stations we have created a network for predicting the pollution value of the test stations with a certain accuracy.

Keywords: Environmental pollution, air quality index, AI driven optimization, monitoring pollution stations, multi-layer perceptron (MLP) classifier.

Introduction

Air pollution is a pressing global issue that significantly impacts public health and the environment. Monitoring and controlling air quality is a crucial issue to mitigate the adverse effects of pollution on human well-being and ecosystem sustainability. Traditional air quality monitoring approaches often involve a large number of measurement stations distributed across vast areas, which can be expensive, resource-intensive, and inefficient. The current work aims to revolutionize pollution monitoring and there after the air quality prediction through advanced machine learning techniques. The primary objective is to find the optimum number of pollution measurement stations using the multi-layer perceptron (MLP) classifier.

---

[a]agnish42@gmail.com, [b]sukumar.c@technoindiaeducation.com, [c]jayanta.p@technoindiaeducation.com

With the help of minimum number of measurement stations, maintain the prediction accuracy is a prime objective of this work. This is motivated by the need for resource optimization, cost-effectiveness, and improved air quality management. By leveraging the capabilities of the MLP classifier, this work seeks to achieve more accurate and efficient air quality predictions, leading to informed decision-making and timely interventions. To achieve these objectives, this work relies on a comprehensive dataset collected from various pollution measurement stations over a defined period of time. The data collection process is thoroughly documented, including any preprocessing steps taken to ensure the data quality and consistency.

The collected dataset is analyzed to gain insights into pollution levels, trends, and patterns. This data analysis serves as a crucial step in preparing the dataset for training the MLP classifier, ensuring that the model can effectively learn from the information present in the data.

Problem Statement

Air pollution poses a significant threat to human health and the environment, demanding effective monitoring and control measures. Traditional air quality monitoring approaches often involve an extensive network of measurement stations deployed across wide geographic regions to capture pollution data. However, this approach can be resource-intensive, expensive, and may not always provide accurate representations of air quality in real-time.

The problem addressed in this work is to optimize pollution monitoring efforts by determining the optimum number of pollution measurement stations required to achieve accurate and reliable air quality prediction without compromising the precision of air quality assessments. In order to satisfy these the focuses are on resource efficiency, cost-effectiveness, and overall environmental sustainability.

The cost of running and maintaining pollution data stations can be significant, and it is one of the key challenges faced in environmental monitoring efforts. Several factors contribute to the high costs:

1. Infrastructure: Setting up and maintaining pollution data stations require substantial infrastructure, including air quality monitoring equipment, sensors, data loggers, communication systems, and power supply. The initial investment and ongoing maintenance costs for this infrastructure can be substantial.
2. Calibration and quality control: Regular calibration and quality control of the monitoring equipment are essential to ensure accurate and reliable data. These processes involve additional expenses for specialized expertise, equipment, and regular checks.
3. Site selection and rent: Identifying appropriate locations for data stations that represent diverse pollution scenarios and securing permissions or renting space for installing the stations can add to the overall cost.
4. Data transmission and storage: Data collected by the stations needs to be transmitted to centralized databases for analysis. This often involves using dedicated communication networks or satellite links, incurring additional operational expenses.
5. Manpower and training: Skilled personnel are required to operate and maintain the data stations. Training and retaining qualified staff can be costly, especially in remote or less developed regions.

6. Data analysis and interpretation: Analyzing and interpreting the vast amount of data collected by the stations require specialized expertise and computational resources, contributing to operational costs.
7. Upgrades and technology refresh: As technology advances, data stations may require regular upgrades or replacements to stay current and maintain data accuracy.

To address this problem, the project leverages the power of the MLP classifier, a deep learning algorithm capable of learning complex patterns and relationships from data. The MLP will be trained on a comprehensive dataset collected from multiple pollution measurement stations, encompassing a diverse range of atmospheric conditions and pollutant concentrations.

The outcome of this work will lead to significant advancements in pollution monitoring and air quality prediction. The optimized deployment of pollution measurement stations will enhance the efficiency of environmental monitoring programs, allowing for better resource allocation and cost savings. Moreover, the project's contributions will extend to informed policy-making, targeted pollution control measures, and sustainable urban planning, ultimately benefiting public health and the overall well-being of communities.

Related Research

There has been numerous works going on globally on various aspects of pollution data analysis. Choudhary et. al. [6] suggested how to find out the optimal number of stations in an urban area depending on various parameters such as the coverage area, the measurable pollutants and their variance in the concentration and most importantly the population density in that area. According to their research for every 100 million population a minimum of four stations are required for measured of suspended particles. Our subject area being Kolkata with a population of approximately 1.5 crores, according to their empirical formula, it will require a minimum of 30 stations to monitor the PM 2.5 concentration, properly. Drewil et. al. [5] in their paper suggested an LSTM based approach for predicting pollution values, where for selecting the hyperparameters they have used a genetic algorithm-based approach. Hoffman and Jasinski [3] suggested a MLP based approach to predict the PM2.5 level using the PM10 concentration values. Maleki et. al. [4] suggested an ANN based approach for prediction of various pollutants on an hourly basis. Mahalingam et. al. [1] suggested two approaches based on neural networks and SVM for prediction of AQI. Singh and Goel, [2], suggested a regression based approach for AQI prediction.

The Dataset

The data for this project was collected from the government website https://www.wbpcb. gov.in/. This website is a reliable source of pollution data provided by the West Bengal Pollution Control Board (WBPCB). The WBPCB is a government regulatory body responsible for monitoring and controlling pollution in the state of West Bengal, India.

Dataset Description

The dataset encompasses pollution data collected from a total of 149 stations strategically positioned across West Bengal. Each station corresponds to a specific location where air quality measurements are routinely conducted. This comprehensive dataset offers an extensive repository of air quality information, covering a significant geographic area.

Spanning from January 1, 2023, to May 23, 2023, the dataset provides a rich temporal perspective with hourly measurements. This level of temporal granularity allows for a detailed exploration of air quality fluctuations throughout the day, as well as seasonal variations. The dataset's fine temporal resolution is pivotal for uncovering intricate patterns and trends within the air quality data.

The Dataset Parameters: Air Pollution Attributes

AQI (Air Quality Index):
The Air Quality Index column represents the overall assessment of air quality at each station. AQI provides a standardized measure that allows comparison of air quality levels across different locations.

PM 2.5 AVG ($\mu g/m^3$):
The PM 2.5 AVG column indicates the average concentration of fine particulate matter (PM 2.5) in micrograms per cubic meter. PM 2.5 particles are small particles suspended in the air that can penetrate deep into the respiratory system and pose health risks.

PM 10 AVG ($\mu g/m^3$):
The PM 10 AVG column represents the average concentration of inhalable coarse particulate matter (PM 10) in micrograms per cubic meter. PM 10 particles are larger particles that can contribute to respiratory issues and reduced visibility.

REL HUMI (%):
The REL HUMI column provides the relative humidity data recorded at each station. Relative humidity is a measure of the amount of moisture present in the air and is expressed as a percentage.

Temperature (°C):
The temperature column denotes the temperature recorded at each station in degrees Celsius. Temperature can affect air quality patterns and influence the behavior of pollutants in the atmosphere.

The dataset presented in Table 24.1 provides information on air quality conditions for this specific time range, enabling the analysis of seasonal variations and trends.

Data Processing

We took on the responsibility of preprocessing the collected data to ensure its quality and consistency. This involved handling missing values, removing outliers, and normalizing or scaling the data to facilitate better model training.

Model development

Here the MLP classifier has developed and fine-tuning its hyperparameters to achieve the optimal performance. This included experimenting with different network architectures

Table 24.1: The sample air pollution data, provided by WBPCB.

Location	District	Date	Ho	Latitude	Longitude	AQI	PM 2.5 AVG ($\mu g/m^3$)	PM 10 AVG ($\mu g/m^3$)	REL HUMI (%)	Temperature (°C)
Ballygunge Campus; C.U	Kolkata	2023-03-01	0	22.5267010	88.3632100	105	56.70	107.76	76.20	27.27
Bidhannagar East Police Station	North 24 Pgs.	2023-03-01	0	22.5874270	88.4335500	200	90.02	166.83	77.12	25.65
Dhapa Lock Pumping Station	Kolkata	2023-03-01	0	22.5580270	88.4099800	168	80.44	155.36	79.00	25.77
East Calcutta Girls College, Lake Town	Kolkata	2023-03-01	0	22.6015830	88.4045560	112	61.28	118.32	78.90	25.84
Flora Fountain	Kolkata	202303-01	0	22.5486280	88.3822560	79	47.26	56.52	83.08	26.67
Karunamoye Bus Stand	North 24 Pgs.	2023-03-01	0	22.5870290	88.4207400	234	100.31	184.45	80.67	25.50
Lady Brabourne College	Kolkata	2023-03-01	0	22.5451541	88.3673286	105	56.08	106.89	77.91	26.86
Leather Complex	Kolkata	202303-01	0	22.4953000	88.5092930	113	63.95	91.89	85.17	24.65
Madhyamgram Municipality	Kolkata	2023-03-01	0	22.6929110	88.4652300	157	77.10	146.67	60.51	25.73
New Tom Kolkata Development Authority (A-l)- HI DCO	North 24 Pgs.	2023-03-01	0	22.6107500	88.4716950	119	64.51	128.61	79.32	23.21
Office of The Assistant Com missioner of Police, Belghoria	North 24 Pgs.	2023-03-01	0	22.6909750	88.3755720	163	78.97	191.57	68.43	24.17
Office of The Commissioner of Police, Barrackpore	North 24 Pgs.	2023-03-01	0	22.7598300	88.3667300	73	4.37	14.63	72.88	22.48
Old Calcutta Road Checkpost	North 24 Pgs.	2023 03-01	0	22.7450030	88.3823900	172	81.57	182.64	68.94	24.12

Location	District	Date	Ho	Latitude	Longitude	AQI	PM 2.5 AVG (µg/m³)	PM 10 AVG (µg/m³)	REL HUMI (%)	Temperature (°C)
Poura Prashikhan Kendra, Barrackpore	North 24 Pgs.	2023-03-01	0	22.7569790	88.3754400	135	70.44	136.13	57.71	26.09
Presidency University	Kolkata	2023-03-01	0	22.5763855	88.3610445	175	82.47	149.21	81.23	26.14
Rajarhat Police Station	North 24 Pgs.	2023 03-01	0	22.6318040	88.4850100	121	64.8	131.86	68.12	25.83
Sarojini Naidu College for Women	Kolkata	2023-03-01	0	22.6265050	88.4183300	103	56.33	104.81	75.30	25.53
Sarsuna College	Kolkata	2023-03-01	0	22.4812700	88.2845540	41	2.38	3.07	88.92	24.37
Sector V Industrial Township Authority, Nabadiganta Bhavan	North 24 Pgs.	2023 03-01	0	22.5760990	88.4342960	196	88.67	167.69	78.46	25.98
Solaris Ideal Riverview	Howrah	2023-03-01	0	22.5544145	88.3096577	83	49.58	78.87	90.53	24.30
Taki Municipality	North 24 Pgs.	2023-03-01	0	22.5819870	88.9349900	105	61.58	107.99	84.69	22.97
Urban Health Center	Howrah	2023 03-01	0	22.6068209	88.3246857	70	22.09	52.41	80.35	25.82
Urbana Housing Complex	Kolkata	2023-03-01	0	22.5113850	88.4123840	31	8.55	20.61	79.84	25.38
VIVEKANANDA COLLEGE FOR WOMEN, BARISHA	Kolkata	2023-03-01	0	22.4885267	88.3141944	50	24.41	41.30	86.72	25.29
WBPCB Office, Saltlake	North 24 Pgs.	2023 03-01	0	22.5622120	88.4093550	152	75.59	146.91	75.91	26.10

and activation functions to arrive at the best configuration for this work. The details of the MLP classifier are describe as follows:

1. Input layer: The input layer represents the features of the data, including latitude, longitude, date time, and the. A total of 13 input nodes, each corresponds to these attributes.
2. Hidden layers: The hidden layers, consisting of three layers with 100 neurons, respectively, are intermediate layers between the input and output layers. Each neuron in the hidden layers performs a weighted sum of the input values and applies an activation function, such as the ReLU function, to introduce non-linearity.
3. Activation function: The ReLU (rectified linear unit) activation function is typically used in the hidden layers. It sets negative values to zero and keeps positive values unchanged, allowing the model to learn complex patterns in the data.
4. Output layer: The output layer predicts the target variable, which is the AQI. In this case, the output layer has a single neuron, as we are predicting a continuous value (AQI) for regression.
5. Training: The model undergoes training using the backpropagation algorithm to adjust the weights and biases of the neurons. During training, the model learns to minimize the prediction errors by comparing the predicted AQI values with the actual AQI values from the training data.
6. Prediction: Once the MLP model is trained, it can be used to predict the AQI values for new input data, which includes latitude, longitude, and date time, enabling real-time air quality prediction.

The flowchart, shown in Figure 24.2 offers a bird's-eye view of the sequential stages involved in air quality prediction model development. The accuracy and reliability of the predictions directly impact environmental decision-making and public health, making this process vital in our quest for cleaner, safer air.

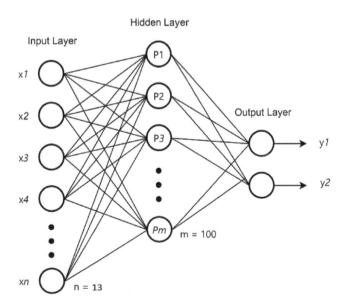

Figure 24.1 Multi-layer perceptron architecture.

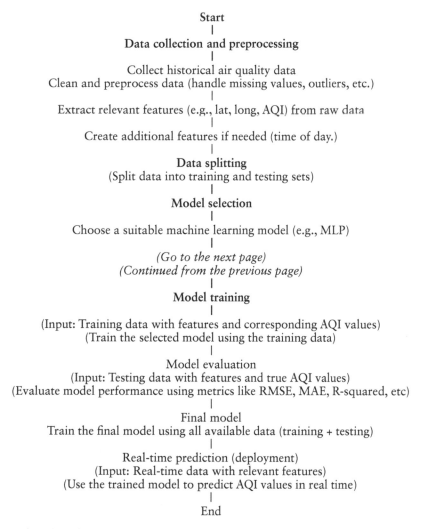

Start
|
Data collection and preprocessing
|
Collect historical air quality data
Clean and preprocess data (handle missing values, outliers, etc.)
|
Extract relevant features (e.g., lat, long, AQI) from raw data
|
Create additional features if needed (time of day.)
|
Data splitting
(Split data into training and testing sets)
|
Model selection
|
Choose a suitable machine learning model (e.g., MLP)
|
(Go to the next page)
(Continued from the previous page)
|
Model training
|
(Input: Training data with features and corresponding AQI values)
(Train the selected model using the training data)
|
Model evaluation
(Input: Testing data with features and true AQI values)
(Evaluate model performance using metrics like RMSE, MAE, R-squared, etc)
|
Final model
Train the final model using all available data (training + testing)
|
Real-time prediction (deployment)
(Input: Real-time data with relevant features)
(Use the trained model to predict AQI values in real time)
|
End

Figure 24.2 Flowchart for air quality model development.

Results

The MLP classifier employed in this project has proven to be a powerful tool for capturing and understanding the complex relationships between various input parameters and pollution levels. By training the classifier with a comprehensive dataset consisting of AQI, PM 2.5 AVG, PM 10 AVG, REL HUMI, and TEMPERATURE data, the model has effectively learned the intricate patterns and correlations within the pollution data.

The accurate predictions generated by the MLP classifier hold great potential for enhancing pollution monitoring, control strategies, and decision-making processes. With the ability to accurately estimate pollution levels based on the input parameters, stakeholders can make informed decisions regarding pollution mitigation measures, resource allocation, and environmental policies.

Table 24.2: Station wise actual and predicted data.

Actual	Predicted	Date	Hour	Stations
283	283	01-01-2023	0	Taki Municipality
277	277	01-01-2023	1	Taki Municipality
253	236	01-01-2023	2	Taki Municipality
236	236	01-01-2023	3	Taki Municipality
391	391	01-01-2023	0	WBPCB Office, Saltlake
404	404	01-01-2023	1	WBPCB Office, Saltlake
438	438	01-01-2023	2	WBPCB Office, Saltlake
426	426	01-01-2023	3	WBPCB Office, Saltlake
453	453	01-01-2023	4	WBPCB Office, Saltlake
437	437	01-01-2023	5	WBPCB Office, Saltlake
383	383	01-01-2023	6	WBPCB Office, Saltlake

In the pursuit of determining the optimal number of pollution measurement stations, a rigorous and systematic analysis was conducted as part of this project. This analysis involved a judicious selection of stations for training and testing the multilayer perceptron (MLP) classifier, a machine learning model used for predicting the AQI. Thirteen stations were strategically chosen for training the model, while an additional two stations were reserved for testing purposes.

The primary objective was to assess the MLP model's performance in predicting the AQI for these specific testing locations. The results of this evaluation were nothing short of remarkable. The model demonstrated a remarkable level of accuracy in its predictions, even when applied to previously unseen stations. This level of accuracy underscores the model's ability to generalize well beyond its training data and make reliable predictions for different geographic locations.

These findings carry profound implications for the field of air quality monitoring and management. The project's results strongly suggest that it is indeed possible to achieve accurate AQI predictions with a reduced number of monitoring stations. This revelation is significant because it opens the door to more streamlined and cost-effective approaches to station distribution and resource allocation.By strategically selecting a smaller number of stations based on considerations such as geographical coverage, pollutant concentration patterns, and spatial distribution, resources can be optimized. This optimization can lead to increased operational efficiency and cost savings in air quality monitoring initiatives. Moreover, it highlights the potential to make monitoring efforts more efficient without compromising the accuracy and reliability of AQI predictions.

Furthermore, we subjected our model to a rigorous evaluation using various metrics, including root mean square error (RMSE), mean absolute error (MAE), and R-squared (R^2) score. These metrics collectively affirmed the model's high accuracy and robustness in predicting AQI values, thereby underlining its reliability for real-world applications. Having successfully fine-tuned the model, we proceeded to train the final version using the entire dataset, combining both the training and testing data. This holistic approach ensures that the model captures the intricate relationships within the air quality dataset.

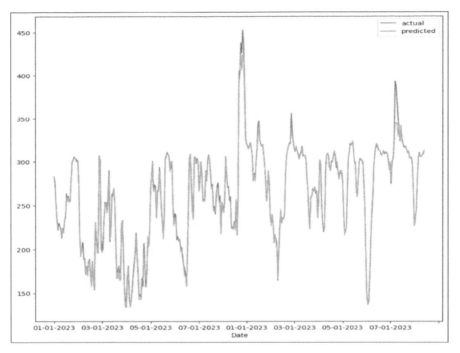

Figure 24.3 Average error estimation for pollution AQI index.

Discussion: The Limitations and Challenges

Despite the positive outcomes, it is essential to acknowledge the limitations and challenges encountered in this project. One limitation is the exclusion of additional factors that may impact pollution levels, such as meteorological conditions, industrial activities, and seasonal variations. Integrating these factors could provide a more comprehensive understanding of the pollution patterns and further enhance the accuracy of the predictions.

Another challenge is the availability and quality of the pollution data. The reliance on data obtained from a specific government website may introduce biases or limitations due to data collection methods or measurement errors. Incorporating data from multiple sources and ensuring data quality could improve the robustness of the analysis and prediction models.

Conclusion and Future Work

In this project, we developed a methodology for finding the optimum number of pollution measurement stations using an MLP classifier. Through the analysis and experimentation, we have obtained several key findings and insights. Firstly, the MLP classifier demonstrated its effectiveness in accurately predicting pollution levels and the air quality index (AQI). The model was trained using data from 13 stations and achieved a high accuracy rate in predicting the AQI for two test stations. This indicates that a reduced number of stations can still provide accurate and reliable pollution predictions, leading to cost savings and improved efficiency in air quality monitoring.

The Future scope of the project involves expanding the implementation to cover a larger number of data points and creating a lowest AQI route across India. Here key aspects to consider in more detail to make the solution more extensive and comprehensive are as follows:

- The possibility of creating a spatio-temporal pollution profile of the city. This will enable us to find out the pollution values of any point in the city, in real time, using its latitude and longitude values.
- Once this is done, then taking this as input, we can extend our work further, by finding out a "Green Route" (the least pollutant route) from one location to the other, at any point of time.

References

1. Choudhary, S. and Kaur, H. (2022). Examining the locational approach towards optimal siting of air quality monitoring stations in India. *Research Square*. https://doi.org/10.21203/rs.3.rs-2079414/v1
2. Drewil, G. I. and Al-Bahadili, R. J. (2022). Air pollution prediction using LSTM deep learning and metaheuristics algorithms. *Journal of Measurement, Sensor,* 24, 100546.
3. Hoffman, S. and Jasiński, R. (2023). The use of multilayer perceptrons to model pm concentrations at air monitoring stations in Poland. *Atmosphere*, 14(1), 96. https://doi.org/10.3390/atmos14010096
4. Mahalingam, U. and Elangovan, K. (2019). A Machine learning model for air quality prediction for smart cities. *2019 International Conference on Wireless Communications Signal Processing and Networking (WiSPNET)* 21–23.
5. Maleki, H. and Sorooshian, A. (2019). Air pollution prediction by using an artificial neural network model. *Clean Technologies and Environmental Policy, SpringerLink*, 1341–1352.
6. Singh, J. K. and Goel, A. K. (2021). Prediction of air pollution by using machine learning algorithm. *7th International Conference on Advanced Computing and Communication Systems (ICACCS 2021)*.

Part 3

Artificial Intelligence & Industrial Applications

25 Energy Efficient Task Execution Through Edge Federation Utilizing Simulated Annealing

Arpita Ray[a], Raksha Prasad[b], Chandreyee Chowdhury[c], and Sarbani Roy[d]

Jadavpur University, Kolkata, India

Abstract

Edge computing is the newest computing paradigm which is distributed in nature, and assists in transferring the computational load of mobile devices to the edge nodes for employing the resources that are under-utilized at the edge nodes. With the emergence of smart technologies and IoT devices, the performance of edge nodes becomes very crucial in minimizing task completion time which is computation intensive and time-sensitive in nature. Due to the limitations of edges, a framework called Edge Federation is proposed, which supervises multiple edge servers. Allocation of tasks to fittest edge servers is an NP-complete problem and is implemented by utilization of an optimizing algorithm, Simulated Annealing (SA). This framework increases the chances of maintaining the quality of service (QoS) requirements by allocating the offloaded tasks, to the edge servers which are fittest, rather than dispatching the tasks over to the cloud. The algorithm is simulated in the PureEdgeSim simulator. According to the simulation outcomes, SA performs better than Fuzzy Logic (FL) or Knapsack (KS) in energy utilization taking minimum task completion time. It is found that the maximum rate of task failure in SA is 10% which is better than the other approaches.

Keywords: Edge federation, federated edge, optimization, simulated annealing.

Introduction

In recent years, ubiquitous computing by using mobile devices have played a crucial role by sensible data collection, analyzing and sharing the sensed information [1]. The Information Technology (IT) sector has progressed at such a monumental rate, it is forecast that the number of devices using Internet of Things (IoT) will jump above and beyond 29 billion by 2030 from mere 9.7 billion in 2020. Moreover, with the generation of vast quantity of data from these IoT devices, which are not only critical but also computation intensive, these data needs to be analyzed for obtaining some significant information. Mobile subscribers currently demand large storage capacity, rapid data exchange, extended battery life, and quick processing capabilities, all of which are challenging to provide with pocket-sized technology. However, the device's limited power and computation capabilities further

[a]arpitaray81@gmail.com, [b]rprasad1480@gmail.com, [c]chandreyee.chowdhury@jadavpuruniversity. in, [d]sarbani.roy@jadavpuruniversity.in

restricts time sensitive and resource limited applications to run on mobile devices [2]. Time sensitive and critical tasks can be offloaded to alternative resourceful devices so that these limitations can be solved and take less time for processing and satisfy the latency requirements. An Edge federation framework has evolved subsequently due to advancement of 5G mobile connectivity and the Internet of Things (IoT), allowing latency-sensitive applications like augmented reality (AR) and autonomous cars to run smoothly. Edge computing [3] is a distributed computational paradigm that has been responsible to broaden the cloud services to the edges where the edges form the logical layer positioned as close to the data sources as possible, and it forms the inter-medial layer joining the edge servers to the cloud servers. There are still few limitations of edge computing (when we consider single standalone edges geographically distributed servicing a number of users) with respect to capacity of the resources and the computational power, when we compare it with cloud computing. In addition to these issues, these individual standalone edge nodes do not share information with the neighboring edge nodes located in nearby areas since the edge nodes are not aware of other edge nodes due to the fact that these nodes are geographically distributed. This results in under-utilisation of the resources. So, an efficient optimization of resource utilization at less time is very difficult to attain. Therefore, the QoS deteriorates due to the lengthy execution and waiting time and interruptions of services.

In, the authors utilized an approach based on fuzzy logic to handle the workload orchestration [8]. Workload orchestration can be described as a process of deciding where and how a task should be computed (edges/ cloud) in an n-tier architecture. The authors proposed to capture the instinct of real-world administrator for automating the management system which performs work orchestration. The authors have not considered task migration between the edges and clouds. In, the authors presented a collaborative scheme of task offloading built on Fuzzy logic in MEC which is distributed in small concentrated networks. The authors have not considered the mobility of the users which restricts the usability of mobile devices. In, the authors proposed a concept of edge infrastructure providers (EIPs) model and termed it as edge federation, where edge and cloud are both part of the federation so that the latency critical tasks can be serviced by resource co-operation. The authors have characterized this resource sharing process as Linear programming problem and transformed it into solvable form by a dynamic algorithm. The authors have considered few fixed length time slots, and have not considered dynamic prediction of the length of time slots.

In, the authors have considered a 3-tier architecture and distinctly differentiating the cloud-fog-edge layers and fog layer is considered to be more resource rich than edge layer, for Vehicular Ad hoc Networks (VANET). A resource management approach was proposed utilizing integrated fuzzy logic that determined the fittest vehicle for task offloading. The authors in explored the issue of tasks scheduling in vehicular edge computing (VEC) and these tasks could be sent off to the road side units (RSU). The aim was to diminish the average time for completing a task. Here, the authors have not considered the communication delay of the tasks and the simulation have been done on small data set. There has been a lot of research initiatives in multi clouds or edge clouds, for scheduling services to decrease energy usage but very few researches have been done on Federated Edges or Edge servers where the issues like QoS and latency has been taken into account for enhancing the performance of the edge nodes. This has motivated us to form an Edge federation framework a novel service-provisioning paradigm for next-generation edge computing networks. The primary objective in our work is to offload tasks that are computationally intensive to the edge servers which are closest to the mobile devices in a more cost-effective manner than the other techniques. SA is being implemented for optimizing the problem of allocation of tasks by the Edge federation to other fittest edge servers, when the closest edge servers to the mobile device fails to complete the offloaded task.

The contribution of our work is proposing an innovative Edge federation framework that provides a transparent and seamless service to the end user, by diminishing the latency and optimizing the energy utilization. This is done by allocating the task to the fittest edge server, based on the principle of SA, in case the nearest edge server fails to provide service to the end user. The proposed work is simulated in PureEdgeSim and the results are compared with state-of-the-art approaches.

The rest of the paper is structured as follows: In Section 2 detailed architecture is described. Section 3 discusses the experimental setup of the work for PureEdgeSim simulator and results are discussed along with comparisons made with two other existing optimization i.e. FL and KS algorithms and finally Section 4 concludes the paper along with future work directions.

Energy Efficient Task Execution Through Edge Federation

Problem Definition

The standalone edge servers distributed geographically over an area cannot use the resources in an optimized way. Hence, a solution is needed to aggregate these standalone edge servers to utilize these edge nodes efficiently while guaranteeing low latency and reducing the energy utilization. To explain the Edge federation framework, a real life scenario is examined where a building in university campus is considered as the total available space for wireless communication, and each floor consists of classrooms and laboratories and has e number of edge servers. It is evident that most of the people in the university building carry smart mobile devices which are equipped with a lot of sensors. The computational intensive tasks/data from these IoT devices are offloaded to its closest edge server. Among the various edge servers, servers with high processing power and resources are grouped in the Edge federation within the building or among various buildings within the university. If the nearest edge server is not capable of performing the task at that given time, then that task is offloaded to the Edge federation. The Edge federation generally takes the decision of sending the tasks to the fittest edge servers, which has the capability of completing that tasks. In case the Edge federation is unable to get the tasks done only then the tasks are sent off to the cloud server for completing the task.

Architecture of Edge Federation

In Edge federation framework, a four-tier architecture is proposed. The Edge federation is an extra layer between the edge servers and the cloud servers. In Figure 25.1, it is seen that the end users are in the lowest layers that have various IoT devices for different applications. Each of the IoT devices, searches for its nearest edge server for offloading the task. A number of edge servers near vicinity forms a group and remain connected to its nearest Edge Federation. This Edge Federation is formed by group of edge servers (in our example it is ES1, ES2, ES3). Edge federation have all the information about all other edge servers, of the whole building. The Edge federation comprises of collection of k edge servers out of e edge servers i.e, Edge federation will be a collection of k numbers of edge servers present in a geographic location, where $k <= e$.

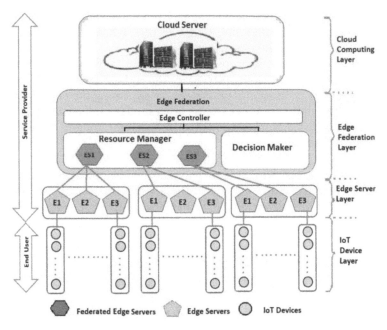

Figure 25.1 Architecture of the proposed edge federation system.

The main functional unit of Edge federation is the edge controller (similar to the concept of edge orchestrator). It has two components namely: i) Resource Manager is in charge of taking care of the various physical resources of the edge servers that are connected to the Edge federation. It keeps a record of the load at each edge servers connected to it along with the information of the tasks that are being processed in the edge servers, ii) Decision maker is responsible for taking decision on behalf of the Edge federation. It dispenses the tasks to the fittest edge servers, depending on the load and the computation power on those edge servers. It also keeps an account of the tasks that are being executed at the edge servers. When there is any failure in task execution, the Edge federation takes the decision of either allotting the tasks to another edge server or to the cloud server. Allocating the task from the Edge federation to the appropriate edge server is performed by implementing the optimization algorithm–SA.

Task Allocation by Edge Federation Implementing Simulated Annealing

Edge federation accomplishes the tasks allocation by implementing a heuristic method. Simulated Annealing for allocation of the tasks to the fittest edge servers so that the task execution time is reduced significantly. The tasks allocation done by the resource manager in Edge federation is an NP-complete problem. In the following segment, first, the algorithm is described and then the mapping of the algorithm is explained i.e., how the algorithm is utilized for solving the tasks allocation problem in Edge Federation.

Globally lowest energy point in an energy distribution environment, can be obtained by the SA optimization algorithm. The algorithm was developed when scientists noticed the nature in which a cooled molten metal gradually may form a regular crystalline structure. So, a metal is heated up till it reaches an annealing temperature, after which it is cooled slowly such that the metal can be changed according to its desired structure thereby minimizing the potential energy of the mass. The algorithm introduces random changes/jumps to explore a possible new and better solution space and avoid local optima. As the algorithm proceeds, these random changes are done in a more controlled and decreasing manner.

The parameters of SA—temperature and energy have been mapped to the task allocation problem. Energy minimization is the objective function and the temperature is used for efficient searching of the solution space avoiding local minima. For an ascending move in the system, temperature (T) is the acceptance probability and the range to which the temperature rises is denoted by delta (Δ). The system either shifts to a reduced energy state with the new solution which results in better fitness. Otherwise, it shifts to the new solution with a probability $P = exp(-\Delta E/kT)$, to a higher energy state, where ΔE is the increase in energy.

Algorithm 1 explains the task allocation implementing SA. Initial solution S_s is randomly selected at the beginning of the algorithm and accordingly, the algorithm calculates the energy/cost $S_s.Energy$ for the present solution S_s. An initial temperature T is set, and then neighbouring solution S_{next} and corresponding energy/cost $S_{next}.Energy$ is computed in reference to the present solution S_s. If the newly calculated energy/cost of any neighbouring solution is less than the existing present cost S_s.Energy, then the new solution S_{next} is accepted, where $S_{next}.Energy$ is the energy/cost of the newly obtained solution. Else, the neighbouring solution is also adopted as the new energy/cost solution, depending on the value of Δ, where $\Delta = S_s.Energy - S_{next}.Energy$ and $exp(-\Delta/T)$ is the probability function used as the deciding factor to accept the new energy/cost solution. On arriving to the thermal equilibrium at temperature (T) for the system, this temperature is reduced by a cooling factor α, and the inner iteration loop is increased by an increment factor β, and the algorithm carries on till it reaches another new thermal equilibrium point with the new diminished temperature. The algorithm stops if it has reached T_{min}, minimum temperature, or the maximum number of iterations. This is stated as the *stopping condition*.

When the algorithm is implemented for finding the optimized solution to a given problem, a number of factors need to be looked into namely, the selection of the energy function, the cooling schedule, the structure of the neighborhood, and annealing parameters. All these parameters are critical as they are deciding factors for the standard of the solution obtained and the speed at which the solution is obtained. a) Energy function: The whole purpose of the algorithm is to reduce the energy that is being used in finding the solution. The energy function is calculated as follows:

$$S_s.EnergyCost = \sum_{i=1}^{|S|} \frac{S_i.power \times S_i.Tasklength}{S_i.MIPS}$$

The energy is calculated in watt-hour and $S_s.EnergyCost$ is the energy that is utilised by the edge servers where the tasks are initially allocated, $S_i.Tasklength$ indicates the tasks allotted to S_i.

b) Neighborhood structure: The set of moves that are implemented to the present solution to obtain a new solution i.e., to shift from one equilibrium point or solution (S_s) to another (S_{next}) in the nearby solution space, is defined as neighbourhood. In this work, the neighbourhood solution is obtained by first selecting a random solution S_s and allocating the task to an edge server, and then based on Δ, where $\Delta = S_s.Energy - S_{next}.Energy$, the next equilibrium point S_{next} (another task allocation schedule) is obtained. c) Cooling schedule

---

Algorithm 1 Task Allocation Implementing Simulated Annealing

Input: Tasks (Kbps, in million instructions, latency), edge servers profile $S.Profile$ (MIPS, max energy $S_s.EC$)(where S is the set of edge servers in Edge federation)

Output: Solution servers S_s (Edge servers to which the tasks will be offload)

Initial temp T, $\alpha = 0.9$ (considered for implementation)($temp\_reducing\_factor$ $\alpha < 1$); $\beta = 1.05$ (considered for implementation)($chain\_increasing\_factor$ $\beta > 1$);

Select: Random initial solution S_s from S (set of edge server), random initial chain variable n_{rep};

$S_s.Energy - Energy\_Cost\ (S_s.Profile, Task)$;

Repeat

for $v = 0$; $v <= n_{rep}$; $v = v + 1$ **do**

     $S_{next} \leftarrow S_s$;   ▷ Neighbouring solution selection

     $S_{next}.Energy - Energy\_Cost(S_{next}.Profile, Task)$ ▷ Calculate the energy of the neighbouring selected server

     $\Delta - S_{next}.Energy - S_s.Energy$ ▷ difference of energy of current server and neighbouring server

     if $\Delta < 0$ **then**

         $S_s - S_{next}$;   ▷ neighbouring server selected

         $S_s.Energy = S_{next}.Energy$;

     else

         $x - rand(0,1)$ ▷ a random number is selected

         if $e^{-\Delta/T} > x$ **then**

             $S_s - S_{next}$;   ▷ neighbouring server selected

             $S_s.Energy = S_{next}.Energy$;

         end

     end

end

Set $T = \alpha \times T$

$n_{rep} = \beta \times n_{rep}$

Until Stopping condition

Return Solution set of servers S_a (Edge servers to which the tasks will offload) $S_s.Energy$ (the energy consumed)

The feature that can be changed and controlled in SA is the temperature T. The strategy that is used to decrease the temperature of the process till the thermal equilibrium state is reached is defined as the cooling schedule. In this work, a cooling schedule which is geometric in nature is taken into consideration where $T = \alpha \times T$, where $\alpha (<1)$ is a constant known as the cooling rate. This cooling schedule is controlled by the number of iterations in the inner loop n_{rep} and the cooling rate α. The number of iterations in the inner loop at every temperature, , is modified according to $n_{rep} = \beta \times n_{rep}$, where β is a constant greater than 1. d) Annealing parameters:

1. Cooling (reducing) rate α: It is defined as the rate at which the temperature (time to complete the task) is decreased. It is found in and that if the cooling rate α is between 0.9 and 0.90 then there are higher chances of success. Here, the rate of cooling is selected to be $\alpha = 0.90$.

2. Increasing rate β: As the temperature T is reduced, the rate at which the number of iterations in the inner loop increases, is defined as the increasing factor β. In this work, the value is chosen to be $\beta = 1.05$. It is decided empirically.

3. Initial temperature T: This is the starting temperature of the system and it takes the decision of the probability to accept a worse solution. Initially, the time is set to a fairly high value to capture a large horizon of the solution space, then slowly reducing this value to obtain the optimal solution.

Experimental Setup and Results

A prototype of the proposed work is simulated using the PureEdgeSim simulator. The system uses Windows 10 as a platform. PureEdgeSim is an object-oriented framework built on CloudSim Plus. It is the framework within which the device-edge-cloud architectures could be effectively simulated. SA is implemented here for task allocation. In the proposed work, three types of tasks have been considered for offloading namely, a) augmented reality, b) healthcare, and c) high computational task. It is assumed that the augmented reality task will need the least and the high computational task will need the most computational requirements. The healthcare system will need average computational requirements. The algorithm is simulated for a large number of tasks generated randomly. Here, 2400 tasks are considered. These tasks have varying levels of computational requirements of the servers, having medium to high computational capabilities, and are allocated within the distributed environment. Five active edge servers are taken into consideration where they are placed at different locations within an area of 100 sq.m. The edge servers are placed in such a way that they are at least 10m apart from each other. After the simulation, all the outputs are saved as a text file (.txt), excel (.csv) file, and the graphs in PNG format. The proposed algorithm is simulated with a sizeable number of tasks that are generated randomly and are allocated in a distributed system. In Figure 25.2 (a), the plotted graph shows the rate at which the system attains an equilibrium state by slowly reducing the temperature. As the algorithm progresses, a steady state is obtained when the number of iterations is more than 20. The cooling factor is kept between 0.90 and 0.95. An experiment is conducted to compare the task allocation success rate of the proposed algorithm with the knapsack algorithm as shown in Figure 25.2(b). Tasks that cannot be executed by the edge server due to issues of mobility are considered failed tasks denoted as 0% success and tasks which are successfully executed after offloading are considered as success, denoted as 100% success. Five edge servers are taken into consideration with tasks. Task 5 and Task 14 failed to be allocated due to a time delay in executing the tasks in the allocated edge servers and the rest tasks are successfully assigned and executed when SA is implemented, but task number 6, 7, 8, 10, 12, 13, 15, 18 and 20 failed to successfully complete when implemented by KS. From the graph, it can be concluded that 18 out of 20 tasks are successfully executed by the edge servers implementing SA while 11 tasks are successfully executed by edge servers with KS. Hence, accuracy of the success rate by the proposed work is 90%. When the standard FL was used instead of SA, it was found that, the success rate was nearly 43.98% due to high delay and the total task execution time was 1074.25 seconds. A comparative analysis of the proposed SA, FL, and KS has been performed in terms of the task completion time as shown in Figure 25.2(c). Task completion in this context refers to the procedure in which tasks are sent off to the edge servers from the IoT devices, executing the tasks on edge servers, and returning the executed tasks with suitable output to the IoT devices.

The task completion time for allotted tasks, initially is the lowest for the proposed approach implementing SA while FL and KS took more time to complete the allotted tasks. But with time, KS algorithm and SA shows comparable performance. However, task failure rate kept on increasing up to 45% with the KS algorithm.

Figure 25.2(d) presents the resulting energy expenditure of the algorithms. It has been found that the consumption of energy was initially lowest for the proposed work. However, as the number of tasks increased, KS algorithm performed comparably with the proposed

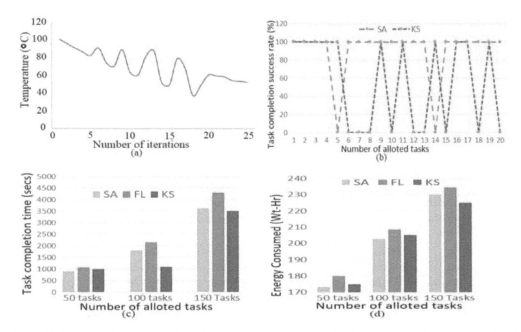

Figure 25.2 (a) Total number of iterations to reach stable state (b) Success rate of allocated tasks of SA & KS (c) Task completion time implementing SA, FL & KS (d) Energy consumed implementing SA, FL & KS.

work though the proposed work resulted in more number of tasks to be allocated and completed.

From the simulation outputs, an inference can be drawn that implementing the SA, the algorithm works most efficiently with 15 to 25 edge servers and 100 to 350 tasks. Therefore, using SA as an optimization algorithm provides better performance with respect to task execution and successful completion of the tasks, utilizing energy in an optimized manner.

Conclusion

In this paper, an energy-efficient task allocation based on SA algorithm is presented, with an integrated service provisioning model, named, Edge federation. A 4-tier architecture comprising edge servers, IoT devices, and edge federation is designed so that energy consumption of the system can be reduced by offloading the tasks to the most suitable edge servers and not offloading the task to a cloud server. An intermediate layer, the Edge federation, takes over the responsibility of allocating the tasks to the fittest edge servers near its vicinity. To swiftly obtain a solution of the task allocation problem, a near-optimal solution is found by implementing an allocation method based on the SA algorithm. When simulated on a sizeable number of randomly generated instances, these instances were used to assess the algorithm's performance. According to the experimentation, the proposed SA-based task offloading by Edge federation results in better performance in contrast to the modern and cutting edge approaches. Finally, this work can be concluded with a next step in the future direction. It has been found that by using constant thermodynamic speed annealing schedules for cooling schedule, better low-energy solutions to problems can be obtained YNou1998. Hence, in the extension of this work, a plan is in progress to implement a thermodynamic speed annealing schedule and perform a comparative analysis between the various cooling schedules of SA.

References

1. Ray, A., Chowdhury, C., Bhattacharya, S., and Roy, S. (2023). A survey of mobile crowdsensing and crowdsourcing strategies for smart mobile device users. *CCF Transactions on Pervasive Computing and Interaction*, 5(1), 98–123.
2. Son, J., Dastjerdi, A. V., Calheiros, R. N., and Buyya, R. (2017). Sla-aware and energy-efficient dynamic overbooking in sdn-based cloud data centers. *IEEE Transactions on Sustainable Computing*, 2(2), 76–89.
3. Shi, W., Cao, J., Zhang, Q., Li, Y., and Xu, L. (2016). Edge computing: vision and challenges. *IEEE Internet Things Journal*, 3(5), 637–646.
4. Kim, S., Park, S., Youngjae, K., Kim, S., and Lee, K. (2017). VNF-EQ: dynamic placement of virtual network functions for energy efficiency and QoS guarantee in NFV. *Journal of ACM, in Cluster Computing*, 20(3), 2107–2117.
5. Nonde, L., El-Gorashi, T. E. H., and Elmirghani, J. M. H. (2015). Energy efficient virtual network embedding for cloud networks. *Journal of Lightwave Technology*, 33(9), 1828–1849. doi: 10.1109/JLT.2014.2380777.
6. Ascigil, O., Phan, T. K., Tasiopoulos, A. G., Sourlas, V., Psaras, I., and Pavlou, G. (2017). On uncoordinated service placement in edge-clouds. In IEEE International Conference on Cloud Computing Technology and Science (CloudCom), (pp. 41–48). doi:10.1109/CloudCom.2017.46.
7. Son, J., and Buyya, R. (2019). Latency-aware virtualized network function provisioning for distributed edge clouds. *Journal of Systems and Software,* 152(C), 24–31. doi.org/10.1016/j.jss.2019.02.030.
8. Sonmez, C., Ozgovde, A., and Ersoy, C. (2019). Fuzzy workload orchestration for edge computing. *IEEE Transactions on Network and Service Management*, 16(2), 769–782.
9. Hossain, M. D., Sultana, T., Nguyen, V., Rahman, W. U., Nguyen, T. D., Huynh, L. N., and Huh, E. N. (2020). Fuzzy based collaborative task offloading scheme in the densely deployed small-cell networks with multi-access edge computing. *Applied Sciences*, 10(9), 3115.
10. Cao, X., Tang, G., Guo, D., Li, Y., and Zhang, W. (2020). Edge federation: Towards an integrated service provisioning model. *IEEE/ACM Transactions on Networking*, 28(3), 1116–1129. doi.org/10.1109/TNET.2020.2979361.
11. Qafzezi, E., Bylykbashi, K., Ampririt, P., Ikeda, M., Matsuo, K., and Barolli, L. (2022). An intelligent approach for cloud-fog-edge computing SDN-VANETs based on fuzzy logic: effect of different parameters on coordination and management of resources. *Sensors,* 22(3), 878.
12. Liu, Y., Wang, S., Zhao, Q., Du, S., Zhou, A., Ma, X., and Yang, F. (2020). Dependency-aware task scheduling in vehicular edge computing. *IEEE Internet of Things Journal,* 7(6), 4961–4971.
13. Deshmukh, S., and Grover, L. (2020). Implementation and analysis of pureedgesim. *International Journal of Engineering Applied Sciences and Technology*, 5(7), 129–133. ISSN No.2455–2143.
14. Santoro, D., Zozin, D., Pizzolli, D., De Pellegrini, F., and Cretti, S. (2017). Foggy: a platform for workload orchestration in a fog computing environment. In IEEE International Conference on Cloud Computing Technology and Science (CloudCom), (pp. 231–234). doi: 10.1109/CloudCom.2017.62.
15. Flores, H., Su, X., Kostakos, V., Ding, A. Y., Nurmi, P., Tarkoma, S., Hui, P., and Li, Y. (2017, March). Large-scale offloading in the Internet of Things. In 2017 IEEE International Conference on Pervasive Computing and Communications Workshops (PerCom Workshops) (pp. 479–484). IEEE.
16. Attiya, G., and Hamam, Y. (2006). Task allocation for maximizing reliability of distributed systems: a simulated annealing approach. *Journal of Parallel and Distributed Computing*, 66(10), 1259–1266.

17. Kirkpatrick, S., Gelatt, C. D., and Vecchi, J. M. P. (1983). Optimization by simulated annealing. *Science*, 220, 671–680.
18. Aarts, E., and Korst, J. (1989). Simulated Annealing and Boltzmann Machines. New York: Wiley.
19. Rutenbar, R. A. (1989). Simulated annealing algorithms: an overview. *IEEE Circuits and Devices Magazine*, 5(1), 19–26.
20. Nourani, Y., and Andresen, B. (1998). A comparison of simulated annealing cooling strategies. *Journal of Physics A: Mathematical and General*, 31(41), 8373.

26 Improving Sentiment Prediction and Opinion Mining using Combined Review Datasets

Kartick Chandra Mondal[1,a], Jayeeta Choudhury[1,b], and Hrishav B. Barua[2,c]

[1]Department of Information Technology, Jadavpur University, Kolkata, India

[2]TCS Research, Tata Consultancy Services, Kolkata, India

Abstract

Sentiment analysis of online opinions is gaining profound attention among the AI and data analytics community across the world. This paper presents an experimental study on sentiment analysis of online reviews of hotels, movies, and airline services. The review data from airline industry about the services people avail, IMDB dataset of movies, Trivago for hotels, and Rotten Tomatoes for movies are considered. In this paper, the classification of the different reviews into respective classes of positive, negative, and neutral, using machine learning algorithms is being studied. A classification engine's architectural backbone and Linear Support Vector Machine (L-SVM) are considered to map the input data into the polarities as mentioned. The results obtained show interesting associations that can certainly help industries to improve their experience and services. The results show that referring to a classification engine's architectural backbone (i.e. forensic identification classification engine) and employing a Naïve Bayes as well as a Linear SVM both give decent accuracy in sentiment analysis. Results from rigorous experiments of a combination of datasets for better insights are presented in the article.

Keywords: Classification engine, data analytics, opinion mining, sentiment analysis.

Introduction

Nowadays, social media and online platforms are flooded with people's opinions and reviews on various services and products that they use or buy. Social media platforms mirror their opinions and thoughts about day-to-day life, services they avail, products they buy, movies they see, etc. People are easily pleased or displeased with the products and services they avail with those opinions. People share opinions of varied sentiments like, positive, negative, and neutral. Every day millions of reviews are generated which amounts to a huge volume of data. This has made the internet an important source of information before availing any services or buying any products.

With the help of sentiment analysis and opinion mining, the organizations can know about the feedback which will help them to enhance their quality of services and products. This will also facilitate the potential customers to know about the advantages and disadvantages of any service or product before availing it. Sentiment analysis basically deals

---

[a]Kartickjgec@gmail.com, [b]jayeeta987@gmail.com, [c]Hrishav.smit5@gmail.com

with the classification of the texts into polarities. The text reflects evaluation, decision, judgment, or statements [1, 2]. Thus, there is a need for effective text mining tools and techniques, to accommodate the huge amount of textual data. Sentiment analysis allows an organization to understand the strengths and weaknesses of the client which can facilitate effective measures to improve their services. Most frequently, people share their reviews based on airline services they avail, movies they see as well as hotels they stay in. The need for sentiment analysis with such diversified datasets is an important research area. It provides an effective way of analyzing the sentiment of the general public.

The techniques have been discussed in this paper which are [3] first inspired from a classification engine architecture and second machine learning-based SVM. The main motive of this paper is to provide a more comprehensive understanding of the sentiments of users using different services. A classification named "forensic identification classification engine" is studied, and based on its architectural background and algorithms used, the classification of sentiments is being done. Similarly, linear SVM architecture is also being studied and implemented for the problem. The main task is to find features such as the most frequent words used, the weightage of those words, the group of words appearing together, using these features and the algorithms being implemented, and trying to analyze the sentiments.

Motivation: The motivation for the classification of sentiments is that we are living in a "data" world today. All the data are used to improve the services and products. There are several algorithms available to perform sentiment analysis. The number of users on social media sites is increasing rapidly. There are several ways for organizations to keep track of their customers and take their feedback about product opinion. Tweets are useful as well as an important source of opinion for businesses, organizations, or companies. Therefore, the method to analyze sentiments needs to be automated so that millions of data can be evaluated without reading them manually. Thus, in this paper, two approaches have been studied for the same. Machine learning model [4] such as the Naive Bayes algorithm used in a classification engine as well as SVM is studied.

Contribution: In this work, a model is developed based on a classification engine's architectural background. Different datasets are taken into account to see if mixing them increases the efficiency of the model. Understanding the architectural background of the classification engine, a vocabulary function is built which splits the input into words, further converts it into lower form, and checks if the words are present in stop-words or length of them is less than two, if so such words are ignored. Otherwise, add them to the vocabulary dictionary and keep count of the frequency of words. The top 1000 words which have a high frequency are taken as features. The input data is again passed and if the word is a part of the 1000 words, it is counted else not counted. There is a fit function that trains the model about the data and later helps in analyzing the sentiments. Linear SVM is also studied and analyzed for the analysis.

Organization of the paper The paper is organized such that there is an introduction at the beginning in section 1 where the need for sentiment analysis is discussed. An approach for sentiment analysis is being introduced, which leads to using a classification engine's architectural idea as well as using SVM for sentiment analysis in section 2. Different datasets are being analyzed and presented in section 3. Conclusions are drawn in section 4.

Proposed System: Overview and Design

The Forensic identification classification engine (as shown in Figure 26.1) is the backbone of the model. This classification engine works atop the concept of M pedigrees and N

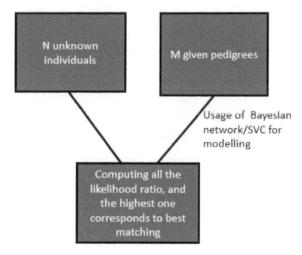

Figure 26.1 Working principle of forensic identification classification engine.

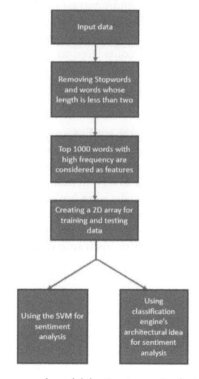

Figure 26.2 Workflow of the proposed model for Sentiment Analysis.

unknown individuals, the classification model constructed is based on the architecture, where it is crucial to find the best matches among the N unknown individuals and M pedigrees. For multiclass classification, generally, SVC (support vector classifier) is used. But in this approach, the Linear SVM is used, which works only on two-class classification, thus OneVsRest Classifier is also stitched into the pipeline.

An overview of the system is shown in Figure 26.2. At first, input data is taken, and further stopwords and words having a length of less than two are removed. The top 1000 words with higher frequency are taken into account. A 2D array is created for the training data. Later based on the approach, it may go through the SVM model else it may go through the classification engine's background idea-inspired model and help in identifying and analyzing sentiments.

Classification Engine for Sentiment Analysis

For the first approach, a classification engine is being studied and considering those steps, sentiment analysis is being implemented where the Naive Bayes algorithm works the best. The working principle of *forensic identification classification engine* are mapped with sentiment analysis process [5]:

- There are N unknown individuals and M pedigree in the architecture, and it is critical to determine the best matches between N unknown and M pedigree. Compute all the likelihood ratios, likewise in sentiment analysis there are N unknown texts and there are three classes (positive, negative, and neutral) and it is important to find the best matches by computing the probability using the Naive Bayes theorem.
- The highest likelihood corresponds to best matching in the classification engine, likewise the highest probability will be considered as the class in the sentiment analysis problem.

Linear SVM for Sentiment Analysis

In the proposed system, Linear SVM is being studied and implemented. Linear SVM is used for linearly separable data, that is if using a straight line dataset can be divided into two classes, such dataset is linearly separable and the classifier is termed as Linear SVM [6, 7]. The SVM or Support Vector Machine is a very popular algorithm. It follows the idea to follow a linear model and to find a linear decision boundary also called a hyperplane that best separates the data. Here the choice of the best hyperplane is the one that represents the largest separation between the two classes. So, the hyperplane is chosen such that the distance from it to the nearest data point on each side is maximized [8].

Experiment and Evaluation

Datasets

Four datasets have been studied and analyzed in the experimentation i.e., Airline, IMDB, Trivago, and Rotten Tomatoes datasets. Detail of the datasets are explained below and presented in Table 26.1.

Airline dataset is a dataset derived from Kaggle, which was created to analyse sentiments. It contains a collection of tweets from six US airlines in sets of positive, neutral, or negative. Users were asked to provide input and give feedbacks and polarize their opinions on the presented data, which was scrapped in February 2015. Around 15,000 rows are present in the dataset [9].

IMDB dataset can be used for natural language processing or text analytics. It contains around 50k movie reviews. It is categorised into class values: Positive and negative, each having 25k rows [10].

Trivago dataset contains sentiments of people who stayed in hotels and shared their reviews refgarding the stay. It contains around 5000 rows having positive and negative categories [11].

Rotten tomatoes dataset is comprised of tab-separated files from Rotten tomatoes dataset. The sentences in the dataset is parsed into many phrases. PhraseID is preserved for each phrase. Sentences also has a sentenceID. The sentiments are labels into five classes: namely, negative, somewhat negative, neutral, somewhat positive, positive [12].

Classification Using a Classification Engine

The algorithm is able to identify positive, negative, and neutral sentiments. It is also analyzed from past findings, that Naive Bayes works much better for sentiment Analysis. It is giving a decent accuracy in all cases being studied. The classification engine's architecture was the backbone of the work and its algorithm was studied and used.

Fit function:
- Fit() function is the function to build the dictionary.
- **It is built in the following format as shown in** Figure 26.3.

Probability function:
- The probability function is used to find the probability of a given X test data. It takes dictionary, and class value as input and returns the probability output for that class given input.

Table 26.1: Detail of the datasets used in the experiment (AWC: Average Word Count).

Dataset		Description
Airline	with stopwords:	Total Fields: 14640; Negative 9178, Neutral 3099, Positive 2363
		Total Words: 258458, AWC per Sentiment: 17.654234972677596
	without stopwords:	Total Words: 153759, AWC per Sentiment: 10.502663934426229
IMDB	With stopwords:	Total Fields: 50000; Negative 25000, Positive 25000
		Total Words: 11558128, AWC per Sentiment: 231.16256
	Without stopwords:	Total Words: 6118738, AWC per Sentiment: 122.37476
Trivago	With stopwords:	Total fields: 4724; Negative 683, Positive 4041
		Total Words: 179554, AWC per Sentiment: 38.00889077053345
	Without stopwords:	Total Words: 99383, AWC per Sentiment: 21.037891617273498
Rotten Tomatoes	With stopwords:	Total Fields: 156060; Negative 7072, Positive 9206, Neutral 79582, Somewhat Negative 27273, Somewhat Positive 32927
		Total Words: 1122624, AWC per Sentiment: 7.1935409457900805
	Without stopwords:	Total Words: 689746, AWC per Sentiment: 4.4197488145585035

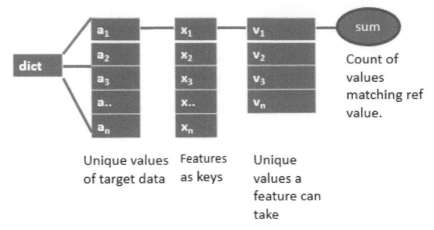

Unique values   Features   Unique
of target data   as keys   values a
feature can
take

Figure 26.3 The dictionary structure used in building the model.

- As Laplace correction is also needed in any good classifier, it is also taken care of.
- Taking log calculations to not get Zero error.

Predict Single Point function:
- initializing the values of best class and prob
- calling function probability() which calculates the probability, if it is for the first time then it is bound to call the function and change the values otherwise if the values are less than the previous values, it won't change. Thus giving the best class for X test.

Classification Using Linear SVM

Linear SVM is another approach used for the experiment. As there are three classes to be classified, OneVsRestClassifier is also imported. Further, the training data is passed to fit the model, and testing data is predicted.

Results and Discussion

Many case studies were being performed. To understand the nature and properties of each of the datasets, extensive experiments are conducted based on different ratios. It also enabled us to infer which one yields the highest accuracy. Different training and test splits have been performed to understand the characteristics of output accuracy on data. In individual datasets, it can be seen that *70:30* ratio yields more accuracy than *80:20* ratio in Airline, IMDB and Trivago datasets. In Rotten Tomatoes dataset, the difference on accuracy is marginal. Overall *70:30* ratio yields a better model than a *80:20* ratio in case of individual datasets.

So it can be understood from the different analysis done above that for individual datasets, *70:30* ratio is giving a higher accuracy value for all the cases in classification engine. In SVM in case of individual datasets, the values are almost similar for both the ratio division. In case of merged dataset, it can be concluded that accuracy while merging datasets were almost constant in all cases except removing somewhere positve and somewhere negative from the Rotten tomatoes dataset, increased the accuracy.

Detailed results generated from the experiments are presented and analyzed in the below tables, Tables 26.2 and 26.3 and Figures 26.4–26.7. The presented work is an early effort in sentiment analysis, many other algorithms need to be studied and analysed, the features generated should also be tuned as per the machine learning algorithm, which is also a

Table 26.2: Accuracy of individual datasets for classification engine (CE) and SVM.

Dataset	Accuracy (CE)	Accuracy (SVM)	Remarks
Airline dataset (80:20)	73.94	75.88	
Airline dataset (70:30)	74.27	76.73	
IMDB dataset (80:20)	82.56	85.71	
IMDB dataset (70:30)	82.75	85.48	
Tivago dataset (80:20)	87.93	90.89	
Trivago dataset (70:30)	89.42	89.84	
Rotten Tomatoes (80:20)	62.35	64.10	Merging negative and somewhat negative together, positive and somewhat positive together
Rotten Tomatoes (70:30)	62.14	63.98	Merging negative and somewhat negative together, positive and somewhat positive together
Rotten Tomatoes (80:20)	86.68	87.25	Removing somewhat positive and negative
Rotten Tomatoes (70:30)	86.60	87.29	Removing somewhat positive and negative
Rotten Tomatoes (80:20)	62.23	63.67	Removing positive and negative
Rotten Tomatoes (70:30)	62.45	63.77	Removing positive and negative

Table 26.3: Accuracy of merging datasets for classification engine (CE) and SVM.

Dataset	Accuracy				Remarks
	CE (70:30)	CE (80:20)	SVM (70:30)	SVM (80:20)	
Airline	74.27	73.94	76.73	75.88	
Airline + IMDB	70.22	70.20	82.74	82.67	
Airline + IMDB + Trivago	71.58	71.64	82.22	82.44	
Airline + IMDB + Trivago + Rotten Tomatoes	58.22	58.38	66.51	66.63	Merging negative and somewhat negative together, positive and somewhat positive together
Airline + IMDB + Trivago + Rotten Tomatoes	76.68	76.53	83.66	83.67	Removing somewhat positive and negative
Airline + IMDB + Trivago + Rotten Tomatoes	60.46	60.63	67.54	67.46	Removing positive and negative

parameter for future work. Sentiment analysis has helped to understand the opinions of the mass public. It is cost effective as well as an easier approach to understand how people react to a particular subject of matter. A major setback is the limit in the number of tweets being analysed. Thus, in future it is important to study other algorithms as well as expand the dataset for a more analytical view.

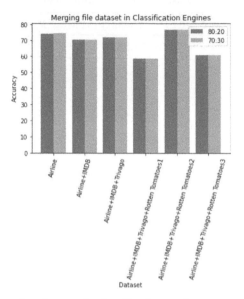

Figure 26.4 Accuracy of merging datasets in classification engine.

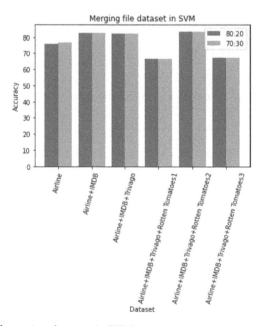

Figure 26.5 Accuracy of merging datasets in SVM.

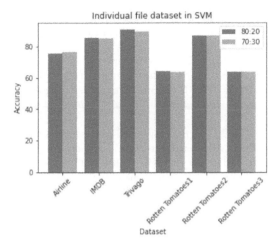

Figure 26.6 Accuracy of individual datasets in SVM.

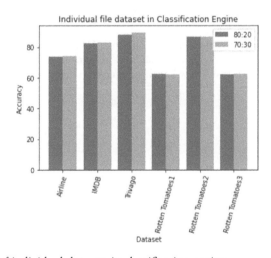

Figure 26.7 Accuracy of individual datasets in classification engine.

Conclusion

In this paper, analysing the sentiments and classifying them into polarities is addressed. One of the main contributions of the work is to express how a classification engine as well as SVM helps in solving this issue. The present study accurately classifies emotions as per human psychology. As per Sentiment analysis, there is still scope for improvement in this analysis as many other classification models are yet to be studied.

In the future big data analysis techniques can be used to classify all the emotions for a large volume of feedbacks and records. Moreover, deep learning models can also be tried to see how they perform in the sentiment analysis problem.

References

1. Das, D. D., Sharma, S., Natani, S., Khare, N., and Singh, B. (2017). Sentimental analysis for airline twitter data. *International Conference on Science Engineering and Technology*, 263, 2–11.

2. Sandaruwani M. D., and Hewapathirana, I. U. (2023). A review of recent trends in sri lankan social media analytics research. In 2023 International Research Conference on Smart Computing and Systems Engineering (SCSE), (Vol. 6, pp. 1–7), 2023.

3. Gupta, A., Pruthi, J., and Sahu, N. (2017). Sentiment analysis of tweets using machine learning approach. *International Journal of Computer Science and Mobile Computing*, 6(4), 444–458.

4. Brar, G. S., and Sharma, A. (2018). Sentiment analysis of movie review using supervised machine learning techniques. *International Journal of Applied Engineering Research*, 13(16), 12788–12791.

5. Siino, V., and Sears, C. (2020). Artificially intelligent scoring and classification engine for forensic identification. Forensic Science International: Genetics, 44, 102162. doi: 10.1016/j.fsigen.2019.102162. Epub 2019 Sep 19. PMID: 31604203.

6. Ahmad, M., Aftab, S., and Ali, I. (2017). Sentiment analysis of tweets using svm. *International Journal of Computer Applications*, 117(5), 25–29.

7. Hasan, A., Moin, S., Karim, A., and Shamshirband, S. (2018). Machine learning-based sentiment analysis for twitter accounts. *Mathematical and Computational Applications*, 23(1), 11. https://doi.org/10.3390/mca23010011

8. Kumar, S., and Zymbler, M. A. (2019). Machine learning approach to analyze customer satisfaction from airline tweets. *J Big Data,* 6, 62. https://doi.org/10.1186/s40537-019-0224-1

9. Twitter us airline sentiment dataset. https://www.kaggle.com/datasets/crowdflower/twitter-airline-sentiment.

10. Imdb dataset. https://www.kaggle.com/lakshmi25npathi/imdb-dataset-of-50k-movie-reviews.

11. Trivago dataset. https://www.kaggle.com/rohitanil/trivago-data.

12. Rotten tomatoes dataset. https://www.kaggle.com/c/sentiment-analysis-on-movie-reviews/data.

27 Identifying Diseased Fish using Improved CNN

Souvik Halder

Research Scholar, Department of I.T., Jadavpur University

Abstract

Fish disease is a major problem for fish farmers, and it can lead to significant production and economic losses. By detecting diseases early, appropriate measures can be taken to contain and manage the infections, safeguarding the aquatic ecosystem and human health. However, in India, detecting fish diseases is done manually, and this traditional method frequently yields inaccurate and misleading results. Therefore, a quick and affordable procedure is crucial and preferred. Recently, the effectiveness of Convolutional Neural Networks (CNNs) has been shown in a range of computer vision and machine learning issues. Therefore, to increase the classification accuracy of datasets containing Fish disease, CNN and transfer learning were applied in this study. Weight acquisition was carried out for five distinct convolutional neural network (CNN) models, namely AlexNet, Lenet-5, VGG16, ResNet50, and MobileNet. Based on Comparative tests showed that transfer learning with ResNet50 demonstrates the best performance among the models AlexNet, MobileNetV2, VGG16, and ResNet50. Therefore, ResNet50 is the preferred model for the fish disease classification task.

Keywords: Convolutional neural networks (CNNs), detection, fish disease.

Introduction

Fish play a crucial role as a global, cost-effective protein source, and the fishing industry's significance extends to food and the economy. With increasing global population and health awareness, the demand for fish is set to rise. However, the vulnerability of fish to diseases poses a challenge to sustaining production in the fishery sector [14, 15], exacerbated by factors such as temperature fluctuations and contaminated water, making infection control vital for fishermen and fish farmers.

Identifying fish diseases holds paramount importance within agricultural industries. Historically, diagnosing afflicted fish relied on manual differentiation through visual cues in images, a laborious and time-intensive process. Developing an automated model for fish disease identification could revolutionize this, ameliorating accuracy and efficiency. While conventional classification algorithms necessitate meticulous manual feature delineation and struggle with extensive data, the emergence of convolutional neural networks (CNNs) like ResNet50 presents a solution. This paper introduces a novel hybrid approach

---

souvikhalder32@gmail.com

by fusing a CNN with transfer learning, addressing the challenge of limited training data. The study's comparative experiments confirm ResNet50's superiority in accuracy, establishing it as the prime candidate for the proposed hybrid model.

When some diseases are not readily identifiable through visual examination alone, our technique remains valuable as it can learn complex patterns and features from substantial datasets. While humans might encounter challenges in recognizing subtle or intricate visual cues, CNNs excel at uncovering concealed patterns that play a role in disease detection. Through training on diverse and extensive datasets, CNNs can capture nuanced differences that might evade human observation. Additionally, CNNs contribute to identifying less apparent features, thereby elevating the precision and effectiveness of disease detection procedures.

Literature Review

Hinton [1] proposed deep learning (DL) [3], with CNNs prominently utilized for image categorization. DL has also been employed in classifying fish diseases, though enhanced training data could improve its performance. Consequently, data augmentation and transfer learning are gaining traction. Similar techniques, such as convolutional neural networks (CNN) [2] and transfer learning (TL) [4, 5], have proven effective in diverse domains, like healthcare material categorization, as shown by Yunyan [21] and the investigation of transfer learning by Amin [24] for establishing decentralized ecosystems.

While previous studies on automated fish disease [10, 11, 12] diagnosis have made valuable contributions, limitations persist. For instance, Park et al. [8] developed a principal component analysis system for feature extraction from microscopic fish images. Yet, limited testing raises questions about its applicability to larger, varied datasets. Matarneh [9] proposed fish image segmentation based on color information but faced limitations due to a small sample size and reliance on a single image processing technique. Similar concerns arise from studies by Divinely et al. [6], Ammar and Neama [7], Waleed et al. [16], Yasruddin [22], and Sharma [23] in terms of dataset size, architecture exploration, and generalization capabilities. Although foundational, these studies point to potential enhancements by addressing limitations and diversifying datasets and architectures for more robust and accurate automated fish disease diagnosis models.

Method

Construction of the Data Set

We gathered a dataset of 1568 fish photos, which we provided to our convolutional neural network (CNN) model. These photos originated from a fish farm in West Bengal and Google Images, collectively featuring infected and healthy fish. This dataset encompassed instances of five distinct fish diseases—anchor worms, carp pox, cotton mouth diseases, clumnaris, and furunculosis for infected fish [13]. We divided the fish image dataset into three segments for data organization: 1280 images for training the model, 188 images for fine-tuning hyperparameters, and 100 images for evaluating model performance. For each type, one examples are given in Figure 27.1.

Model

Convolutional Neural Networks (CNNs) represent a powerful form of deep learning technology widely known for their efficacy across various image processing tasks. Figure 27.2 shows

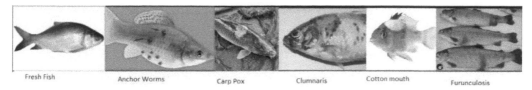

Fresh Fish   Anchor Worms   Carp Pox   Clumnaris   Cotton mouth   Furunculosis

Figure 27.1 Samples of fresh fish and diseased fish.

the Convolutional Neural Network (CNN) structure. They capitalize on local connectivity and parameter sharing to extract features from images, leading to superior learning efficiency and generalization compared to other deep learning models. Comprising convolution and pooling layers, CNNs excel in feature extraction by reducing parameters and enhancing feature quality. Five CNN configurations were chosen for model training: AlexNet [17], VGG16 [18], ResNet50 [18], Lenet-5 [20], and MobileNet [19], each employing different kernels in the convolution layer to extract diverse image features. Pooling layers summarized feature map regions, reducing image size and computational requirements. The automated parameter tuning and feature extraction followed a sequential process: forward calculations were executed, gauging the error between predicted and actual outputs. Subsequently, model parameters underwent adjustments via backpropagation (BP), wherein network weights were updated by propagating error backward and optimizing weights in the direction of the steepest descent. This iterative process aimed to reach the minimum of the loss function and attain optimal parameter values.

AlexNet: AlexNet marked a significant milestone by achieving leading performance on the ImageNet dataset as the inaugural CNN to do so. With eight layers, this architecture delves relatively deep and is an early adopter of ReLU activation functions within CNNs.

VGG16: VGG16, comprising 16 layers, is a CNN of greater depth. Resembling AlexNet, it employs smaller convolutional filters, resulting in heightened computational efficiency while also introducing a trade-off in accuracy.

ResNet50: ResNet50 stands as an extensively deep CNN, boasting 50 layers, and distinguishes itself through the incorporation of residual connections to mitigate the vanishing gradient challenge. This architectural choice enhances its accuracy beyond that of AlexNet or VGG16, although it does entail a higher computational cost.

MobileNetV2: MobileNetV2 is a CNN tailored for mobile devices, featuring a compact and efficient design, in contrast to the larger architectures of AlexNet.

LeNet5: With a modest depth of merely five layers, it assumes a relatively shallow architectural profile. While its accuracy might not parallel the CNN as mentioned above architectures, it retains its potency as a robust instrument for image classification.

Optimization

The Adam optimizer amalgamates the benefits of AdaGrad and RMSProp, two distinct optimization algorithms. AdaGrad adapts the learning rate via parameter gradients to avert its extreme diminution or escalation, thereby enhancing algorithm convergence. Meanwhile, RMSProp employs a rolling average of squared gradients to update parameters, thereby diminishing slope noise and augmenting algorithm convergence.

Dropout

During training, dropout involves randomly deactivating nodes in a CNN, compelling the network to depend on all nodes instead of a limited subset, preventing excessive reliance on

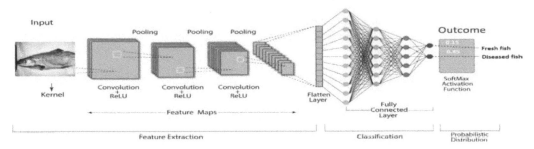

Figure 27.2 CNN architecture.

specific features, and ultimately enhancing the network's generalization capabilities. The dropout layer's equation is as follows:

$$y = x(1 - p) + 0.p$$

y represents the dropout layer's output and input is x. The dropout rate, abbreviated as p, ranges from 0 to 1. By randomly changing some of the input units to 0 with probability p, the dropout layer operates. The sum of the remaining input units is kept constant by scaling them up by a factor of $1 - p$. As a result, the network is prevented from relying too heavily on any one set of input units, which aids in avoiding overfitting.

Performance Metrics

Precision, recall, and the F1 measure are assessment metrics used in the classification of fish disease models. Precision evaluates the accuracy of positive predictions, computed as the ratio of correct predictions to all positive predictions. Recall measures the model's ability to identify true positive instances by calculating the ratio of true positives to all actual positives. The F1 measure combines precision and recall to offer a holistic performance indicator, particularly useful when dealing with uneven class distributions.

$$Precision(P) = \frac{TP}{TP + FP} \, , \, Recall = \frac{TP}{TP + FN}, \, F1score = \frac{2.Precision.Recall}{Precision + Recall}$$

Transfer Learning

Within the realm of CNNs, transfer learning entails employing a pre-trained CNN, previously trained on an extensive image dataset, as a foundational basis for training a fresh CNN on a smaller image dataset. This approach optimizes training efficiency and resource allocation when establishing a new CNN.

Pipeline of our Method

The comprehensive sequence of steps in our method is illustrated in Figure 27.3. This entails inputting data, performing image processing, constructing a CNN model, and evaluating the model's performance, constituting the entirety of the workflow.

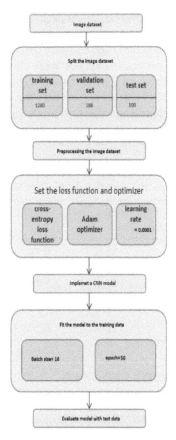

Figure 27.3 The diagram illustrates the training process of CNN model.

1. Datasets were generated for training, validation, and testing purposes.
2. Employing the initial image data, a series of image preprocessing steps were executed, encompassing actions like image scaling, normalization, standardization, and converting image tags into one-hot coding.
3. CNN models were built, with parameters initialized at random. Following that, an optimizer, learning rate, and cross-entropy loss function were chosen based on past performance.
4. The network model underwent training using the partitioned training dataset, while the validation set was employed to assess classification accuracy after each epoch. The training process encompassed a total of 50 epochs.

Result and Analysis

The experiments were conducted on a system comprising an Intel i7 processor and Nvidia RTX 2060 GPU, employing the PyTorch 2.0 deep learning toolbox on a Windows 10 platform. 100 fish disease images, including infected and healthy fish, were tested. The accuracy outcomes were used to construct a confusion matrix, illustrated in Figure 27.4. The diagonal pattern in blue hues on this matrix represented accurate identifications, while the white diagonal pattern indicated misclassifications. The results from training four distinct convolutional neural networks (CNNs) using varying learning rates are outlined as follows: Utilizing a learning rate of 0.0001 led to swift convergence to an optimal solution. In contrast, divergence and poor processing speeds were caused by learning rates of 0.001 and 0.00001, respectively. A batch size 16 was chosen, with a random dropout rate of 0.5 applied based on prior experience. Table 27.1 presents the corresponding outcomes after 50 iterations, excluding transfer learning. The findings suggested that the CNN, encountered delayed convergence when learning parameters from initial inputs, likely due to the limited training dataset size.

Table 27.1: Results for four convolutional neural network (CNN) models.

Model	Input size	Validation accuracy	Test accuracy	Precision	Recall	F1-measure
AlexNet	227×227×3	80%	81%	0.7872	0.8043	0.7956
LeNet-5	32×32×3	87%	85%	0.8888	0.8421	0.8648
VGG-16	64×64×3	84%	87%	0.9402	0.8750	0.8740
ResNet50	110×110×3	85%	88%	0.8793	0.9444	0.9107

Table 27.2: Outcomes obtained from four CNN models.

Model	Validation accuracy	Improvements (Validation)	Test accuracy	Improvement (Test)	Precision	Recall	F1-measure
AlexNet	93%	13%	97%	16%	0.9824	0.9655	0.9739
VGG-16	92%	5%	94%	7%	0.9803	0.90	0.9433
MobileNet	89%	-	91%	-	0.9090	0.9259	0.9174
ResNet50	97%	12%	98%	13%	0.9833	0.9833	0.9833

The previously trained model's parameters served as initial values for transfer learning. This process involved a learning rate 0.0001 and a dropout rate 0.5, with fine-tuning applied to all layers' parameters. To enhance transfer learning, additional data were incorporated into the training set. Table 27.2 illustrates this data, with the third column highlighting the enhanced test accuracy between the second and first experiments. These improvements underscored the positive impact of transfer learning, where pre-learned knowledge adapted from initial training improved accuracy in the new task. Figure 27.5 showcased training and validation accuracy curves for four models—AlexNet, MobileNetV2, VGG16, and ResNet50—allowing a visual performance comparison. The confusion matrix emphasized ResNet50's superiority. In general, ResNet50 emerged as the optimal CNN architecture for most tasks due to its accuracy, efficiency, and ease of training. Based on the depicted accuracy curves in Figure 27.5, ResNet50 exhibited the most favorable performance among the models, thereby being selected as the preferred choice for classifying fish diseases. Its outstanding accuracy and robustness rendered it the most suitable candidate for accurate fish disease classification within this context.

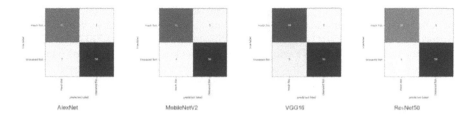

Figure 27.4 Illustrates the confusion matrix depicting the performance of AlexNet, MobileNetV2, VGG16, and ResNet50.

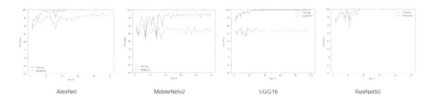

Figure 27.5 Displays the accuracy curves for training and validation of AlexNet, MobileNetV2, VGG16, and ResNet50.

Conclusions

CNNs are increasingly sophisticated methods used in image recognition, offering enhanced accuracy with lower computational complexity than other techniques. CNNs' robust fault tolerance improves accuracy even with less clear images. This technology holds significant advantages for detecting fish diseases, as demonstrated in this study. While the current accuracy isn't perfect, further refinements can enhance disease control strategies. The model's potential for identifying diverse animal diseases warrants further research to confirm its applicability.

References

1. Hinton, G. E., and Salakhutdinov, R. R. (2022). Reducing the dimensionality of data with neural networks. *Science*, 313(5786), 504–507.
2. Krizhevsky, A., Sutskever, I., and Hinton, G. E. (2017). ImageNet classification with deep convolutional neural networks. *Communications of the ACM*, 60, 84–90. https://doi.org/10.1145/3065386.
3. Hou, Y., Quan, J., and Wang, H. (2017). Review of deep learning development. *Ship Electronic Engineering*, 37(5–9), 111.
4. Wang, Y., Wang, C., Luo, L., and Zhou, Z. (2019). Image classification based on transfer learning of convolutional neural network. In 2019 Chinese Control Conference (CCC), 2019.
5. Anjomshoaa, A., and Curry, E. (2021). Transfer learning in smart environments. *Machine Learn Knowledge Extract*, 3, 318–332. https://doi.org/10.3390/make3020016.
6. Divinely, S. J., Sivakami, K., and Jayaraj, V. (2019). Fish diseases identification and classification using machine learning. *Int. J. Adv. Res. Basic Eng. Sci. Technol.(IJARBEST)*, 5(6), 46–51.
7. Ammar, A., and Neama, Y. (2019). Detection of freshwater fish white spots disease using the machine learning LR classifier and ACO. *Beni-Suef University Journal of Basic and Applied Sciences*, 78–87.

8. Park, J. S., Oh, M. J., and Han, S. (2007). Fish disease diagnosis system based on image processing of pathogens' microscopic images. In Proceeding Frontiers in the Convergence of Bioscience and Information Technologies FBIT, 2007, (pp. 878–883). doi: 10.1109/FBIT.2007.157.

9. Lyubchenko, V., Matarneh, R., Kobylin, O., and Lyashenko, V. (2016). Digital image processing techniques for detection and diagnosis of fish diseases. *International Journal of Advanced Research in Computer Science and Software Engineering,* 6(7), 79–83.

10. Zhang, X., Fu, Z., Cai, W., Tian, D., and Zhang, J. (2009). Expert systems with applications applying evolutionary prototyping model in developing FIDSS: an intelligent decision support system for fish disease/health management. *Expert Systems with Applications,* 36(2), 3901–3913. doi:10.1016/j.eswa.2008.02.049.

11. Han, C., Lee, S., Han, S., and Park, J. (2011). Two-stage fish disease diagnosis system based on clinical signs and microscopic images. *ICCSA,* 2, 635–47.

12. Hitesh, C., Rituraj, P., and Prodipto, D. (2015). Image processing technique to detect fish disease. *International Journal of Computer Science and Security,* 9(2), 121–131.

13. Hu, J., Li, D., Duan, Q., Chen, G., and Si, X. (2016). Preliminary design of a recognition system for infected fish species using computer vision. *Computer and Computing Technologies in Agriculture,* 5, 530–534.

14. Jovanovic´, V., Risojevic´, V., Babic´, Z., Svendsen, E., and Stahl, A. (2016). Splash detection in surveillance videos of offshore fish production plants. In International Conference on Systems, Signals and Image Processing, doi:10.1109/IWSSIP.2016.7502706.

15. Gujjala, J., Sujatha, K., Ponmagal, R. S., Anand, M., and Srividhya, V. (2019). Diagnosis of fish disease using UKF and Elman neural networks. *International Journal of Engineering and Advanced Technology,* 8(3), 438–441.

16. Waleed, A., Medhat, H., Esmail. M., Osama, K., Samy, R., and Ghanim, T. M., (2019). Automatic recognition of fish diseases in fish farms. In Proceeding International Conference on Computer Engineering and Systems, (pp. 201–206). doi:10.1109/ICCES48960.2019.9068141.

17. Alom, M. Z., Taha, T. M., Yakopcic, C., Westberg, S., Sidike, P., Nasrin, M. S., Van Esesn, B. C., Awwal, A. A., and Asari, V. K. (2018). The history began from alexnet: a comprehensive survey on deep learning approaches. arXiv preprint arXiv:1803.01164.

18. Theckedath, D., and Sedamkar, R. R. (2020). Detecting affect states using VGG16, ResNet50 and SE-ResNet50 networks. *SN Computer Science,* 1, 1–7.

19. Qin, Z., Zhang, Z., Chen, X., Wang, C., and Peng, Y. (2018). Fd-mobilenet: Improved mobilenet with a fast downsampling strategy. In 2018 25th IEEE International Conference on Image Processing (ICIP), pp. 1363–1367. IEEE, 2018.

20. Wei, G., Li, G., Zhao, J., and He, A. (2019). Development of a LeNet-5 gas identification CNN structure for electronic noses. *Sensors,* 19(1), 217.

21. Lu, Y., and Tian L. (2022). Global, regional, and national burden of hypertensive heart disease during 1990–2019: an analysis of the global burden of disease study 2019. *BMC Public Health,* 22(1), 841.

22. Yasruddin, M. L., Ismail, M. A., Husin, Z., and Tan, W. K. (2022). Feasibility study of fish disease detection using computer vision and deep convolutional neural network (dcnn) algorithm. In 2022 IEEE 18th International Colloquium on Signal Processing & Applications (CSPA) (pp. 272–276). IEEE.

23. Tyagi, A., Sharma, C., Srivastava, A., Kumar, B.N., Pathak, D., and Rai, S. (2022). Isolation, characterization and complete genome sequencing of fish pathogenic Aeromonas veronii from diseased Labeo rohita. *Aquaculture,* 553, 738085.

24. Amin Ul, H., Ping Li, J., Ahmad, S., Khan, S., Alshara, M. A., and Alotaibi, R. M. (2021). Diagnostic approach for accurate diagnosis of COVID-19 employing deep learning and transfer learning techniques through chest X-ray images clinical data in E-healthcare. *Sensors,* 21(24), 8219.

28 Robot Learning and Adaptation using Artificial Intelligence (AI) and Machine Learning (ML): Review

K. Viswanath Allamraju[1,a], K. Satyanarayana[2,b], T. Ravi Kiran[3,c], and Manepalli Sailaja[2,d]

[1]Department of Mechanical Engineering, Institute of Aeronautical Engineering, Hyderabad, Telangana state, India

[2]Department of Mechanical Engineering, ANITS, Visakhapatnam, Andhra Pradesh, India

[3]Department of Mechanical Engineering, Rabindranath Tagore University, Bhopal, Madhya Pradesh, India

Abstract

Robot Learning and Adaptation using AI and ML is a field of study that focuses on developing intelligent robotic systems capable of acquiring new knowledge and skills through interaction with their environment. This abstract provides an overview of the key concepts and applications in this emerging field. AI & ML techniques are employed to enable robots to learn from their experiences, make informed decisions, and adapt their behaviour based on changing circumstances. One of the fundamental aspects of robot learning and adaptation is the integration of discernment and accomplishment. Robots utilize sensors to perceive the environment and collect data, which is then processed using AI and ML techniques to derive meaningful information. This information is used to guide the robot's actions and decision making processes, allowing it to interact effectively with its surroundings. Robot learning and adaptation have numerous practical applications across various domains. In industrial settings, robots can learn to perform complex tasks, such as assembly line operations or quality control, by observing human demonstrations or through trial and error. In healthcare, robots can learn to assist in surgical procedures or provide personalized care to patients by adapting their behaviour based on individual needs. Furthermore, robot learning and adaptation find applications in autonomous vehicles, where AI and ML enable vehicles to learn from real-life driving scenarios and acclimatise to changing conditions. In the field of robotics research, learning and adaptation are crucial for developing robots that can operate in unstructured and dynamic environments, such as search and rescue missions or exploration tasks. However, there are several challenges associated with robot learning and adaptation. In this article , presented the AI and ML adaptation to robots in comprehensive picture.

Keywords: Artificial intelligence (AI), machine learning (ML), robot.

Introduction

Robot Learning and Adaptation (RLA) using AI & ML is an exciting and rapidly growing research that combines the disciplines of robotics, AI, and ML to enable robots to acquire novel aids and awareness, make informed decisions, and adapt their behaviour based on

[a]k.viswanathallamraju@iare.ac.in, [b]ksnarayana.me@anits.edu.in, [c]travi.kiran@aisectuniversity.ac.in, [d]saisailaja2k@gmail.com.

changing circumstances. This introduction provides an overview of the fundamental concepts and motivations behind this field. Robots have traditionally been programmed with explicit instructions to perform specific tasks. However, this approach has limitations when it comes to handling complex, unstructured environments and dynamic situations. Robot Learning and Adaptation aim to address these limitations by allowing robots to learn and improve their performance through interaction with the environment. The assimilation of AI and ML techniques in robot learning and adaptation empowers robots to acquire knowledge from data, perceive their surroundings, make sense of sensory inputs, and generate appropriate responses. AI provides the foundation for intelligent decision-making, while ML algorithms enable robots to learn patterns, recognize objects, predict outcomes, and adapt their behaviour based on observed data. The motivation behind robot learning and adaptation is to create robots that can operate in real-world scenarios with minimal human intervention. By equipping robots with learning capabilities, they can adapt to new situations, acquire new skills, and autonomously navigate complex environments. This enables robots to be more versatile, flexible, and useful in a broad area of applications, from mechanical and medical care to autonomous vehicles and space exploration. The key components of robot learning and adaptation include perception, learning algorithms, decision-making, and action. Perception involves the robot's ability to sense and perceive its environment using sensors such as cameras, lidar, or tactile sensors. Learning algorithms enable the robot to extract meaningful information from sensory data, learn models of the environment, and acquire new skills through supervised, unsupervised, or reinforcement learning techniques. Decision-making involves using the acquired knowledge to make informed choices and generate appropriate actions. Finally, action refers to the physical execution of tasks and interactions with the environment based on the decisions made. There are several challenges in robot learning and adaptation that researchers are actively addressing. These challenges include designing efficient and scalable learning algorithms, dealing with uncertainty and noise in sensory data, ensuring safety and ethical considerations, and achieving seamless integration of learning and real-time control. In conclusion, Robot Learning and Adaptation using AI and ML is a dynamic and interdisciplinary field that aims to create intelligent robotic systems capable of learning, adapting, and interacting effectively with their environment. By supporting robots to acquire novel knowledge, improve their abilities, and autonomously adapt to changing situations, we can unlock the potential for robots to become valuable companions, assistants, and collaborators in various domains, enhancing productivity, safety, and human-machine interaction.

Classification of Robot Learning and Adaptation

Reinforcement Learning for Robot Control

Reinforcement learning (RL) algorithms permit robots to acquire from interactions with their environment and optimize their actions based on rewards or feedback. Researchers explore RL algorithms for different robotic errands such as grasping objects, movement, manipulation, and navigation. RL has extended significant responsiveness in the areas of robotics as a promising approach for robot control. RL enables robots to learn and optimize their behavior through interactions with their environment, without relying on explicit programming or predefined rules. By formulating robot control as a sequential decision-making problem, RL algorithms learn to select actions that maximize a cumulative reward signal. One of the key advantages of RL for robot control is its capacity to handle complex-dynamic environments. RL algorithms have been efficaciously applied to innumerable robotic tasks, including grasping and manipulation, locomotion, navigation, and autonomous driving. By learning from trial and error, robots can adapt their control policies to achieve desired objectives in different scenarios. Several research studies have focused on developing RL algorithms specifically tailored for robot control. Proximal Policy Optimization (PPO) was proposed by Schulman et al. [1]. PPO has shown promising results in training robots for tasks such as robotic grasping and locomotion. Another popular RL algorithm is DDPG proposed by Lillicrap et al. [2]. DDPG combines deep neural networks with the actor-critic framework and has been widely used for continuous control tasks in robotics, including robotic arm manipulation and quadcopter control. Furthermore, there has been research in incorporating additional techniques into RL for better performance. For instance, Hindsight Experience Replay (HER) introduced by Andrychowicz et al. [3] enables robots to learn from failure by replaying experiences with different goals. This approach has been effective in training robots for tasks involving sparse rewards, such as pushing and stacking objects. Moreover, transfer learning has been explored in RL for robot control. Tan et al. [4] proposed Domain Randomization (2018), which involves training robots in simulated environments with varying dynamics to improve their adaptability in real-world scenarios. This technique has been successfully applied to tasks like robotic grasping and manipulation. Recent research has also focused on combining RL with other learning methods, such as imitation learning and meta-learning, to leverage human demonstrations and prior knowledge for faster and more efficient learning.

Transfer Learning and Generalization

Transfer learning aims to leverage knowledge absorbed from one assignment or environment to progress learning and adaptation in a new task or environment. This research focuses on developing techniques that allow robots to generalize their learning across different scenarios, tasks, or even robot platforms. Transfer learning and generalization techniques in the context of robots using AI and ML have gained significant attention as they enable robots to leverage knowledge from previous tasks or domains to improve learning and adaptation in new scenarios. These techniques aim to enhance the efficiency and effectiveness of robot learning, reduce the need for extensive data collection, and facilitate the deployment of robots in various real-world applications. One of the prominent research areas in transfer learning for robots is domain adaptation. For instance, in the domain of robotic grasping, an algorithm trained on simulated data can be adapted to real-world environments through domain adaptation techniques. Several approaches, such as adversarial domain adaptation done by Ganin and Lempitsky [5] and self-supervised domain adaptation, Saito et al. [6], have been proposed to address this challenge. Furthermore, research has explored transfer learning techniques that enable robots to transfer knowledge across different tasks or environments. One popular approach is to learn a shared representation or embedding space that captures common features across tasks. By doing so, the robot can transfer knowledge from one task to another. For example, research has demonstrated the effectiveness of shared embedding spaces in enabling robots to learn object manipulation skills, James et al. [7] or locomotion behaviors by Rajeswaran et al. [8]. Meta-learning, also known as "learning to learn," is another area of research that complements transfer learning in robots. Meta-learning algorithms aim to enable robots to quickly adapt to new tasks by leveraging prior knowledge or experience from related tasks.

These algorithms learn a meta-policy or meta-learner that can efficiently adapt to new tasks with minimal data or interaction. Meta-learning has shown promising results in a variety of robotic domains, including grasping, manipulation, and navigation done by Finn et al. [9]. In addition to these approaches, research has also explored techniques to enhance generalization in robot learning. Generalization refers to the ability of robots to adapt and perform well in novel or unseen scenarios. Techniques such as data augmentation, regularization, and uncertainty estimation have been investigated to improve the generalization capabilities of robot learning models. These techniques aim to enhance the robustness of learned policies and models to variations and uncertainties in the environment.

Online Learning and Continuous Adaptation

Kober and Peters [10] explains about robots often operate in dynamic and changing environments. Online learning algorithms enable robots to continuously adapt their behavior and update their models based on new data. This research investigates techniques to facilitate online learning and adaptation in real-time robotic systems. The research investigates methods such as Gaussian processes and local models for modeling and updating the policies. It addresses the challenges of sample efficiency and exploration in online learning for robots. The authors also discuss the application of these techniques to real robotic systems and present experimental results.

Human-Robot Interaction (HRI) and Learning from Demonstration (LfD)

Learning from human demonstration involves the robot observing and imitating human actions to acquire new skills or behaviors. Researchers aim to improve processes and techniques that make robots to gain complex tasks from human interaction, making human-robot collaboration more intuitive and efficient. Argall et al [11] introduces a comprehensive review of robot LfD techniques in the context of HRI. The paper explores various AI and ML approaches used for learning from human demonstrations to make robots to gain novel skills and behaviors. The authors discuss different paradigms and algorithms for LfD, including methods based on supervised learning, apprenticeship learning, and inverse reinforcement learning. They highlight the challenges and considerations in designing effective interaction and learning frameworks for robots that rely on human demonstrations.

Multi-Robot Systems and Collective Learning

In scenarios where multiple robots cooperate or collaborate, collective learning techniques enable robots to learn from the experiences and knowledge of other robots in the system. This research focuses on distributed learning algorithms and mechanisms for efficient knowledge sharing among robots. Matignon and Gribovskiy [12] and Brambilla et al. [13] provides a comprehensive review of swarm robotics, a field that focuses on the coordination and collective behavior of large-scale multi-robot systems. The paper explores the principles of swarm intelligence and the application of AI and ML techniques for collective learning in robot swarms. The authors discuss various algorithms and methodologies used for collective learning in multi-robot systems, including techniques inspired by social insects and evolutionary computation. 2. Christensen and Lilienthal [14] investigated methods for self-organization, task allocation, cooperative decision-making, and information sharing among robots in a swarm. The paper also examines the challenges and open research questions in swarm robotics, such as scalability, robustness, and adaptability. It presents case studies and applications of swarm robotics in diverse domains, including exploration, surveillance, and environmental monitoring. Pinto and Gupta [15] explained a unique work on supersizing self-supervision.

Active Learning and Exploration

Akrour et al. [16], Poupart et al. [17] and Cully et al. [18] explained about model based exploration, analytic solutions and also about quality and diversity of robots but provides a comprehensive survey of active learning and exploration strategies for robot learning. The paper explores different AI & ML methods used to allow robots to actively acquire information and explore their environment to improve their learning capabilities. The authors discuss different active learning approaches. They examine exploration strategies such as curiosity-driven exploration, Bayesian optimization, and information gain-based exploration. The paper also investigates the exploration techniques with reinforcement learning, imitation learning, and other learning paradigms. It explores their application in various robotic domains, including object recognition, robot grasping, and autonomous navigation. Active learning algorithms allow robots to actively seek out informative data or experiences to improve their learning and adaptation. This research investigates techniques for intelligent exploration, where robots can autonomously select actions to gather relevant information and learn more efficiently.

Robustness, Safety, and Ethical Considerations

Learning and adaptation in robots should consider safety and ethical aspects. Research in this area explores methods to ensure robust and safe behavior of robots, prevent unde[1]sirable adaptations, and address ethical considerations such as fairness, transparency, and accountability in robot decision-making. Quillen et al. [19] addressed the challenges and considerations related to robustness, safety, and ethical aspects of AI and ML systems, including those used in robotics. While the paper focuses on AI safety in general, it provides valuable insights into the specific concerns and considerations for robots utilizing AI and ML techniques. The authors discuss concrete problems and potential risks associated with AI systems, such as specification gaming, reward hacking, and unintended consequences. Zhu et al. [20] explore the need for robust and reliable AI systems that can handle uncertainties, adversarial inputs, and system failures in real-world scenarios. Furthermore, the paper delves into the safety concerns related to autonomous systems and the potential risks they pose. Christiano et al. [21], Hester et al. [22] emphasized the importance of designing AI and ML algorithms that prioritize safety and mitigate hazards in human-robot interaction and decision-making. Levine et al. [23] worked on research related to deep learning concepts of robots learning and adaptation using AI and ML.

Conclusion

In conclusion, Robot Learning and Adaptation using AI & ML is a quickly growing field that aims to enhance the capabilities of robots by enabling them to learn, make informed decisions, and adapt their behaviour based on changing circumstances. Through the combination of AI and ML techniques, robots can grasp knowledge, extract meaningful information from data, and increase their efficiency over period of time. The applications of robot learning and adaptation are diverse and span various domains. In industrial settings, robots can learn to perform complex tasks, increasing efficiency and productivity. In healthcare, robots can provide personalized care and assist in medical procedures, enhancing patient outcomes. Autonomous vehicles can learn from practical driving situations to improve navigation and safety. Additionally, in robotics research, learning and adaptation enable robots to operate effectively in unstructured and dynamic environments, such as search and rescue missions or exploration tasks. However, challenges exist in the field of robot learning and adaptation. Efficient and scalable learning algorithms, robust perception systems, safety considerations, and ethical implications are areas that require continued research and development. Interdisciplinary collaboration between AI, ML, robotics, and cognitive science is crucial to overcome these challenges and unlock the full potential of intelligent robotic systems. The future of robot learning and adaptation looks promising. Advancements in deep learning, reinforcement learning, transfer learning, and human-robot interaction are paving the way for more sophisticated and capable robots. Ongoing research aims to address the limitations of current approaches and develop algorithms (AL) and methodologies that make robots to develop with skills, generalize knowledge across tasks and environments, and interact seamlessly with humans. In conclusion, Robot Learning and Adaptation using AI and ML holds immense potential for revolutionizing robotics and enabling robots to operate autonomously and effectively in complex and dynamic environments. By continuously advancing the field and addressing its challenges, we can create intelligent robotic systems that have a wide range of applications, ultimately improving productivity, safety, and human-machine collaboration in various domains.

References

1. Schulman, J., Wolski, F., Dhariwal, P., Radford, A., and Klimov, O. (2017). Proximal policy optimization algorithms. arXiv, 1707.06347, 1–12. https://doi.org/10.48550/arXiv.1707.06347
2. Lillicrap, T. P., Hunt, J. J., Pritzel, A., Heess, N., Erez, T., Tassa, Y., Silver, D., and Wierstra, D. (2016). Continuous control with deep reinforcement learning. arXiv preprint arXiv, 1509.02971, 1–14. https://doi.org/10.48550/arXiv.1509.02971
3. Andrychowicz, M., Wolski, F., Ray, A., Schneider, J., Fong, R., Welinder, P., McGrew, B., Tobin, J., Abbeel, O. P., and Zaremba, W. (2017). Hindsight experience replay. In Advances in Neural Information Processing Systems, (pp. 5048–5058, 1–12).
4. Tan, J., Zhang, T., Coumans, E., Iscen, A., Bai, Y., Hafner, D., and Abbeel, O. P. (2018). Sim-to-real: learning agile locomotion for quadruped robots. In Conference on Robot Learning, (pp. 409–419). arXiv:1804.10332.
5. Ganin, Y., and Lempitsky, V. (2015). Unsupervised domain adaptation by backpropagation. In International Conference on Machine Learning, (Vol. 37, pp, 1180–1189).
6. Saito, K., Yamamoto, S., Ushiku, Y., and Harada, T. (2018). Maximum classifier discrepancy for unsupervised domain adaptation. In Proceedings of the IEEE Conference on Computer Vision and Pattern Recognition, (pp. 3723–3732).
7. James, S., Bloesch, M., and Siegwart, R. (2017). Sim-to-real via sim-to-sim: Data-efficient robotic grasping via randomized-to-canonical adaptation networks. *IEEE Robotics and Automation Letters*, 2(2), 437–444.
8. Rajeswaran, A., Kumar, V., Gupta, A., Schulman, J., and Levine, S. (2017). Learning complex dexterous manipulation with deep reinforcement learning and demonstrations. In IEEE International Conference on Robotics and Automation, (pp. 3306–3313).

9. Finn, C., Abbeel, O. P., and Levine, S. (2017). Model-agnostic meta-learning for fast adaptation of deep networks. In International Conference on Machine Learning, (Vol. 2, pp. 1126–1135).

10. Kober, J., and Peters, J. (2010). Policy search for motor primitives in robotics. *Machine Learning*, 84(1–2), 171–203.

11. Argall, B. D., Chernova, S., Veloso, M., and Browning, B. (2009). A survey of robot learning from demonstration. *Robotics and Autonomous Systems*, 57(5), 469–483.

12. Matignon, L., and Gribovskiy, A. (2012). A survey on multi-robot systems: Applications, classifications, and challenges. *International Journal of Robotics and Automation*, 27(1), 1–17.

13. Brambilla, M., Ferrante, E., Birattari, M., and Dorigo, M. (2013). Swarm robotics: a review from the swarm engineering perspective. *Swarm Intelligence*, 7(1), 1–41.

14. Christensen, A. L., and Lilienthal, A. J. (2016). Autonomous exploration and mapping of unknown environments—a review. *Journal of Autonomous Robots*, 41(4), 637–662.

15. Pinto, L., and Gupta, A. (2016). Supersizing self-supervision: learning to grasp from 50k tries and 700 robot hours. In Proceedings of the IEEE Conference on Computer Vision and Pattern Recognition, (pp. 3376–3385).

16. Akrour, R., Abdolmaleki, A., Peters, J., and Neumann, G. (2012). Model-based active exploration for learning inverse dynamics models. In IEEE International Conference on Robotics and Automation (ICRA), (pp. 2065–2072).

17. Poupart, P., Vlassis, N., Hoey, J., and Regan, K. (2006). An analytic solution to discrete bayesian reinforcement learning. In Proceedings of the 23rd International Conference on Machine Learning, (pp. 697–704).

18. Cully, A., Demiris, Y., and Korkinof, D. (2015). Quality and diversity optimization: a unifying modular framework. *IEEE Transactions on Evolutionary Computation*, 19(1), 138–157.

19. Quillen, D., Janner, M., Impinna, F., and Levine, S. (2020). Deep dynamic reaching in cluttered environments. *Science Robotics*, 5(45), 2400.

20. Zhu, Y., Wang, Z., Merel, J., Rusu, A., Erez, T., Cabi, S., Tunyasuvunakool, S., Kramár, J., Hadsell, R., Freitas, N., and Heess, N. (2018). Reinforcement and imitation learning for diverse visuomotor skills. *Science Robotics*, 3(18), 9354.

21. Christiano, P., Leike, J., Brown, T.B., Martic, M., Legg, S., and Amodei D., (2017). Deep reinforcement learning from human preferences. In Advances in Neural Information Processing Systems, (pp. 4299–4307).

22. Hester, T., Vecerik, M., Pietquin, O., Lanctot, M., Schaul, T., Piot, B., Horgan, D., Quan, J., Sendonaris, A., Osband, I., Dulac-Arnold, G., Agapiou, J., Leibo, J., and Gruslys, A. (2018). Deep Q-learning from demonstrations. In Proceedings of the AAAI Conference on Artificial Intelligence, 32(1), (pp. 2924–2931). https://doi.org/10.1609/aaai.v32i1.11757.

23. Levine, S., Pastor, P., Krizhevsky, A., and Quillen, D. (2016). Learning hand-eye coordination for robotic grasping with large-scale data collection. In Proceedings of the IEEE International Conference on Robotics and Automation (ICRA), (pp. 3342–3349).

29 Physics Informed Neural Network (PINN) for Load Reconstruction for Vibrating Pipes in Process Plants

Subrata Saha

Associate Professor, Mechanical Engineering Dept.,Techno-India University, Sector V, EM-Block, Salt Lake, Kolkata, India

Abstract

The objective of this paper is to present a method based on Physics Informed Neural Network (PINN) and Theory of InverseProblems for estimating the dynamic loads on a vibrating piping in a Process plant. The loads are essential for calculating the dynamic stresses which determine the mechanical integrity of the piping system. However, in view of the complexity of the physics of the phenomenon, the estimation of the forcing function is practically impossible and the system of governing partial differential equation (PDE) is ill-posed. In such a situation a data driven approach on a theoretical frame-work of Inverse Theory has been investigated.The forcing function has been modelled by a Deep Neural Networks (DNN) and the displacement function is obtained as a solution to the PDE. In-situ vibration readings fromthe field constitute the input reference data. The problem can be reduced to an optimization problem for the minimization of the loss functional which is essentially a mean square error (MSE) of the measured vis-a-vis the calculated displacements. The motivation behind the use of DNN has been the wideproliferation in application of the Machine Learning (ML) in the multi-disciplinary field of Artificial Intelligence (AI), physical sciences and engineering. The availability of open-source software and hardware for high performance computing resourceshas made the use of ML tools quite attractive. To the best knowledge of the author,such an application of PINN is not found in literature till date. The problem is important from the perspective of condition monitoring of plant equipment.

Keywords: Inverse theory, machine learning, neural networks, optimization, partial differential equations, vibrations.

Introduction

Vibrationfailures of piping are a serious problem and a safety hazard for plant operations [10, 8]. An operating piping carrying service fluid experiences various mechanical loads throughout its life cycle. Hence, the engineering design should provide safeguardsfor the piping system against such failures. The design procedure is to calculate the induced dynamic stresses and keep them within the allowable limits by suitable modification of the design parameters. The stresses are computed from the displacements which are obtained as a solution to the governing PDE for a vibrating beam (Craig, 1981).

subrata.s@technoindiaeducation.com, subrsaha@yahoo.co.in

The nature of vibration is flow induced and the forces are induced on the piping through pressure acting on the inside walls of the pipes. However due to the complexity of the phenomenon, no closed form analytical expressions are available for the forcing function. Thus, the direct or the forward problem is ill posed. In such a scenario it becomes imperative to invoke the theory of Inverse Problems (Nicaise, Yamamoto) in conjunction with the vibration data of the operating piping captured from the field to reconstruct the forcing function. Instruments like portable vibrometers are normally used for data acquisition.

The problem of determination of point forces has been studied by several researchers Moussa and AbdelHamid [6], Nicaise and Zair [7] and Yamamoto [9]. Moussa and AbdelHamid Hamid [6] have used FEA in the time domain to estimate the unknown forces based of displacement observations. The approach was mainly heuristic, but practically simple to adopt. The mathematical aspects of existence and uniqueness of solution and load reconstruction for hyperbolic systems have been studied by the other researchers. With some internal or boundary observations of the primary variable or its derivative, the existence and uniqueness of the solution could be established. Saha, [8] applied the reconstruction strategy to solve some real-life problems.

With the advent of high-power computing resources and advancement in the machine learning techniques the data-driven approach using PINNs Lagaris and Likas [4], Raissi et al. [5] is gaining vast popularity. The introduction of PINNs in solving PDE's is relatively new. The mathematical basis for the ability of the DNN to solve PDE'scan be attributed to the Uniform Approximation Theorem [2]. By virtue of this property, it is possible for the DNN model to satisfy the PDE as a solution of constrained optimization problem. In this respect PINN's are more efficient for solving inverse problems rather than direct problems as for the latter, relatively large number of parameters have to be optimized.

The remaining part of the paper is comprised of sections commencing with the mathematical formulation, wherein the PDE is stated along with thedefinitions of all the variables and the physical parameters. In the sequel the Forward problem, Inverse problem are described.In a logical sequence, DNN model and loss functions are presented. As a part of validation of the proposed method some numerical experiments are presented in the section under Numerical Simulation. Finally, the summary and the concluding remarks are provided in the section Conclusion.

Mathematical Formulation

Figure 29.1 depicts the physical model of a pipe simply supported at the ends A and B. The length of the span AB is L. Concentrated forces act at P points $x_1, x_2, \ldots x_p$. The equilibrium equation for the vibration of a uniform beam is given in Eq.1 (Craig, 1981). The mass per unit length is m, the damping coefficient c and flexural rigidity is EI. The displacement variable is denoted by u, with dot referring to the to the time differentiation D^n and as the nth order partial derivative with respect to the space variable x. The boundary conditions correspondto that of a simply supported beam and the initial conditions to that of a beam initially at rest.

The forcing function $F(x, t)$ is expressed as the product of the weighted delta functions in space and the function representing the temporal variation of the forcing function as shown in Eq.2. Here H_k denotes the weight of the Dirac delta function at the point x_k.

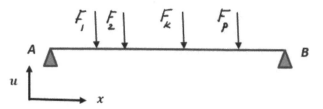

Figure 29.1 Simply supported pipe with point forces acting on it.

$$m\ddot{u}(x,t) + c\dot{u}(x,t) + EID^4u(x,t) = F(x,t) \tag{1}$$

for $0 \leq x \leq L$

$$F(x,t) = \sum_{i=1}^{P} H_i \delta(x - x_i) f_i(t) \tag{2}$$

Boundary Conditions (B.C's):

$$u(0,t) = 0 u(L,t) = 0 \tag{3}$$

$$EID^2u(0,t) = 0 \ EID^2u(L,t) = 0 \tag{4}$$

Initial Conditions (I,C's):

$$u(x,0) = 0 \dot{u}(x,0) = 0 \tag{5}$$

Forward Problem

Eq.1 is termed as a well posed forward or direct problem if $F(x, t)$, BC and IC's are completely defined. For the proof of the existence of the solution of problem we can follow the procedure in the papers by Nicaise and Zair [7] and Yamamoto [9]. It has been shown that for some suitable function spaces the forward problem has a unique solution. Each function $f_i(t)$ and its first order derivative is continuous (i.e. $f_i(t) \in C^1[0,T]$), where $[0,T]$ is the time interval) with an additional condition $f_i(0) \neq 0$.

Using the spectral expansion or mode superposition principle (Craig, 1981), the displacement can be expressed in terms of the modal components as Eqn. (6).

$$u(x,t) = \sum_{n=1}^{N} \phi_n(x) q_n(t) \tag{6}$$

With this expression, Eq.1 can be recast into a problem in the time domain for a particular mode of vibration (Eq. 7). For details one may refer to (Craig, 1981). The number of modes is denoted by N, the undamped natural frequency for mode n by w_n, the mass normalized eigen-value by $\Phi_n(x)$, modal damping by ς_n, and the modal displacement by $q_n(t)$.

$$\ddot{q}_n(t) + 2\omega_n \zeta_n \dot{q}_n(t) + \omega_n^2 q_n(t) = \sum_{k=1}^{P} \Gamma_{nk} f_k(t) \tag{7}$$

The modal participation factor $\Gamma_{nk} = \int^{L} \Phi_n(x) H_k \delta(x - x_k) dx$ (8)

The initial conditions are

$$q_n(0) = 0 \dot{q}_n(0) = 0 \tag{9}$$

The closed form solution to Eqs.7 to 9 can be obtained from Duhamel's Integral as

$$q_n(t) = \left(\frac{1}{w_{dn}}\right) \int_0^t \mathcal{F}_n(\tau) e^{-\zeta_n w_n(t-\tau)} \sin(w_{dn}(t-\tau)) d\tau \tag{10}$$

Here the following notations have been used

$$\mathcal{F}_n(t) = \sum_{k=1}^{P} \Gamma_{nk} f_k(t) \tag{11}$$

The damped natural frequency $w_{dn} = \sqrt{1 - \zeta_{sn}^2}$

Inverse Problem

For the inverse problem, $F(x, t)$ is the unknown function which has to be estimated. The forcing function has to be reconstructed such that the analytical solution of the PDE (Eq. 1) would match with the data of displacement observations taken from the field. The existence and uniqueness of the load reconstruction strategy have been studied by various researchers Nicaise and Zair [7] and Yamamoto [9]. Their approach has been to transform the PDE to a Volterra integral equation of the first kind. The unknown weights of the Dirac delta function are obtained as a solution and thus the forcing function can be reconstructed. However, conditions on the forcing functions are restrictive. For example, the time variation of all the functions is identical and at time $t = 0$, their values are not equal to zero.

In the present study, Neural Network Theory has been used to model the forcing function. Subsequently, the problem is formulated as that of a multi-variable optimization. The objective function is the loss functional calculated from the difference between the measured and calculated displacements. The PDE (Eq.1)is satisfied as a constraint.

PINN–Neural Network Model

A schematic diagram of the PINN Model is shown in Figure 29.2. The input layer consists of a single node which takes time variable as input. The output layer consists of P nodes, each of which corresponds to the value of the forcing function f_i at point x_i.

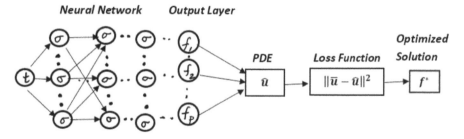

Figure 29.2 Schematic of DNN Model.

The Neural Network model is defined as follows:

Input Layer: $\mathcal{N}^0(t) = t \in \mathbb{R}$

Hidden Layers: $\mathcal{N}(t) = \sigma\left(W^l \mathcal{N}^{l-1} + b^l\right) \in \mathbb{R}^l$ for $1 \leq l \leq L-1$

Output Layer: $\mathcal{N}^L(t) = \sigma\left(W^L \mathcal{N}^{L-1} + b^L\right) \in \mathbb{R}^P$

The model represents a single input and multiple output system. The output is a vector consisting of the elements of the forcing functions at P points. Here L denotes the number of layers W^l, the weights, b^l the bias for the layer l and σ the activation function.

$$H_k f_k(t) = \mathcal{N}_k^L(t) \text{ for } 1 \leq k \leq P \tag{12}$$

From Eq. 10, the modal displacement can be expressed as

$$q_n^k(t) = G_n^k\left(\mathcal{N}_k^L(t)\right) \tag{13}$$

The approximation for the displacement function is given as

$$\hat{u}(x,t) = \sum_{k=1}^{P} \sum_{n=1}^{N} \phi_n(x) \mathcal{N}_k^L(t) \tag{14}$$

Loss Functional

The displacement u is the solution of Eq,1 and $\|(.)\|$ denotes the L^2 norm (Kreyszig, 1978). The residual R_{ki} for the kth observation $\bar{u}(x_k, ti)$ at time t_i is shown in Eq.15.

$$R_{ki} = \left\| \bar{u}(x_k, t_i) - u(x_k, t_i) \right\| \tag{15}$$

The loss functional is given as

$$Loss = (0.5) * \left(\frac{1}{MP}\right) \sum_{i=1}^{M} \sum_{k=1}^{P} R_{ki}^2 \tag{16}$$

where M is the number of observation points.

$$loss(\theta) = 0.5\left(\frac{1}{NT}\right) \sum_{i=1}^{NT} \sum_{j=1}^{M} \left(\bar{u}(x_j, t_i) - \hat{u}(x_j, t_i)\right)^2 \tag{17}$$

(Where θ is the vector of the training parameters)

Problem is to minimize the loss function with respect to the optimization parameters.

$$\theta^* = \text{Arg Minimum } loss(\theta) \tag{18}$$

Numerical Simulation

The problem studied was for an application of a single point force. The physical parameters are given in Table 29.1. The forward problem is solved for the displacements for a given force. These synthetic displacements are treated as observations for the inverse

Table 29.1: Physical data (SI units have been used for the quantities).

L	m	EI	c	Location of Force	Location of Obs. Point P1	Location of Obs. Point. P2
6	49	6860655	100	3	3	2.75

problem. Two observation points (*P1* and *P2*) are considered. Random noise has been added to the displacement data to simulate the real-life scenario. The maximum variation is kept at 5% of the peak displacement amplitude. The number of modes used is two, which has been found to be fairly accurate. The time interval (in secs)is [0, 0.5] with the number of time steps $NT = 1024$.

Results and Discussion

Tensorflow software has been used to build the DNN model. The parameters of the DNN model are: Number of hidden layers = 5 Number of neurons per layer = 130. These parameters have been finalized after training the model for a large number of epochs. The number of trainable variables = 68511.

The displacement-time plot of point *P1* is shown in Figure 29.3. The plot of the calculated vis-à-vis the applied force time-history is given in Figure 29.4. A close match is observedwith the final loss = 1.0E-05.This demonstrates the effectiveness of the method. Although this example involves a single force, the results for the application of two-point forces havealso been found to be sufficiently accurate.

However, the major challenge lies in the initialization of the training variables. If they are chosen arbitrarily then convergence becomes difficultand even not guaranteed. Hence it is extremely important to have some understanding of the physics of the problem in order to make a proper choice. Our approach has been heuristic and initial values were finalized by scaling the parametersthrough several trials.

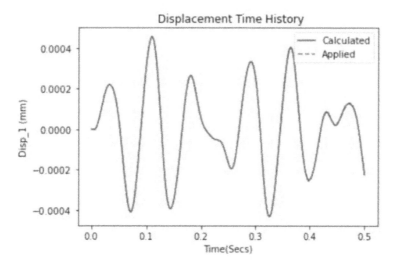

Figure 29.3 Displacement data: observations vs. calculated.

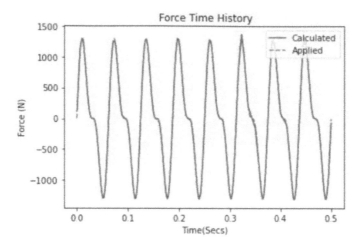

Figure 29.4 Force time history: applied vs. calculated.

Conclusion

A novel method based on Inverse Theory and PINN has been presented. The factors behind the motivation are the rapid development of ML techniques, availability of open-source software and advanced computing resources. The success of the method has been amply demonstrated through the numerical experiments. However, the most challenging step is the initialization of the training parameters. This critically governs the convergence rate and the solution time. In this study the approach for the initialization has been heuristic. There is a need to explore other methods with proper mathematical basis which can be a subject for further study.

The problem is of practical importance from the aspects of safety and mechanical integrity of an operating plant. However, no standard analytical methodology is available apart from empirical ones [10]. To the knowledge of the author this study is the first of its kind and it is a vindication of the effectiveness of machine learning in engineering applications.

References

1. Roy, Jr C. R. (1981). Structural Dynamics. USA: John Wiley.
2. Hornik, K. (1989). Multilayer feedforward networks are universal approximators. *Neural Networks,* 2, 359–366.
3. Kreyszig, E. (1978). Introductory Functional Analysis with Applications. India: John Wiley.
4. Lagaris, I. E., and Likas, A. (1998). Artificial neural networks for solving ordinary and partial differential equations. *IEEE Transactions on Neural Networks*, 9(5), 987–1000.
5. Raissi, M., Perdikaris, P., and Karniadakis, G. E. (2019). Physics-informed neural networks: a deep learning framework for solving forward and inverse problems involving nonlinear partial differential equations. *Journal of Computational Physics*, 378, 686–707.
6. Moussa, W. A., and AbdelHamid, A. N. (1999). On the evaluation of dynamic stresses in pipelines using limited vibration measurements and FEA in the time domain. *Journal of Pressure Vessel Technology*, ASME, 121, 37–41.

7. Nicaise, S., and Zair, O. (2004). Determination of point sources in vibrating beams by boundary measurements: identifiability, stability and reconstruction results. *Electronic Journal Differential Equations*, 20, 1–17.
8. Saha, S. (2008). Estimation of point vibration loads from field measurements for industrial piping. *Journal Pressure Vessel Technology, Transactions ASME*, 131(031206), 1–9.
9. Yamamoto, M. (1996). Determination of forces in vibrations of beams and plates by point wise and line observations. *Journal of Inverse and Ill-Posed Problems*, 4(5), 437–457.
10. Wachel, J. C. (1981). Piping Vibration and Stress. In Proceedings Machinery Vibration Monitoring & Analysis, Vibration Institute, New Orleans, USA, (pp. 1–20).

30 Dense Recognition of Offline Unconstrained Devanagari Handwriting

Bappaditya Chakraborty[1,a], Soumitra Roy[2,b], and Ujjwal Bhattacharya[3,c]

[1]MAKAUT, WB, Kolkata, India

[2]Dr. Sudhir Chandra Sur Institute of Technology & Sports Complex, Kolkata, India

[3]Indian Statistical Institute Kolkata, India

Abstract

Deep convolutional network (DCNN) based feature extraction has been popularly used in various classification tasks. Though the features computed at a deeper layer are found to be more efficient, requirement of the number of training samples as well as computational resources increase significantly with the depth of the network. In view of the same, here we present an end-to-end offline devanagari handwriting recognizer based on a DenseNet, a lightweight convolutional feature extractor. Although it's dense architecture performs equivalently in terms of recognition metric as the commonly used state-of-the-art deeper networks, but it involves comparatively less network parameters and simulation infrastructure. The performance of DenseCNN-RNN hybrid architecture with spatial attention is competitive based on both simulation time and recognition accuracy. The experimental results show that our approach achieves 10.73% word error rate(WER). Importantly, it consumes affordable training and testing time in a simple infrastructure despite the large volume and inherent challenges of devanagari script.

Keywords: Deep learning, handwriting recognition, segmentation free recognition, sequence to sequence.

Introduction

Over time, the need for efficient offline handwriting recognition has grown significantly, driven by the increasing demand for digitizing handwritten manuscripts. Extensive literature exists that explores this expansive research field. Recent advancements have prominently utilized segmentation-free strategies, with LSTM-based architectures being a prominent choice for decades. Researchers have integrated offline and online features to incorporate temporal information, yielding notable accuracy [1, 2]. Addressing the scarcity of comprehensive handwritten Devanagari datasets until recently, this paper focuses on Devanagari handwriting. Devanagari is widely used in Northern India, with over 400 million writers. In previous work Dutta et al. [3], real-time data augmentation (RTDA) dynamically boosted training data volume, employing an 18-layer Residual Network (ResNet-18) for recognition. Similar tasks utilizing deep hybrid networks include Bengali,

[a]bappa.chakraborty84@gmail.com, [b]mosinapur.sou@gmail.com, [c]ujjwal@isical.ac.in

Malayalam, Hindi, Arabic, and more highly cursive scripts. The offline handwritten text recognition (HTR) task is divided into two phases: an efficient feature extraction module for positional features and utilizing high-level features to map a target distribution. Feature extraction methods have transformed with the rise of convolutional neural networks (CNNs), which provide learnable feature extractors (Rania Maalej et al., 2018) [10]. The CNN-BLSTM-CTC-based HTR approach introduces segmentation-free recognition but with resource consumption and inference time limitations. This paper introduces a DenseNet-based approach for feature extraction, aiming for lower parameters and faster execution in both training and inference. The goal is to maintain performance quality while mitigating the resource-intensive nature of deep architectures, making them more applicable for production scenarios.

IIT-HW-Dev Unconstrained Handwritten Devnagari Dataset

Description of the Datasets

IIIT-HW-Dev dataset Dutta et al. [3] is one of the large publicly available dataset containing 95381 handwritten word images with 9540 distinct devanagari words. It contains 108 characters including numerals, vowels, consonants, vowel modifiers, special characters and end of line character. Presence of large variations in the data and non-uniform length of the word images, made it further complex large class recognition problem to address.

Complexity of Devanagari Script

Devanagari script consists of groups of characters identified as basic characters, vowel modifiers and conjunct characters. There are 48 basic characters of which 11 are vowels and another 37 are consonants. The concept of Shirorekha (a horizontal line on the upper part of the word) is one common and important feature of devanagari script. The dataset used in this experiment contains not only the character groups mentioned above, but also ten devanagari numerals and special characters. Presence of different unicodes for characters with almost similar shapes have further increased the complexity of the dataset. Figure 30.2 illustrates few such similarities present in dataset. However, high variation in width of the word images in terms of dimension and number of characters present in the label, makes it more generalized dataset to the researchers. Figure 30.1 is referencing such variations present in the corpus.

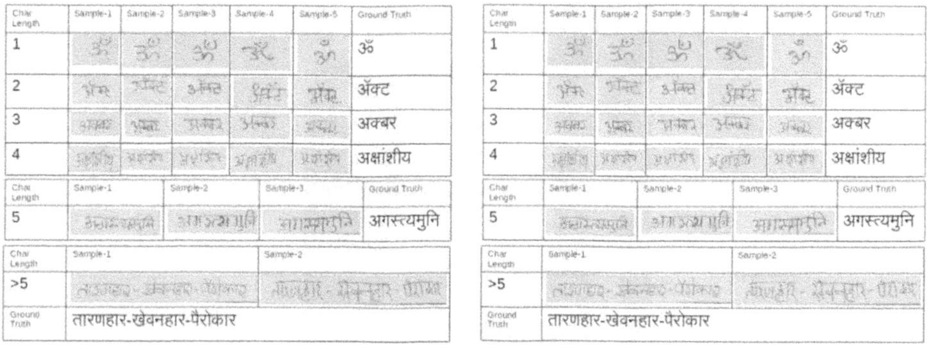

Figure 30.1 The figure depicts the presence of high width variations in the dataset. Inter class variations in writing styles can be visualized column wise.

1	ऍ ऎ ए ऐ	5	ळ ऴ
2	ऑ ऒ ओ औ अँ अ	6	न ऩ
3	ग ज ड ग़ ज़ ड़ ग ज ड	7	र ऱ
4	क ख क़ ख़	8	ऋ ॠ

Figure 30.2 Clusters of characters present in the dataset that exhibit shape similarities.

Pre-Processing

The samples are converted into grayscale images and height normalized with height H = 32 pixels. We apply white pixel padding to gain uniformity. Each image is resized to 32 × 512 dimensional grayscale intensity vector. The choice of H and W are made after experimenting with other higher and lower dimensions. No other image level processing is used. However, groundtruth level unicode reordering is used to map the visual position of the vowel modifiers to the position of their unicodes in groundtruth [11].

Network Architecture

The network architecture presented is centered around dense convolutional connections and bottleneck features, incorporating a parallel convolution block with three distinct convolutional kernels. A visual depiction of the architecture can be observed in Figure 30.3.

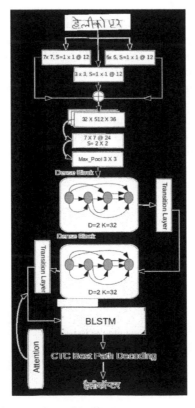

Figure 30.3 Dense CNN based network used in the experiment.

This parallel convolution layer is designed to capture fine to coarse convolutional features from raw images. Specifically, 3 × 3, 5 × 5, and 7 × 7 convolutional kernels are employed with a stride (S) of 1. Extracted features from these kernels are concatenated and then input into sequential DenseCNN blocks.

Dense Convolutional Feature Extractor

The Dense Convolutional Feature Extractor follows the principles of DenseNets, introduced by Forrest Iandola et al. in [12]. DenseNets contain multiple densely connected blocks, each housing numerous layers producing k features, where k is the growth rate. A key aspect is that each layer's input is the concatenated feature vector originating from prior layers within the same dense block. This approach optimizes gradient flow and minimizes feature map redundancies, as represented by Equation 1.

Where $x_i(r)$ is the i th input of region r. k_{ij} is the convolution kernel corresponding to ith input. $b_j(r)$ is bias. ReLU is a non linear activation function given by ReLU (x) = max(0, x) and out_j (r) is the jth output for region r.

$$x_l = H_l[x_0, x_1, x_2 \ldots x_{l-1}] \tag{1}$$

Where H_l indicates concatenation operation of the layer outputs and x_i indicates output features from ith layer.

A convolution operation is articulated as:

$$out_j(r) = \text{ReLU}\,(b_j(r) + \Sigma k_{ij}(r) * x_i(r)) \tag{2}$$

Furthermore, each dense block is followed by a transition layer utilizing a 1 × 1 convolution followed by 2 × 2 average pooling with a stride of 2. This reduction step is the sole means of dimension reduction in the DenseNet's feature vectors. Each layer within a dense block consists of initial batch normalization and a rectifier linear unit (BN-Relu) layer, followed by a 1 × 1 convolution, another BN-Relu, and a final 3 × 3 convolutional layer. The output of this final layer is concatenated with the input vector along the feature dimension. This concatenated output undergoes a similar layer iteratively k times within the dense block. For this implementation, k (the growth rate) is set to 32 for both dense blocks. The feature extractor generates a feature vector denoted as B × H × W × F, where B represents batch size, H signifies height, W denotes width, and F represents the number of features. The architecture aims to optimize feature extraction while minimizing resource usage.

Attention Based BLSTM

Attention-based networks are gaining traction in the domains of sequence classification and sequence-to-sequence mapping. The utilization of attention mechanisms isn't novel within document processing, as evidenced by the works of Patrick Doetsch et al. (2016) and Wenpeng Yin et al. [13]. The fundamental concept behind an attention mechanism involves selectively emphasizing intermediate representations of a sequence learner, facilitating improved alignment with the desired output. Despite the modest increase in computational overhead attributed to attention mechanisms, their implementation doesn't significantly impact the overall network size. This rationale spurred our adoption of attention-based decoding within our experiment.

The role of the attention mechanism is to synthesize encoded image features for every horizontal position. This operation is expressed mathematically in Equation 3:

$$F_{(i,j),t} = \text{Att}(\text{enc}, \alpha_{(i,j),t-1}, s_{t-1}) \tag{3}$$

Here, (i, j) denote the coordinates within the feature maps, enc signifies the encoded image derived from our proposed encoder, st–1 represents the state of the LSTM node at the previous time step, and α(i,j),t stands for the attention map at time t as given by Equation 4:

$$\alpha_{(i,j),t-1} = e^{z(i,j),t} / \sum_{i',j'} e^{z(i',j'),t} \tag{4}$$

The ultimate bidirectional LSTM (BLSTM) with the integrated attention mechanism generates hidden states amounting to (256 + 256) = 512 at each time step. The incorporation of a BLSTM is strategic in comprehending feature maps in both forward and backward directions, as detailed by Graves et al. [14]. As a result, the BLSTM network yields an output vector of dimensions 31 × 128, which is subsequently subject to decoding using the CTC (Connectionist Temporal Classification) best path decoding algorithm, as proposed by Graves et al. [14]. Figure 30.3 depicts the overall architecture of Densenet and Figure 30.4 is a detailed architecture of one single Dense block.

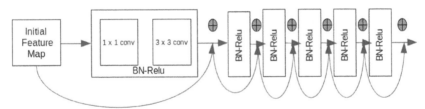

Figure 30.4 Architecture of a single dense block. A transition layer shown in diagram is composed of a Batch Normalization followed by a rectifier linear unit layer. Input from preceding layer is concatenated in forward direction. The ⊕ symbol indicates layer concatenation.

Experiments

The IIIT Devanagari dataset contains 95,381 word images of varying sizes and writing styles. These samples are divided into training, testing, and validation sets in a 6:2:2 ratio. To improve network generalization, Random Targeted Data Augmentation (RTDA) and random shuffling are applied during training. Due to limited memory, training data is read in batches of 25,000 samples from disk. This leads to longer convergence times as not all samples are seen within a single epoch, mimicking resource-constrained training conditions. The experiment is conducted on a moderate home computer with a 4-core 3.60GHz CPU and 4GB RAM. For comparison, recognition is also performed on higher-spec machines. Due to hardware limitations, deep architecture training is done on a higher-configured machine with GPU support. Pre-trained models are tested on the lower-spec machine for recognition time. A performance analysis is presented in Table 30.1, comparing computation and recognition metrics. Additionally, the study contrasts results with two other prevalent deep networks trained on the same dataset. The ResNet-based network's recognition performance is detailed in Dutta et al.'s [3] work. Further, the evaluation extends to the VGG-16-BLSTM hybrid architecture for a comprehensive assessment.

Evaluation Metrics

The evaluation of our proposed approach encompasses two key aspects. Firstly, the training and recognition times required by the network in a cost-effective, basic hardware environment. Secondly, the accuracy in word and character recognition. The metrics employed to gauge performance are elaborated below.

Training time per epoch (TTE) is computed mathematically by equation 5.

$$t_{epoch} = (t_s - t_e) \tag{5}$$

Where ts and te indicates start and end timestamp in seconds. Recognition time per sample (RTS) is computed by equation 6.

$$t_{recog} = (t_{st} - t_{et}) / TNS \tag{6}$$

Where t_{st} and t_{et} indicates start and end timestamp of test batch recognition in seconds, and TNS is the Total Number of Test Samples in batch. Recognition time per sample is computed as average inference time per batch. The second set of metrics are used to note the recognition performance of system. We use the traditional word error rate (WER) and character eror rate (CER) to evaluate our model. CER and WER are computed by equation 7 and 8.

$$CER = \Sigma \ ED(GT, RT)/C \tag{7}$$

Where ED= Edit Distance, GT= Ground Truth and RT=Recognized Text, C= Total Number of Characters present in test set.

$$W \ ER = M \ W / T \ N \ W \tag{8}$$

Where MW= Total Number of Mich-recognized words, TNW= Total Number of Word present in the test set. Detailed performance based quantitative results can be understood from Table 30.1.

The qualitative assessment of our proposed recognizer is demonstrated in Figure 30.5. The outcomes highlight our model's proficiency in handling diverse input image widths. Notably, character frequencies exhibit non-uniformity, ranging from the most frequent "aa" (Frequency-52046) to the least frequent "ll" (Frequency-10). Special characters generally have lower frequencies compared to consonants and select vowel modifiers. As anticipated,

Table 30.1: Experimental results providing comparative performance of various network architectures.

Feature Extractor	# of Layers	Params (Millions)	TTE(H)	RTS(Sec)	WER	CER
VGG16 + BLSTM	18	15.122 M	8.56	6.35	11.23	4.02
ResNet18 + BLSTM [6]	20	11.469 M	6.2	4.15	10.78	3.85
DenseCNN + BLSTM (Proposed Model)	6	4.473 M	0.5	0.8	10.73	3.82

Figure 30.5 Qualitative performance of our model on test set. Few samples and recognized ground truth are shown. Red colors indicate misrecognized words.

the model struggles in cases involving low-frequency or visually similar characters, as evident from Figure 30.5. The misrecognition can be attributed to shape resemblances or the scarcity of training instances for less frequent characters.

Conclusion

The paper propose a lightweight CNN-RNN hybrid architecture with spatial attention that can perform equivalently with well known deeper architectures without compromising the recognition accuracy. The structural complexity and visual similarities of characters have been an open challenge to address in future. Addition of more dense blocks and/or inclusion of more samples per word class may increase the recognition performance further. However, the purpose of the literature with the available devanagari datasets is well established. Experimental results show that the proposed lightweight architecture have improved the result presented in Dutta et al. [7] by a moderate fraction.

References

1. Mukherjee, S., Kumar, P., and Roy, P. P. (2019). Fusion of spatio-temporal information for indic word recognition combining online and offline text data. *ACM Transactions on Asian and Low-Resource Language Information Processing (TALLIP)*, 19(2), 1–24.
2. Valikhani, S., Abdali-Mohammadi, F., and Fathi, A. (2019). Online continuous multi-stroke persian/arabic character recognition by novel spatio-temporal features for digitizer pen devices. *Neural Computing and Applications*, vol. 32, 1–20.
3. Dutta, K., Krishnan, P., Mathew, M., and C. V. Jawahar. Offline handwriting recognition on Devanagari using a new benchmark dataset. In 2018 13th *IAPR international workshop on document analysis systems* (DAS), pp. 25–30. IEEE, 2018.
4. Dutta, K., Krishnan, P., Mathew, M., and Jawahar, C. V. (2017). Towards accurate handwritten word recognition for hindi and bangla. In National Conference on Computer Vision, Pattern Recognition, Image Processing, and Graphics, (pp. 470–480). Springer, 2017.
5. Chakraborty, B., Mukherjee, P. S., and Bhattacharya, U. (2016). Bangla online handwriting recognition using recurrent neural network architecture. In Proceedings of the Tenth Indian Conference on Computer Vision, Graphics and Image Processing, (pp. 1–8), 2016.
6. Iandola, F., Moskewicz, M., Karayev, S., Girshick, R., Darrell, T., and Keutzer, K. (2014). Densenet: implementing efficient convnet descriptor pyramids. arXiv preprint arXiv:1404.1869, 2014.
7. Yin, W., Schütze, H., Xiang, B., and Zhou, B. (2016). Abcnn: attention-based convolutional neural network for modeling sentence pairs. *Transactions of the Association for Computational Linguistics*, 4, 259–272.

8. Graves, A., Fernández, S., and Schmidhuber, J. (2005). Bidirectional lstm networks for improved phoneme classification and recognition. In International Conference on Artificial Neural Networks, (pp. 799–804). Springer, 2005.

9. Graves, A., Fernández, S., Gomez, F., and Schmidhuber, J. (2006). Connectionist temporal classification: labelling unsegmented sequence data with recurrent neural networks. In Proceedings of the 23rd International Conference on Machine Learning, (pp. 369–376), 2006.

10. Doetsch, P., Zeyer, A., and Ney, H. Bidirectional decoder networks for attention-based end-to-end offline handwriting recognition. In 2016 15th International Conference on Frontiers in Handwriting Recognition (ICFHR), pp. 361–366. IEEE, 2016.

31 Comparative Analysis of Electric and Hybrid Vehicles: A Review

Shweta Goyal[1,a], Abhishek Chakravorty[2,b], Vrij Mohan Vidhyarthi[2,c], Nanse[2,d], and Sheetal kapoor[2,e]

[1]Assistant professor, Graphic era deemed to be University, Dehradun

[2]Assistant professor, Tulas Institute Dehradun, Dehradun

Abstract

When cars were initially introduced, engineers and designers made a lot of effort to make them more affordable, inexpensive to run, and environmentally friendly. Fuels like gasoline and diesel are often used to power these cars. Internal combustion engines, one of the causes of pollution, have historically powered cars. The study seeks to increase knowledge of alternative internally combusted engine solutions that are both economically and environmentally sound, such as electric vehicles (EVs), in light of the rising cost of gasoline and the environmental issues connected with cars.

The development and use of electric and hybrid cars have significantly increased as a result of the shift towards more environmentally friendly transportation options. In this study, electric cars (EVs) and hybrid vehicles (HVs) are thoroughly reviewed and compared, with an emphasis on their technical developments, environmental effects, performance measures, market penetration, and future prospects. Battery systems, electric drivetrains, energy management systems, and charging infrastructure are just a few of the important parts and technologies that are extensively explored in this examination of EVs and HVs. To comprehend the benefits and constraints of EVs and HVs in different circumstances, performance indicators including as acceleration, range, and efficiency are rigorously assessed. The present level of adoption and future estimates may be understood by comparing market trends, consumer preferences, governmental incentives, and industry investments.

The paper also explores difficulties that EVs and HVs encounter, including limits in battery technology, accessibility to charging infrastructure, and manufacturing scalability. The report emphasizes current initiatives in research and innovation focused at solving these problems.

Keywords: batteries, electric propulsion, electric vehicles, engines, hybrid electric vehicles, motors.

Introduction

Electric vehicles employ motors that are powered by batteries that are connected to the motors, unlike conventional automobiles. Since electric automobiles don't produce exhaust gas, they are less harmful to the environment [1]. These cars' rechargeable batteries may be charged at any outlet. Reusable batteries may be used without worrying about appropriate

[a]shwetugoyal@gmail.com, [b]abhishek.chakravorty@tulas.edu.in, [c]vmvidyathi@tulas.edu.in, [d]na nse@tulas.edu.in, [e]sheetal.kapoor@tulas.edu.in

disposal. One of the key advantages of the car is its great efficiency, which comes from its internal electric motor [2]. Electric cars contribute to lowering pollution and greenhouse gas emissions both today and in the future since they don't emit any exhaust gases. There are many different types of cars, including conventional and hybrid electric ones [3].

Conventional Vehicles

These automobiles are equipped with an engine that generates power by burning gasoline. An internal combustion engine that uses either gasoline or diesel to power the wheels [4]. These are less costly than electric cars since they don't consume any energy for propulsion— just for radios and headlights [5]. It is believed that the car's engine is a substantial source of pollution. The majority of the cars on the road are propelled by these engines. They are affordable for many individuals wants to buy these cars because of the strength that fuel-burning engines provide.

Hybrid Electric Vehicles

In hybrid electric automobiles, the two energy sources combined are an internal combustion engine and an electric motor. A hybrid electric car is one that combines an ICE engine with an electric motor [6]. This technology can help with a variety of optimization tasks, including boosting power and fuel efficiency. Hybrid electric cars are commonplace. Electric and diesel engines are regularly combined in large mining vehicles. Submarines, which run on both nuclear and diesel power, may be considered to be similar [7]. Any vehicle having two or more sources of power that may be utilized to produce driving force either directly or indirectly is said to have two or more sources of power.

Series hybrid and parallel hybrid are the two halves of a HEV [8]. The battery is connected to the power-generating engine in a series hybrid system. The vehicle's power comes from the battery. Energy propulsion makes use of both motors and combustion engines, much as parallel hybrid systems [9]. Both the motor and the engine contribute to your overall torque. The motor engine, which also functions as a generator, absorbs the energy from the internal combustion engine. Both the parallel and series hybrid systems absorb the energy generated when breaks are needed [10]. Figure 31.1 shows the series and parallel path for HEVs.

Parts in Electric Vechiles

The EV's construction is rather straightforward. The gearbox part is quite important. The battery has a lot of power [11]. To recharge the battery by converting the electric current

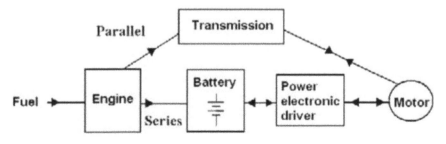

Figure 31.1 HEV-Series or parallel hybrid path.

from the source, the automobile must have a battery charger [12]. An electric current inverter converts the battery's DC voltage into a modified form of signal that powers the engine. The battery voltage of extra electrical car parts may be reduced to 10V–15V using a DC–DC converter.in Figure 31.2 we can see the basic key elements for EV.

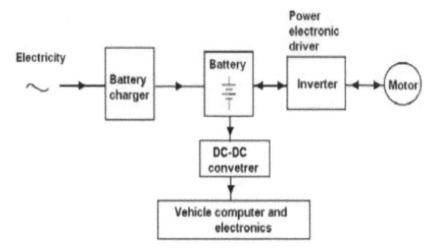

Figure 31.2 Key elements of EV.

Motors: Some Different Types of Motors that are Used in the Electric Vehicles

DC Motors Drive
Due to the torque-speed characteristics' low power consumption and simplicity in controlling the speed of the DC motor drive, this drive is in the forefront of electric thrust. The commutator in the DC drive means that it needs periodic repair [13]. Commutator is now in an era when great motor efficiency is being developed as a result of recent technological advancements.Induction motor is widely used AC motor. This motor also has a wide range of usability because of variable speed drive [14]. It can be used in air conditions, elevators etc. Lots of high power electric cars with more than 5kw, use induction motors.

Brushless Motor Drive
Because the drive of a conventional DC motor rotates at a low power while the windings rotate at a high power, the drive is inefficient [15]. The excitement is produced by a permanent magnet, and both the engine and the field are at risk from the engine's stable end's maximum voltage. Despite costing a bit more, the engine tends to survive longer than a DC motor. With the right driver, a brushless motor may take the place of most DC motors. It is used in low-powered electric cars [16].

Fuel Cells

The fuel cell is an electrochemical apparatus that converts electrochemical reaction energy into electrical energy. Fuel cells create electricity instead of storing it, and they may continue to do so for as long as fuel is available [17]. Fuel cells' low-emission capacity to convert fuel into electrical energy, durability, and heat recovery allow noiseless operation.

Ultra-Capacitor

Fast transitions between the battery's charging and discharging phases are possible because to the flexibility of the ultra-capacitors used in electric vehicles [18]. Because of their short-term energy storage, which helps with acceleration and mountain climbing, and their high average power to peak power ratio, electric cars perform better. Rapid charging or short-term energy storage are both possible uses. The high current charging method used by ultra-capacitors allows for fast charging.To charge the battery, a lot of power must be applied, and the charger's design is determined by how quickly or slowly the battery charges. High efficiency H-bridge power converters are used to convert electricity while the battery is being charged as shown in Figure 31.3.

Energy System

Battery: There are three Main Types of Charging Levels to Charge the Battery

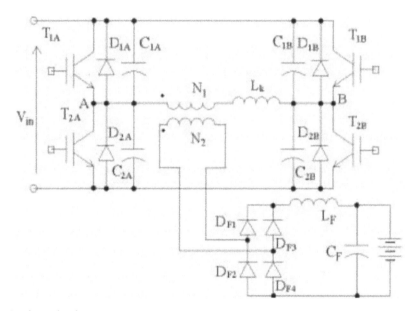

Figure 31.3 The H-bridge converter.

- **Regular charging current:** A regular or slow charger, which uses 15A, takes more than six hours to completely charge an electric vehicle. Because of the charger's sluggish charging rate, the best charger for an electric vehicle provides outstanding efficiency and extended battery life.
- **Medium current for charging:** The charging current for the batteries of electric cars ranges between 30 and 60 A. The battery may be charged with reasonable speed compared to a standard charger. The medium charger's installation cost is more than the usual charger's because of the equipment upgrade and the medium charger's higher power than the normal charger.
- **Current for quick charging:** Electric vehicle batteries are quickly charged using fast charging infrastructure that employs high charging currents of 150–400A. Fast

charging infrastructure is more expensive to develop than standard and medium charging infrastructure since a quicker charge demands a higher charging current and less efficient charge.

The electric car's batteries may be charged overnight at home using a regular charging method to increase battery efficiency. Although there is infrastructure for rapid and medium charging in homes and public places. Rapid charging infrastructure is usually used by the general people because of the high current demand and limited time available. The battery and the cars' efficiency are more significantly impacted by this high power charging.

- **Effects of battery rapid charging:** Electric car batteries are affected by fast charging in both good and bad ways. The electric vehicle battery may be charged overnight or while it is half charged to fulfill `the car's powered level and improve the power system. Chargers are non-linear devices that introduce harmonics into the power supply, whilst the normal charging of an electric vehicle battery puts additional demand on the charging infrastructure.

- **Minimizing Harmonic:** Scientists and engineers have suggested a variety of strategies to lessen harmonics in the power unit system. It is split into two sections: device levels and system levels.

- A new international study has been launched to lower the harmonic current in battery chargers and get rid of distortion. It is possible to achieve this by creating novel, affordable battery chargers with lower harmonic distortion.

- It is further divided into passive and active filter types at the system level. Passive filters make use of a simple phase-shifting transformer to minimize low-frequency harmonics and harmonics from inductors and capacitors that are unnecessary for the system. While a highly developed power electronics system called an active filter system level is able to more rapidly and correctly identify harmonics of the same magnitude. However, it was discovered that they needed more electricity to operate. Reducing current demand:

There has been a proposed concept to minimize maximum power usage to charge the electric vehicle [19]. Its main idea is to control and coordinate between the charging current and time to charge number of electric vehicles at the same time in the same charging station. By this the system can achieve the maximum current demand and batteries to charge quickly at the charging station.

- **Ultra-capacitor charger:** The power output in the ultracapacitor will fluctuate to minimum from maximum voltage if the energy stored in the power unit drops from highest to lowest. The system functions substantially differently from a battery because of the 25% power variation [20]. This voltage capacitor is not visible to customers since it is concealed within the power unit. When converting power, a tapped converter is employed because it has a higher efficiency than a transformer isolated version, which has a lower efficiency.

- **BMS (Battery management system):** It is called battery management system. This system is a formation of many cells connected either in parallel or in series as per the system is designed and has to checked and monitored. This system looks after the voltage, current and temperature of the system as this helps system control and protect the parameter of the system. To charge the battery a cable has to plug in to the charging point. A dedicated charging station are installed to charge the electric vehicles and the tariff is calculated by the unit it consumes during charging. It usually take 2.8kW of power to charge a private four wheeler electric car.

There are two parameter in the system. The state of charge (SoC) which gives the system the charging condition if the battery by calculating the information provided by the voltage and current [21]. The other is state of health (SoH) which is used to look after battery's health in the state of charge and its life period in Figure 31.4 we can clearkt see the BMS system of EVs.

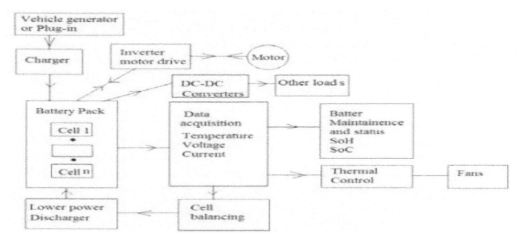

Figure 31.4 Structure of Battery Management System.

- **EMS(Energy management system):** It's also called energy management system. The energy stored in the ultra-capacitor uses number of capacitor to store the energy or uses other energy storage equipment like battery. This type of structure helps the system to monitor and control the system at the same time [22].

Charging Requirement

Network for Charging Charging Stations

Similar to gas pumps, electric vehicle charging stations operate similarly, and certain power outlets include rapid charging connectors for batteries. Since a significant current is required to swiftly charge the battery, the fast charging station is run on three-phase line energy. There are several ways to power the charging station. Magnetic contactless charging transmits electricity using magnetic induction since it is not connected to any metal. Safeguards are required because the high power voltage and current being delivered pose a risk to people [23].

Future Technologies for Batteries

Since the battery is the primary electrical component of an electric vehicle and necessary for the operation of all other electrical components, it is exposed to severe pressure and eventually loses efficacy over time. In order to solve these issues, scientists and engineers are always working to improve batteries and create new battery kinds that are highly effective yet have minimum detrimental effects on the health of vehicles.

Our era has been nicknamed the "next gold rush," and new technologies will make batteries capable of producing a great deal of power and efficiency. New technologies may help to tackle issues like range anxiety and battery charging times as they may reduce charging times while increasing range anxiety. Because batteries are so expensive, electric cars are expensive. A few issues must be resolved before electric cars are permitted to be driven on public roads. While the operating systems developed for power savings and enhanced performance are not spectacular and the lithium-ion batteries now used in electric vehicles have substantial shortcomings, new technologies will overcome these problems and increase the number of EVs on the road. Researchers at Chalmers University of Technology are changing the way electric vehicles are made to take advantage of recent breakthroughs in battery technology. The weight of the autos will be lowered as a consequence. To increase their durability, these batteries' negative and positive electrodes are constructed of lithium iron phosphate and carbon fiber, respectively. Recently, new battery designs and technologies, as well as certain battery-related research, have been created.

- Nanotube-based carbon electrodes
  The ultrafast carbon electrode is a ground-breaking innovation from NAWA Technologies that might enhance the current battery. They are using a unique electrode material that blends Nanotechnology and Clean Technology to create the Ultrafast Carbon Battery. Compared to other batteries, this one is more power-efficient and has a larger energy capacity. Additionally, it can charge more fast. As the battery's condition becomes better, charging will be quicker—just five minutes to reach 80% capacity.
- Cobalt-free batteries are a
  Cobalt, the most expensive material used as a cathode in lithium-ion batteries, raises the price of batteries for electric automobiles by 35 to 45 percent. Cobalt will be used instead, bringing down the price of batteries while also making them more ecologically friendly. Cobalt-free batteries with high energy density and a range anxiety of more than 500 kilometers are produced by a Chinese company under the name of SVOLT.
- Battery for seawater
  The development of new kinds of batteries that will improve both the health of the cars and the batteries and exacerbate the range anxiety that vehicles suffer when fast charging is being advanced by several other growing businesses and technology. Making batteries more environmentally friendly is one of the objectives of developing new technologies and enhancing current ones. Due to the lightning-fast pace of battery development and the success of new technologies with regard to electric vehicles, this industrial era is analogous to the following gold rush. Several manufacturers, including Tesla, Mercedes-Benz, and others, are working on new battery technologies.

EV or HEV

When in use, electric and hybrid automobiles have a number of potentials. Just batteries are used in electric cars, whereas an internal combustion engine is also used in hybrid vehicles. EVs are small and light since they just have a motor and battery. As contrast to hybrid electric vehicles, which include an engine that emits exhaust gas and a battery that may be used when required, they don't discharge any toxic exhaust gases into the environment, making them more environmentally friendly. HEVs have the benefit of being able to operate the engine and charge the batteries at the same time. Electric cars need to be charged at a charging station with the right amount of electricity in order to avoid battery damage and degraded battery performance. The driving range of electric cars is impacted by battery charging, which is a major problem. Several research and development projects are under process to address these concerns. New theories and technologies are created to address these problems, extending battery life and improving battery efficiency, enabling the vehicle to go further and farther. As technology advances, electric car range and battery efficiency must increase.

Conclusion

The research concludes by offering a thorough examination of electric and hybrid cars with an emphasis on battery and energy management systems. These two crucial elements have fundamentally altered the effectiveness, performance, and all-around viability of electric and hybrid cars.Energy storage units are now operated and maintained at their best for a longer period of time thanks to the development of battery management systems. Improved safety, range, and dependability have been made possible by advanced monitoring, thermal management, and cell balancing approaches. Additionally, current battery chemistry research indicates even higher energy densities, quicker charging speeds, and increased environmental sustainability. As a vital interface between power sources, storage facilities, and vehicle propulsion, energy management systems have arisen. Energy is distributed and used intelligently to improve both vehicle economy and the seamless integration of renewable energy sources and smart grid technologies. These developments provide a way forward for lowering transportation's carbon impact in addition to lowering operating expenses. However, issues with batteries and energy management still exist. Obstacles persist in the form of battery deterioration, limited charging infrastructure, and the need for defined communication protocols. Similar to this, it takes ongoing research and innovation to improve energy management algorithms to account for various driving situations and customer preferences.

The conclusions drawn from this study highlight the value of ongoing multidisciplinary research cooperation as the automobile industry moves toward electrification. It will be crucial to take use of developments in battery chemistry, materials science, electronics, and software engineering in order to get beyond present obstacles and introduce electric and hybrid cars into the general public's consciousness. The development of battery and energy management systems for electric and hybrid cars, in conclusion, has enormous potential. Accelerating the worldwide shift to a more sustainable and effective transportation paradigm will depend on addressing the outstanding issues and taking advantage of new possibilities. Electric cars are an alternative to other kinds of vehicles and are more environmentally friendly as a result of rising gasoline prices. However, there are certain benefits and drawbacks to electric cars as well that will be resolved in the future.

References

1. Cheng, K. W. E. (2009). Recent development on electric vehicles. In 2009 3rd International Conference on Power Electronics Systems and Applications, PESA 2009, (pp. 1–5).
2. Jones, W. D. (2003). Hybrids to the rescue [hybrid electric vehicles]. *IEEE Spectrum*, 40(1), 70–71. doi: 10.1109/MSPEC.2003.1159736.
3. Jones, W. D. (2005). Take this car and plug it [plug-in hybrid vehicles]. *IEEE Spectrum*, 42(7), 10–13. doi: 10.1109/MSPEC.2005.1460339.
4. Chan, C. C. (2002). The state of the art of electric and hybrid vehicles. *Proceedings of the IEEE*, 90(2), 247–275. doi: 10.1109/5.989873.
5. Charadsuksawat, A., Laoonual, Y., and Chollacoop, N. (2018). Comparative study of hybrid electric vehicle and conventional vehicle under New European driving cycle and Bangkok

driving cycle. In 2018 IEEE Transportation Electrification Conference and Expo, Asia-Pacific (ITEC Asia-Pacific), 2018, (pp. 1–6). doi: 10.1109/ITEC-AP.2018.8432599.

6. Cheng, K. W. E. (2009). Recent development on electric vehicles. In 2009 3rd International Conference on Power Electronics Systems and Applications, PESA 2009, (pp. 1–5).

7. Gao, Z., LaClair, T., Ou, S., Huff, S., Wu, G., Hao, P., Boriboonsomsin, K., and Barth, M. (2019). Evaluation of electric vehicle component performance over eco-driving cycles. *Energy*, 172, 823–839. https://doi.org/10.1016/j.energy.2019.02.017

8. Yaşar, H., and Canbolat, G. (2019). Performance comparison for series and parallel modes of a hybrid electric vehicle. *Sakarya University Journal of Science*, 23(1), 43–50. https://doi.org/10.16984/saufenbilder.369149

9. Fayaz, F., and Ilahi Bakhsh, F. (2018). Electrical vehicle, Parallel Hybrid electrical vehicle, Battery electrical vehicle, Plug-in Hybrid electrical vehicle. *International Journal of Engineering Research in Electrical and Electronic Engineering (IJEREEE)*, 4(1). https://www.researchgate.net/publication/323240047

10. Bansal, A. (2018). Comparison of electric and conventional vehicles in indian market: total cost of ownership, consumer preference and best segment for electric vehicle. https://doi.org/10.21275/ART20181202

11. Paykani, A., and Shervani-Tabar, M. T. (2011). A comparative study of hybrid electric vehicle fuel consumption over diverse driving cycles. *Theoretical and Applied Mechanics Letters*, 1(5), 052005, ISSN 2095–0349. https://doi.org/10.1063/2.1105205.

12. Singh, R., Kumar, N., Kumar, V., and Bindal, R. K. (2021). Design and implementation of hybrid solar e-rickshaw (pp. 133–138). https://doi.org/10.1007/978-981-15-7504-4_14

13. Katoch, S., and Rahul, R. K. B. (2019). Design and implementation of smart electric bike eco-friendly. *International Journal of Innovative Technology and Exploring Engineering (IJITEE)*, 8(6S4), pp. 965–967. (Scopus indexed), ISSN: 2278-3075.

14. Bindal, R. K., and Kaur, I. (2020). Torque Ripple Reduction of Induction Motor using Dynamic Fuzzy Prediction Direct Torque Control. has been published in ISA Transactions (Elsevier), (Vol. 99, pp. 322–338).

15. Anjum, R., Yar, A., and Yousufzai, I. K. (2019). Dual loop speed tracking control for torque management of gasoline engines. In Proceeding 18th European Control Conference, (pp. 3084–3089).

16. Feng, M., and Jiao, X. (2017). Double closed-loop control with adaptive strategy for automotive engine speed tracking system. *Internation Journal Adaptive Control Signal Process*, 31, 779–790.

17. Roy, J. V., and Indulal, S. (2018). Hybrid electric three-wheeler with ann controller. In 2018 International Conference on Circuits and Systems in Digital Enterprise Technology (ICCSDET), (pp. 1–5).

18. Kwon, Y., Kim, S., Kim, H., and Byun, J. (2020). What attributes do passengers value in electrified buses. *Energies*, 13(10), 2646.

19. Yang, Z., Shang, F., Brown, I. P., and Krishnamurthy, M. (2015). Comparative Study of Interior Permanent Magnet, Induction, and Switched Reluctance Motor Drives for EV and HEV Applications. *IEEE Transactions on Transportation Electrification*, 1, 245–254.

20. Kwon, Y., Kim, S., Kim, H., and Byun, J. (2020). What attributes do passengers value in electrified buses. *Energies*, 13(10), 2646.

21. Nugraha, Y. U., Asfani, D. A., Negara, I. M. Y., Aziz, M., and Yuniarto, M. N. (2021). Technology review of electric motor for hybrid-electric vehicle. In Region 10 Conference (TENCON) TENCON 2021-2021 IEEE, (pp. 777–781), 2021.

22. Ye, J., Feng, H., Xiong, W., Gong, Q., Xu, J., and Shen, A. (2021). A real-time model predictive controller for power control in extended-range auxiliary power unit. *IEEE Transactions on Vehicular Technology*, 70(11), 11419–11432.

23. Feng, H., Ye, J., Xiong, W., Gong, Q., and Xu, J. (2021). Delay-dependent MIMO robust control for power following system of auxiliary power unit in series hybrid electric vehicles. *IEEE Transactions on Vehicular Technology*, 70(5), 4353–4365.

32 Hallucination Reduction in Long Input Text Summarization

Tohida Rehman[1,a], Ronit Mandal[1,b], Abhishek Agarwal[1,c], and Debarshi Kumar Sanyal[2,d]

[1]Department of Information Technology, Jadavpur University, Kolkata, India

[2]School of Mathematical and Computational Sciences, Indian Association for the Cultivation of Science, Kolkata, India

Abstract

Hallucination in text summarization refers to the phenomenon where the model generates information that is not supported by the input source document. Hallucination poses significant obstacles to the accuracy and reliability of the generated summaries. In this paper, we aim to reduce hallucinated outputs or hallucinations in summaries of long-form text documents. We have used the PubMed dataset, which contains long scientific research documents and their abstracts. We have incorporated the techniques of data filtering and joint entity and summary generation (JAENS) in the fine-tuning of the Longformer Encoder-Decoder (LED) model to minimize hallucinations and thereby improve the quality of the generated summary. We have used the following metrics to measure factual consistency at the entity level: precision-source, and F1-target. Our experiments show that the fine-tuned LED model performs well in generating the paper abstract. Data filtering techniques based on some preprocessing steps reduce entity-level hallucinations in the generated summaries in terms of some of the factual consis tency metrics.

Keywords: Data filtering, hallucination, JAENS, LED, summary-worthy entities, text summarization.

Introduction

With the exponential growth of textual data, the need for effective summarization techniques becomes crucial to extracting relevant and concise information from lengthy documents. Text summarization plays a vital role in various domains, including news articles, legal documents, and scientific papers. However, when it comes to handling long input texts, such as research papers or legal documents, the task becomes even more challenging. The input documents of such tasks are often significantly longer than the maximum context lengths of most standard transformer models. This has motivated researchers to explore changes in model architecture and training strategies. For instance, to avoid the quadratic growth in memory consumption of the attention computation in transformers, many memory-efficient transformer variants have been proposed in [1]. To handle long

---

[a]tohidarehman.it@jadavpuruniversity.in, [b]ronitmandal21@gmail.com, [c]abhi2001agarwal@gmail.com, [d]debarshi.sanyal@iacs.res.in

inputs, Beltagy et al. [2] have added a long input pre-training stage to the transformer while Chalkidis et al. [3] have only fine-tuned the models with long inputs without any pre-adaptation.

In the context of long input text summarization, one common issue is the presence of hallucinations, that is, the generated summary includes factual inconsistencies or introduces information not present in the source document. Hallucinations can be categorized as intrinsic and extrinsic hallucinations [4]. Intrinsic hallucinations occur when the model interprets information from the input text incorrectly but uses the terms or concepts that occur in the source document. Extrinsic hallucinations occur when the model generates text that does not match the input text, that is, uses terms and concepts not even present in the source document. Hallucinations can undermine the reliability and accuracy of the summarization process, potentially leading to misinformation or misleading interpretations. These contradictions in fact can exist at the entity or phrase level. A model-generated summary may include named-entities that were not included in the source document. This is known as the entity hallucination problem [5].

The main contributions of this paper are:

1. We use Longformer Encoder-Decoder (LED) model [2] to generate summary of scientific articles in the PubMed dataset [6]. In addition, we explore two techniques, namely, data filtering and JAENS (Join sAlient ENtity and Summary generation) [5] to study their effect on the factual consistency of the generated summaries.
2. We analyze the factual consistency of the output summary at the entity level using the following metrics: precision-source and F1-target, introduced by [5]. We also use the traditional metrics, namely, ROUGH [7], METEOR [8], and BERTScore [9], to evaluate the performance of the models. The entity-based data filtering technique improves the precision-source but the other metrics achieve higher values when fine-tuning with LED is done without the other two techniques. Our code and results are available on github[1].

Literature Survey

Early research efforts in text summarization predominantly focused on extractive methods, which involve selecting the most significant sentences or phrases from the source document to form the gist. While extractive summarization approaches achieved reasonable success, it was hard to modify these methods to handle information that required rephrasing or merging content from multiple sentences. This limitation led to research in abstractive summarization techniques, which aim to generate summaries by understanding the source text and producing new sentences that capture the essential information. The emergence of recurrent neural networks (RNNs) that are capable of processing and producing text has significantly improved abstractive summarization, but they sometimes exhibit undesirable behavior such as incorrectly reproducing factual details, an inability to deal with out-of-vocabulary (OOV) words, and repetitive themselves [10]. The pointer-generator model with a coverage mechanism helps to resolve the problem of out-of-vocabulary (OOV) words, and repeating phrase generation [11–15].

Large pre-trained transformer models have proven to be exceptionally capable of dealing with natural language tasks [16, 17]. Handling extended textual sequences, on

---

1   https://github.com/tohidarehman/Hallucination-Reduction-Text-Summarization

the other hand, remains a considerable issue for these models. These challenging input documents are often substantially longer than the maximal context lengths of typical transformer models, necessitating both specialized model architectural adjustments and unique training regimes to accommodate. For example, numerous memory-efficient transformer variations have been proposed to prevent the quadratic escalation in memory consumption of the attention estimation in transformers. Another severe issue is the inability of current abstractive summarization methods to generate faithful results. These systems frequently struggle to verify that the generated summaries only include information extracted from the source document and do not include manufactured or hallucinated statements. These hallucinations can occur for a variety of causes, including biases in the training data, a lack of context perception, or model overoptimization. Cao et al. [18] and Krys´cin´ski et al. [19] reported that approximately 30% of the summaries generated by seq2seq models suffer from the issue of hallucination. As a result, as noted in the works of the NLP community, attention has been drawn more and more to the faithfulness and factual components of abstractive summarization [19–21]. Many recent works study entity-level and relation-level hallucination problems in the generated text. Nan et al. [5] address entity hallucination by applying a filter on the training data and multi-task learning. Goyal and Durrett [20] study relation hallucination, that is, whether the semantic relationships manifested by the individual dependency arcs in a generated sentence are entailed by the source sentence. One notable work by Narayan et al. [22] incorporates entity chain content planning to guide faithful summary generation. There has been growing interest in quantitatively measuring the faithfulness of text generation models. Most widely-adopted evaluation metrics for text generation, such as ROUGE [7] and BERTScore [9] correlate poorly with the human perceived faithfulness of the generated text [19]. Recent studies explore categorical and content-based analysis for measuring the faithfulness of summaries [20].

Methodology

To handle long input sequences, we utilized the pre-trained checkpoints of the Longformer Encoder Decoder (LED) model [2], which incorporates a sliding window and dilated sliding window attention mechanisms. It consists of both the encoder and decoder Transformer stacks, but instead of using full self-attention in the encoder, it employs the Longformer's efficient local+global attention pattern.

The decoder applies full self-attention to all encoded tokens and previously decoded locations. Because pre-training LED is expensive, authors in [2] have used BART parameters to initialize LED parameters and adhered to BART's exact design in terms of the number of hidden sizes and layers. This allows it to effectively process lengthy inputs. We performed fine-tuning of the pre-trained LED model to adapt it specifically for text summarization of scientific documents. To ensure the accuracy of the summaries, we implemented *scispaCy-based* Named Entity Recognition (NER) on the ground truth summaries. We applied the JAENS (Jointly Aligned Entity Names and Summaries) approach to augment salient entities in front of the abstracts. Training the model to recognize summary-worthy named-entities aims to enhance the precision and recall related to named-entities in the generated summaries.

We have performed experiments with 3 variants with the LED model: (1) fine-tuned on the LED model, (2) fine-tuned LED model with the filtered dataset, and (3) fine-tuned LED model using the JAENS approach on the filtered dataset.

Fine-Tuning LED

Pre-trained models like LED learn rich language representations from a large corpus. Fine-tuning customizes these models for specific tasks. It initializes the model with pre-trained weights, then fine-tunes it on a task-specific dataset using backpropagation. Fine-tuning leverages the model's language understanding saves time and resources, and requires less labeled data. This approach enhances text summarization by adapting the model to task-specific data while leveraging its pre-trained knowledge.

Entity-Based Data Filtering

As demonstrated successfully by [5], the training dataset's quality has a significant impact on the amount of entity-level hallucinations present in the generated summary. With that in mind, we applied *scispaCy-based* Named Entity Recognition (NER) to the gold summary for the PubMed dataset. This allows us to identify all the named-entities present in the gold summary. Our objective is to ensure that these named-entities have corresponding *n-gram* matches within the source document. For unigram matching, we avoid matching any stop words. Therefore, if any named-entity of a sentence in the summary cannot be found within the source document, we decided to exclude that sentence from the summary. If the number of sentences in the summary is one and using the filtering technique, we need to remove that sentence, then the entire article-summary pair has been removed from the dataset.

Joint sAlient ENtity and Summary Generation (JAENS)

The JAENS (Joint sAlient ENtity and Summary generation) approach, originally introduced by Nan et al. [5], is an alternative generative approach aimed at enhancing entity-level precision, and recall metrics. JAENS trains the LED model to construct a sequence that contains summary-worthy named-entities, a special token, and the summary itself, as opposed to typical summarization approaches. This approach enables the model to simultaneously learn the identification of summary-worthy named-entities while generating summaries, similar to the multitask learning approach. By prioritizing the generation of salient named-entities in the decoder, JAENS ensures that the summaries incorporate and highlight these important entities through decoder self-attention. By incorporating the JAENS approach into our project, we aim to mitigate entity-level summary hallucinations and improve the overall quality of the generated summaries.

Experimental Setup

Datasets

We used a dataset collected from a scientific repository, PubMed[2], and was introduced in [6]. We chose scientific papers as our dataset because they are examples of long documents with a standard discourse structure. Furthermore, scientific papers are rich in domain-specific terminology and technical information, which makes them an important source of information for researchers and practitioners alike. PubMed is a biomedical literature database that contains over 30 million citations and abstracts of research articles. The

---

2   https://pubmed.ncbi.nlm.nih.gov/

dataset contains almost 19,000 scholarly publications on diabetes from the PubMed database, which are categorized into one of three categories. In our experiment, we choose, for training 2000 examples, validation 250 examples, and testing 250 examples.

The size of the used dataset after applying the entity-based filtering procedure was 1798 examples for training, 232 examples for validation, and 236 examples for testing. The average number of sentences in summary before applying the entity-based data filtering technique was 7.33, 7.04, and 7.51 for training, validation, and test datasets. The average number of sentences in a summary after applying the entity-based data filtering technique is 4.34, 4.11, and 4.58 for training, validation, and test datasets.

Data Processing

We eliminated all punctuation, numerals, special characters, mathematical formulas, and citation markers from the documents and lowercase the entire corpus. When we were going through documents, we made sure they were the right length and had the right structure. If something was too long, like a thesis, or too short, like a tutorial announcement, we removed it. We also looked for documents that did not have an abstract or a clear structure. To understand the structure, we used the section headings as clues. Sometimes, documents had figures or tables that did not help us understand the text. We got rid of those, keeping only the words. In our model, the maximum number of allowed input tokens is 8192, that of output tokens is 512, and the minimum number of output tokens is 100 only.

In line with the JAENS approach, we used the *scispaCy model(en_core_sci_sm)*[3] library to generate summary-worthy named-entities and augmented the list of comma-separated named-entities before the ground truth summary (abstract) for each sample of the dataset. The sequence of named-entities is followed by a special token, which helps separate the entities from the abstract. This special token is chosen from the model's vocabulary such that it is not commonly occurring and can help the model learn to recognize the named-entities separately from the actual abstract. This helps in training the model as now the model will apply special attention to these entities while generating the summary.

Implementation Details

We conducted our experiments using *Google Colab Pro+,* which provided us with an *NVIDIA A100 GPU.* For all experiments, we used the base variant of the pre-trained LED model *led-base-16384*[4], due to resource limitations. Firstly, we fine-tuned the LED model on the original 2000-sample PubMed dataset. Secondly, we utilized a filtered version of the taken dataset by removing article-abstract pairs with a $prec_s$ score (to be defined in the next subsection) less than 1 (i.e., we ensure that the abstract – which is the ground-truth summary contains almost no hallucinations of entities) and performed fine-tuning on this filtered dataset. Finally, we incorporated the JAENS approach into the fine-tuning process by augmenting summary-worthy named-entities in front of the abstract for each example of the filtered train dataset, aiming to enhance entity-level precision, recall, and F1 metrics in the generated summaries and thus reduce the entity-level hallucinations. For all the models, we fine-tuned up to 10 epochs. To evaluate the models, we used the same test dataset that was obtained after the entity-based data filtering technique.

---

3   https://allenai.github.io/scispacy/
4   https://huggingface.co/allenai/led-base-16384/

Evaluation Metrics

We employ a comprehensive set of widely used text summarization evaluation metrics, including ROUGE [7], METEOR [8], BERTScore [9], to assess the quality and effectiveness of the generated summaries. Unfortunately, these metrics are inadequate to quantify factual consistency [19]. Hence, we have also used three new metrics, introduced by [5], to evaluate the factual consistency of the generated summaries.

We define $N(t)$ as the count of named-entities in the target (ground truth or gold summary) and $N(h)$ as the count of named-entities in the hypothesis (generated summary). To determine the number of entities in the hypothesis that have corresponding matches in the source document, we use $N (h \cap s)$. In circumstances when a named-entity in the summary spans many words, we consider it a match if any component of the named-entity can be identified in the original document, permitting partial matching based on n-grams. **Precision-source**, defined as, $prec_s = N (h \cap s)/N(h)$ is a metric that is used to determine the intensity of hallucination in relation to the source. Note that $prec_s$ represents the percentage of entities mentioned in the generated summary that can be retrieved from the source. Low $prec_s$ indicates that hallucination is possibly present in the generated text. However, $prec_s$ does not capture the computed summary's entity-level correctness in relation to the ground-truth summary. Entity-level accuracy of the generated summary is calculated using the **precision-target** as $prec_t = N (h \cap t)/N (h)$; the **recall-target** as $recall_t = N (h \cap t)/N (t)$; and **F1 score** as $F1_t = \dfrac{2 * (recall_t * prec_t)}{recall_t + prec_t}$. Here $N (h \cap t)$ represents the number of matched named-entities in the generated summary and the ground truth summary.

Note that the above precision and recall scores can be calculated in two ways. One is to consider the entity mentioned in each document (which may be the source s or target t or hypothesis h) as a set so that multiple occurrences of an entity in a document are equivalent to a single occurrence. The other is to consider the entity mentioned in a document as a list; here, if a metric is defined as $\mu = N (x \cap y) /N (x)$, then for each entity mention in x, we check if it occurs in y, and if so, increment the intersection count $N (x \cap y)$ by unity. The second approach is followed in [5], In the first approach, we denote the metrics as $prec_t^U$, $recall_t^U$, and $F1_t^U$ (U indicates that only unique entity mentions are considered). In the second, we represent them as $prec_t^{NU}$, $recall_t^{NU}$, and $F1_t^{NU}$.

Results

Comparison of the Models

In this sub-section, we report the results of the variations of fine-tuning the LED model on PubMed dataset. Table 32.1 shows the F1-scores for ROUGE-1 (R-1), ROUGE-2 (R-2), ROUGE-L (R-L), BERTScore, and METEOR metrics along with values of the entity-level factual consistency metrics $prec_s^U$, $prec_s^{NU}$, $F1_t^U$, and $F1_t^{NU}$ on the filtered test dataset.

The LED model fine-tuned on the filtered dataset achieves the highest $prec_s$ scores. However, when fine-tuning with LED is done without additional techniques like filtering or JEANS, the values of ROUGE, METEOR, BERTScore, and even $F1_t$ are the highest. This shows that not only n-gram matches and cosine similarity of embeddings are higher for the plain LED model, but the entity-level hallucination is also lower for it. Nan et al. [5] also observed a reduction in ROUGE scores when data filtering and JEANS were applied, and remarked that it could be due to the increased complexity during decoding. Surprisingly, we find that data filtering and JEANS do not improve the $F1_t$ scores. In future, we intend

Table 32.1: Evaluation of the models: F1-scores for ROUGE, METEOR, BERTScore, along with the $prec_s^U$, $prec_s^{NU}$, $F1_t^U$, and scores are used for evaluating the factual consistency of the generated summaries for the PubMed dataset. All score in percentage (%).

Model Name	R-1	R-2	R-L	R-LSum	METEOR	BERTScore	$prec_s^U$	$prec_s^{NU}$	$F1_t^U$	$F1_t^{NU}$
Fine-tuned LED	**35.12**	**14**	**21.57**	**29.96**	**32.08**	**84.96**	93.38	94.76	**43.76**	**46.14**
Fine-tuned LED + Filtered dataset	33.18	12.04	19.93	28.48	27.43	84.74	**96.04**	**96.83**	40.15	43.27
Fine-tuned LED +Filtered dataset + JAENS	30.21	09.13	18.26	25.87	23.55	84.35	92.16	89.36	40.15	36.34

to conduct a detailed study of this behavior and try to decipher its reason. This could be related to the inaccuracy in entity recognition that we observed for the dataset; for example, on manual review, we found that many phrases detected as entities do not appear to be very important, but their match/mismatch between the generated and golden summary do impact the $F1_t$ scores. In contrast, in [5], standard entities are detected which could be achieved with high accuracy. Another difference with [5] is that in our case, the dataset is much smaller and the summaries longer.

Case Study

Figure 32.1 shows sample outputs generated by the fine-tuning of the LED model without the filtered dataset, with the filtered dataset, and with both the filtered dataset and JAENS.

Ground-truth summary (abstract): "in recent years, evidence has accumulated that a dedicated plant cell wall integrity (cwi) maintenance mechanism exists. the available data suggest that it may represent a component of the stress response mechanisms underlying biotic and abiotic stress responses, which has not been identified previously as a distinct mechanism."

entities='years', 'evidence', 'plant', 'cell wall', 'integrity', 'cwi', 'maintenance', 'mechanism', 'data', 'stress response', 'mechanisms', 'biotic', 'abiotic stress responses'

Fine-tuned LED: "plant cells need to maintain the functional integrity of their walls during cell morphogenesis and exposure to biotic / abiotic stress. the available evidence suggests that a dedicated plant cell wall integrity (cwi) maintenance mechanism exists. while our understanding of the mechanisms regulating stress responses and morphogenesis has increased significantly, our knowledge regarding the processes maintaining cwi is still limited. in the last years, a number of yeast strains have been identified and the available knowledge regarding cwi maintenance during plant development and in yeast has increased substantially. in this review, we will focus on the role of particular cell wall defects in the perception of and response to botic stress responses.

materials and methodsa yeast cwi monitoring and maintenance network is quite complex, providing an indication of the possible complexity of the interaction and the direct impact of these differences on the biological role and function of the plant cwi."

entities='plant cells', 'functional', 'integrity', 'walls', 'cell morphogenesis', 'exposure to', 'biotic', 'abiotic stress', 'evidence', 'plant', 'cell wall', 'cwi', 'maintenance', 'mechanism', 'mechanisms', 'stress responses', 'morphogenesis', 'increased', 'knowledge', 'processes', 'years', 'yeast', 'strains', 'identified', 'plant development', 'review', 'defects', 'perception', 'response', 'botic stress', 'methodsa', 'monitoring', 'network', 'quite complex', 'indication', 'complexity', 'interaction', 'impact', 'biological role', 'function', 'plant cwi'

Fine-tuned LED + Filtered dataset: "plant cells need to maintain the functional integrity of their walls during cell morphogenesis and exposure to biotic / abiotic stress responses. the available evidence suggests that a dedicated plant cell wall integrity (cwi) maintenance mechanism exists. while the yeast cwi monitoring and maintenance network is quite complex, certain important differences exist that affect the biological role and function of the plant cwi maintenance mechanism. in this review, we will focus on the effects of particular cell wall defects on pathogen resistance and the mode of action of specific signaling cascades."

entities= 'plant cells', 'functional', 'integrity', 'walls', 'cell morphogenesis', 'exposure to', 'biotic', 'abiotic stress responses', 'evidence', 'plant', 'cell wall', 'cwi', 'maintenance', 'mechanism', 'yeast', 'monitoring', 'network', 'quite complex', 'biological role', 'function', 'review', 'effects', 'defects', 'pathogen resistance', 'mode', 'signaling cascades'

Fine-tuned LED + Filtered dataset + JAENS: "plant cells need to maintain the functional integrity of their walls during cell morphogenesis and exposure to biotic / abio stress responses. the available evidence suggests that a dedicated plant cwi maintenance mechanism exists, but what about the role of dedicated cells wall integrity (cwi) maintenance during process."

entities='plant cells', 'functional', 'integrity', 'walls', 'cell morphogenesis', 'exposure to', 'biotic', 'abio stress responses', 'evidence', 'plant', 'maintenance', 'mechanism', 'cells', 'cwi', 'process'

Figure 32.1 Comparison of ground-truth and generated summary from the test set of the PubMed dataset. Hallucinations are highlighted. Input and ground-truth summary (abstract) taken from https://pubmed.ncbi.nlm.nih.gov/22629279/.

The entities detected in each summary are also shown. The original abstract consists of 5 sentences, but after using entity-based filtering technique it consists of only 2 sentences. In this case study, yellow color represents an incorrect representation of entity during summary generation. In this case study, cyan color denotes a correct entity mention that was incorrectly generated by the fine-trained LED model.

Conclusion

We applied the Longformer Encoder-Decoder model on scientific research papers to generate summaries and used data filtering along with the JAENS approach to reduce entity hallucinations. We found that the simple fine-tuned LED model performs the best in terms of ROUGE, METEOR, and BERTScore but entity-based data filtering improves the scores of some of the factual consistency metrics. In the future, we would like to investigate in detail the reason behind the low performance of the JEANS approach. We also noticed that entities are not always identified with high recall and precision in the summary. We would like to analyze this issue in detail and improve the entity recognition module. Finally, we would like to study the reduction in ROUGE, METEOR, and BERTScore values that we observed in all the hallucination-mitigating designs.

References

1. Huang, L., Cao, S., Parulian, N., Ji, H., and Wang, L. (2021). Efficient attentions for long document summarization. In Proceedings NAACL-HLT. ACL, 2021, (pp. 1419–1436).
2. Beltagy, I., Peters, M. E., and Cohan, A. (2020). Longformer: the long-document transformer. arXiv preprint arXiv:2004.05150.
3. Chalkidis, I., Dai, X., Fergadiotis, M., Malakasiotis, P., and Elliott, D. (2022). An exploration of hierarchical attention transformers for efficient long document classification. arXiv preprint arXiv:2210.05529.
4. Maynez, J., Narayan, S., Bohnet, B., and McDonald, R. (2020). On Faithfulness and factuality in abstractive summarization. In Proceedings ACL. ACL, 2020, (pp. 1906–1919).
5. Nan, F., Nallapati, R., Wang, Z., Dos Santos, C. N., Zhu, H., Zhang, D., McKeown, K., and Xiang, B. (2021). Entity-level factual consistency of abstractive text summarization. In Proceeding EACL. ACL, 2021, (pp. 2727–2733).
6. Cohan, A., Dernoncourt, F., Kim, D. S., Bui, T., Kim, S., Chang, W., and Goharian, N. (2018). A discourse-aware attention model for abstractive summarization of long documents. In Proceeding NAACL-HLT. ACL, 2018, (pp. 615–621).
7. Lin, C. Y. (2004). ROUGE: a package for automatic evaluation of summaries. In Text Summarization Branches Out. ACL, 2004, (pp. 74–81).
8. Banerjee, S., and Lavie, A. (2005). METEOR: an automatic metric for MT evaluation with improved correlation with human judgments. In Proceedings of the ACL Workshop on Intrinsic and Extrinsic Evaluation Measures for Machine Translation and/or Summarization. ACL, 2005, (pp. 65–72).
9. Zhang, T., Kishore, V., Wu, F., Weinberger, K. Q., and Artzi, Y. (2020). BERTScore: evaluating text generation with BERT. In Proceeding ICLR. 2020.
10. Nallapati, R., Zhou, B., Dos Santos, C., Gulc̦ehre, C., and Xiang, B. (2016). Abstractive text summarization using sequence-to-sequence RNNs and beyond. In Proceeding CoNLL. ACL, 2016, (pp. 280–290).
11. See, A., Liu, P. J., and Manning, C. D. (2017). Get to the point: summarization with pointer-generator networks. In Proceeding ACL. 2017, (pp. 1073–1083).

12. Rehman, T., Sanyal, D. K., Chattopadhyay, S., Bhowmick, P. K., and Das, P. P. (2021). Automatic generation of research highlights from scientific abstracts. In Proceeding EEKE Workshop at JCDL 2021. CEUR-WS.org, 2021, (pp. 69–70).

13. Rehman, T., Sanyal, D. K., Majumder, P., and Chattopadhyay, S. (2022). Named entity recognition based automatic generation of research highlights. In Proceeding SDP Workshop at COLING 2022. ACL, 2022, (pp. 163–169).

14. Rehman, T., Sanyal, D. K., and Chattopadhyay, S. (2023). Research highlight generation with ELMo contextual embeddings. *Scalable Computing: Practice and Experience,* 24(2023), 181–190.

15. Rehman, T., Sanyal, D. K., Chattopadhyay, S., Bhowmick, P. K., and Das, P. P. (2023). Generation of highlights from research papers using pointer-generator networks and scibert embeddings. *IEEE Access,* 11(2023), 91358–91374.

16. Devlin, J., Chang, M. W., Lee, K., and Toutanova, K. (2019). BERT: pre-training of deep bidirectional transformers for language understanding. In Proceeding NAACL-HLT. ACL, 2019, (pp. 4171–4186).

17. Rehman, T., Das, S., Sanyal, D. K., and Chattopadhyay, S. (2022). An analysis of abstractive text summa rization using pre-trained models. In Proceeding International Conference Computational Intelligence, Data Science and Cloud Computing. Springer Nature Singapore, 2022, (pp. 253–264).

18. Cao, Z., Wei, F., Li, W., and Li, S. (2018). Faithful to the original: fact-Aware neural abstractive summarization. In Proceeding AAAI. AAAI Press, 2018.

19. Kryscinski, W., McCann, B., Xiong, C., and Socher, R. (2020). Evaluating the factual consistency of abstractive text summarization. In Proceeding EMNLP. ACL, 2020, (pp. 9332–9346).

20. Goyal, T., and Durrett, G. (2020). Evaluating factuality in generation with dependency-level entailment. In Findings EMNLP. ACL, 2020, (pp. 3592–3603).

21. Zhu, C., Hinthorn, W., Xu, R., Zeng, Q., Zeng, M., Huang, X., and Jiang, M. (2021). Enhancing factual consistency of abstractive summarization. In Proceeding NAACL-HLT. ACL, 2021, (pp. 718–733).

22. Narayan, S., Cohen, S. B., and Lapata, M. (2018). Don't give me the details, just the summary! topic-aware convolutional neural networks for extreme summarization. In Proceeding EMNLP. ACL, 2018, (pp. 1797–1807).

33 Detection of Diseases in Mango Leaves Based on Different Classification Algorithms and their Comparisons using Machine Learning

Pintu Das[1,a], Mausumi Maitra (Mazumdar)[2,b], Sudarshan Chakraborty[3,c], and Rajarshi Sanayal[4,d]

[1]IT, MCKVIE, MAKAUT, Howrah, India

[2]IT, GCECT, MAKAUT, Kolkata, India

[3]ECE, Techno India University, Kolkata, India

[4]ECE, MCKVIE, MAKAUT, Howrah, India

Abstract

This article presents an extended investigation regarding the identification of ambiance of the leaf illness. In this regard, the investigation has been carried out using 4000 mango leaf samples. On the basis of this investigation, the classification grade of unhealthy leaves and the rate of infection have been obtained through peripheral extraction and contour segmentation procedure. The aim of this study was to explore the detection of foliar diseases using different foliar infection detection algorithms, and to compare them. This classification is absolute based on these criteria, and it automatically extracts features that are very useful. As a result of the SVM, the methodology presented is the most accurate for datasets at 97.88%. The obtained experimental results prove that among all the algorithms used for classification, SVM is the most efficient in detecting mango leaf disease. Vision-based leaf disease presented adequate results and excellent working efficiency.

Keywords: Classification, random forest, segmentation, support vector machine.

Introduction

The global mango market volume is expected to rise approximately from $57,332.9 million in 2021 to $77,942.1 million in 2022. India has estimated exports at three billion rupees in 2022 as a result of the growing demand for fresh mangoes. Natural mango production has been continuously decreasing due to the significant degradation of mango trees in the last few years. Therefore, the recognition of mango leaf disease becomes a focusing issue among the researchers. It adversely predicts disease at the initial stage. Therefore, the leaf quality graduation is a core aspect of the investigation. he previous literature mainly focused on the binary decision of the leaf, such as whether the leaf belongs to a healthy or unhealthy group. In this regard, A. Nage et al. has discussed the fact that the leaf disease is in its early stages [1]. If properly analysed and timely measures are taken, this infection can be easily avoided. Y. Kurmi et *al.* has presented a potential investigation. However, the image classification has been tabulated into two sections: healthy and diseased leaves [2]. Furthermore, D. Akanjuna *et al.* has developed a mobile application [3]. These applications

[a]daspintu2001@gmail.com, [b]mou1232005@yahoo.com, [c]sudarshan.c@technoindiaeducation.com, [d]rajarshi.sanyal@mckvie.edu.in

revealed the detection of different types of diseases in the leaf. N. K. Trived *et. al.* have suggested using the Convolutional Neural Network (CNN) algorithm to filter extracted images. It is used for training data, and each different type of layer is connected. Performs an accurate series of computations on the input test set and training data. Based on its results, the proposed model predicts a figure of 98.49% [4]. Moreover, [5] K. Sarode et al. has mentioned various types of features such as energy, entropy, homogeneity, dissimilarity, and contrast [6]. S. Mahajan *et.al.* has mentioned that image segmentation is the way to distinguish many important parts of the image section.

The Literature on Survey Work

Agricultural productivity has an important role that can radically change the economic system of any country. In this research paper, a lot of studies have been done to use machine learning techniques and different approaches for automatically detecting leaf diseases. The Authors of the paper [1] inspected using the illness may show symptoms in different places. Essentially, the proposed method localises the leaf region before identifying if the image is healthy or diseased [2]. Divide the source photos into many binary images by thresholding the source image using thresholds starting at the lowest threshold (minThreshold). The Authors of the paper [3] advised disease identification by different types of names, which provide useful information [4]. A pre-competent network model has been suggested for identifying and analysing mango illnesses [5]. The most commonly used features among the many that have been considered include energy, entropy, homogeneity, and dissimilarity [6]. Also, a technique for segmenting images that combines the objective function and the number of thresholds (th) is provided by histogram-based thresholding [7]. In this study, to investigate Convolutional neural network (CNN) models are created for segmented image data to investigate a potential fix for this problem. Histogram-based thresholding makes this possible [8]. When compared to the traditional CNN model, the lightweight technique had a better accuracy rate for mango leaf disease. An irregular leaf fall with spots can be difficult to diagnose [9]. The preferred and most popular machine learning technique is deep learning (DL) [15, 16]. In this paper, a hybrid random forest Multiclass SVM model is proposed for the detection of plant foliar diseases (SVM). According to a survey of the most common machine-learning algorithms, support vector machines (SVMs) are widely used for disease classification [17, 18]. Inspected the paper, the SVM test, the linear kernel is more efficient than the other three algorithms.

Research Methodology

The models SVM, CM, RF, and GLCM were created on Machine Learning (ML) accesses for the finding of leaf diseases that are discussed in this portion. The method that is suggested consists of four basic steps: collection, image pre-processing and segmentation, feature extraction using Dataset GLCM, model training and testing, and image prediction based on input. The Flowchart is shown in Figure 33.1.

Dataset Collection

The Leaves are collected from both healthy and unhealthy mango plants. The leaf images are saved in jpg format. The Kaggle dataset served as the source for the images. The folder contains 4000 images for training data, 3200 and 800 images for testing data. It is a collection of image pixels arranged in rows and columns. Pixels are image elements that provide

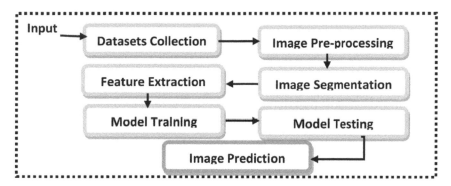

Figure 33.1 Flowchart of image prediction.

information about color and intensity. This is identified as an RGB image. The Flowchart is shown in Figure 33.1.

Pre-Processing of Images

Image preprocessing is an important part of image retrieval systems. If each mango leaf image is not resized properly, the operation time will increase during disease detection and image classification. The first task is to convert a color image to a grayscale image; as well, each image is scaled down to 128 × 128 from 256 × 256 images of mango leaves.

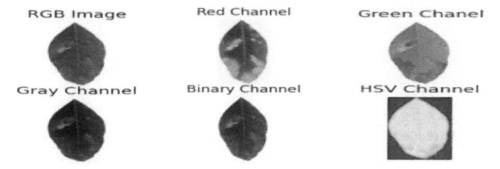

Figure 33.2 Image generate different scale.

The various techniques are applied at the time of image fetching, such as RGB (Red, Green, Blue), Gray, Binary, HSV (Hue Saturation Value), resizing, filtering, elimination of noise, color metamorphosis, etc. The color is the most significant component of image extraction. It is subjected to appropriate pixel and image models. The selection of color pixel has been defined in color features such as RGB, Gray, Binary, HSV, etc. The Image generated from different scales is shown in Figure 33.2.

The natural points are mean and standard deviation and the calculation can be presented as follows:

$$\mu_i = \frac{1}{N} \sum_{j=1}^{N} f_{ij} \tag{1}$$

$$\sigma_i = \left(\frac{1}{N} \sum_{j=1}^{N} (f_{ij} - \mu_i)^2 \right)^{\frac{1}{2}} \tag{2}$$

SVM Classification

The major desire of the SVM algorithm is to operate both in image classification and regression. These mango leaves have three types of disorders and are expected to evaluate their accuracy and perceive the leaf as healthy or unhealthy. This result is a sample of 4000 mango leaf images from the datasets. After that, the models are grown with a ratio of 80:20, and the test division is used to achieve the best results regarding the image classification. The Architecture is shown in Figure 33.3.

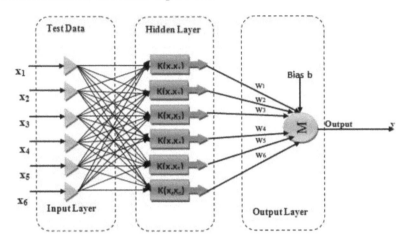

Figure 33.3 SVM hidden layer.

Now, development to measures w and b, which take a selection at the most efficient classifier: The hypothesis function is denoted by h as:

$$h(xi) = f(x) = \begin{cases} w \cdot x + b \geq 0 \, for + 1 \\ w \cdot x + b < 0 \, for - 1 \end{cases} \tag{3}$$

Feature Extraction GLCM

The recommended model GLCM uses machine learning to recognize infected mango leaf images. The Haralick texture algorithm implements (Gray Level Co-occurrence Matrix (GLCM) in statistical texture analysis. The Gray Level Co-occurrence Matrix is calculated to obtain the statistical texture features. GLCM is a current and innovative computational system for evaluating textural patterns.

Random Forest and Decision Tree Classifier

Random Forest (RF) is an ensemble that uses and combines many decision tree classifiers. It is used for image classification. The decision tree takes the votes and calculates their average to enhance accuracy. The majority vote is predicted as the final result. A Decision Tree is a supervised method. In order to comprehend what random forests endeavour, it is crucial to become familiar with decision trees. The possibility of classification features is expected to be. The possibility of classification features is expected to be

$$P(x_i \, |y) = \frac{1}{\sqrt{2\mu\sigma^2}} \exp\left(-\frac{(x_4 - \mu_y)^2}{2\sigma_y^2} \right) \tag{4}$$

Confusion Matrix

The classification results are evaluated using the accuracy percentage. The equation provides it with a mathematical representation and uses the whole sample size that was provided to the model as input to generate the precise model prediction.

$$\text{Accuracy} = \frac{TP + TN}{TP \mid TN \mid FP \mid FN} \tag{5}$$

$$\text{Precision} = \frac{TP}{TP + FP} \tag{6}$$

$$Recall = \frac{TP}{TP + FN} \tag{7}$$

$$F1 = \frac{2 \times \text{Precision} \times \text{Recall}}{\text{Precision} + \text{Recall}} \tag{8}$$

Comparison Study of SVM, RF, CM and DT

In this literature, SVM, RF, CM, and DT are shown for comparative research on image visualization. In this dataset, the data is sparse and can be grouped very easily; therefore, SVM works faster and gives better results. However, Random Forest also shows good results, but SVM is much better for this appropriate dataset. The SVM is used for binary classification. On the other hand, Decision Tree is used for Boolean classification. A confusion matrix is a matrix that works with a machine learning model on a group of datasets. RF and DT are very complicated when dealing with high-dimensional datasets. Hence, it is very difficult to implement and impractical to understand the model in such cases. This is a weakness of Confusion Matrix. Reviewing the results of the three models discussed, it can be seen that SVM accuracy is higher than others.

Result Analysis

In this research paper, one of the most important points is to analyze each algorithm in detail. RF is a very robust classifier that incorporates the adaptability of many decision tree models into a single model. The total mango leaves are divided into eight groups. It confirmed that SVM produced the highest classification accuracy for datasets, RF, DT, and CM (97.88%, 95.72%, 94.01%, and 93%, respectively) which is as shown in Figure 33.4. Finally, from this research analysis, it can be concluded that the SVM algorithm revealed the highest accuracy compared to others, and references for different types of classification and reference reports have been shown in Table 33.3. The model successfully identifies 4000 samples with average accuracy in SVM, RF, DT, and CM of 95.48%, 93.42, 84.75%, and 82.5% in Table 33.2.

Leaf diseased region optimization as:

$$PIR = [((\beta * .75) + (\gamma * .25) + (\delta *.15) + (\varepsilon * 0.0)) - (\alpha * 1)] \tag{9}$$

Where, PIR is the Predicted Infected Region. The Weight factor value is shown in Table 33.1.

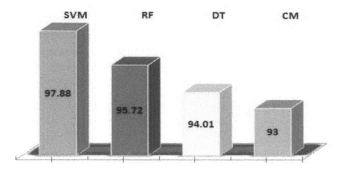

Figure 33.4 Comparisons accuracy chart.

Table 33.1: Weight factor.

Name	Value	Symbol
Green	1.0	α
Brown	0.75	β
Yellow	0.50	γ
Black	0.25	δ
White	0.0	ε

Table 33.2: Different leaf disease accuracy.

Sl. No.	Disease Name	SVM (%)	RF (%)	DT (%)	CM (%)
1.	Anthracnose	97	89	82	77
2.	Bacterial Canker	96.5	88.4	88	83
3.	Cutting Weevil	89	95	84	73
4.	Die Back	97.45	94.6	76	82
5.	Gall Midge	93	95.88	79	88
6.	Healthy	96.88	94.5	83	85
7.	Powdery Mildew	97.20	96	91	78
8.	Sooty Mould	96.87	94	95	94
Average		95.48	93.42	84.75	82.5

Table 33.3: Research reference shows on disease classification results.

Reference Paper	Investigation Algorithms	Performance Report (%)			
		SVM	RF	DT	CM
Ref. [11]	SVM	100	NA	NA	NA
Ref. [10]	SVM and DT	100	NA	93.1	NA
Ref. [12]	RF	NA	91	NA	NA
Ref. [13]	SVM	96.77	NA	NA	NA
Ref. [14]	RF	NA	94	NA	NA
Proposed Work	SVM, RF, DT and CM	95.48	93.42	84.75	82.5

Conclusion

The agricultural sector always faces losses due to plant diseases, which greatly affect the economy of that country. The technology of computer vision has greatly advanced over the past few decades. A noticeable classification mechanism for infected leaves has been introduced in this article. SVM, RF, DT, and CM optimization techniques have been adopted to classify the mango leaves according to their infection rate. An empirical formulation of infection rate is developed, and we can accurately determine the percentage of infected and healthy, with this new formula. The accuracies found among the given models are: SVM produces the highest accuracy of 97.88%, followed by Random Forest, DT CM, and 95.72%, 94.01%, and 93%, respectively. In conclusion, there are a lot of impressive subjects that would be studied in future research work. Future work can be done to include new techniques and more plant species with different diseases and structural characteristics. This will increase the accuracy more. The predictions of a given model can be further improved by extracting many distinct features from plant leaves, thereby allowing us to provide more early warnings.

References

1. Nage, A., and Raut, V. R. (2019). Detection and identification of plant leaf diseases based on Python. *International Journal of Engineering Research and Technology,* 8(5), 296–299.
2. Kurmi, Y., and Gangwar, S. (2022). A leaf image localization based algorithm for different crops disease classification. *Information Processing in Agriculture,* 9(3), 456–474.
3. Akujana, D., Akankwasa, D., Atukunda, S., Mukasa, A., and Nsabagwa, M. Using machine learning to determine the level of infection of cercospora leaf spot disease in cowpeas. Available: https://www.academia.edu/download/81621160/
4. Trivedi, N. K., Gautam, V. and Anand, A. (2021). Early detection and classification of tomato leaf disease using high-performance deep neural network. *Sensors,* 21(23), 7987.
5. Khushal, S., Savedkar, R., and Choudhury, T. (2022). Texture feature Analysis of an image using gray level co-occurance matrix. *International Journal of Novel Research and Development,* 7(2), 139–143.
6. Mahajan, S., and Pandit, A. K. (2022). Image segmentation and optimization techniques. *Medicon Engineering Themes,* 2(3), 47–49.
7. Sharma, P., Berwal, Y. P. S., and Ghai, W. (2020). Performance analysis of deep learning CNN models for disease detection in plants using image segmentation. *Information Processing in Agriculture,* 7(4), 566–574.
8. Zeng, T., Li, C., and Zhang, B. (2022). Rubber leaf disease recognition based on improved deep convolution neural networks with a cross-scale attention mechanism. *Frontiers in Plant Science,* 13, 1–12.
9. Srivstava, S., Divekar, A. V., Anilkumar, C., Naik, I., Kulkarni, V., and Pattabiraman, V. (2021). Comparative analysis of deep learning image detection algorithms. *Journal of Big Data,* 8(66), 1–27.
10. Balasundaram, A., and Arunkumar, S. (2022). Deep learning and computer vision based model for detection of diseased mango leaves. *International Journal on Recent and Innovation Trends in Computing and Communication,* 10(6), 70–79.
11. Arivazhagan, S., and Ligi, S. V. (2018). Mango leaf diseases identification using convolutional neural network. *International Journal of Pure and Applied Mathematics,* 120(6), 1067–1079.
12. Pham, T. N., Tran, L. V., and Dao, S. V. T. (2020). Early disease classification of mango leaves using feed-forward neural network and hybrid metaheuristic feature selection. *IEEE Access,* 8, 189960–189973.

13. Kaur, L., and Laxmi, V. (2015). Detection of unhealthy region of plant leaves Using neural network. *International Journal of Latest Engineering Research and Applications*, 1(5), 34–42.
14. Basavaiah, J., and Anthony, A. A. (2020). Tomato leaf disease classification using multiple feature extraction techniques. *Wireless Personal Communication*, 115, 633–651.
15. Sahu, S. K., and Pande, M. (2023). An optimal hybrid multiclass SVM for plant leaf disease detection using spatial fuzzy C-means model. *Expert systems with applications*, 214, 118989.
16. Goel, L., and Nagpal, J. (2023). A systematic review of recent machine learning techniques for plant disease identification and classification. *IETE Technical Review*, 40(3), 423–439.
17. Javidan, S., M., Banakar, A., and Vakilian, K. A. (2023). Diagnosis of grape leaf diseases using automatic K-means clustering and machine learning. *Smart Agriculture Technology*, 3(1), 2772–3755.
18. Sarkar, C., Gupta, D., and Gupta, U. (2023). Leaf disease detection using machine learning and deep learning: review and challenges. *Applied Soft Computing*, 145, 110534.

34 An Extensive Review on the Controller Design Strategies for Non-Linear Systems

Prashant Nigam[1,a], Nidhi Singh[2,b], and Lillie Dewan[3,c]

[1]Research Scholar, Electrical Engineering Department, Gautam Buddha University, Greater Noida, U.P, India

[2]Assistant Professor, Electrical Engineering Department, Gautam Buddha University, Greater Noida, U.P, India

[3]Professor, Electrical Engineering Department, National Institute of Technology, Kurukshetra, Haryana, India

Abstract

Most systems in real life are nonlinear., despite the fact that many of them are designed to 'behave' linearly in the vicinity of a specific operating point at moderate speeds, at least within a wide operating range under specified conditions. Nonlinear models can reflect a various physical phenomena. Many scholars in fields, such as process control, biomedical engineering, robotics, aviation and spacecraft control have recently expressed a keen interest in the creation and study of nonlinear control strategies. Most real-world issues inevitably face nonlinearities. Therefore, this survey aims to make a review on 30 research articles that concern on design strategies for non-linear systems. Consequently, the systematic analyses on the proposed methods are made from each reviewed paper. Likewise, the performance measures and the highest achievements in each research paper are also analyzed in this work. Moreover, the chronological assessment and different tools used for implementing the non-linear systems were considered and reviewed in this work. At last, the survey depicts various research gaps and challenges that are more supportive for researchers to implement novel controller design methods for non-linear systems.

Keywords: algorithm, controller design, machine learning, non-linear system, optimization.

Nomenclature

Abbreviation	Description	Abbreviation	Description
PID	Proportional Integral Derivative	DP	Dynamic Positioning
PSO	Particle Swarm Optimization	AII	Average Impulsive Interval
OLAs	Online Approximators	FTS	Finite Time Stability
NNs	Neural Networks	NTPZSD	Nonlinear Two-Player Zero-Sum Differential
MIMO	Multiple-Input Multiple-Output	SDRE	State-Dependent Riccati Equation

[a]prashantnigs108@gmail.com, [b]nidhi@gbu.ac.in, cl_dewan@nitkkr.ac.in

Abbreviation	Description	Abbreviation	Description
NLPID	Nonlinear PID	PD	Proportional Derivative
GA	Genetic Algorithm	RESs	Renewable Energy Sources
OPI	Output Performance Index	BESSs	Battery Energy Storage Systems
LPID	Linear PID	BH	Black Hole
RL	Reinforcement-Learning	RISE	Robust Integral of The Sign of The Error
DOF	Dynamic Output Feedback	CQLF	Common Quadratic Lyapunov Function
SMC	Sliding Mode Control	PQLFs	Piece-Wise Quadratic Lyapunov Functions
ISpS	Input-To-State Practical Stability	SMSDO	Sliding Mode State And Disturbance Observer
SMC	Sliding Mode Control	PSS	Power System Stabilizers
VRFT	Virtual Reference Feedback Tuning	IMC	Internal Model Control
T-S	Takagi–Sugeno	ANN	Artificial Neural Network
LMI	Linear Matrix Inequalities	LFC	Load Frequency Control
ADP	Adaptive Dynamic Programming	CS	Cuckoo Search
HJB	Hamilton–Jacobi–Bellman	ARX-NN	ARX-Neural Network
PI	Policy Iteration Algorithm	MCH	Methylcyclohexane
IMC	Internal Model Control	CFNSs	Complex Flexible Nonlinear Systems
PLS	Partial Least Squares	FSPM	Fuzzy Singularly Perturbed Model

Introduction

For a long period, linear models were used to create excitation regulators for synchronous generators. Linear control approaches were extensively used applications because of their simple strategy and convenient real-time execution. Linear PSS are used to improve dynamic stability limits through high quality suppression via excitation control. The establishment of the nonlinear decentralized control system is to handle the issue of generic nonlinear interconnectivity constraints. A nonlinear decentralized excitation controller with adaptive back stepping technique was developed to boost multi machine power systems' transient stability performance. In contrast to previous dispersed excitation control techniques, this strategy eliminates the need for specified connectivity parameter limits [1]. PID tuning and adaptation methods for linear time-invariant systems have been extensively studied [6]. The addition of strained input and lack of ability accurately identify the uncertainty which promotes the formation of a stabilizing controller based on RL methods [8]. Numerous methods for optimal management of nonlinear systems have been suggested. Among the numerous approaches, DP provides an optimal control for nonlinear dynamic systems [3]. A conic-type nonlinear system is a form of nonlinear system that survives within a hyper sphere with a center linear system. The radius is strained by other linear systems, as well as nonlinear systems with conic-type nonlinear effects have finite-time stability. However, time delays are common in dynamic systems and often result in unwanted behaviors, such as poor performance or severe fluctuations [12]. Controlling fractional-order nonlinear systems is primarily concerned with stability analysis. Nonlinear fractional-order systems may suffer from system variables, such as unknown external disruptions, modeling mistakes, and plant parameter changes which may degrade the system's management performance if they are not handled properly [15]. Iterative learning control was used to create a self-tuning control for optimizing the control settings of turbocharged motors. VRFT has also been used by neural operators in nonlinear systems and MIMO linear systems [16]. To address problems with nonlinear control, many academics use linear methods. Via local linearization. Nevertheless, use of linearization techniques has limitations in regulating complex chemical processes. Model-based control system IMC was created by Gacia and Morari that has been usefully employed in industrial processes. Few researchers handled robustness and stability problems by extending linear IMC to a nonlinear scheme due to its simple and transparent arrangement. The incorporation of ANN in the IMC strategy offers a more comprehensive and flexible method of dealing with nonlinearity. Several nonlinear system problems, such as delay and stability, can be investigated in order to obtain outstanding results under the NN based IMC paradigm [21].

Literature Survey

In 2010, Rui Yan et al. [1] have developed a new excitation control for enhancement of power system transient stability. The control technique is built on recursive adaptive back stepping without linearizing the system model. When a significant fault happens, the controller is designed using the Lyapunov function technique to guarantee the generator's power, speed in relation to it, and power angle convergence. In contrast to current nonlinear decentralized control methods, the suggested controller has no necessity for power system interconnection bounds. Furthermore, the presence of a solution to a designed algebraic Riccati equation is not required by the novel method. Moreover, the performance of transient power systems' stability can be improved by using the suggested management approach. In 2010, Wei-Der Chang, Shun-Peng Shih [2] has presented an enhanced PSO for finding the nonlinear system's ideal PID controller gains. The suggested method aims to improve the searching effectiveness by modifying the PSO velocity formulae. The PID control gains such as k_p, k_i, and k_d are combined to generate an input vector known as a particle which is the fundamental module of PSO networks, in which a population is made up of many such particles. The updated velocity and position modified formulae are used to shift all position particles in the population which helps to determine the optimum nonlinear PID gains. At the same time, a common nonlinear system with an inverted pendulum tracking control are demonstrated to validate the suggested method's control performance. In 2012, Qinmin Yang and Sarangapani Jagannathan [3] have suggested broad unknown multi-input, multi-output, affine discrete temporal systems with adaptive critic controllers based on output feedback and reinforcement learning with limited disruptions using OLAs. The suggested controller design consists of two components: an action network which intended to generate optimum signals as well as critical network which assesses the action network's performance. The critical network calculates the cost-to-go function, which is adjusted in real time with recursive formulae drawn from heuristic dynamic programming. To determine the missing system states for the equivalent output-feedback, an extra NN is assigned to the observer, and thus the separation principle is not needed. Using Lyapunov theory, tuning of NN weights for the controller algorithm is derived, which ensure unchanging final limitedness of the closed-loop system. In 2015, Yan-Jun Liu and Li Tang [4] has presented a nonlinear MIMO system with adaptive neural network tracking control. Furthermore, the MIMO systems under consideration are made up of N components, each of which includes unknown functions and exterior disturbance. To resolve the noncausal issue, the investigated systems are converted into a specific shape appropriate for the back-stepping design by specifying the coordinate transformations. The basic function of NN is used to estimate the system's undetermined roles. The changed arrangements are used to create the adaptation rules and controllers. The Lyapunov technique is used to substantiate the closed-loop system's stability in which the unchanging bound set with each signal and tracking errors converge to a bounded compact set. In 2015, A.Y. Abdelaziz, E.S. Ali [5] have designed a novel optimization method known as the CS for optimal PI controller calibration for LFC. To achieve the most optimistic outcomes, an objective function based on the time domain is developed. for tuning the constraints of a PI-based LFC that is resolved by using CS algorithm. In order to validate the efficacy of the proposed approach, a test system with system nonlinearity is taken into consideration. Thus, the implemented findings demonstrate the improved performance of the established CS-based controllers in contrast to GA, PSO, and traditional integral controllers. These findings indicate that the suggested controllers outperform others in terms of settling time as well as reliability. In 2014, Aydogan Savran, Gokalp Kahraman [6] have developed a new adaptive tuning technique for classical PID controllers to manage nonlinear processes to modify PID gains, a challenging issue to solve classical PID controllers. A technique for controlling nonlinear processes that incorporates well-known conventional PID control is suggested. Controller design in this technique is not considered as the main model because it is very challenging to acquire. Apart from this, it is based on a fuzzy model process which is built from the observed input-output data. Furthermore, a soft limiter with control input has its indus-trial boundaries. Hence, the system's performance was effectively evaluated in the bioreac-tor, which is an extremely nonlinear process with uncertainties. Various experiments demonstrated the proposed effectiveness in tracking, robustness to noise, and adaptation characteristics. In 2019, Aws Abdulsala Najm, Ibraheem Kasim Ibraheem [7] have estab-lished a NLPID controller to steady a 6-DOF quadrotor system's translational and rota-tional motion to force it to travel a specific path with the least amount of energy and inaccuracy. Using the Euler-Newton formalism, the design method makes use of the 6-DOF quadrotor technology, resulting in an equation that is more precise and closer to the real system 6-DOF quadrotor model. Six NLPID controllers are developed, with their settings are adjusted using a GA optimization. The stability of the 6-DOF UAV components was investigated using the Hurwitz stability theorem under various constraints based on the NLPID controller gains. Thus, the simulations demonstrated the efficacy of the suggested NLPID controller. In 2015, Derong Liu et al. [8] have developed a novel resilient category of nonlinear continuous-time systems with uncertainty along with input limitations can be controlled using an adaptive control approach based on RL. By selecting appropriate value functions, the restricted optimum control problem is created from the robust control prob-lem. When compared to the typical action-critical dual networks used in RL, the estimated optimum control is determined by only one critical NN. Meanwhile, unlike stabilizing control, the RL with initial control has no particular requirements. Therefore, the closed-loop optimal control system employs Lyapunov's direct method. And the predicted weights of the critic NN are constrained uniformly. As a result, the simulation demonstrates the efficacy and utility of the current strategy. In 2012, Qing Gao et al. [9] have created Fuzzy dynamic Takagi-Sugeno (T-S) models for generic nonlinear systems management. It is first demonstrated that by using a generalised T-S fuzzy system, one can calculate a stabilization problem using approximation mistakes as the uncertainty component. The broad nonlinear system's robust semiglobal stabilisation and H∞ control is expressed as linear matrix equations based on a piecewise quadratic Lyapunov function. In 2017, Bo Fan et al. [10] have released a continuous control technique that integrates the NN and rising signal of MIMO uncertain nonlinear systems with bounded disturbances in order to compel the system state to adhere to user-defined time-variable boundaries. The criteria for understanding system uncertainties, external disruptions, and rebuilding errors have been eased. Strict theoretical analysis has also ensured the asymptotic tracking efficiency. All other closed-loop system outputs have been proven in which the suggested controller can enhance the efficiency of the transient state even more. Though, the suggested controller can guarantee the transient output performance. In 2010, Jiuxiang Dong et al. [11] have looked into the construction of output feedback controllers for local nonlinear submodels in discrete-time fuzzy systems. The observer and a controller with nonlinear feedback laws can be created individually. Furthermore, a two-step method using CCL method is provided for developing H∞ controllers. Similarly using the nonlinear information of local sub models, a convex H∞ control synthesis condition is supplied. Compared to standard methods, the current approaches based on fuzzy controllers not only minimize the processing load but also favors engineering applications. In 2017, Shuping He et al. [12] have created an SMC technique for a conic nonlinear system with time delay and mismatched external perturbation that is finite in time. The slack matrices method is employed to develop a delay-dependent adequate condition that ensures the FTB of the SMC system within a finite time span. The suggested SMC scheme has the advantage of being able to calculate the dynamics paths into the specified sliding surface within given time period. Two simulations demonstrated the efficacy and relevance of the suggested finite-time SMC strategy in the presence of conic-type nonlinearity and time delays. In 2014, Zhi Liu et al. [13] have introduced an ISpS issue involving nonlinear systems with unmodeled dynamics in the face of unstructured uncertainties and dynamic disruptions. The dynamic conflicts are determined by the system state, observed outputs and its statement circumstances remain eased in comparison to typical constraints. Based on an input-driven filter, the unknown fuzzy logic system and desired control signals along with an integrated backstepping technique suggested an adaptive output-feedback controller which certifies fitness in terms of unidentified constraints and uncertain nonlinearities. Thus, this article demonstrates that the suggested method ensures uniformly bounded closed-loop system. In 2019, Wenqiang Ji et al. [14] have investigated a dynamic fuzzy output feedback SMC technique fuzzy affine systems with mismatched external disruptions under uncertain T-S. A descriptor system characterizes the dynamic property of the sliding motion using a state—input enhancement method. Based on CQLF and PQLFs, novel outcomes on asymptotic stability analysis for sliding motion was presented. Moreover, a convex optimization setup with sliding surface gain can be obtained. Formerly, a dynamic output fuzzy feedback, is advised to use the SMC method to force the closed-loop system states onto the planned sliding surface. This may also be important to note that the suggested controller may deviate from the idea that all local affine design has a common input control channel as well as control input matrices with uncertain parameters. In 2018, Heng Liu et al. [15] have suggested an adaptable fuzzy controller for unidentified fractional-order nonlinear systems. An undefined nonlinear function is determined using a fuzzy logic system. Additionally, utilizing the fractional-order Lyapunov stability criterion, an adaptive fuzzy controller for both the Caputo and Riemann-Liouville fractional-order systems is performed. Then, the fuzzy constraints were upgraded using fractional-order differential equations, and the corresponding parameter finiteness is assured. In 2016, Pengfei Yan et al. [16] have developed a new data-driven multivariate nonlinear controller design approach for MIMO nonlinear systems using VRFT and neural networks. Moreover, this is one of the first theoretical applications of nonlinear VRFT to MIMO systems. When compared to the conventional VRFT, it is restated that the VRFT objective function is made simpler by reference control problem with time-domain model. without the use of a linear filter. The target value of VRFT then achieves its lowest value at the same point as the model reference control optimisation problem, and a connection between the bounds of the two optimisation problems linked to model reference control and VRFT is shown. The devised technique is implemented with a three-layer neural network. In 2010, Cheung-Chieh Ku et al. [17] have created a passive fuzzy controller design approach with multiplicative noises for nonlinear system. The energy creativity and the stochastic differential equation can be used to evaluate external disruption and multiplicative noise features using Ito's formula. The recommended T-S fuzzy controller design was taken into consideration by resolving a BMI stability condition using the PDC thoughts. The BMI stability condition was transformed into an LMI form by using the convex optimal programming method. To demonstrate the utility and efficacy of the suggested fuzzy control methodology, mathematical scenarios for synchronous generator systems and two-link robot systems has been provided. In 2012, Thomas Gubner et al. [18] have suggested a novel design approach for nonlinear systems with input saturation. The suggested approach uses feedback linearization to reformulate the nonlinearity system as a linear system with state-dependent input saturation. Therefore, convex optimization issues can be expressed with respect to unequal linear matrix and sum of square conditions. Moreover, a stabilizing algorithm for a single link manipulator with a flexible joint is created to show the advantage of this design approach. In 2014, Changzhu Zhang et al. [19] have investigated the issue of a H∞ output feedback controller with medium access constraints, in which networked nonlinear systems are modeled by T-S fuzzy models and the control technique. A collection of spatially dispersed sensors measures the numerous system outputs, which are then relayed to the remote controller via the common communication network. Moreover, it requires only one sensing node which is allowed to contact the network stochastically and transmit data within a sampling period. Similarly, a strategy based on the design of a H∞ feedback output controller with respect to linear matrix inequality is suggested. In 2015, Jilie Zhang et al. [20] have suggested two novel adaptive dynamic programming-based implementation methods with on-line model-free optimum control algorithm. The HJB problem was solved using ADP, which incorporates the NN approximator, the PI algorithm, and the least-squares technique. The primary concept behind the approach is to use the least—squares methodology to update the NN weights by sampling data from the state, state derivative, and input. The PI framework is used to perform the update procedure. Thus, certain instances have demonstrated the efficacy of the suggested method. In 2012, Bin Hu et al. [21] have introduced a new multi-loop nonlinear IMC approach for MIMO systems based on the PLS framework, which dynamically divides the system into a number of univariate parts. Moreover, there offered an ARX-NN cascaded structure which is integrated with the PLS paradigm to develop a nonlinear dynamic PLS framework. Based on the ARX-NNPLS model, a gradient-based optimization method is introduced, which reduces the plant-model mismatch. Further, it demonstrates that the closed loop system decreases to a basic IMC filter that is linear and uses the original system delay. Thus, the findings shows the efficacy of the method with respect to disorder rejection and monitoring performance. In 2014, Jinxiang Chen et al. [22] have suggested CFNS controller design and modelling using the fuzzy singularly perturbed approach. To begin with, a conventional discrete-time FSPM is constructed to forecast CFNSs. Moreover, a static output feedback controller is developed based on a matrix spectral norm method. However, the controller gains are achieved by cracking a collection of independent LMIs, allowing the poorly-conditioned issues induced can be easily prevented. To demonstrate its efficacy, the given technique is used to model and control a Van der Pol circuit and CFSs. The time required for the closed-loop device to reach steady state is considerably decreased and small control inputs are required, which means that during controller design, the amplitudes of the control inputs are not restricted. Thus, the findings demonstrate that the method is also capable of overcoming external disruptions and noise. In 2018, Xiaogong Lin et al. [23] have established DP systems for ships using a input saturation, unknown time-varying environmental disturbances, unknown dynamic model parameters, and high-gain observer in a nonlinear adaptive fuzzy output feedback controller. Moreover, the suggested control strategy is only dependent on observations of the ship's location and direction because it helps to calculate the ship's missing speed. It has been demonstrated that the suggested control strategy arbitrarily reduces position error by ensuring uniform boundedness closed loop signals in the DP system. Hence, the suggested DP control scheme's efficacy and reliability have been proven by simulation findings and comparisons. In 2017, Xiaoxiao Lv, Xiaodi Li et al. [24] have investigated the FTS for nonlinear explosive sampled-data systems. A nonlinear impulsive sampled-data systems can be determined based on LMIs using a suitable Lyapunov function and the AI technique, and it can be confirmed using the LMI toolkit. Using theknown LMIs, an impulsive controller and a sampled-data controller are combined to create a hybrid controller. In addition, the impulse impact studied in this article includes both stabilizing and destabilizing impulses. Furthermore, the impulse impact studied in this article includes both stabilizing and destabilizing impulses. Thus, the established findings are lower conservative than former research study. In 2017, Yazdan Batmani [25] has developed a novel technique for addressing the H∞ tracking issue in nonlinear systems. Moreover, an NTPZSD game is specified using the discounted cost function. The latter NTPZSD game is subjected to an SDRE method which helps to determine an estimated solution. The tracking error between the system state and its intended trajectory is thus displayed to converge asymptotically to constant under moderate discount factor circumstances. As a result, the simulation findings reveal that the suggested H∞ tracking controller is extremely active for solving the problem of following a changing desired trajectory in nonlinear dynamical systems. In 2021, Jeang-Lin Chang and Tsui-Chou Wu [26] have created a robust SMSDO capable of estimating both system status and unexpected disruption. Furthermore, if the first-order of the disturbance is supposed to be limited, the state estimation error and instabilities are ultimately constrained in an area. Such that, the suggested nonlinear PD controller can accomplish minimal overshoot and fast reaction. A variable damping ratio nonlinear PD controller along with prediction state and disruptions should have less overshoot and a quick reaching time. Thus, the simulated findings validate the efficacy and benefits of the design technique. In 2021, HamedeKarami and Reza Ghasemi [27] have developed a controller based on neuro-observation for both traditional and clever methods. With guaranteed steadiness, an adaptive NN observer is created for the nonlinear dynamics of a hovercraft. Moreover, it is controlled by a nonsingular super-twisting terminal sliding mode technique. Due to system's Lyapunov stability, the backpropagation method is used to resolve the weighting factors. Thus, the super-twisting method lowers the chattering problem. Furthermore, the closed-loop system's Lyapunov stability, tracking convergence and the controller can be made resilient to uncertainty and shocks when observer errors decrease to zero. In 2014, M. A. Mahmud et al. [28] have developed a robust nonlinear distributed controller architecture for islanded microgrid operation with active and reactive power. However, the suggested distributed controller's design process is straightforward when variables are considered. As the control method is accomplished of withdrawing the innate nonlinearity by keeping stable internal dynamics. Then, the designed controller accomplishes the control goals in a more rapid and robust manner than an LQR-based controller. The simulation findings plainly show that the designed controller responds more quickly under various operating circumstances. In 2020, Hazem I. Ali et al. [29] have established a novel optimum nonlinear controller for three types of nonlinear systems: nonaffine nonlinear strict-feedback systems, time-varying (TV) nonlinear systems, and uncertain nonlinear systems. The suggested controller design verified the steadiness and optimum performance for various issues in nonlinear systems. After implementing the suggested controller, the system became asymptotically steady. Following that, the optimal settings of the suggested controller are optimized utilizing the BH algorithm to achieve the best performance in every instance. Lastly, after certain hypotheses and a proper analysis, the suggested controller balance many nonlinear systems with the associated issues. In 2011, Tufan Kumbasar et al. [30] have implemented an inverse model controller, which provide an effective method to regulate nonlinear processes using an inverse controller based on type-2 fuzzy model. Despite the fact that different fuzzy inversion techniques for type-1 fuzzy logic systems have been developed, there is no method for type-2 fuzzy logic systems. Based on a pure analytical technique, a systematic strategy for forming the inverse of the interval type 2 T-S fuzzy models was suggested in this research. The inverse model is computed using basic manipulations of the fuzzy model's antecedent and consequence components. Furthermore, to provide effective and robust control performance, the type-2 fuzzy model is incorporated in a nonlinear internal model control structure.

Problem Statement of Reviewed Approaches

The reviews on proposed techniques in the 30 review papers are determined and it is represented in Table 34.1. On analyzing the review, it is clearly shown that the Adaptive decentralized controller method was used in [1] possess better efficiency with improved stability and it suffers with disturbances which is the major drawback. PSO algorithm was adopted in [2] which offer excellent tracking performance and is highly feasible. It has a major drawback of low-quality system and needs memory to update the velocity. Moreover, OLA's model was adopted in [3], which is highly feasible but low stability and computational cost is high. Furthermore, MIMO approach was suggested in [4], which has tracking errors with high efficacy. Anyhow, it has additional complexities with low stability analysis. Moreover, CS based PI controller technique was designed in [5] which offers high performance and it has a major drawback of high convergence time and easily trapped in local minimal solution. Nevertheless, Adaptive tuning method was built in [6], which minimizes the error with high accuracy; but it is less compensate. DOF based UAV model was proposed in [7], which offers high speed with more stability; but it is inaccurate and takes more time to execute. Likewise, RL method was developed in [8], which provide less computation complexity. But, the system has system instability. Furthermore, T-S fuzzy dynamic model was developed in [9, 19], which provide better performance and is very complex with transmission delays. Additionally, Adaptive control strategy was investigated in [10], which offers high tracking performance. However, it may affect the high frequency with more disturbances. Moreover, Non-linear discrete time fuzzy system was developed in [11], which is used for engineering implementations with less computational burden. The major drawback is less conservativeness of the model. SMC approach was adopted in [12], which is highly effective but, it may undergo time delay with external disturbances. The ISpS model was implemented in [13], which minimizes the computation burden and achieves better performance. But it may suffer with computation problem. Yet, output feedback SMC method was suggested in [14], which is robust with good transient performance; also, it suffers with external disturbances. Fractional-order nonlinear system was adopted in [15], which provide tracking error reduction with high stability analysis. Unfortunately, the model may undergo modelling errors and plant parameter variations. Data-driver controller design was introduced in [16], which helps to solve practical problems and provide noiseless results. A major drawback is its cost and is very complex. The Linear Matrix Inequality was suggested in [17], which is highly stable and provides passive performance. But it may cause disturbance and is highly complex. Convex optimization was used in [18], which is highly flexible with less complexity. Here, the input signal is only utilized for large initial values. Moreover, ADP method was suggested in [20], which reduces the complexity of NN framework and is highly effective; anyhow it may suffer with a major drawback of time-delay. The ARX-NN approach was adopted in [21], which may support disturbance rejection and provide better tracking performance. Nevertheless, it has high complexity and time delay. CFNSs strategy was introduced in [22], which is highly flexible and enhances the control performance. Moreover, it may undergo a major disadvantage of noises well as external disturbances. Adaptive fuzzy with auxiliary dynamic strategy was created in [23], which avoids damages and is highly accurate. But, it has low-frequency positions and has dynamic disturbance. AII technique was proposed in [24], which provide easy installation and is highly reliable; but it has a time-invariant performance. Likewise, SDRE strategy was introduced in [25], which is very effective and it may lead to complex computation. Similarly, PD controller was suggested in [26], which provide high disturbance rejection control as well as gain control. But, it may cause less sensitivity and state peaking phenomenon. Neuro-observer based robust controller was designed in [27], which offers high accuracy with more flexibility; but it may cause disturbance effects. Partial feedback linearization was built in [28], which provide tracking error performance and is very slow with low bandwidth. BH optimization algorithm was suggested in [29], which possess high efficacy with better performance. A major drawback of this method is its external disturbance. Finally, interval type 2 T-s fuzzy model was adopted in [30], which provide simple calculations and is highly effective. A major disadvantage occurred in this method is model mismatch and its disturbances.

Table 34.1: Review on controller design strategies for non-linear systems.

Author Name	Method	Advantages	Drawbacks
Rui Yan et al. [1]	Adaptive decentralized controller	• better efficiency • Improved stability.	• Disturbance occurs.
Wei-Der Chang and Shun-Peng Shih [2]	PSO approach	• Excellent tracking performance. • More Flexible.	• Low quality system. • Needs memory to update velocity.
Qinmin Yang and Sarangapani Jagannathan [3]	Online approximators (OLAs)	• High feasibility	• Low stability. • High computational cost
Yan-Jun Liu, Li Tang [4]	multiple-input multiple-output (MIMO)	• Less Tracking errors • High effectiveness	• Additional complexities. • Less stability analysis.
A.Y. Abdelaziz& E.S. Ali [5]	CS+PI controller	• High performance	• High computation time • Easily trapped in local optimal solution.
AydoganSavran & GokalpKahraman [6]	adaptive tuning method	• Minimizes the error • High accuracy	• less compensate
Aws Abdulsalam Najm, Ibraheem Kasim Ibraheem [7]	DOF & UAV	• High speed • More stability	• Inaccurate • It takes more time
Derong Liu et al.[8]	reinforcement-learning (RL) methods	• Computation complexity reduces.	• system instability
Qing Gao et al. [9]	T-S fuzzy dynamic models	• Good performance	• very complex
Bo Fan et al. [10]	adaptive control strategy	• High tracking performance.	• It affects the high frequency. • Disturbance occurs
Jiuxiang Dong et al. [11]	Nonlinear discrete time fuzzy system.	• Used for engineering implementations. • lighter computational burden	• Less conservative.
Shuping He et al. [12]	Sliding mode control (SMC)	• Effectiveness	• Time delay • External disturbance occurs.
Zhi Liu et al. [13]	input-to-state practical stability (ISpS)	• reduce the computation burden • Improved performance	• computation problem
Wenqiang Ji et al. [14]	output feedback sliding mode control (SMC)	• strong robustness • good transient performance	• External disturbance

(continued)

Table 34.1: Continued

Author Name	Method	Advantages	Drawbacks
Heng Liu et al. [15]	fractional-order nonlinear systems control	• Guarantee reduced tracking errors. • Stability analysis.	• modeling errors • plant parameter variations
Pengfei Yan et al. [16]	data-driven controller design	• Solve practical problems. • Noiseless	• more complex • costly
Cheung-Chieh Ku et al. [17]	Linear Matrix Inequality (LMI)	• stability • active performance	• disturbance occurs • more complex
Thomas Gubneret al. [18]	convex optimization	• It is flexible • Less Complexity	• Input signal is utilized here.
Changzhu Zhang et al. [19]	akagi–Sugeno (T–S) fuzzy models	• less conservative • effectiveness	• transmission delays • packet dropouts
Jilie Zhang et al. [20]	adaptive dynamic programming (ADP)	• reduces the burden of neural-network framework • Effectiveness	• time-delay
Bin Hu et al. [21]	ARX-neural network (ARX-NN)	• disturbance rejection • tracking performance	• complexity • time delay
Jinxiang Chen et al. [22]	complex flexible nonlinear systems (CFNSs)	• flexible • control performance is greatly improved	• noise occurs • external disturbance
Xiaogong Lin et al. [23]	Adaptive fuzzy system and auxiliary dynamic system.	• Accuracy • avoids damaging	• Low frequency positions. • dynamic disturbance
XiaoxiaoLv&Xiaodi Li [24]	Average impulsive interval (AII)	• easy installation • high reliability	• time-variant
Yazdan&Batmani [25]	state-dependent Riccati equation (SDRE)	• very effective	• computational complexity • tracking error
Jeang-Lin Chang and Tsui-Chou Wu. [26]	PD controller	• disturbance rejection control • high gain control	• state peaking phenomenon • less sensitivity
HamedeKarami and Reza Ghasemi [27]	neuro-observer-based robust controller	• flexibility • high accuracy	• disturbance influences • robustness
M. A. Mahmud et al. [28]	partial feedback linearization	• tracking error	• low-bandwidth • very slow
Hazem I. Ali et al. [29]	Black hole (BH) optimization method.	• Efficiency • better performances	• external disturbance is the major drawback
Tufan Kumbasaret al. [30]	interval type2 Takagi– Sugeno fuzzy model	• Simple calculations • Effectiveness	• Efficiency Reduction occurs

Analysis on Performance

The performance analysis from the reviewed paper was based on controller design strategy for nonlinear system and the analysis is represented. On observing, the terminal voltage in [1] obtains maximum values of 1.05p.u and 102p.u and convergence rate of [2, 5] has attained a value of 0.0002747 and 23.3, accordingly. Additionally, the joint angle measure has obtained better value of 0.05 and tracking trajectory has attained a value of 1V, 1.2V and 2.2 V in [4], respectively. Moreover, the IAE, ITAE, ISE, ITSE has attained a maximum value of 4.1453, 28.9463, 0.4882 and 2.7914 and it is examined in [5]. However, the mean square error has attained a value of 4.98×10^4 as indicated in [6]. Similarly, steady state error in [7] obtained a maximum value of 0.56. Likewise, the weight convergence has attained a transverse value of [2.5849, 2.0037, 0.6158, 1.1825, 1.5860, –0.1390, 0.6583, –0.5108, 0.6364, 0.6695, –0.1333, 0.3175, –0.1374, 0.3578, –0.2267, 0.3878, 0.2951, –0.0874, 0.1738, –0.0517, 0.1243, –0.0437, 0.1497, –0.0854]. Although, the performance in [9, 11, 14, 15] obtained a maximum value of 1.0336, 16.5304, 3.9993 and 0.001 respectively. Moreover, the measures such as sampling period, output error, disturbance, torque, attenuation, state trajectory, efficiency, tracking error, settling time and estimation error have attained better values of 0.01 sec, 0.01 for error 1 and 0.05 for error 2, 2.9791. Furthermore, the constraint measures in [10, 12] attains a value of 0.2 sec and 2.2 sec respectively. Finally, the measures such as randomized set-point and set-point change have attained 0.7–0.99 and 0.98–0.88 values in [21] respectively. In [22], the controller gains and efficiency has attained 0.8 and 0.5 values in which the error and response time has attained a zero value in 150 sec. In [26], the settling time and overshoot have attained a low value while in [28], the attained active and reactive power is 1 kW and 1.62kVAr. Hence, the obtained sampling time and computational time in [30] have achieved a value within 2 sec and 8.115ms, correspondingly.

Research Gaps and Challenges

The issue with linear systems is that they may not be able to accommodate modern and advanced technology, such as multi-degree of freedom, etc. and require a wider range of application. The choice between linear and nonlinear control for a specific application can be difficult. Linear management has been thoroughly tried, and industry professionals trust it. For linear systems, there are numerous good analysis tools accessible. Nonlinear systems, on the other hand, require complicated mathematical analysis, such as the Lyapunov stability criterion, the Popov criterion, and singular perturbation techniques. Nonlinear systems may also make mathematical modeling difficult. Limit cycle, instability, and splitting can all occur in nonlinear systems. Most schemes can only assure local stability and cannot promise worldwide stability. Furthermore, due to the high initial expenditure required, processing capacity may be one of the primary constraints to implement sophisticated nonlinear control methods. Nonlinear control development model is still very expensive. No low-cost, portable, and competent technology is currently available to market these sophisticated methods. To minimize computational overhead, more effective programming may be needed. Future study will concentrate on how to adapt the techniques to nonlinear systems with delays, switched systems, and practical issues. The approach must specifically examine model-free optimal control for time-delay systems. Furthermore, the nonlinear controller must be improved so that the control system is asymptotically steady.

Conclusion

A complete review of available technology solutions is conducted in this paper based on controller design strategies for non-linear systems and is given by reviewing the contributed 30 papers. Each solution's concept is extensively studied, and its viability is assessed in light of the current limits in nonlinear systems. Furthermore, the disadvantages and benefits of each strategy are discussed. It reviews a bunch of 30 research papers and analyses various controller design strategies for non-linear systems. For addressing various groups of nonlinear systems, it conducts analysis on various controllers, including intelligent, adaptive, and hybrid controllers. The performance measures evaluated in each paper has been reviewed and its maximum achievements were analyzed. At last, this survey presented different research gaps and challenges that were useful for researchers to perform the upcoming work based on the control of non-linear systems.

References

1. Yan, Rui, Dong, Zhao Yang, Saha, Tapan, and Majumder, Rajat. (2010). A power system non-linear adaptive decentralized controller design. *Automatica, 46*(2), 330–336.
2. Chang, W. D., and Shih, S. P. (2010). PID controller design of nonlinear systems using an improved particle swarm optimization approach. *Communications in Nonlinear Science and Numerical Simulation*, 15(11), 3632–3639.
3. Yang, Q., and Jagannathan, S. (2011). Reinforcement learning controller design for affine non-linear discrete-time systems using online approximators. *IEEE Transactions on Systems, Man, and Cybernetics, Part B (Cybernetics), 42*(2), 377–390.
4. Liu, Y. J., Tang, L., Tong, S., and Chen, C. L. P. (2014). Adaptive NN controller design for a class of nonlinear MIMO discrete-time systems. *IEEE Transactions on Neural Networks and Learning Systems, 26*(5), 1007–1018.
5. Abdelaziz, A. Y., and Ali, E. S. (2015). Cuckoo search algorithm based load frequency controller design for nonlinear interconnected power system. *International Journal of Electrical Power and Energy Systems, 73*, 632–643.
6. Savran, A., and Kahraman, G. (2014). A fuzzy model based adaptive PID controller design for nonlinear and uncertain processes. *ISA Transactions, 53*(2), 280–288.
7. Najm, A. A., and Ibraheem, K. I. (2019). Nonlinear PID controller design for a 6-DOF UAV quadrotor system. *Engineering Science and Technology, an International Journal* 22(4), 1087–1097.
8. Liu, D., Yang, X., Wang, D., and Wei, Q. (2015). Reinforcement-learning-based robust controller design for continuous-time uncertain nonlinear systems subject to input constraints. *IEEE Transactions on Cybernetics, 45*(7), 1372–1385.
9. Gao, Q., Zeng, X. J., Feng, G., Wang, Y., and Qiu, J. (2012). T–S-fuzzy-model-based approximation and controller design for general nonlinear systems. *IEEE Transactions on Systems, Man, and Cybernetics, Part B (Cybernetics)*, 42(4), 1143–1154.
10. Fan, B., Yang, Q., Jagannathan, S., and Sun, Y. (2017). Asymptotic tracking controller design for nonlinear systems with guaranteed performance. *IEEE Transactions on Cybernetics, 48*(7), 2001–2011.
11. Dong, J., Wang, Y., and Yang, G. H. (2010). Output feedback fuzzy controller design with local nonlinear feedback laws for discrete-time nonlinear systems. *IEEE Transactions on Systems, Man, and Cybernetics, Part B (Cybernetics)*, 40(6), 1447–1459.
12. He, S., Song, J., and Liu, F. (2017). Robust finite-time bounded controller design of time-delay conic nonlinear systems using sliding mode control strategy. *IEEE Transactions on Systems, Man, and Cybernetics: Systems*, 48(11), 1863–1873.

13. Liu, Z., Wang, F., Zhang, Y., Chen, X., and Chen, C. L. P. (2013). Adaptive fuzzy output-feed-back controller design for nonlinear systems via backstepping and small-gain approach. *IEEE Transactions on Cybernetics*, 44(10), 1714–1725.

14. Ji, W., Qiu*, J., Wu, L., and Lam, H. K. (2019). Fuzzy-affine-model-based output feedback dynamic sliding mode controller design of nonlinear systems. *IEEE Transactions on Systems, Man, and Cybernetics: Systems*, 51(3), 1652–1661.

15. Liu, H., Li, S., Li, G., and Wang H. (2018). Adaptive controller design for a class of uncertain fractional-order nonlinear systems: an adaptive fuzzy approach. *International Journal of Fuzzy Systems*, 20, 366–379.

16. Yan, P., Liu, D., Wang, D., and Ma H.(2016). Data-driven controller design for general MIMO nonlinear systems via virtual reference feedback tuning and neural networks. *Neurocomputing*, 171, 815–825.

17. Ku, C. C., Huang, P. H., and Chang, W. J. (2010). Passive fuzzy controller design for nonlinear systems with multiplicative noises. *Journal of the Franklin Institute*, 347(5), 732–750.

18. Gußner, T., Jost, M., and Adamy, J. (2012). Controller design for a class of nonlinear systems with input saturation using convex optimization. *Systems and Control Letters*, 61(1), 258–265.

19. Zhang, C., Feng, G., Qiu, J., and Zhan, W-A. (2014). T–S fuzzy-model-based piecewise H∞ output feedback controller design for networked nonlinear systems with medium access constraint. *Fuzzy Sets and Systems*, 248, 86–105.

20. Zhang, J., Zhang, H., Liu, Z., and Wang, Y. (2015). Model-free optimal controller design for continuous-time nonlinear systems by adaptive dynamic programming based on a precompensator. *ISA Transactions*, 57, 63–70.

21. Hu, B., Zhao, Z., and Liang, J. (2012). Multi-loop nonlinear internal model controller design under nonlinear dynamic PLS framework using ARX-neural network model. *Journal of Process Control*, 22(1), 207–217.

22. Chen, J., Sun, F., Sun, Y., and Yu, L. (2014). Modeling and controller design for complex flexible nonlinear systems via a fuzzy singularly perturbed approach. *Information Sciences*, 255, 187–203. doi:10.1016/j.ins.2013.08.031.

23. Lin, X., et al. (2018). Nonlinear adaptive fuzzy output—feedback controller design for dynamic positioning system of ships. *Ocean Engineering*, 158, 186–195.

24. Lv, X., and Li, X. (2017). Finite time stability and controller design for nonlinear impulsive sampled-data systems with applications, *ISA Transactions*, S0019057817305104,doi:10.1016/j.isatra.2017.07.025

25. Batmani, Y. (2017). H∞ suboptimal tracking controller design for a class of nonlinear systems. *International Journal of Control, Automation and Systems*, 15, 2080–2087.

26. Chang, J. L., and Wu, T. C. (2021). Robust output feedback sliding mode state and disturbance observer-based controller design for nonlinear systems. *International Journal of Control, Automation and Systems*, 19(4), 1550–1559.

27. Karami, H., and Ghasemi, R. (2021). Adaptive neural observer-based nonsingular super-twisting terminal sliding-mode controller design for a class of hovercraft nonlinear systems. *Journal of Marine Science and Application* 20(2), 325–332.

28. Mahmud, M. A., et al. (2014). Robust nonlinear distributed controller design for active and reactive power sharing in islanded microgrids. *IEEE Transactions on Energy Conversion*, 29(4), 893–903.

29. Ali, H. I., and Hadi, M. A. (2020). Optimal nonlinear controller design for different classes of nonlinear systems using blackhole optimization method. *Arabian Journal for Science and Engineering*, 45(8), 7033–7053.

30. Kumbasar, T., et al. (2011). Interval type-2 fuzzy inverse controller design in nonlinear IMC structure. *Engineering Applications of Artificial Intelligence*, 24(6), 996–1005.

35 Analytical Detection of Volatile Organic Compound by Supervised Learning Methods

Paramita Chowdhury[1,a], Sunipa Roy[2,b], and Utpal Biswas[3,c]

[1]Electronics & Communication Engineering Department, Netaji Subhash Engineering College, Kolkata & Research Scholar, University of Kalyani, West Bengal

[2]Electronics and Communication Engineering Department, Guru Nanak Institute of Technology, Kolkata

[3]Computer Science and Engineering Dept, University of Kalyani, West Bengal

Abstract

This paper reports the sensing of thin films sensors to address two volatile organic compounds; acetone and formaldehyde.Highest responses obtained at the concentration of 3600 ppm. Analytical detection of volatile organic compound using supervised learning methods has been reported here. In this case, the Random Forest algorithm has been used to classify VOC vapors onZnO surface and the 10 second response time is used to provide the necessary data for classification. By Random Forest Algorithm, the present sensor can be transformed into a more efficient one with improved detection capabilities. Among all classification algorithms, Random Forest reaches the highest accuracy of 88%.

Keywords: Acetone, formaldehyde, random forest, supervised learning, SVM, ZnO.

Introduction

Selective detection of vapors of volatile organic compounds (VOC) is of great importance in many fields, from home, factories to diagnosis of hazarders' diseases [1–5]. In this sense, gas sensors offer many advantages, including low cost, high response and recovery times, and low energy consumption [6]. Among the various types of gas sensors based on different operating principles, such as acoustic wave, calorimetric, capacitance, optical, 98, and so on. Among the various metal oxides, zinc oxide (ZnO) is one of the proven materials in the gas sensor family. Since it has unique properties such as higher band gap, direct band gap, porosity, smaller grain size, etc., the morphology and patterning of ZnO is multidimensional. A ZnO gas sensor consists of a sensing element and a transducer capable of converting energy from one species to energy from another. The high sensitivity of ZnO and its derivatives in terms of measurable changes in their conductivity when exposed to reducing or oxidizing gasses makes them a natural choice for VOC vapor sensing. ZnO as sensing elements in resistive gas sensors has been intensively studied because of its open coordination sites for axial ligations [7]. Due to their conductivity, they exhibit excellent sensing properties for various gasses such as H_2 [8], CO_2 [9], SO_2 [10], VOCs [11], etc. Indeed, the main drawback of ZnO-based sensors is their lack of selectivity towards VOC

---

[a]paramita.chowdhury@nsec.ac.in, [b]sunipa_4@yahoo.co.in, [c]utpal01in@yahoo.com

vapors, which severely limits their use in sensing applications. This is a problem that needs to be solved, especially when it comes to making high-tech devices such as an electronic nose. An essential component of an electronic nose is a system that establishes a relationship between the sensor response and the gas species in order to selectively detect gasses. This can be achieved through machine learning, where feature extraction and classification are two important steps. Since the feature plays a key role in classification, it should efficiently represent the features of the original high-dimensional gas sensor dataset. If this is the case, it is usually easier to find a successful classification algorithm. Therefore, many previous works have made contributions based on steady-state response in various gas sensor applications [12–16].

In this work, to detect two different VOC vapors with different concentrations from 50 ppm to 3600 ppm, an experiment was conducted using a metal oxide based gas sensor on ZnO surface. It was found that the interaction between the vapors and the surface is not selective enough. To solve this problem, it is aimed that the system can discriminate vapors using machine learning algorithms. As in many other studies where robust information was extracted from the sensor characteristic curve, we also used the steady-state characteristic curve. However, the results show that the classification accuracy was very low when only the steady-state response was used in our work. Therefore, it is considered that not only one response but also a few seconds of responses from the steady-state could be used to utilize more information from the response curve. The performance of this feature was tested with multiple AI algorithms such as decision tree, support vector machine (SVM), K-nearest neighbor (KNN) and Random Forest ensemble method.

VOC Sensing Experimental

All of the chemicals were analytical reagent grade and were used directly out of the package. Diethanolmine, hexamethylenetetramine [$(CH2)6N4$], deionized water (resistivity: 18.2 M-cm), and zinc acetate dihydrate (Merck, 99.9%). The sample was created via the Roy et. al. recommended method. The specifics of the vapor measuring configuration have already been disclosed [17–18].

Sensing Results

Figure 35.1 depicts the transient response of the sensor and illustrates how the sensor resistance was originally high since there were no free electrons present on the ZnO sensing surface. When formaldehyde vapor was introduced, the sensor resistance decreased because free electrons were released; when the supply of vapor was cut off, the sensor resistance increased because oxygen species from the air were recapturing the adsorption site, which resulted in a decrease in the concentrations of surface electrons that were available. The formaldehyde sensor has high response magnitude (94.8%) at 2020 ppm concentration with a 20 s response time and 88 s recovery time, however at lower concentrations (190 ppm), the sensor delivers a response magnitude of 70.4% with a high response time of 45 s. Due to the sluggish desorption rates of the formaldehyde molecules from the ZnO surface, the reaction time drops at greater concentrations while the recovery time increases. Due to the saturation of all the surface's adsorbent sites at higher concentrations, the formaldehyde sensor has a propensity toward saturation. The minimal baseline shift can also be attributed to the fact that some target vapor species remained adsorbed long after the vapor pulse was shut off and that their amount increased as the vapor concentration rose.

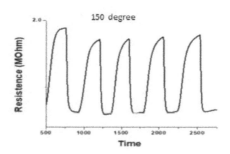

Figure 35.1 Transient response of sensor at 100°C.

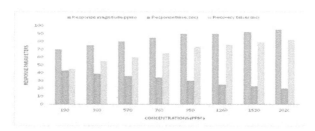

Figure 35.2 Comparison chart of different parameter of sensor at different ppm level.

The bar chart representation in Figure 35.2 depicts the variance in response magnitudes as a function of formaldehyde concentrations (190, 380, 570, 760, 950, 1260, 1530, and 2020 ppm) at a certain optimal operating temperature (100°C).

Sensor Data Extraction

The ZnO thin film's response recovery properties after exposure to two distinct amounts of organic vapours, formaldehyde and acetone. We noticed that after being exposed to all of the VOC fumes, ZnO film shows a rise in current. For all concentrations of the tested VOC vapors, formaldehyde vapors showed the greatest increase in sensor current, followed by acetone. The following equation has been used to compute the sensitivity (S) of the ZnO thin film towards VOC vapors from the measured dynamic properties of the sensor:

$$S = 1/C_v . \Delta I/I_0 \tag{1}$$

Where I_0 is the baseline current of the ZnO thin film prior to exposure to the VOC vapours, and C_v is the concentration of the VOC vapor under investigation [19–21]. The current sensor has low selectivity, making it impossible to accurately identify the vapors being considered using conventional techniques. Therefore, as described in the next sections dimensionally, we used a AI mechanism to provide the system with a means of identifying the vapors on its own.

The collected data should be preprocessed before the feature extraction step. Since data processing in a machine learning system directly determines the input, it is an indispensable step before training a model. There are no general guidelines to determine the appropriate data preprocessing technique, so the technique to be used varies from application to application. Among different techniques, in this research, On the unprocessed data, the initial state removal technique was initially applied to mitigate the impact of the initial state on the information and guarantee data dependability. We deducted the sensor readings from the initial state values to negate the impact of the initial state on the information. As illustrated in Equation (2), the dynamic response is given by (t), while the initial state is given by ((0), which corresponds to the minimum sensor reading when exposed to a reference gas. Thus, the preprocessed data corresponds to the discrepancy between the initial state and the response.

$$YS = (t) - (0) \tag{2}$$

Feature extraction step helps to reduce the amount of redundant data from the original high dimensional gas sensor data set. in our approach, we use 10 s response values of the steady state region, In order to obtain successful classification results, only a few seconds of sensor data after the response curve reaches 90% of its maximum value were directly used without taking into account the other information in the whole response curve.

Data Classification

Since each gas has two different concentrations ranging from 50 ppm to 3600 ppm, there are a total of 324 samples in the sub-data set for categorization. 150 data for formaldehyde and 174 data for acetone were extracted from 324 data.In addition, a second data set comprised of the highest value for each concentration location was produced for comparison purposes.

The classification stage deals with the issue of establishing a connection between the sensor responses and the various gas types utilizing the steady state 10 s responses. The dataset, which comprises of 55 samples for each type of gas, is randomly divided into 70% training sets and 30% test sets.

Random forests combine several decision trees to make predictions or classifications. Integrating the results of these trees produces a consolidated and more accurate result. The effectiveness of several categorization algorithms was evaluated using categorisation accuracy as a performance metric. Counting the samples and dividing by the number of samples that were successfully identified determines the number of samples. The training set and test set were randomly selected from the original dataset of gas sensor data, therefore we repeated the train-test methodology ten times to prevent bias in the classification process. Then, the final classification accuracy of each classifier was calculated by averaging the outcomes of 10 iterations. The same training set was used to train all classification algorithms, and the same test set was used to assess their classification accuracy. Therefore, the categorization method with the highest accuracy is unquestionably the best.

Table 35.1 shows the classifiers used in this study and their descriptions. In Table 35.2, the accuracy of these algorithms is evaluated to determine how well they function. Comparing the results of the considered classification algorithms, the Random Forest algorithm performs the best. One of the most important characteristics of the Random Forest Algorithm is its ability to handle data sets containing continuous variables, such as in regression, and categorical variables, such as in classification. As far as classification and regression are concerned, it performs better. Using a classification job as an example, we will examine the random forest's operation. With the Random Forest technique, 88% accuracy is achieved in this work. Additionally, the SVM method also produced good classification accuracy, but it guards against over fitting and makes predictions that are more precise.

Table 35.1: Classification algorithms used in this work.

Classifier	Description
Logistic Regression	Preset: Logistic Function
K Nearest Neighbor(KNN)	Preset: Weighted KNN Number of neighbor: 3 DistanceMatrix: Euclidean
Random Forest Classifier	Preset: Fine Tree Splits maximum:200 Indicator: entropy
SVM	Kernel: rbf

Table 35.2: Classification performance of the algorithms.

Classifiers	Accuracy	F1-score
Logistic Regression	66%	55%
KNN	66%	57%
Random Forest Classifier	88%	80%
SVM	88%	80%

Result Analysis

Between the proposed classifiers SVM, KNN, Nave Bayes, and Neural Network, the accuracy of the results is compared. This comparison aims to determine which gas diagnosis classifier performs at the highest level. Random Forest has a comparatively high specificity of 88.0% compared to other classifiers. The comparison table is displayed in Figure 35.3 down below.

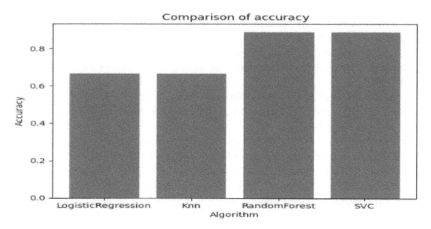

Figure 35.3 Comparison chart of accuracy of different algorithm.

Conclusion

In this study, the sensing capabilities of a VOC gas sensor on a thin ZnO screen were examined at two different concentrations of VOC vapor, ranging from 50 to 3600 ppm.

To address the issue of selectivity, machine learning methods were used. from the gas sensor. In order to achieve high classification accuracy and quick classification times, only 10 s of steady state sensor curve data is required. With a single mental oxide based gas sensor, Random Forest, one of the quickest and simplest to use algorithms, achieved the best accuracy (88%) which is an impressive accomplishment. In this investigation, the proposed feature's performance was also contrasted with that of the conventional steady state response feature, and the proposed feature offered much higher classification accuracy. As a result, a few seconds of steady state reactions might offer the most crucial data to distinguish between the various kinds of VOC vapors. One advantage of the suggested strategy in this study is its speed and simplicity. In this investigation, excellent classification accuracy was achieved by using only one feature based on a single sensor, even though in many cases a single feature can't adequately capture the properties of sensor responses. Using a single gas sensor to resolve the selectivity issue is a great practical solution to the selectivity problem.

References

1. Yamazoe, W. N., Fuchigami, J., Kishikawa, M., and Seiyama, T. (1979). Interactions of tin oxide surface with O2,H2O and H2. *International Journal of Surface Science*, 86, 335–344.
2. Banerjee, N., Bhowmik, B., Roy, S., Sarkar, C. K., and Bhattacharyya, P. (2013). Anomalous recovery characteristics of Pd modified ZnO nanorod based acetone sensor. *Journal of Nanoscience and Nanotechnology*, 13(10), 6826–1634.
3. Deng, H., Li, H., Wang, F., Yuan, C., Liu, S., Wang, P., Xie, L., Sun, Y., and Chang, F. (2016). A high sensitive and low detection limit of formaldehyde gas sensor based on hierarchical flower-like CuO nanostructure fabricated by sol–gel method. *Journal of Materials Science: Materials in Electronics*, 27(7), 271–350.
4. Li, Y., Guo, Z., Su, Y., Jin, X., Tang, X., Huang, J., Huang, X., Li, M., and Liu, J. (2017). Hierarchical morphologydependent gas-sensing performances of threedimensional SnO_2 nanostructures, *ACS Sensors*, 2(1), 102–110.
5. Roy, S., Sarkar, C. K., and Bhattacharyya, P. (2012). Low temperature fabrication of a highly sensitive methane sensor with embedded Co-Planar nickel alloy microheater on MEMS platform. *Sensor Letters*, 10, 1–10.
6. Pandya, H. J., Chandra, S., and Vyas, A. (2011). MEMS based ethanol sensor using ZnOnanoblocks, nanocombs and nanoflakes as sensing layer. *Sensors and Transducers Journal*, 134(11), 85–94.
7. Roy, S., Banerjee, N., Sarkar, C. K., and Bhattacharyya, P. (2013). Development of an ethanol sensor based on CBD grown ZnO nanorods. *Solid-State Electronics*, 87, 43–50. K. Elissa, "Title of paper if known," unpublished.
8. Prajapati, C. S., and Sahay, P. P. (2011). Alcohol-sensing characteristics of spray deposited ZnO nano-particle thin films. *Sensors and Actuators B*, 160(1), 1043–1049.
9. Roy, S., Banerjee, N., Sarkar, C. K., and Bhattacharyya, P. (2014). Butanone sensing characteristics, mechanism and equivalent circuit model of Pd decorated ZnO nanorod based resistive sensors. *Sensor Letters*, 12(1), 89–96.
10. Lenaerts, S., Roggen, J., and Maes, G. (1995). FT-IR characterization of tin dioxide gas sensor materials under working conditions. *Spectrochimica Acta Part A*, 51(5), 883–894.
11. Yamazoe, N., Fuchigami, J., Kishikawa, M., and Seiyama, T. (1979). Interactions of tin oxide surface with O2, H2O and H2. *International Journal of Surface Science*, 86, 335–344.
12. Chang, S. C. (1980). Oxygen chemisorption on tin oxide: correlation between electrical conductivity and EPR measurements. *Journal of Vacuum Science and Technology*, 17, 366–369. (Nanorod Based Resistive Sensors, Sensor Letters, 12(1), 2014, pp. 89–96.)

13. Weng, X., and Kong, C. (2021). Detection of volatile organic compounds (VOCs) in livestock houses based on electronic nose characterization of tin dioxide gas sensor materials. *Applied Sciences*, 11(5), 2337, https://doi.org/10.3390/app11052337.
14. Banerjee, N., Roy, S., Sarkar, C. K., and Bhattacharyya, P. (2013). High dynamic range methanol sensor based on aligned ZnO nanorods. *IEEE Sensors Journal*, 13(5), 1669–1676.
15. Banerjee, N., Bhowmik, B., Roy, S., Sarkar, C. K., and Bhattacharyya, P. (2013). Anomalous recovery characteristics of Pd modified ZnO nanorod based acetone sensor. *Journal of Nanoscience and Nanotechnology*, 13(10), 6826–1634.
16. Roy, S., Sarkar, C. K., and Bhattacharyya, P. (2012). Low temperature fabrication of a highly sensitive methane sensor with embedded Co-planar nickel alloy microheater on MEMS platform. *Sensors Letters*, 10, 1–10.
17. Pandya, H. J., Chandra, S., and Vyas, A. (2011). MEMS based ethanol sensor using ZnOnanoblocks, nanocombs and nanoflakes as sensing layer. *Sensors and Transducers Journal*, 134(11), 85–94.
18. Harikrishna, N. B. (2019). Confusion matrix, accuracy, precision, recall, F1 score. *Analytics Vidhy* 2019, https://medium.com/analytics-vidhya/confusion-matrix-accuracy-precision-recall-f1-score-ade299cf63cd.
19. Brownie, J. (2022). How to calculate precision, recall, F1, and more for deep learning models. *Machine Learning Mastery*, 2021 https://machinelearningmastery.com/how-to-calculate-precision-recall-f1-and-more-for-deep-learning-models/.
20. Lovell, D., McCarron, B., Langfield, B., Tran, K., and Bradley, A. (2021). Taking the confusion out of multinomial confusion matrices and imbalanced classes. 2021, 10.1007, 978-981-16-8531-6_2.
21. Joseph, V. R. (2022). Optimal ratio for data splitting. *Statistical Analysis and Data Mining: The ASA Data Science Journal*. 15.10.1002, sam.11583.
22. Khalid, M., and Peter, E. (2021). Algorithmic Splitting: A Method For Dataset Preparation. *IEEE Access*. 1–1. 10.1109/ACCESS.2021.3110745

36 Fake News Detection in Social Media: A Literary Review

Anudeepa Gon[1,a] and Sangeeta Bhattacharya[2]

[1]Assistant Professor, Brainware University, Kolkata

[2]Associate Professor, Guru Nanak Institute of Technology, Kolkata

Abstract

The development of technology and the internet has made the entire world accessible online. Everybody today relies on a range of online news sources resulting to how pervasive the internet has become in our contemporary culture. The news quickly reaches millions of people at a short span of time without being verified, along with Facebook, Twitter and other social media platforms are becoming quite common. False articles created with the goal to mislead are the most prevalent definition of fake news on the internet. To increase reading, social media and news organizations create fake news. Profiting from clickbait is typically the aim. To precisely identify fraudulent postings, a few basic, carefully chosen elements of the title and content are used. Malicious spam or email attachments may cause the same issue, and people might be deceived by such fraudulent information. This review paper's objective is to offer information that can be used to spot false or misleading material. This study examines the motivations behind fact-checking efforts, such as maintaining credibility, preventing damage, and maintaining a healthy information ecosystem. It explores the methods used by fact-checkers, from manual verification by experts to the development of automated tools employing Machine learning and Natural Language Processing. Challenges, such as the rate at which false information spreads and the possibility of algorithmic bias, are discussed alongside possible solutions. In addition, we can implement how to explore the implications of fact-checking, including its role in cultivating critical thinking, collaborative partnerships between platforms and fact-checkers, and the evolving legal and regulatory landscape. Understanding the intricate interactions between technology, information sharing, and the quest for truth is improved by this study and trustworthy digital communication by investigating the multifaceted approach to combating false news in social media.

Keywords: Deep learning, extracting features, fake news identifying, machine learning, natural language processing, online social media.

Introduction

With the popularity of social media and other mobile technologies growing, information is now more easily accessible. Social media platforms and mobile apps have displaced print media as the primary means of news and information transmission. Digital media's ease

---

[a]gonanudeepa@gmail.com, ang.cs@brainwareuniversity.ac.in

and speed force people to use it for their daily information needs. Customers may readily access a variety of data, for-profit organizations benefit from a solid platform for growing their reach. The forum seems to have trouble identifying authentic news from bogus news. False information is regularly spread with the aim of misleading people or developing prejudice in order to gain politically or economically from it. As a result, in order to attract more visitors, it could offer amusing news articles or other content. Several news stories that supported specific candidates and their political platforms during the most recent US elections have been the subject of intense debate regarding their validity. Due to this rising worry and an effort to lessen its negative effects on individuals and society, the study of false news is gaining popularity. Fake news detection systems commonly use a variety of machine learning models, including Random Forests, Logistic Regression, Decision Trees, Stochastic Gradient Descent, and others Lee, et al. [14]. A random forest classifier based model that classifies news as authentic or not must be put into practice in this project. Even aesthetically displayed news can be categorized by it.

The goal of this study is to use machine learning to identify fake news so that you can get what is legitimately yours—the truth—and ensure and increase the news' trustworthiness. Recent elections in India and other nations have demonstrated the development of misleading information that is routinely communicated in an effort to alter students' political opinions or worldviews, particularly during the Covid-19 virus's global spread Eklund et al. [9]. Because of the increased availability of faulty information, including deep fakes, biased reporting, and sources that are only partially credited, news writers find it more and more challenging to distinguish between what is true and what is false. The bulk of current rumors regarding social media include social media, even if fake news is not a new problem and can be found in any media, including books, television, radio, and the internet. False news routinely spreads on social media platforms despite several companies' efforts to identify and remove it.

These offer a comprehensive understanding of the motivations, challenges, methodologies, and implications of false news detection in the context of social media. It integrates multiple perspectives to provide a comprehensive comprehension of the issue's complexity.

Motivations and Objectives: The survey sheds light on why individuals, organizations, and platforms engage in efforts to identify and combat false information by examining the motivations behind fake news detection. This understanding contributes to establishing the ethical, societal, and technological foundations for these initiatives.

Methodologies and Approaches: The survey investigates the variety of methodologies and approaches used for detecting false news, such as the human fact-checking of experts and the application of cutting-edge technology like machine learning and natural language processing. This analysis illustrates the diversity of counter-misinformation strategies.

By addressing obstacles associated with the pace of information dissemination, algorithmic biases, and the ever-evolving nature of false news, the survey highlights the obstacles encountered in this field. In addition, it describes the prospective solutions and innovations that researchers and practitioners are devising to address these obstacles.

Implications and Results: The survey outlines the broader implications of false news detection, including its role in fostering critical thinking, nurturing collaborations between platforms and fact-checkers, and aligning with legal and regulatory efforts. This understanding offers a road map for shaping the future of information dissemination and consumption. Eventually, the contribution of the survey on detecting fake news rests in its capacity to consolidate knowledge, provide insights, guide decision-making, and influence ongoing efforts to combat misinformation on social media platforms. It fosters a more informed and proactive approach to addressing the challenges posed by false news and functions as a valuable resources for academics, practitioners, policymakers, and the general public.

Motivation

- **Spread of Misinformation:** Through social media platforms, false information can travel quickly and quickly reach a huge number of individuals. This may result in factual distortion, inaccurate portrayal of events, and misunderstanding among the general audience.
- **Impact on Society:** The dissemination of false information has the power to sway public opinion, mould views, and even have an impact on important decision-making processes including elections, public health, and social issues. In order to sustain an informed and healthy society, information accuracy must be ensured.
- **Credibility and Trust:** Information and news are frequently accessed from social media platforms. If users consistently encounter false information, it can erode their trust in both the platform and the information they receive.
- **Preventing Harm:** False information can lead to harmful consequences, such as the spread of medical misinformation, causing panic, and promoting dangerous behaviors. Fact-checking helps prevent such harm by providing accurate information and dispelling false narratives.
- **Ethical Responsibility:** Social media platforms often see themselves as responsible for the content that is shared on their platforms. This includes addressing the dissemination of misinformation, which can have real-world consequences.

Literature Review

Fake news was identified using Naive Bayes in (Manag.Proc.2020). The testing was conducted using a dataset of Facebook news posts. Buzz feed's news dataset was also utilized. An estimated 74% classification accuracy was achieved on the test set.

A simple fake news detector with good classification accuracy was created and applied network analysis and linguistic clues techniques to it. Both strategies use machine learning to train classifiers that match the analysis and 72% accuracy rate was found that can be higher. This might be achieved by using cross-corpus analysis to the classification models and shrinking the size of the input feature vector [13].

It was possible to ascertain whether a news report was true or false using an ensemble approach and recurrent neural network (LSTM and GRU). For assessing the integrity of a news piece, an Android application was also created and performed the tests using a sizable dataset that was created. This method has the drawback that it needed the item to be of specific size.

Extensive checking and impartiality were used in [1, 6] as a crucial factor to increase the efficiency of identifying false news. Three distinct datasets were used. Along with cosine similarity, Bi-grams and Tri-grams approaches was implemented and count and Tf-Idf vectorizer was used. Both Naive Bayes and Random forest models were used to train the model. The reported accuracy was 81.6%.

A rumor detection system that assesses the veracity of the information and categorizes it as a rumor or not was proposed in [12] and Twitter API was used to gather data.

In order to identify fake news, multiple Machine Learning and Deep Learning based approaches are implemented using representative and open datasets in [15].

Using fivefold cross-validation, [16] evaluated twelve performance indicators and established the accuracy of the data collected. The evaluation's findings show how well-researched ground truth data are, how valid they are, and how well they work to build models that can spot false information.

Methodology Used

False, deceptive, or whose source cannot be confirmed information is referred to as fake news. This content might have been created with the goal of deceiving others, harming someone's reputation, gaining attention, or making money. The 2016 US Presidential Elections saw a spike in the use of the phrase. According to reports, economic factors most likely affected the election's outcome due to fake news [11].

Fake news spreading techniques in heterogeneous medium
This can be considered in two parts:

- In order to ascertain whether the statement as a whole refers to all the material that is accurate and authentic with proof (content based), one must look at the language used [7].
- To examine the statement for any hidden motives that can impede the reader from understanding the news's genuine meaning and expose it as fake (context based).

The two methods for identifying fake news are:

- **Machine learning approach:** Machine learning has been utilized in various research as a stand-alone method to identify how to discern between real and fake news, despite being a component of deep learning. The study by Georgios Gravanis et al. is the only one utilized as a reference to understand this concept has proposed an approach that uses content-based features and machine learning techniques to identify bogus news.
- **Deep Learning methodology:** Using a variety of deep learning concepts, including neural networks and artificial intelligence, is the main principle of this approach. Although machine learning is a component of deep learning, it has been chosen to compare it using a different approach. In [8], the described method of how the deep learning is used to determine if the news is authentic or fraudulent by its attitude in anyone of the primary researches that is evaluated for understanding this strategy.

Methodology Used in Fake news Identification
The methodology consists of various steps which is explained in Figure 36.1:

1. **Dataset Preparation:** The three stages listed below make up the majority of this step:
   - **Data Collection:** At first, information needs to be collected from several social media networks. For this reason, various data gathering technologies are available, such as Facebook, Twitter, etc., which offer a number of APIs to collect posts containing fake news via web scraping [5].

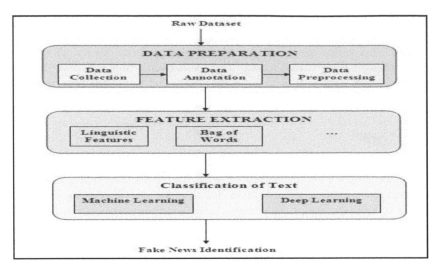

Figure 36.1 Shows the methodology used in fake news identification.

- **Data Annotation:** Data annotating comes after data gathering. Data classification and tagging are both components of data annotation. If fake news is present in the data that has been obtained, it is primarily how the categorization technique is built.
- **Data Pre-processing:** Data pre-processing follows the completion of data annotation. Pre-processing entails filtering input text to improve a proposed system's accuracy by eliminating redundant features that analyze raw data. Tokenization, which divides a post into individual tokens [2], the removal of URL addresses, brackets, dashes, colons, stop words, and all newline symbols, the replacement of redundant white spaces with a single white space, and other steps are required as part of the pre-processing of the data [3].
2. **Feature Extraction:** The goal of feature extraction is to transform unprocessed data into the inputs needed by a particular machine learning or deep learning algorithm. Features must best satisfy the needs of the algorithm that will be used to identify fake news by representing the information in the data. Some of the often utilize characteristics are statistical features, syntactic features, linguistic features, bag of words, word frequency features, topic features, n-gram features [10], and word embedding feature.
3. **Classification of Texts:** After creating the dataset with the extracted features, it is supplied as an input to the classification algorithms to determine whether or not false news intentions are present in that dataset. Numerous deep learning [4] and classification algorithms using machine learning are employed for this goal.
   A comparison of the various approaches used to identify fake news across various social network platform is provided in Table 36.1.

Conclusion and Future Work

Although various studies on the subject of fake news identification in online social networks have been undertaken, there are still a number of areas for investigation in this field.

Table 36.1: A comparative analysis of different methodologies.

Reference	Methodology	Drawback	Dataset
[14]	Several machine learning techniques, including random forest, logistic regression, multilayer perceptron, and support vector machines.	The sole variation from the configuration used for the larger dataset	Word2vece m
[9]	• Tweets pre-processing • Features extraction • Classifiers Evaluation	Need to release a publicly available large-scale COVID-19 benchmark sentiment analysis dataset	Tweepy, an official Python Twitter API library
[13]	Features extracted mainly from the text content of news tweets	Absence of fact automated fact checking system.	Sina Weibo
[6]	A graphical social context representation and learning framework for fake news detection	Machine learning techniques not used.	Fake news dataset collected
[12]	NLI models as well as BERT.	Ensemble model can be used.	Fake news dataset.
[11]	Used fact-checking systems	Need to experiment each feature type.	Data is collected from news websites
[10]	Deep Learning based Fact checking system	Can be applied new model EchoFakeD, when new features are added	Real world Fake-news dataset
[3]	Natural Language Processing (NLP) techniques.	Accuracy need to be better by using new classifiers.	False and Manipulate d contents

- Fake news detection can be done extremely well using historical or temporal data. The majority of publications, however, have not examined this; rather, few works have used merely data from a single post without any context. Studying the connections between fake news and factors like financial and political advantage, etc. would help the detection tools work better.
- Automatic labeling of the data is preferred to manual labeling, which introduces bias into the annotation process due to established annotation norms.
- In order to identify rumors, multi-modalities in the data, such as pictures, videos, and hyperlinks, also need to be analyzed. To provide the researchers with a standard testing environment to gauge the value of the proposed model, a dataset made up of features derived from multiple types of postings, such as text, image, and video, needs to be generated utilizing various social network data.
- In order to decrease the amount of false positives and increase the precision of detecting fake news or fraudulent ideation, a layer expert-based suggestion can be implemented to the utilized model in place of supervised learning classification techniques.

Acknowledgement

I would like to take this opportunity to record my deep sense of gratitude to my supervisor and guide, Prof. Dr. Sangeeta Bhattacharya of Department of CSE, Guru Nanak Institute of Technology (GNIT) for her supervision, advice, guidance and constant inspiration throughout the period of my work. A special thanks to my family for inspiring me, guiding me, helping me. Last but clearly not the least I would thank the almighty for giving strength to do the paper.

References

1. Alexandrov, D., Baly, R., Glass, J., Karadzhov, G., and Nakov, P. (2018). Predicting factuality of reporting and bias of news media sources. In Proceedings of the 2018 Conference on Empirical Methods in Natural Language Processing, EMNLP, 2018, (pp. 3528–3539) (2020).

2. Alonso-Galbán, P., and Alemañy-Castilla, C. (2022). Curbing misinformation and disinformation in the Covid-19 era: a view from cuba. *MEDICC Review*, 22, 45–46. https://doi.org/10.37757/MR2020.V22.N2.12.

3. Andersen, J., and Søe, S. (2020), Communicative actions we live by: the problem with fact-checking, tagging or flagging fake news-the case of Facebook. *European Journal of Communication*, 35(2), 126–139. https://doi.org/10.1177/0267323119894489.

4. Baptista, J. P., and Gradim, A. (2022). A working definition of fake news. *Encyclopedia*, 2(1), 632–645. https://doi.org/10.3390/encyclopedia2010043.

5. Bastick, Z. (2021). Would you notice if fake news changed your behavior? An experiment on the unconscious effects of disinformation. *Computers in Human Behavior*, 116, 106633. https://doi.org/10.1016/j.chb.2020.106633.

6. Cao, J., Zhang, Y., Jin, Z., Zhou, J., and Tian, Q. (2020). Novel visual and statistical image features for microblogs news verification. *IEEE Transactions on Multimedia*, 19(3), 598–608.

7. Cao, J., Qi, P., Sheng, Q., Yang, T., Guo, J., and Li, J. (2020). Exploring the role of visual content in fake news detection. In Disinformation, Misinformation, *Fake News Social Media*, (pp. 141–161) (2020).

8. Tianyu, G., Xingcheng, Y., and Danqi, C. (2021). In Proceedings of the 2021 Conference on Empirical Methods in Natural Language Processing, pp. 6894–6910

9. Eklund, P. W., Khushi, M., Kim, J., Naseem, U., and Razzak, I. (2021). Covidsenti: a large scale benchmark Twitter data set for COVID-19 sentiment analysis. *IEEE Transactions on Computational Social Systems*, 8(4), 976–988.

10. Goswami, A., Kaliyar, R. K., and Narang, P. (2021). EchoFakeD: improving fake news detection in social media with an efficient deep neural network. *Neural Computing and Applications*, 33(14), 8597–8613. https://doi.org/10.1007/s00521-020-05611-1.

11. Honghao, Z.,Tinghuai, M., Rong, H., Qian, Y., Tian, Y., and Al-Nabhan, N. (2022). MDMN: multi-task and domain adaptation based multi-modal network for early rumor detection. *Expert Systems with Applications*, 195, 116517.

12. Kan, M. Y., Nguyen, V. H., Sugiyama, K., and Nakov, P. (2020). FANG: leveraging social context for fake news detection using graph representation. In International Conference on Information and Knowledge Management Proceeding, 2020. https://doi.org/10.1145/3340531.3412046.

13. Lee, D., Liu, H., Mahudeswaran, D., Shu, K., and Wang, S. (2020). Fake news net: a data repository with news content, social context, and spatiotemporal information for studying fake news on social media. *Big Data*, 8, 171–188. https://doi.org/10.1089/big.2020.0062.

14. Lee, D., Liu, H., Mahudeswaran, D., Shu, K., and Wang, S. (2020). Fake news net: a data repository with news content, social context, and spatiotemporal information for studying fake news on social media. *Big Data*, 8, 171–188. https://doi.org/10.1089/big.2020.0062.

15. Linmei, H., Siqi, W., Ziwang, Z., and Bin, W. (2022). Deep learning for fake news detection: a comprehensive survey, *AI Open*, 3, 133–155. ISSN 2666-6510. https://doi.org/10.1016/j.aiopen.2022.09.001.

16. Elhadad, M. K., Li, K. F., and Gebali, F. (2019). Fake news detection on social media: A systematic survey. In 2019 IEEE Pacific Rim Conference on Communications, Computers and Signal Processing (PACRIM), Victoria, BC, Canada, 2019, (pp. 1–8). doi: 10.1109/PACRIM47961.2019.8985062.

37 Optimization of a Trapezoidal Fuzzy and Time-based Multi-objective Optimization Problem for y_i-out-of-n_i Systems

Satyajit De[1,a] and Anil Bikash Chowdhury[2,b]

[1]Department of Computer Science, Maheshtala College, Kolkata, India

[2]Department of Computer Application, Techno India University, Kolkata, India

Abstract

Time is an important parameter to analyse the cost and reliability of a reliability redundancy allocation model (RRAM). In this study a trapezoidal fuzzy and time based multi-objective reliability redundancy allocation problem (TFT-MORRAP) is proposed for the m level $y_i - out - of - n_i$ series parallel system. Maximization of system reliability with minimization of system cost according to time through optimization of level wise redundant components by maintaining some constraints are the main objectives of this study. Maximum entropy constraints with bounded redundant components in each level follows to achieve the objectives. Trapezoidal Fuzzy Numbers (TRFN) come into play to manage the vagueness surrounding both component cost and reliability, effectively addressing their uncertainties. The correlation between component cost and reliability reduction corresponds to the arc length of the inverse logarithmic spiral as it varies with the angle, akin to the concept of time in this context. Over-speed protection system of a gas turbine (OPSGT) is used to analyse our introduced problem. Satisfaction level of peak solutions achieved through the utilization of the well-known algorithms, Multi-objective particle swarm optimization (MOPSO) and Non-dominated sorting genetic algorithm-II (NSGA-II), are compared with a focus on time. Several performance metrics are employed to evaluate diverse algorithmic performances, aiming to identify the optimal solutions for the previously mentioned issue. In the majority of instances, MOPSO exhibits superior performance compared to NSGA-II.

Keywords: Inverse logarithmic spiral (ILS), multi-objective reliability redundancy allocation problem (MORRAP), non-dominated solutions, performance metric indicators, trapezoidal fuzzy number (TRFN).

Introduction

Here an m level $y_i - out - of - n_i$ series-parallel RRAM De et al. [7] is considered. In this model every ith (i = 1, 2, ...m) level contains n_i redundant components, out of which y_i components are operational. Reliability always improved in a multi stage series parallel RRAM, by distributing indistinguishable duplicated components within every level. Different stages contains different kind of components.

---

[a]satyajitde973@gmail.com, [b]abchaudhuri007@gmail.com

A TFT-MORRAP is introduced for the m level $y_i - out - of - n_i$ series parallel RRAM. The issue at hand involves a pair of contradictory goals: the enhancement of system reliability with the reduction of system cost. The solution space of the objectives is managed by imposing an entropy constraint and restricted quantity of extra components in every level and throughout the entire system. Entropy constraints measure the scattering of allocation among various levels in a reliability redundancy allocation problem (RRAP) Roy et al. [21]. Cost and reliability of a system may change with respect to time. The main objective of the said problem is to maximize the system reliability and minimizing the system cost on time by maintaining the said constraints. The value of component cost and reliability is not rigid, so their uncertain information are maintained by using trapezoidal fuzzy number (TRFN) Barros et al. [2]. In this case MORRAP becomes more flexible and more realistic. Components of a system becomes old according to time. Hence, we have taken into account the simultaneous reduction of component cost and reliability, both scaled by a consistent geometric progression factor from their initial values. The diminishing characteristic adheres to the ILS's radial dimension. The intersection points of the spiral with the radius of the ILS exhibit diminishing distances in an inverse geometric progression Weisstein, [23]. Here time is seen as the angular measure of the spiral.

A meta-heuristic technique MOPSO introduced by Coello et al. [4] and a fast elitist multi-objective evolutionary algorithm NSGA-II introduced by Deb et al. [8] are used to solve the aforementioned problem. Recently various types of NSGA-II and MOPSO are applied to solve various types of multi-objective optimization problems (MOOP) [5, 7, 16, 25, 29]. In this work the personal and global best solution set in MOPSO and Pareto front sets in NSGA-II are updated by applying non-dominated sorting techniques Deb et al. [8] on membership values of optimal solutions. Here the best solution from Pareto optimal set is always chosen depending on Topsis approach Kumar and Yadav [12]. A five valves OPSGT Kumar and Yadav [11] model is used to illustrate the TFT-MORRAP on time. A comparative study has been shown between NSGA-II Deb et al. [8] and MOPSO Coello et al. [4] algorithms to solve the TFT-MORRAP of protection system. MOPSO shows the better optimal solutions and greater degree of contentment compared to NSGA-II across the majority of time values. Performance metrics *MID* (Mean Ideal Distance) Mobin et al. [18], Spacing Mobin et al. [18] and Spreading Deb et al. [8] are used to measure the performances of the algorithms as accuracy, diversity and convergence rate to solve the TFT-MORRAP. MOPSO shows the best performance in all cases.

Literature Review

Recently, a variety of RRAM types and RRAP have been introduced for k-out-of-n series parallel systems and other system configurations. In their research, Zhang et al. [28] introduced a novel RRAP tailored for complex systems. In their study, Zarei et al. [27] formulated a MOOP aimed at minimizing pressure deficit and cost across the entire network, which they subsequently addressed using NSGA-II and MOPSO algorithms. Kumar and Yadav [11] developed a fuzzy based MORRAP for an OPSGT, and their proposed hybrid NSGA-II algorithm was employed to solve the problem effectively. In their investigation, Maneckshaw and Mahapatra [17] employed a multi-objective RRAM with constraints and utilized evolutionary algorithms to optimize the identified objectives. Li et al. [13] formulated a global reliability model with a RRAP, which they successfully solved using an advanced swarm optimization technique. Roy et al. [21] assessed the optimal system reliability and minimal system expenses in a MORRAP while considering an entropy constraint.

In recent times, multiple MOOPs have been tackled using a variety of MOPSO algorithms and NSGA-II algorithms. De et al. [7] proposed a novel fuzzy rank oriented MOPSO approach for effectively addressing a fuzzy MORRAP concerning xj – out – of – m_j series parallel systems. Mahapatra et al. [16] applied an MOPSO algorithm to effectively solve a MORRAP in the scenario of a hesitant uncertain framework. Gan et al. [9] adapted the conventional technique MOPSO to address an optimization problem related to centrifugal pumps. In the study conducted by Li et al. [14], improvements were made to the search efficiency and effectiveness involving the MOPSO algorithm. Davoudi et al. [6] introduced an innovative MOPSO algorithm aimed at efficiently addressing the model of optimization they posed. In their investigation, Liu et al. [15] constructed a model for optimization of multi-objectives to enhance the greenhouse light environment. The model was effectively solved by employing an improved NSGA-II algorithm. Zand et al. [26] employed a hybrid approach, combining NSGA-II with MOPSO algorithm, to effectively address their introduced two-objective non-linear redundant allocation problem. In their research, Nath and Muhuri [20] defined a MORRAP with diverse structures and successfully tackled it through NSGA-II and NSGA-III algorithms. Brentan et al. [3] employed NSGA-II to discover the optimal solutions of the MOOP related to water quality sensor placement. Muaddi and Singh [19] integrated MOPSO into their study to optimize both cost and reliability aspects of a MOOP.

TFT-MORRAP for yi – out – of – ni System with Spiral Model

In this section we describe the proposed problem TFT-MORRAP containing two multi-objective functions with some constraints on time t.

m Stage yi – out – of – ni Series Parallel System with Inverse Logarithmic Spiral

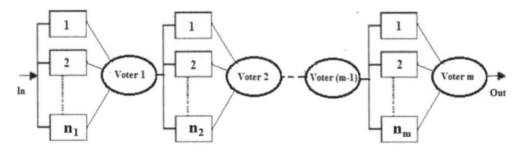

Figure 37.1 *m* level y_i – *out* – *of* – n_i series parallel system.

Figure 37.1 depicts a *m* level y_i – *out* – *of* – n_i series parallel system utilizing a k– out – of - n Xie et al. [24] system. Within this system, there are *m* levels connected in series, where every ith (i = 1, 2, ... n) level incorporates parallel connections of y_i components. From the entirety of y_i components, $(y_i – 1)$ elements are designated as redundant. Optimization of the count of redundant components y_i at ith level (for i = 1, 2, ...n), while adhering to the previously mentioned objectives and constraints is the main goal. $r = ae^{-b\theta}$ Weisstein, [23] be the polar form of the ILS. Here a and b are constants that determine the characteristics of the spiral. The distance r from the origin to a point on the spiral decreases in geometric progression when the spiral angle θ increases. In general, both the cost and reliability of components tend to decrease proportionally with respect to time t reflecting a similar factor reduction from their previous values. These factors undergo changes according to the radial progression of the ILS $r = ae^{-bt}$. In this context, the angle θ is equated to the variable time t. The parameter b governs the modulation of the rate of decrease. In this analysis, we take a = 1 while keeping all other parameters unaffected.

System Reliability and System Cost Functions on Time

In $y_i - out - of - n_i$ system the component cost and reliability of the ith level are denoted as R_i and C_i respectively. The total reliability at the ith level is computed using the binomial distribution defined in Xie et al. [24] as

$$\sum_{j=y_t}^{n_t} \frac{n_i!}{j!(n_i-j)!} R_i^j (1-R_i)^{(n_t-j)}$$

The system reliability function for the m level $y_i - out - of - n_i$ system can be written from Xie et al. [24] as

$$R(y_1, y_2, \ldots y_m) = \prod_{i=1}^{m} \left\{ \sum_{j=y_t}^{n_t} \frac{n_i!}{j!(n_i-j)!} R_i^j (1-R_i)^{(n_t-j)} \right\}$$

Over time, the system's components age, causing a decline in their reliability, a pattern akin to the ILS's radius. So the component reliability at ith level becomes $R_i e^{-bt}$. Equation 1 represents the system reliability function on time t.

$$R(y_1, y_2, \ldots y_m) = \prod_{i=1}^{m} \left\{ \sum_{j=y_t}^{n_t} \frac{n_i!}{j!(n_i-j)!} R_i^j (1-R_i)^{(n_t-j)} \right\} \tag{1}$$

Component reliability holds a direct proportionality to the associated component cost. So the complete cost of y_i elements at ith level with characteristic factor (shape parameter) x_i can be calculated as $C_i y_i R^{x_i}$. Characteristic factors represents the different characteristics of elements in various levels. The cost function of the system mentioned earlier at time t is represented by equation 2.

$$C(y_1, y_2, \ldots y_m, t) = \sum_{i=1}^{m} C_i e^{-bt} y_i \left(R_i e^{-bt} \right)^{x_i} \tag{2}$$

Development of MORRAP under Crisp and Fuzzy Conditions over Time

The MORRAP containing the said objectives and constraints for the m level $y_i - out - of - n_i$ system with series-parallel configuration in a crisp setting on time is shown in (3) based on the (1) and (2).

$$Maximize \ R(y_1, y_2, \ldots y_m) = \prod_{i=1}^{m} \left\{ \sum_{j=y_t}^{n_t} \frac{n_i!}{j!(n_i - j)!} R_i^j \left(1 - R_i\right)^{(n_t - j)} \right\}$$

$$C(y_1, y_2, \ldots y_m, t) = \sum_{i=1}^{m} C_i e^{-bt} y_i \left(R_i e^{-bt} \right)^{x_i}$$

$$Subject \ to : entropy \ constraint - \sum_{i=1}^{m} \left\{ \frac{y_i}{\Sigma_j y_j} \ln \left(\frac{y_i}{\Sigma_j y_j} \right) \right\} \leq S,$$

$$L \leq \sum_{i=1}^{m} y_i \leq U \ and \ B_l \leq y_i \leq n_i \ for \ i = 1, 2, 3, \ldots, m. \tag{3}$$

Entropy constraint are depicted in Roy et al. [21]. The trapezoidal fuzzy and time based MORRAP (TFT-MORRAP) for the m level $y_i - out - of - n_i$ series-parallel system is depicted in (4) based on (3).

$$Maximize \ \tilde{R}(y_1, y_2, \ldots y_m, t) = \prod_{i=1}^{m} \left\{ \sum_{j=y_t}^{n_t} \frac{n_i!}{j!(n_i - j)!} \left(\tilde{R}_i e^{-bt} \right)^j \left(1 - \tilde{R}_i e^{-bt} \right)^{(n_t - j)} \right\} \tag{4}$$

$$Minimize \ \tilde{C}(y_1, y_2, \ldots y_m, t) = \sum_{i=1}^{m} \tilde{C}_i e^{-bt} y_i \left(\tilde{R}_i e^{-bt} \right)^{x_i}$$

Subject to same constraints as in equation 3

Approach for Resolving the Presented TFT-MORRAP

In our approach, we systematically progress through the subsequent four sections to address the formulated TFT-MORRAP outlined in (4), employing the NSGA-II Deb et al. [8] and MOPSO Coello et al. [4] algorithms.

Boundary Values of Individual Objective Function on Time

All objective functions are individually solved while adhering to the constraints specified in (4), treated as distinct single objective optimization problems for time parameter t. Suppose V^1 and V^2 consecutively represent the optimal solution of the reliability and cost objective functions. Then the boundary values interval for the system cost and system reliability are $[C_{min}, C_{max}] = [C(V^2), C(V^1)]$ and $[R_{min}, R_{max}] = [R(V^1), R(V^2)]$. In this study, the particle swarm optimization (PSO) Song et al. [22] algorithm is employed to evaluate the extreme values of the individual objective function for different time values $t = 0, 1, \ldots 10$. Parameters of the PSO approach are configured as follows: a cognitive acceleration value is 0.5, a global acceleration value is 0.5, an upper limit and lower limit of inertia weight are at 0.9 and 0.2.

Membership Functions of TRFN Based Reliability and Cost

In this investigation, when applying the membership functions, the reliability $\tilde{R} = (r_1, r_2, r_3, r_4)$ and cost $\tilde{C} = (c_1, c_2, c_3, c_4)$ in the Trapezoidal fuzzy number (TRFN) form, are transformed into their respective crisp values using the center of gravity (COG) Kim et al. [10] technique as defined in equation 5.

$$R^* = \frac{\sum_{i=4}^{4} r_i \mu_{\tilde{R}_s}(r_i)}{\sum_{i=4}^{4} \mu_{\tilde{R}_s}(r_i)} \qquad C^* = \frac{\sum_{i=4}^{4} C_i \mu_{\tilde{C}_s}(C_i)}{\sum_{i=4}^{4} \mu_{\tilde{C}_s}(C_i)} \tag{5}$$

In this context, the system cost and system reliability, expressed in the TRFN form, are denoted as 4-tuples $\tilde{R}_s = (R_{min}, R_{mid1}, R_{mid2}, R_{max})$ and $\tilde{C}_s = (C_{min}, C_{mid1}, C_{mid2}, C_{max})$ respectively. The benefit of utilizing TRFN lies in the flexibility it offers to decision-makers, as they can strategically set the modal values R_{mid1}, R_{mid2} within the respective uncertain interval $[R_{min}, R_{max}]$ and the modal values C_{mid1}, C_{mid2} within the respective uncertain interval $[C_{min}, C_{max}]$. This allows decision-makers to achieve an optimal level of satisfaction that aligns with their specific requirements and preferences. In order to attain an enhanced level of gratification for the stated problem, decision-makers have the option to position R_{mid1} and R_{mid2} closer to R_{max}, while moving C_{mid1} and C_{mid2} towards C_{min}. Equation 6 outlines the membership functions that characterize the precise (crisp) values of system reliability and system cost.

$$\mu_{\tilde{R}_s}(y) = \begin{cases} \dfrac{y - R_{min}}{R_{mid1} - R_{min}} & \text{if } R_{min} \le y \le R_{mid1} \\ 1 & \text{if } R_{mid1} \le y \le R_{mid2} \\ \dfrac{R_{max} - y}{R_{max} - R_{mid2}} & \text{if } R_{mid2} \le y \le R_{max} \\ 0 & \text{otherwise} \end{cases} \quad \mu_{\tilde{C}_s}(y) = \begin{cases} \dfrac{y - C_{min}}{C_{mid1} - C_{min}} & \text{if } C_{min} \le y \le C_{mid1} \\ 1 & \text{if } C_{mid1} \le y \le C_{mid2} \\ \dfrac{C_{max} - y}{C_{max} - C_{mid2}} & \text{if } C_{mid2} \le y \le C_{max} \\ 0 & \text{otherwise} \end{cases} \tag{6}$$

Here, the variable y denotes the defuzzified value of a TRFN, computed utilizing equation 5. In the proposed problem we set

$$R_{mid1} = R_{min} + \frac{(R_{max} - R_{min})}{8} * 5, \quad R_{mid2} = R_{max} + \frac{(R_{max} - R_{min})}{8} * 2,$$

$$C_{mid1} = C_{min} + \frac{(C_{max} - C_{min})}{8} * 2, \quad C_{mid2} = C_{max} + \frac{(C_{max} - C_{min})}{8} * 5,$$

Construction of Proposed Problem with Fuzzy Objectives

The highest level of satisfaction is attained by optimizing the membership functions corresponding to the objective functions Kumar and Yadav [11]. Equation 7 represents the construction of a fuzzy MOOP based on time, derived from (4). In this specific issue, achieving the peak satisfaction level entails optimizing the membership functions of all the objective functions.

$$Maximize \ \mu(f) = \left[\mu_{\tilde{R}_s}\left(R^*\right), \ \mu_{\tilde{R}_s}\left(C^*\right) \right] \tag{7}$$

Subject to same constraints as in equation 4

In this context, C^* and R^* sequentially denote the deconstructed values corresponding to the cost and reliability fuzzy objective functions.

Utilizing the MOPSO Algorithm for Problem Solving

Algorithm 1: Application of MOPSO Algorithm to Solve TFT-MORRAP

Input: A set of M arbitrary solutions of dimension m at time point t.

Output: Acquire the Pareto optimal set, including the peak alternative, at time point t.

Step 1: Set the solution dimension m along with the lower and upper limit L and U respectively for the total permissible components. Additionally, initialize the lower boundary B_1 and an array of upper boundaries $[n_1, n_2, n_3...n_m]$ of the sample space, an array of shape parameters $[x_1, x_2, x_3, ...,x_m]$, and trapezoidal fuzzy-shaped components for reliability ($\widetilde{R1}, \widetilde{R2}, \widetilde{R3}, ... \widetilde{Rm}$) and cost ($\widetilde{C1}, \widetilde{C2}, \widetilde{C3}, ... \widetilde{Cm}$). Furthermore, establish the population size M, the upper limit for iterations *maximum_ite*, the maximum count of independent trials *maximum_trial*, and the highest allowable time limit *maximum_time*.

Step 2: Set the maximum entropy $S \leftarrow 1.8$, the spiral parameter $b \leftarrow 0.028$, Inertia weight 0.4 and the mutation rate as 0.5.

Step 3: $t \leftarrow 0$ (Initialize time variable)

Step 4: while $t \leq maximum\_time$ do

Step 5: Initialize the trial counting variable as $trl\_count \leftarrow 1$

Step 6: Initialize the Pareto optimal set as $Pr\_opt\_set \leftarrow \varnothing$

Step 7: while $trl\_count \leq maximum\_trial$ do

Step 8: Set up global best solution set $Gbest \leftarrow \varnothing$

Step 9: Generate M random chromosomes (particles) of dimension m, ensuring they satisfy the constraints specified in (7). Independently establish a repository for the personal best of each particle, initializing it with the particle's own data. Utilize the non-dominated solutions technique to update the global best solution set *Gbest* using the current particles in the *Gbest* set with the particles from the personal best set.

Step 10: $ite\_count \leftarrow 0$ (Initialize the loop iteration counter)

Step 11: while $ite\_count < maximum\_ite$ do

Step 12: Set mutation probability $mut\_prob = \dfrac{ite\_count}{max\_ite}$

Step 13: The MOPSO algorithm is applied to the problem described in (7). We incorporate mutation with mutation probability and mutation rate to update the positions of M particles within the search space. Non-dominated solutions are stored within a hypercube, and the selection process is conducted through the implementation of the roulette-wheel method. The application of the non-dominated sorting technique is also employed to derive enhanced positions for the particles within the swarm and to refine the global best set *Gbest*.

Step 14: $ite\_count \leftarrow ite\_count + 1$

Step 15: end while

Step 16: $Pr\_opt\_set \leftarrow Pr\_opt\_set \cup Gbest$

Step 17: $trl\_count \leftarrow trl\_count + 1$

Step 18: end while

Step 19: Display the Pareto optimal solutions at time point t, extracted from the Pr_opt_ set, limited to solutions where the membership vector components (evaluated from (7)) are non-zero.

Step 20: Utilize the Topsis method to choose the most suitable solution from the Pareto optimal solutions with non-zero membership vector components, and present it along with the corresponding time t.

Step 21: (Increment time variable) $t \leftarrow t + 1$.

Step 22: end while

Utilizing the NSGA-II Algorithm for Problem Solving

Algorithm 2: Application of NSGA-II Algorithm to Solve TFT-MORRAP

Input: A set of M arbitrary solutions of dimension m at time point t.

Output: Acquire the Pareto optimal set, including the peak alternative, at time point t.

Step 1: Same as step 1 of algorithm 1.

Step 2: Set the maximum entropy $S \leftarrow 1.8$, the spiral parameter $b \leftarrow 0.028$, the weighted parameter for whole arithmetic crossover Ali el al. [1] as $\alpha \leftarrow 0.7$, the decision index parameter for polynomial mutation [30] as $\eta_m \leftarrow 75$, the mutation probability as 0.1, and crossover probability as 0.9.

Step 3: $t \leftarrow 0$ (Set up time variable)

Step 4: while $t \leq maximum\_time$ **do**

Step 5: Initialize the trial counting variable as $trl\_count \leftarrow 1$

Step 6: Initialize the Pareto optimal set as $Pr\_opt\_set \leftarrow \varnothing$

Step 7: while $trl\_count \leq maximum\_trial$ **do**

Step 8: Create a set Pop_set by generating M random chromosomes of dimension m, ensuring they satisfy the constraints specified in equation 7.

Step 9: $ite\_count \leftarrow 0$ (Initialize the loop iteration counter)

Step 10: while $ite\_count < maximum\_ite$ **do**

Step 11: The problem stated in equation 7 is tackled using the NSGA-II algorithm by utilizing whole arithmetic crossover technique for crossover, the polynomial mutation technique for mutation, crowding distance to choose diverse solution. The fast non-dominated sorting method is used to obtain the Pareto fronts from the Pop_set.

Step 12: Substitute the M solutions in Pop_set with the newly gathered M solutions extracted from the Pareto fronts.

Step 13: $ite\_count \leftarrow ite\_count + 1$

Step 14: end of while

Step 15: $Pr\_opt\_set \leftarrow Pr\_opt\_set \cup Pop\_set$

Step 16: $trl\_count \leftarrow trl\_count + 1$

Step 17: end of while

Step 18: Display the Pareto optimal solutions at time point t, extracted from the Pr_opt_ set, limited to solutions where the membership vector components (evaluated from (7)) are non-zero.

Step 19 Utilize the Topsis method to choose the most suitable solution from the Pareto optimal solutions with non-zero membership vector components, and present it along with the corresponding time t.

Step 20: (Increment time variable) $t \leftarrow t + 1$

Step 21: end of while

Illustrative Example for TFT-MORRAP

The benchmark model depicted in Figure 37.2 vividly demonstrates the effectiveness and practicality of the proposed TFT-MORRAP.

OPSGT with Five Valves

The model of a OPSGT, encompassing five control valves (from V_1 to V_5), is presented in Figure 37.2, Kumar and Yadav [11]. The continuous speed detection is achieved through a combined mechanical and electrical mechanism. Upon detecting an over-speed condition, the five control valves actuate to promptly cut off the fuel supply to the turbine. The configuration of this OPSGT exhibits a series-parallel arrangement, which is depicted in Figure 37.1. To enhance its reliability, (y_i - 1) extra valves are interconnected in parallel way at every ith level within this system. To enhance the adaptability of the OPSGT, the submitted information regarding the cost and reliability of individual valve has been formulated in the form of TRFN.

Equation 8 outlines the precise definition of the TFT-MORRAP for this particular protection system. Main goal of this system is to maximize the count of operational valves y_i (for i = 1, 2,...,5) within the designated exploration range, while adhering to the criteria and limitations stipulated in Equation 8.

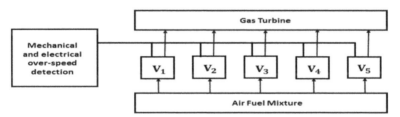

Figure 37.2 Gas turbine over-speed protection mechanism.

TFT-MORRAP of OPSGT

The TFT-MORRAP for the OPSGT, formulated in Equation 8, is derived from Equation 4 with a consideration of m = 5.

$$Maximize \; \tilde{R}(y_1, y_2, \ldots y_5, t) = \prod_{i=1}^{5} \left\{ \sum_{j=y_t}^{n_t} \frac{n_i!}{j!(n_i - j)!} \left(\tilde{R}_i e^{-bt} \right)^j \left(1 - \tilde{R}_i e^{-bt} \right)^{(n_t - j)} \right\}$$

$$Minimize \; \tilde{C}(y_1, y_2, \ldots y_5, t) = \sum_{i=1}^{5} \tilde{C}_i e^{-bt} y_i \left(\tilde{R}_i e^{-bt} \right)^{x_i}$$

$$Subject \; to: entropy \; constraint - \sum_{i=1}^{5} \left\{ \frac{y_i}{\sum_j y_j} \ln \left(\frac{y_i}{\sum_j y_j} \right) \right\} \le S,$$

$$L \le \sum_{i=1}^{5} y_i \le U \; and \; B_l \le y_i \le n_i \; for \; i = 1, 2, 3, \ldots, 5.$$

(8)

Analogous to (7), the fuzzy MOOP based on time for the OPSGT is developed using (8) as a foundation, ultimately defined in (9).

$$\text{Maximize } \mu(f) = \left[\mu_{\tilde{R}_s}\left(R^*\right), \ \mu_{\tilde{R}_s}\left(C^*\right) \right] \tag{9}$$

Subject to same constraints as in equation 8

Findings and Analysis

This section presents a comprehensive comparison and analysis of the satisfaction levels achieved by the Pareto optimal solutions obtained from the optimization problem outlined in (9), utilizing the methodologies defined above. The performance evaluation of these algorithms is conducted and compared using a set of performance metrics, including *MID* (Mean Ideal Distance), Spacing and Spreading.

Numerical Findings

The values of crisp, shape parameters and TRFN representations of component cost and reliability for Equation 8 are detailed in Table 37.1. Table 37.1 offers an elaboration of the crisp values, shape parameters and TRFN representations associated with the component cost and reliability values for Equation 8. Here the crisp form of component reliability is changed to the TRFN form by considering 1% uncertainty for middle values and 5% uncertainty for extreme values. Also the crisp form of component cost is changed to the TRFN form by considering 5% uncertainty for middle values and 15% uncertainty for extreme values. The decision maker has the flexibility to modify these values according to their specific requirements.

Table 37.1: Values of the parameters employed in Equation 8.

i	n_i	x_i	R_i	C_i	$\tilde{R}_i$	$\tilde{C}_i$
I	9	0.75	0.9	40	[0.855. 0.891, 0.909, 0.945)	[34.0, 38.0, 42.0. 46.0]
2	8	0.6	0.85	35	[0.8075, 0.8415, 0.8585, 0.8925]	[29.75. 33.25. 36.75. 40.25]
3	7	0.55	0.75	30	[0.7125, 0.7425, 0.7575, 0.7875]	[25.5, 28.5,31.5, 34.5]
4	8	0.65	0.82	42	[0.779, 0.8118, 0.8282, 0.861]	[35.7, 39.9. 44.1, 48.3]
5	10	0.7	0.93	36	[0.8835, 0.9207, 0.9393, 0.9765]	[30.6. 34.2, 37.8, 41.1]

The following values are attributed to the remaining parameters of equation 8: $b = 0.028$, $S = 1.8$, $L = 10$, $U = 35$, $B_1 = 2$. The parameter values employed in the algorithms mentioned (algorithm 1, algorithm 2, and PSO Song et al. [22] are as follows: maximum time limit *maximum_time* = 10, maximum limit of iteration *maximum_ite* = 200, population size $M = 60$ and maximum autonomous trials *maximum_trial* = 15. The algorithm PSO is implemented according to the guidelines provided in above section for various time instances $t = 0, 1,... 10$. The resulting boundary values of cost and reliability of the system are recorded in the following Table 37.2. Utilizing the formulas specified in section above, we assess the modal values R_{mid1}, R_{mid2}, C_{mid1}, and C_{mid2}, and subsequently, record them in Table 37.2.

Table 37.2: Extreme system reliability and system cost values along their corresponding modal values.

t	0	1	2	3	4	5	6	7	8	9	10
R_{min}	0.98426	0.98031	0.97247	0.96272	0.95086	0.94425	0.92345	0.90895	0.88510	0.85701	0.83762
R_{max}	0.99854	0.99776	0.99553	0.99490	0.99266	0.98978	0.98606	0.97134	0.96915	0.96229	0.96020
R_{mid1}	0.993185	0.991216	0.986882	0.982832	0.976985	0.972706	0.962581	0.947944	0.937631	0.92281	0.914232
R_{mid2}	0.99497	0.993397	0.989765	0.986855	0.98221	0.978398	0.970408	0.955742	0.948138	0.93597	0.929555
R_{min}	356.515	348.522	330.599	310.382	303.131	289.356	276.205	262.337	252.905	240.243	224.665
R_{max}	473.536	451.535	430.961	436.784	421.125	397.970	381.712	367.350	347.194	331.311	315.185
R_{mid1}	385.77	374.275	355.69	341.982	332.63	316.51	302.582	288.59	276.477	263.01	247.295
R_{mid2}	400.398	387.152	368.235	357.783	347.379	330.086	315.77	301.717	288.263	274.394	258.61

Table 37.3: Analyzing the optimal solutions derived from solving (9) (Emphasized values indicate superior outcomes).

Time t	MOPSO									NSGA-II								
	y_1	y_2	y_3	y_4	y_5	R^*	C^*	$\mu_{\overline{R}^*}(R^*)$	$\mu_{\overline{C}^*}(C^*)$	y_1	y_2	y_3	y_4	y_5	R^*	C^*	$\mu_{\overline{R}^*}(R^*)$	$\mu_{\overline{C}^*}(C^*)$
0	2	3	2	3	2	0.99503	309.645	0.98319	1	2	2	2	4	2	0.99308	101.128	0.98824	0.99001
1	2	3	2	3	2	0.99282	387.071	1	1	3	3	2	4	2	0.99213	388.129	1	0.98482
2	2	3	2	3	3	0.98873	367.772	1	1	2	2	2	3	3	**0.98968**	367.973	1	1
3	4	2	2	3	3	0.98518	359.292		0.9809	3	3	2	3	3	0.98499	360.761		0.9623
4	2	2	2	3	3	0.98165	344.498	1	1	4	4	2	2	3	0.97928	350.028	1	0.96408
5	2	3	2	3	2	0.97802	326.781	1	1	3	2	2	3	2	0.96804	333.816	0.85711	0.94506
6	2	2	2	3	2	0.96797	313.577	1	1	3	3	2	3	3	0.96378	319.09	1	0.94966
7	2	2	2	2	2	0.95518	297.761	1	1	2	2	2	3	3	0.95317	299.222	1	1
8	2	3	2	2	2	0.94783	285.214	1	1	2	2	3	3	2	0.94606	286.279	1	1
9	2	3	2	2	2	0.9357	271.529	1	1	2	2	2	3	2	0.93177	272.638	1	1
10	3	3	2	2	2	0.9252	256.632	1	1	3	2	2	3	3	0.90805	260.643	0.91931	0.96407
Average value						0.9684827	328.1611	0.9984718	0.9982636						0.9645118	330.5843	0.9786054	0.9781818

The outcomes of solving the MOOP described in (9) using two methodologies, namely, MOPSO (Algorithm 1) and NSGA-II (Algorithm 2), at different time steps t = 0, 1 ...10, are presented in Table 37.3.

Analysis of the data showcased in Table 37.3 indicates that, for the majority of time values, the MOPSO algorithm exhibits high system reliability and low system cost. Furthermore, it can be observed that, on average, the algorithm MOPSO outperforms the algorithm NSGA-II in discovering optimal outcomes for the equation 8-defined problem. The visual depiction of membership values, as displayed in Figure 37.3, aligns with the data presented in Table 37.3, indicating predominant strong satisfaction levels for the algorithm MOPSO across the majority of time point values. The visual depiction of fitness values, as displayed in Figure 37.4, aligns with the data presented in Table 37.3, indicating better performance for the MOPSO technique across the majority of time values.

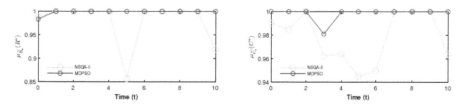

Figure 37.3 Analyzing optimal solution membership values over time.

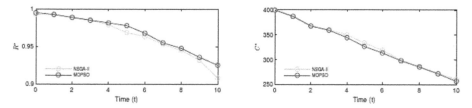

Figure 37.4 Analyzing optimal solution fitness values over time.

Performance Comparison

In this context, the performance metrics *MID*, Spacing and Spreading are applied to evaluate the precision, diversity with other characteristics of the algorithm MOPSO and the algorithm NSGA-II when addressing the TFT-MORRAP problem in the aforementioned system. With no loss of generality, we present the performance comparison for arbitrary time points $t = 2, 4, 6$ to demonstrate the results. The *MID* is employed to assess the proximity of each element within the algorithm-generated Pareto-optimal set towards the ideal point. Within this investigation, $(C_{mid1} + C_{mid2})/2$ and $(R_{mid1} + R_{mid2})/2$ represent the ideal points for the cost and reliability functions, respectively. Here, N represents the count of objective functions, while Q signifies the collection of algorithm-attained Pareto-optimal solutions.

$$MID = \frac{1}{|Q|} \sum_{j=1}^{|Q|} \left(\sqrt{\sum_{Z=1}^{N} \left| f_z^j - x_z \right|} \right)$$

Additionally, suppose x_z stands for the ideal point corresponding to the zth objective function, also let f_z^j represent the zth objective function's fitness value for the jth solution in the set Q. Figure 37.5 provides a graphical comparison of the *MID* indicators computed on the Pareto optimal front evaluated by the algorithms NSGA-II and MOPSO for the time points $t = 2, 4, 6$. In Figure 37.5, it is evident that the MOPSO algorithm exhibits lower *MID* values, indicating superior performance compared to the algorithm NSGA-II.

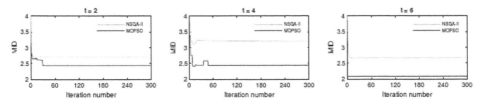

Figure 37.5 Analyzing MID indicator values across iterations.

The *Spacing* metric indicator characterizes the evenness of the distribution of Pareto-optimal solutions yielded by an algorithm. A decrease in the Spacing value signifies an enhanced diversity in the Pareto-optimal front Mobin et al. [18].

$$Spacing = \sqrt{\frac{1}{|Q|} \sum_{p=1}^{|Q|} \left(d_p - \bar{d}\right)^2}.$$

Where $d_p = \min_{j \in Q \text{ and } j \neq p} \sum_{z=1}^{N} \left|f_z^P - f_z^j\right|$ and $\bar{d} = \frac{1}{|Q|} \sum_{p=1}^{|Q|} d_p$

Here f_z^p denote the zth objective function's fitness value for the pth solution in the set Q. In Figure 37.6, we visually compare the *Spacing* indicator values across iterations along the Pareto-efficient frontier attained through the algorithms at time instances $t = 2, 4, 6$. In Figure 37.6, it is evident that the MOPSO algorithm exhibits lower *Spacing* values, indicating superior performance relative to the algorithm NSGA-II.

The subsequent metric, *Spreading*, quantifies the dispersion or spread of the solutions within the algorithm-derived Pareto optimal set. A decrease in the *Spreading* value signifies an enhanced diversity within the Pareto-optimal front Mobin et al. [18].

$$Spreading = \frac{\sum_{i=1}^{N} d_i^e + \sum_{p=1}^{|Q|} \left|d_p - \bar{d}\right|}{\sum_{i=1}^{N} d_i^e + |Q|\bar{d}}$$

Here the parameters d_p, $\bar{d}$, N and Q are defined above. The parameter d_i^e represents the Euclidean distance separating the extreme solutions of the ith objective function.

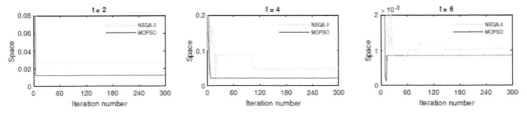

Figure 37.6 Analyzing *Spacing* values across iterations.

In Figure 37.7, we visually present a comparative analysis of the *Spreading* indicator values across iterations within the Pareto front achieved through the algorithms at time instances $t = 2, 4, 6$. In Figure 37.7, it is evident that the MOPSO algorithm exhibits lower *Spreading* values, indicating superior performance compared to the NSGA-II algorithm.

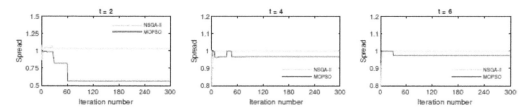

Figure 37.7 Analyzing Spreading values across iterations.

Conclusions

In this research work, MOPSO approach and NSGA-II approach are employed to tackle the proposed TFT-MORRAP associated with a 5-stage OPSGT. The Pareto optimal solutions' membership scores and fitness values are presented in both numeric and visual forms over time. For the majority of time parameter values, MOPSO demonstrates superior results compared to NSGA-II. The MOPSO technique exhibits better average performance across all time parameter values. Therefore, the MOPSO algorithm proves to be a more appropriate choice for tackling the fuzzy multi-objective optimization problem (MOOP). In this situation, the metrics *MID*, *Spacing* and *Spreading* are employed to evaluate the effectiveness of the MOPSO approach and NSGA-II approach in addressing the TFT-MORRAP. The visual depiction of the performance metrics clearly illustrates that the MOPSO outperforms the NSGA-II in all instances.

As reliability and cost are subject to fluctuations over time, the temporal factor continually plays a significant role in any given system. In our future endeavors, we plan to explore various fuzzy numbers such as Gaussian fuzzy numbers, Hesitant fuzzy numbers, and others, as well as investigate different spiral models like Fermat's spiral, Archimedean spiral, and more.

References

1. Ali, M. Z., Awad, N. H., Suganthan, P. N., Shatnawi, A. M., and Reynolds, R. G. (2018). An improved class of real-coded genetic algorithms for numerical optimization. *Neurocomputing, 275*, 155–166.
2. Barros, de L. C., Bassanezi, R. C., and Lodwick, W. A. (2017). A First Course in Fuzzy Logic, Fuzzy Dynamical Systems, and Biomathematics: Theory and Applications. Springer.
3. Brentan, B., Carpitella, S., Barros, D., Meirelles, G., Certa, A., and Izquierdo, J. (2021). Water quality sensor placement: a multi-objective and multi-criteria approach. *Water Resources Management, 35*, 225–241.
4. Coello, C. A. C., Pulido, G. T., and Lechuga, M. S. (2004). Handling multiple objectives with particle swarm optimization. *IEEE Transactions on Evolutionary Computation*, 8(3), 256–279.
5. Dash, S., De, B., Das, R., Samanta, P. K., Bhowmik, W., Kar, R., and Bakshi, A. (2023). Optimal design of current starved oscillator using mopso. In 2023 International Conference on Communication, Circuits, and Systems (ic3s), (pp. 1–4).

6. Davoudi, M., Jooshaki, M., Moeini-Aghtaie, M., Barmayoon, M. H., and Aien, M. (2022). Developing a multi-objective multi-layer model for optimal design of residential complex energy systems. *International Journal of Electrical Power and Energy Systems*, 138, 107889.

7. De, S., Roy, P., and Chowdhury, A. B. (2023). Multi-objective fuzzy reliability redundancy allocation for xj-out-of-mj system using fuzzy rank-based multi-objective pso. In Applied Computing for Software and Smart Systems: Proceedings of ACSS, 2022, (Vol. *555*, p. 145).

8. Deb, K., Pratap, A., Agarwal, S., and Meyarivan, T. (2002). A fast and elitist multiobjective genetic algorithm: Nsga-ii. *IEEE Transactions on Evolutionary Computation*, 6 (2), 182–197.

9. Gan, X., Pei, J., Wang, W., Yuan, S., and Lin, B. (2023). Application of a modified mopso algorithm and multi-layer artificial neural network in centrifugal pump optimization. *Engineering Optimization*, 55 (4), 580–598.

10. Kim, D., Choi, Y. S., and Lee, S. Y. (2002). An accurate cog defuzzifier design using lamarckian co-adaptation of learning and evolution. *Fuzzy Sets and Systems*, 130 (2), 207–225.

11. Kumar, H., and Yadav, S. P. (2019a). Hybrid nsga-ii based decision-making in fuzzy multi-objective reliability optimization problem. *SN Applied Sciences*, 1(11), 1–14.

12. Kumar, H., and Yadav, S. P. (2019b). Nsga-ii based decision-making in fuzzy multi-objective optimization of system reliability. In Deep, K., Jain, M., and Salhi, S. eds. Decision Science in Action: Theory and Applications of Modern Decision Analytic Optimisation, (pp. 105–117). Singapore: Springer Singapore.

13. Li, S., Chi, X., and Yu, B. (2022). An improved particle swarm optimization algorithm for the reliability redundancy allocation problem with global reliability. *Reliability Engineering and System Safety*, 225, 108604.

14. Li, Y., Zhang, Y., and Hu, W. (2023). Adaptive multi-objective particle swarm optimization based on virtual pareto front. *Information Sciences*, 625, 206–236.

15. Liu, T., Yuan, Q., Ding, X., Wang, Y., and Zhang, D. (2023). Multi-objective optimization for greenhouse light environment using gaussian mixture model and an improved nsga-ii algorithm. *Computers and Electronics in Agriculture*, 205, 107612.

16. Mahapatra, G., Maneckshaw, B., and Barker, K. (2022). Multi-objective reliability redundancy allocation using mopso under hesitant fuzziness. *Expert Systems with Applications*, 198, 116696.

17. Maneckshaw, B., and Mahapatra, G. (2022). Multi-objective evolutionary algorithm on reliability redundancy allocation with interval alternatives for system parameters. *Neural Computing and Applications*, 34(21), 18595–18609.

18. Mobin, M., Mousavi, S. M., Komaki, M., and Tavana, M. (2018). A hybrid desirability function approach for tuning parameters in evolutionary optimization algorithms. *Measurement*, 114, 417–427.

19. Muaddi, S. A., and Singh, C. (2022). Reliability constrained optimal sizing and examining capacity credit and alternatives for renewable energy sources. *IEEE Access*, 10, 71133–71142.

20. Nath, R., and Muhuri, P. K. (2022). Evolutionary optimization based solution approaches for many objective reliability-redundancy allocation problem. *Reliability Engineering and System Safety*, 220, 108190.

21. Roy, P., Mahapatra, B., Mahapatra, G., and Roy, P. (2014). Entropy based region reducing genetic algorithm for reliability redundancy allocation in interval environment. *Expert Systems with Applications*, 41(14), 6147– 6160.

22. Song, B., Wang, Z., and Zou, L. (2021). An improved pso algorithm for smooth path planning of mobile robots using continuous high-degree bezier curve. *Applied Soft Computing*, 100, 106960.

23. Weisstein, E. W. (2006). Multiset. from mathworld–a wolfram web resource. Retrieved July 19, 2006.

24. Xie, M., Dai, Y. S., and Poh, K. L. (2004). Computing System Reliability: Models and Analysis. Springer Science and Business Media.

25. Yadav, A., Mishra, S., and Sairam, A. S. (2022). A multi-objective worker selection scheme in crowdsourced platforms using nsga-ii. *Expert Systems with Applications*, 201, 116991.
26. Zand, A. D., Khalili-Damghani, K., and Raissi, S. (2022). An evolutionary approach with reliability priority to design scada systems for water reservoirs. *Evolving Systems*, 13(3), 499–517.
27. Zarei, N., Azari, A., and Heidari, M. M. (2022). Improvement of the performance of nsga-ii and mopso algorithms in multi-objective optimization of urban water distribution networks based on modification of decision space. *Applied Water Science*, 12(6), 1–12.
28. Zhang, Z., Yang, L., Xu, Y., Zhu, R., and Cao, Y. (2023). A novel reliability redundancy allocation problem formulation for complex systems. *Reliability Engineering and System Safety*, Jun 23, 109471.
29. Zhao, D., Chen, M., Lv, J., Lei, Z., and Song, W. (2023). Multi-objective optimization of battery thermal management system combining response surface analysis and nsga-ii algorithm. *Energy Conversion and Management*, 292, 117374.
30. Zhao, Y.-T., Li, W.-G., and Liu, A. (2020). Optimization of geometry quality model for wire and arc additive manufacture based on adaptive multi-objective grey wolf algorithm. *Soft Computing*, 24(22), 17401–17416.

38 Pneumonia Detection from Chest X-ray Plates using Attention-based CNN Architecture

Maheak Dave[a], Saurabh Kumar Sharma[b], and Abhishek Majumdar[c]

Department of Computer Science and Engineering-Artificial Intelligence, Techno India University, West Bengal, India

Abstract

Detecting diseases from chest X-ray images has been a significant area of research, particularly utilizing deep learning techniques. However, several existing models seem to prioritize background details excessively, overlooking crucial diagnostic information. In this study, an attention-based Convolutional Neural Network (CNN) model is introduced specifically for detecting Pneumonia in chest X-ray images. The approach we employ integrates the Convolution Block Attention Module (CBAM) into the CNN architecture, effectively harnessing attention mechanisms. These mechanisms have played a crucial role in enhancing the capabilities of deep learning models, particularly transformers. By incorporating CBAM within CNN models for image processing, this study demonstrates its adaptability and effectiveness.

Keywords: Attention-based convolutional neural network, convolution block attention module, deep learning, pneumonia disease, x-ray plates.

Introduction

Pneumonia is a disease that affects the alveoli, which are tiny air sacs in the lungs. Dry cough, chest pain, fever, and difficulty breathing are common symptoms. To increase the accuracy and speed of pneumonia diagnosis, researchers are currently exploring the use of machine learning models, specifically deep learning approaches such as convolutional neural networks (CNNs). Deep learning has shown encouraging results in a variety of medical image-processing tasks, making it a viable choice for this application. In particular, CNNs have emerged as a widely used deep learning model in the medical industry for accurate image categorization and interpretation.

Pneumonia detection using X-ray imaging utilizing machine learning models has its own set of challenges because the same models may be distracted by irrelevant information in the images. Critical information in images can be reinforced by models using attention-based architectures. Although attention-based designs are not extensively employed in medical imaging, they have showed promise in other areas, including natural language

[a]maheakdave@gmail.com, [b]imesskayesss@gmail.com, [c]hod.aiml@technoindiaeducation.com

processing. These architectures have the potential to improve diagnosis accuracy and effectiveness as research progresses.

Attention-based techniques have showed considerable promise for improving pneumonia detection using X-ray images. The model can enhance accuracy in identifying pneumonia by focusing on critical locations such as the lungs and airways. Irrelevant elements, such as background noise, can be learned to ignore by the model. With data from multiple regions pooled, this strategy shows potential in providing a more accurate and streamlined pneumonia diagnosing process. Attention-based architectures, for example, have shown encouraging results when used to reliably detect pneumonia while simultaneously increasing efficiency. Attention-based architectures may one day replace standard diagnostic techniques in medical diagnosis.

Proposed here is an innovative method combining CBAM in CNN designs for pneumonia diagnosis. By emphasizing specific areas of images, CBAM functions as an attention-focused component of neural networks. The proposed approach was analyzed using data consisting of X-ray images that show patients with pneumonia. Our technique far exceeded the performance of typical CNNs regarding precision. We managed to hit a remarkable high of 93.02% accuracy.

Related Works

The literature presented highlights the effectiveness of deep learning methods, particularly CNNs with attention to mechanisms, in detecting pneumonia from chest radiographs. These proposed methods have demonstrated high accuracy across various datasets. As ongoing research in this field progresses, it is anticipated that deep learning techniques will further enhance their efficiency in identifying not only pneumonia but also other diseases.

Woo et al. [1] in 2018 proposed CBAM, a fundamental however powerful attention module for feedforward convolutional neural networks. it really works by means of first generating a channel interest map that indicates the significance of each channel in the feature map. It similarly computes a spatial interest map that demonstrates the significance of each spatial place inside the feature map. those interest maps are then expanded with the function map, which facilitates the version awareness on critical statistics.

Varshni et al. [2] put forward an approach for identifying pneumonia from chest radiographs using a convolutional neural network (CNN) in 2019. A dataset of chest radiographs with and without pneumonia was used to train the CNN. When examined using chest radiographs, this approach had an accuracy of 88.9%.

Zhou et al. [3], in their own research publication in 2019, presented a strategy for recognizing cardiomegaly among chest X-ray images utilizing an established technique called transfer learning. This method capitalizes on reusing models previously primed for specific tasks to tackle new tasks at hand. In this instance, the researcher tapped into a CNN model pre-trained on an extensive stockpile of natural images and duly refitted it with a smaller assortment of chest X-ray snaps. The upshot: an impressive accuracy rating reaching the heights of 79.70%.

Sharma et al. [4] presented a novel strategy to detect pneumonia in X-ray images of the chest employing convolutional neural networks (CNNs) in their 2020 publication. They used a vast dataset of 5,800 chest X-ray views, of which 2,400 were diagnosed as pneumonia and the remaining 3,400 as normal. They used a CNN model trained on this dataset to evaluate its performance against a testing sample set of 1,000 photos. Notably, their model had an accuracy rate of 90.68%.

On the other hand, Garstka and Strzelecki [5] proposed their unique procedure for detecting pneumonia from chest X-ray images via CNNs in addition to data augmentation techniques. Data augmentation involves artificially expanding datasets by generating brand-new images using existing ones. In this specific instance, these authors employed various augmentation methods like rotation translation and flipping. They trained their CNN using an extensive 10,000-image dataset of chest X-rays with an equal split of 5,000 labeled as pneumonia cases and 5,000 categorized as normal cases resulting in an impressive accuracy score of 85%.

Singh's [6] groundbreaking proposal during 2021 unfolded its novel way to detect pneumonia leveraging technological components deep learning principles offer through processing chest X-ray imagery. Their chosen arsenal to ignite results is none other than AlexNet—an uncomplicated yet potent CNN architecture that brings serious potential once unleashed against sizable datasets such as their holdings containing solid numbers namely: 5,000 X-ray scans—half were plagued with pneumonia indicators while the rest appeared normal by nature—yielding an accurate result of remarkable percentages at 92.63%.

Jassam et al. [7] introduced a novel approach to identify pneumonia from X-ray images in the thorax employing DenseNet, a more intricate CNN model compared to AlexNet. They trained the model on a dataset containing 10,000 X-ray images of the thorax, evenly distributed between 5,000 diagnosed as pneumonia and 5,000 categorized as normal. The trained model achieved remarkable accuracy levels clocking in at 90%.

In 2022, Chen [8] reviewed progress made in deep learning and attention mechanisms discussing how these mechanisms can enhance the performance of deep learning models for diverse tasks including pneumonia detection.

In the year 2023, a research group led by Chen [9] presented a technique that detected pneumonia from chest radiographs. This technique utilized an advanced deep learning architecture, employing both Convolutional Neural Networks (CNNs) and Multi-level Attention Transformers. The attention transformers enabled inference of the most important features in these aforementioned radiographs for diagnosis accuracy optimization. Notably, their method demonstrated impressive results with a remarkable accuracy of 92.3% on a test set comprising various chest radiographs.

Moving forward, another scientific team headed by An [10] put forth an innovative method to bolster object detection performance in YOLOv5 using what is known as the Convolution Block Attention Module (CBAM). This CBAM application is designed to enhance the capability of CNNs in reproducing attention towards vital features within images. The proposed solution showed an increase of 2.5% in object detection performance over the existing YOLOv5 model.

Proposed Architecture

The proposed strategy is composed of two primary components: The CBAM block depicted in Figure 38.1 and the CNN + CBAM model which are illustrated below.

CBAM Block

The CBAM block is an attention module convolutional block that could be used to enhance the effectiveness of convolutional neural networks (CNNs). It is divided into two sub-blocks: the channel attention module (CAM) and the spatial attention module (SAM).

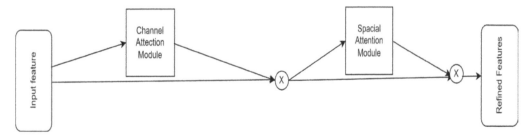

Figure 38.1 Working of CBAM module.

a. *CAM block*
    The CAM block shown in Figure 38.2 consists of an Average Pooling layer and a Max Pooling Layer, input feature is passed to each of the layers individually and then passed through a shared Multi-Layer Perceptron after each layer. The resultant features are then added and then passed through a Sigmoid layer. The refined features are then multiplied with the input features.

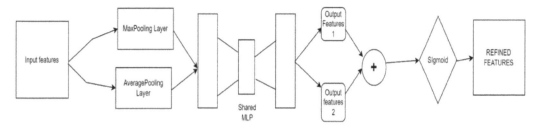

Figure 38.2 Architecture and flow of CAM block.

b. *SAM block*
    A MaxPooling layer, an Average Pooling layer, a 2D Convolution layer, and a sigmoid layer comprise the SAM block. The input characteristics are separated into two channels, each representing average and maximum channel pooling. The resultant data are then merged and fed via a 2D Convolution layer before being passed through a sigmoid layer. After that, the generated features are multiplied by the input features. The complete architecture of the SAM block is depicted in Figure 38.3. below.

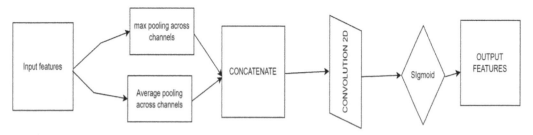

Figure 38.3 Architecture and flow of SAM block.

The CBAM block is formed by combining the CAM and SAM blocks. The output of the CBAM block is then multiplied by the supplied features map, and the resulting feature representation is passed on to the CNN's next layer.

CNN + CBAM Model

The CNN + CBAM mechanism happens to be an intricate convolutional neural network design, wherein the CBAM block remarkably enhances its overall performance. To start with, the CNN architecture that powers this model is none other than the LeNet architectural setup. The LeNet architecture, being a straightforward yet highly efficacious convolutional neural network setup, has been fruitfully employed over a wide array of image classification undertakings.

Fundamentally, this architecture incorporates two convoluted layers paired with two fully connected layers. Following these initial layers, the innovative addition of the CBAM block is introduced immediately after the primary convoluted layer.

Now let's dive into how this cutting-edge CNN + CBAM model was cultivated and honed for optimal results. This particular model underwent rigorous training utilizing Mendeley Data's invaluable "Labelled Optical Coherence Tomography (OCT) and Chest X-Ray Images for Classification" dataset. This extensive dataset comprises a whopping 10,000 chest X-ray images, which are helpfully divided into two equal halves—one set of 5,000 images are labelled as pneumonia while the remaining 5,000 are classified as normal scans.

To ensure top-notch performance during training sessions, the Adam optimizer was meticulously employed alongside a cross-entropy loss function to keep track of any discrepancies between predicted outcomes and expected target classifications. The proposed architecture of CNN + CBAM model is depicted in Figure 38.4 below.

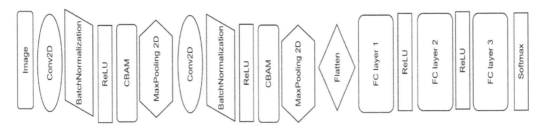

Figure 38.4 Architecture of the proposed model.

Result and Discussions

The model was trained for 50 epochs, with the learning rate decreasing exponentially with each epoch. The model was able to learn from a wide range of data and, as a result, improved its performance over time. The data was divided into three categories viz. training, validation, and testing. The training set was used to train the model and the test set was used to determine the final findings. We calculated the loss of model output due to dataset imbalance using weighted binary cross-entropy loss. When miscategorising minority groups, the classification model is supported by increased weighting of weighted binary cross-entropy loss. Equation (1) for the weighted binary cross-entropy loss is given below:

$$\Theta = -\sum_{i=1}^{N}\sum_{j=1}^{M}\left(\lambda_j{}^1 Y_{ij} log Z_{ij} + \lambda_j{}^0 \left(1 - Y_{ij}\right)\left(1 - log Z_{ij}\right)\right) \tag{1}$$

Where, $\lambda_j^1 = N \div count(Y_j = 1)$ and $\lambda_j^1 = N \div count(Y_j = 0)$ are the loss weights for the jth label's positive and negative samples, respectively. M is the number of labels, while N represents the number of samples.

Table 38.1: Accuracy, precision, recall and F1-score of the model, for different partitions of the dataset.

Dataset partition used	Accuracy (%)	Precision (%)	Recall (%)	F1-Score
Training Set	97.062	87.97	65.78	75.27
Validation Set	93.598	73.27	80.48	76.70
Test Set	93.015	71.15	83.69	76.91

Table 38.1 displays the model's final outcomes in terms of accuracy, precision, recall, and F1-score, depending on how the dataset was partitioned. The model performed well on the test set, with an accuracy of 93.02%. The training data model does not overfit, as evidenced by high recall and precision. A higher precision score would result in a better model. Various loss control or regularization strategies may aid in achieving this goal.

Conclusion and Future Scope

Finally, the work adds to the expanding corpus of research on using deep learning models for medical picture processing. The attention-based technique used in this study yielded excellent findings, encouraging continued research and improvement in the field of medical picture analysis. In conclusion, our study emphasises the importance of attention processes in deep learning models for detecting pneumonia from chest X-ray pictures. The suggested CNN with CBAM solution is strong and effective, supporting developments in computer-aided diagnosis systems and, eventually, contributing to the betterment of healthcare practices. Future research might build on these findings and investigate new improvements to improve the accuracy and efficiency of pneumonia diagnosis and other medical image processing jobs as the field evolves.

References

1. Woo, S., Park, J., Lee, J. Y., and Kweon, I. S. (2018). CBAM: convolutional block attention module. In Proceedings of the European Conference on Computer Vision (ECCV), (pp. 3–19). 2018.
2. Varshni, D., Thakral, K., Agarwal, L., Nijhawan, R., and Mittal, A. (2019). Pneumonia detection using CNN based feature extraction. In 2019 IEEE International Conference on Electrical, Computer and Communication Technologies (ICECCT), (pp. 1–7). IEEE. doi: 10.1109/ ICECCT.2019.8869364.
3. Zhou, S., Zhang, X., and Zhang, R. (2019). Identifying cardiomegaly in Chest x-ray 8 using transfer learning. In MEDINFO 2019: Health and Wellbeing E-Networks For All, (pp. 482– 486), IOS Press.
4. Sharma, H., Jain, J., Bansal, P., and Gupta, S. (2020). Feature extraction and classification of chest x-ray images using cnn to detect pneumonia. In 2020 10th International Conference

on Cloud Computing, Data Science and Engineering (Confluence), (pp. 227–231), IEEE. doi: 10.23919/SPA50552.2020.9241305.

5. Garstka, J., and Strzelecki, M. (2020). Pneumonia detection in x-ray chest images based on convolutional neural networks and data augmentation methods. In 2020 Signal Processing: Algorithms, Architectures, Arrangements, and Applications (SPA), (pp. 18–23), IEEE. doi: 10.23919/SPA50552.2020.9241305.

6. Singh, S. (2021). Pneumonia detection using deep learning. In 2021 4th Biennial International Conference on Nascent Technologies in Engineering (ICNTE), (pp. 1–6). IEEE. doi: 10.1109/ICNTE51185.2021.9487731.

7. Jassam, I. F., Elkaffas, S. M., and El-Zoghabi, A. A. (2021). Chest x-ray pneumonia detection by dense-net. In 2021 31st International Conference on Computer Theory and Applications (ICCTA), (pp. 176–179). IEEE. doi: 10.1109/ICCTA54562.2021.9916637.

8. Chen, X. (2022). The advance of deep learning and attention mechanism. In 2022 International Conference on Electronics and Devices, Computational Science (ICEDCS), (pp. 318–321). IEEE. doi: 10.1109/ICEDCS57360.2022.00078.

9. Chen, S., Ren, S., Wang, G., Huang, M., and Xue, C. (2023). Interpretable CNN-multilevel attention transformer for rapid recognition of pneumonia from chest x-ray images. *IEEE Journal of Biomedical and Health Informatics*, doi: 10.1109/JBHI.2023.3247949. https://ieeexplore.ieee.org/document/10050021

10. An, J., Putro, M. D., Priadana, A., and Jo, K. H. (2023). Improved YOLOv5 network with CBAM for object detection vision drone. In 2023 IEEE International Conference on Industrial Technology (ICIT), (pp. 1–6), IEEE. doi: 10.1109/ICIT58465.2023.10143118.

39 Statistical Approaches and Framework for Meta-analysis: A Brief Review

Tanushree Chakraborty[1,a], Kartick Chandra Mondal[1,b], and Abhijit Sarkar[2,c]

[1]Department of Information Technology, Jadavpur University, Kolkata, India

[2]Department of Botany, University of Gour Banga, West Bengal, India

Abstract

Meta-analysis is a potent statistical method that enables researchers to combine the results from various studies on a specific research subject. To estimate an overall effect size across several related investigations, a meta-analysis may be utilized. This paper offers a succinct and tabular overview of the well-known frameworks and statistical methods employed in meta-analysis. Frameworks address specific challenges in meta-analysis, and their use enhances the reliability, transparency, and credibility of meta-analytic research. By implementing these frameworks, researchers can ensure that their meta-analyses adhere to best practices, leading to more accurate and informative conclusions. Also, various statistical methods used for meta-analysis are explained to give an overall idea that everyone can understand, including those with no prior expertise in statistics and the field of meta-analysis.

Keywords: Framework, meta-analysis, review, statistical approaches.

Introduction

The meta-analysis first familiarized by Karl Pearson in the early 20th century, has become an epidemiological tool [1] for evidence-based decision-making and scientific progress. In the areas of scientific research, it is being employed more frequently for combining data and findings from multiple studies and provides a powerful approach to systematically synthesizing the findings on a particular research question [2]. It is widely used in various fields of science like medical sciences, social sciences, and environmental sciences where data are combined from independent studies to draw more robust and generalizable conclusions. The goal of the meta-analysis (MA) is to compute the direction and/or magnitude of an effect across all related studies, both published and unpublished [3].

The benefits of meta-analysis are numerous. Firstly, meta-analysis offers a systematic method for managing a large number of investigations. Secondly, the procedure is organized and thoroughly recorded, enabling readers to assess the researchers' judgments and findings. Third, compared to a qualitative summary, meta-analysis enables researchers to investigate an effect within a group of studies in a more comprehensive way. To ease the process of meta-analysis, we have seen multiple frameworks proposed by the researchers.

[a]tc4research@gmail.com, [b]kartickjgec@gmail.com, [c]abhijitbhu@gmail.com

Meta-analysis normally involves two stages. To uniformly describe the observed intervention impact across all studies, a summary statistic is first produced for each trial. For example, summary statistics may be a risk ratio if the data are dichotomous; else we considered differences in mean. A weighted average of the intervention effects estimated in each study is used to create a summary intervention effect estimate in the second step.

Motivations: The motivation behind this research paper has been inspired by the recognition of the growing importance of meta-analysis as a crucial tool in evidence-based research. Meta-analysis has become increasingly popular across various disciplines, including medicine, psychology, education, social sciences, agricultural sciences, and more. The paper seeks to address the following key factors:

1. As meta-analysis gains popularity, there is a growing need for researchers, practitioners, and policymakers to have a comprehensive understanding of the statistical approaches and frameworks involved in this technique. The paper aims to bridge this knowledge gap and provide a clear and concise review of the essential statistical methods used in meta-analysis.
2. A thorough understanding of statistical methods used in meta-analysis is essential for researchers aiming to advance the field of evidence synthesis. By reviewing recent developments and emerging statistical approaches, this paper seeks to encourage researchers to explore and apply innovative techniques in their meta-analytic studies, further enhancing the methodological landscape of meta-analysis.
3. The paper's motivation is also to raise awareness about the importance of transparent and comprehensive reporting in meta-analytic research.

Objectives: The objective of this research paper is to provide a brief and concise overview of the statistical approaches and frameworks commonly employed in meta-analysis. The paper explains the concept of meta-analysis and its significance in evidence-based research first. It also highlights the advantages of meta-analysis in aggregating study findings, resolving discrepancies, and increasing statistical power [4]. It comprehensively reviews the various statistical methods and techniques used in meta-analysis and includes the presentation of effect size measures, calculation of standard errors, and weighting methods to appropriately combine study results.

Contributions: The novelty of this research paper lies in its specific focus on providing a concise overview of the statistical methods and frameworks used in meta-analysis. Several key aspects contribute to the paper's uniqueness and novelty are,

1. While there are existing literature studies [3, 5, 6, 7] and book [8] on meta-analysis are available, this paper stands out by presenting a thorough review of statistical approaches and frameworks in a concise format in one single place. It aims to cover the essential concepts and methods without being overwhelmed with excessive technical details, making it accessible to a broader audience, including researchers, practitioners, and policymakers with varying levels of statistical expertise.
2. The paper makes an effort to present complicated statistical ideas in simple language that everyone can understand, including those with no prior expertise in statistics and meta-analysis. It avoids unnecessary jargon and provides intuitive explanations to make it easier for readers to understand the ideas and uses of the statistical approaches effectively.

3. While some papers may focus solely on one theoretical aspect, the novelty of this paper lies in its emphasis on providing an overall framework and its extensions for conducting meta-analyses.

Organization of the paper: The paper begins with the introduction of meta-analysis with its importance in evidence-based research, objectives, and novelty of this paper in section 1. In section 2, various frameworks used for conducting and reporting meta-analysis are briefly explained chronologically. In section 3, brief summary of statistical approaches is presented in a tabular form. Finally, the article concludes at the end in section 4.

Frameworks for Meta-Analysis

In the context of meta-analysis, a framework refers to a set of guidelines, principles, or tools that provide a structured approach for conducting and reporting the meta-analytic study. These frameworks help researchers ensure that their meta-analyses are conducted in a rigorous and transparent manner, leading to more reliable and valid results. Frameworks may address various aspects of the meta-analysis process, such as study selection, data extraction, statistical analysis, and interpretation of findings. It also aims to provide a structured and transparent way to report the methods, results, and interpretations of a meta-analysis study. The Preferred Reporting Items for Systematic Reviews and Meta-Analyses (PRISMA) is the most widely recognized and used reporting guideline for meta-analyses [9]. Early methodological frameworks like QCA, MR meta-analysis, and Cochrane have limited statistical methods, and the inclusion of poor-quality studies may introduce bias and reduce the reliability of meta-analytic results [10, 11, 12]. The difficulty in assessing the risk of bias in non-randomized studies has been overcome by ROBINS-I [13]. Later, frameworks like MOOSE, PICOS aid researchers in providing essential information about study selection, data extraction, and statistical methods [14]. Bayesian meta-analysis [7] allows researchers to incorporate prior knowledge (prior distributions) into the analysis. It overcomes the limitation of traditional frequentist meta-analysis by explicitly modeling uncertainty and providing credible intervals, which offer a different perspective on the effect size estimation [6]. Different extensions of PRISMA [9] [15, 16] provides a standard format and checklist for authors to report the methodology, data, and results of their meta-analyses, ensuring transparency and completeness. Table 39.1 presents a compact list of meta-analysis frameworks present in the literature.

Table 39.1: Summary collection of meta-analysis frameworks.

Sl No.	Framework Name	Proposed Year	Description	Methods Used
1	QCA [10]	1987	It recognizes necessary and/or sufficient conditions and derives generalization from patterns of configurations across multiple studies	Configurational analysis, Qualitative data analysis
2	Cochrane [12]	1993	Developed by the Cochrane Collaboration and it emphasizes systematic reviews and meta-analyses of healthcare interventions.	Systematic review, Quality assessment

(continued)

Table 39.1: Continued

Sl No.	Framework Name	Proposed Year	Description	Methods Used
3	PICOS [17]	1995	It is used in evidence-based medicine to formulate research questions and define study inclusion criteria.	Population, Intervention, Comparison, Outcome, Study design
4	MOOSE [18]	2000	The MA Of Observational Studies in Epidemiology provides guidelines for MA for observational studies.	Study selection, Bias assessment
5	Meta-regression [19]	2002	Identify probable sources of heterogeneity and assess the impact of study-level factors (covariates) on the overall results.	Regression analysis, Heterogeneity analysis
6	QUADAS [20]	2003	Provides criteria for evaluating various aspects of study design, conduct, and reporting to ensure the reliability and accuracy of diagnostic test evaluations.	Diagnostic accuracy assessment, Quality assessment
7	MASEM [18]	2005	It combines MA and structural equation modeling (SEM) to inspect complex relationships among variables across multiple studies. It integrates effect sizes and latent variable models to investigate structural relationships.	Structural equation modeling, Integration of effect sizes
8	Scoping review [21]	2005	It aims to identify key concepts, sources, and types of evidence, providing an overview of the existing literature and highlighting areas for further research.	Literature mapping, Summary of evidence
9	Overlapping meta-analysis [7]	2009	It addresses the issue of duplicate or redundant data and accounts for the correlation structure between effect sizes from overlapping studies.	Correlated samples, Data overlap analysis
10	PRISMA (Original) [9]	2009	This framework provides a checklist of items to include when reporting for transparency in reporting.	Reporting guidelines
11	PRISMA-Equity [22]	2012	An extension of PRISMA specifically focused on systematic reviews and MA for health equity issues.	Checklist of items with a focus on health equity considerations
12	Bayesian network meta-analysis [3]	2013	It combines network MA with Bayesian statistical methods. It integrates multiple treatment comparisons and the estimated treatment effects while incorporating uncertainty through the use of prior distributions.	Bayesian statistics, Network analysis

(continued)

Table 39.1: Continued

Sl No.	Framework Name	Proposed Year	Description	Methods Used
13	MR meta-analysis framework [11]	2015	It is explicit to Mendelian randomization studies and incorporates the principles of genetic instrumental variable analysis between exposures and outcomes.	Genetic instrumental variables, Causal inference
14	PRISMA-P [15]	2015	Reporting planned or ongoing reviews and protocols.	Checklist of items given to report review protocols and ongoing reviews
15	PRISMAIPD Framework [14]	2015	individual-level data from multiple studies are collected and combined for meta-analysis.	Individual-level data integration
16	PRISMANMA [23]	2015	Focuses on network meta-analyses (NMA) and multiple treatment comparisons	Guideline for reporting network meta-analyses
17	ROBINS-I [13]	2016	The ROBINS-I framework is used in non-randomized studies for assessing the risk of bias.	Risk of bias assessment
18	PRISMAScR [24]	2018	Aim to map the available evidence on a broad research topic	Guideline for reporting scoping reviews
19	PRISMADTA [25]	2018	PRISMA diagnostic test accuracy provides transparent reporting and enhances replicability of reviews and makes DTA results more useful. Assist in the evaluation of validity and applicability.	27-item PRISMA diagnostic test accuracy
20	PRISMAA [26]	2019	PRISMA for Acupuncture Specify study characteristics (e.g., PICOS, length of follow-up) and report characteristics (e.g., years considered, language, publication status) used as criteria for eligibility, giving rationale.	Guideline for reporting rapid reviews for Acupuncture
21	PRISMA 2020 for Abstracts [27]	2020	New reporting guidelines are added to PRISMA 2020 statement and replace the 2009 statement. This mainly reflects advances in methods to identify, select, appraise, and synthesize studies.	27-item checklist has details of reporting recommendations for different components
22	PRISMA-S [28]	2021	Extension to PRISMA Statement for literature search reporting	checklist contains a total of 22 essential reporting items.
23	PRISMA-EcoEvo [16]	2021	PRISMA for Ecology and Evolutionary Biology	Summarize the main findings in terms of the precision of effects (e.g. size of confidence intervals, statistical significance)

Statistical Approaches for Meta-Analysis

For dichotomous results, there are four common meta-analysis techniques. Three techniques with fixed effects i) Mantel-Haenszel ii) Peto iii) Inverse variance and one random-effects method is DerSimonian and Laird inverse variance [29]. A meta-analysis helps the researchers to aggregate different indices of effect sizes like standardized mean difference(s), correlation(s), or odds ratio(s). This results in estimates of the overall effect size and their confidence intervals. Researchers can combine effect sizes from various studies using meta-analysis approaches, such as standardized mean differences, correlations, or odds ratios [30]. A brief collection of statistical methods and their applications are summarized in Table 39.2.

Table 39.2: Brief summary of statistical methods and indices for meta-analysis.

Sl No	Statistical Application	Measures/Indices	Description
1	Effect Size [31]	• Cohen's d-Standardized Mean Difference (SMD) • Odds Ratio (OR) • Risk Ratio (RR) • Correlation Coefficient • Hazards Ratio	• Quantify the magnitude of the treatment effect or association in each individual study. • Practical or clinical importance of the studies is not provided by the p-value. Moreover, the effect size does not vary with the sample size. • Cohen's d is commonly used when comparing the means of two groups. • SMD Effect Size (Hedges' g) is used when comparing the means of two groups with unequal standard deviations. • OR and RR for binary outcomes, and r for correlation between variables. • OR measures the association between two categorical variables. • When examining the relationship between two continuous variables, the correlation coefficient is commonly used as the effect size measure. • The common correlation coefficient is Pearson's correlation coefficient (r) to measure the linear association between the variables.
2	Weighting [5]	• Inverse Variance (IV) • Weighting • Sample Size (SS) Weighting • Quality Score Weighting	• Assign different weights to each study based on various factors. • IV weighting gives more weight to studies with smaller variances, sample size weighting assigns weights based on the study sample sizes, and quality score weighting assigns weights based on study quality assessments. • SS weighting assigns weights to studies based on their sample sizes. Larger studies, with larger sample sizes, are given more weight as they provide more precise estimates of the treatment effect. • Weighting ensures that studies with more precise estimates or higher quality contribute more to the overall effect estimate in meta-analysis.
3	Forest Plot [5]	• A graphical representation of the effect sizes and their confidence intervals from individual studies.	• Shows estimated results from a number of scientific studies addressing the same question, along with the overall results Provides a visual summary of the results, displaying the point estimates and confidence intervals of each study. • Includes a summary effect estimate, typically represented by a diamond symbol, which represents the pooled effect size across all studies.

(continued)

Table 39.2: Continued

Sl No	Statistical Application	Measures/Indices	Description
4	Heterogen [4]	• e-ity Cochran's Q test • I-squared (I^2) statistic	• Refers to the variability in effect sizes across studies. Cochran's Q • test assesses the statistical significance of heterogeneity by comparing the observed variation to the expected variation. • The I^2 statistic quantifies the proportion of total variation in effect sizes that is due to heterogeneity. Higher values of I^2 indicate greater heterogeneity. • Assessing and quantifying heterogeneity is crucial in determining the appropriate statistical model for meta-analysis (e.g., fixed-effects or random effects).
5	Fixed-effects Model [4]	• Assumes the underlying effect size is the same across all studies (e.g., Mantel and Haenszel).	• It also assumes that any observed differences in effect sizes between studies are due to random error. • It gives equal weight to each study and provides a single pooled effect size estimate. • This model is appropriate when there is no significant heterogeneity among the included studies.
6	Random-effects Model [4]	• Assumes heterogeneous effects across studies (e.g., DerSimonian and Laird)	• Assumes the true effect size varies across studies due to both random error and genuine differences between studies. • It accounts for both within-study and between-study variability. • Provides a pooled effect size estimate along with a confidence interval. • This model is suitable when there is heterogeneity among the included studies.
7	Subgroup Analysis (SA) and Meta-regression (MR)	• Participant characteristics • Intervention characteristics • Study design features • Outcome measures • MR allows exploration of the relationship between study-level characteristics (e.g., age, publication year, study quality) and effect sizes.	• SA can be conducted based on various study-level characteristics, such as Participant characteristics (e.g., age, gender), Intervention characteristics (e.g., dosage, duration, type), Study design features (e.g., randomized controlled trials, observational studies), Outcome measures (e.g., different scales or assessments) • Subgroup analysis involves exploring potential sources of heterogeneity by dividing studies into subgroups based on specific study characteristics or participant characteristics. • MR allows exploration of the relationship between study-level characteristics (e.g., age, publication year, study quality) and effect sizes. • Unlike subgroup analysis, meta-regression treats the characteristics as continuous variables rather than categorical groupings. • It is used to explore the relationship between the effect size (or outcome) and one or more study-level covariates. • It helps identify potential sources of heterogeneity and explain variability in effect sizes across studies.
8	Publication Bias Assessment [32]	Funnel Plot • Egger's Regression Test • Begg's Rank Correlation Test	Publication bias refers to the selective publication of studies based on the direction or significance

Conclusion

This survey paper provides a comprehensive overview of the statistical approaches and frameworks used in meta-analysis. By following best practices and addressing potential challenges, researchers can employ meta-analysis effectively to advance knowledge in their respective domains and contribute to evidence-based decision-making. Meta-analysis represents a summary or a base for data and results. It determines the path and research direction of the researchers. Decision makers and policymakers can take the help of this to put their hands on a large number of results and conclusions.

The evolving landscape of meta-analysis opens avenues for innovative approaches and enhanced frameworks to address emerging challenges. To improve the effectiveness and accuracy of the meta-analysis process, machine learning algorithms will be essential to automating data extraction, identifying study biases, and detecting publication bias. Hierarchical models and machine learning algorithms can be integrated to explore sources of heterogeneity and improve precision in decision-making.

References

1. Vetter, M. P. H., and Thomas, R. (2019). Systematic review and meta-analysis: sometimes bigger is indeed better. *Anesthesia and Analgesia*, 128(3), 575–583.
2. Kang, H., and Ahn, E. (2018). Introduction to systematic review and meta-analysis. *Korean Journal of Anesthesiology*, 71(2), 103–112.
3. Wilson. L. C. (2014). Introduction to meta-analysis: a guide for the novice. *APS Observer*, 27, 2014.
4. Schmid, E. J., Koch, G. G., and LaVange, L. M. (1991). An overview of statistical issues and methods of meta-analysis. *Journal of Biopharmaceutical Statistics*, 1, 103–120.
5. Bansal P., Tushir M., Emilia V., and Srivastava B. R. Advances in intelligent systems and computing 1164. In Proceedings of International Conference on Artificial Intelligence and Applications ICAIA 2020.
6. Nguyen, P. Y., Kanukula, R., McKenzie, J. E., Alqaidoom, Z., Brennan, S. E., Haddaway, N. R., Hamilton, D. G., Karunananthan, S., McDonald, S., Moher, D., Nakagawa, S., Nunan, D., Tugwell, P., Welch, V. A., and Page, M. J. (2022). Changing patterns in reporting and sharing of review data in systematic reviews with meta-analysis of the effects of interventions: cross sectional meta-research study. *BMJ*, 22(379), 2022.
7. Stroup, D. F., Berlin, J. A., Morton, S. C., Olkin, I., Williamson, G. D., Rennie, D., Moher, D., Becker, B. J., Sipe, T. A., and Thacker, S. B. (2000). Meta-analysis of observational studies in epidemiology: a proposal for reporting. (moose) group. *JAMA*, 19, issue 283(15), 2008–2012.
8. Chandler, J., Cumpston, M., Li, T., Page, MJ., and Welch, VA. (2019). (editors) Higgins JPT, Thomas J. Cochrane handbook for systematic reviews of interventions version 6.4 (updated august 2023), 2nd edition,: chichester (UK): John Wiley Sons, 2019.
9. Moher, D., Liberati, A., Tetzlaff, J., and Altman, D. G. (2009). Prisma group. preferred reporting items for systematic reviews and meta-analyses: the prisma statement. *Annals of Internal Medicine*, 264–269.
10. Ragin, C. C. (2014). The Comparative Method: Moving Beyond Qualitative and Quantitative Strategies. University of California Press.
11. Burgess, S., and Thompson, S. G. (2015). Mendelian Randomization: Methods for Using Genetic Variants in Causal Estimation. CRC Press.
12. Chalmers, I. (1993). The cochrane collaboration: preparing, maintaining, and disseminating systematic reviews of the effects of health care. *Annals of the New York Academy of Sciences*, 31(703), 156–163.

13. Sterne, J. A., Hernán, M. A., Reeves, B. C., Savović, J., Berkman, N. D., Viswanathan, M., Henry, D., Altman, D. G., Ansari, M. T., Boutron, I., Carpenter, J. R., Chan, A-W., Churchill, R., Deeks, J. J., Hróbjartsson, A., Kirkham, J., Jüni, P., Loke, Y. K., Pigott, T. D., Ramsay, C. R., Regidor, D., Rothstein, H. R., Sandhu, L., Santaguida, P. L., Schünemann, H. J., Shea, B., Shrier, I., Tugwell, P., Turner, L., Valentine, J. C., Waddington, H., Waters, E., Wells, G. A., Whiting, P. F., and Higgins, J. PT. (2016). Robins-i: a tool for assessing risk of bias in non-randomised studies of interventions. *BMJ*, 355:i4919.

14. Riley, R. M., Simmonds, M., Stewart, G., Tierney, J. F., Stewart L. A., and Clarke M. (2012). Referred reporting items for systematic review and meta-analyses of individual participant data: the prisma-ipd statement. 313, 16 (2015): 1657–1665, 2012.

15. Moher, D., Shamseer, L., Clarke, M., Ghersi, D., Liberati, A., Petticrew, M., Shekelle, P., and Stewart, L. A. (2015). Preferred reporting items for systematic review and metaanalysis protocols (prisma-p) 2015 statement. Syst Rev., 4(1), 2015.

16. Jennions, MD., Koricheva, J., Noble, DWA., Parker, TH., Gurevitch, J., Page, MJ., Stewart, G., Moher, D., Nakagawa, S., O'Dea, RE., and Lagisz M. (2021). Referred reporting items for systematic reviews and meta-analyses in ecology and evolutionary biology: a prisma extension. *Biological Review*, 96(5), 1695–1722. Biol Rev., 2021.

17. Thompson, S. G., and Higgins, J. PT. (2002). How should meta-regression analyses be undertaken and interpreted. *Statistics in Medicine* 21.11, 1559–1573.

18. Olkin, I., and Gleser, L. (2009). Stochastically dependent effect sizes. In introduction to meta-analysis. 2, 357–376.

19. Arksey, H., and O'Malley, L. (2005). Scoping studies: towards a methodological framework. *International Journal of Social Research Methodology* 8.1: 19–32.

20. Westwood, ME., Mallett, S., Deeks, JJ., Reitsma, JB., Leeflang, MM., Sterne, JA., Bossuyt, PM., Whiting, PF, and Rutjes, AW. (2011). Quadas-2: A revised tool for the quality assessment of diagnostic accuracy studies. *Annals of Internal Medicine*, 155, 8: 529–36.

21. Sutton, AJ., Caldwell, DM., Lu, G., Ades, AE., Dias, S., Welton, NJ. Evidence synthesis for decision making 4: inconsistency in networks of evidence based on randomized controlled trials. *Med Decis Making*. Jul; 33(5):641–56., 2013.

22. Tugwell, P., Moher, D., O'Neill, J., Waters, E., White, H., Welch, V., Petticrew, M. Prismaequity bellagio group. prisma-equity 2012 extension: reporting guidelines for systematic reviews with a focus on health equity. PLoS Med, Vol 9(10), 2012.

23. Caldwell, DM., Chaimani, A., Schmid, CH., Cameron, C., Ioannidis, JP., Straus, S., Thorlund, K., Jansen, JP., Mulrow, C., Catalá-López, F., Gøtzsche, PC., Dickersin, K., Boutron, I., Altman, DG., Moher, D., Hutton, B., and Salanti, G. The prisma extension statement for reporting of systematic reviews incorporating network meta-analyses of health care interventions. *Ann Intern Med*. 2015 Jun 2;162(11):777–84.

24. Tricco, A. C., Lillie, E., Zarin, W., O'Brien, K. K., Colquhoun, H., Levac, D., Moher, D., Peters, M. D., Horsley, T., Weeks, L., Hempel, S., et al. (2018). Prisma extension for scoping reviews (prisma-scr): checklist and explanation. *Annals of Internal Medicine*, 169(7), 467–473.

25. McGrath, TA., Thombs, BD., Hyde, CJ., Macaskill, P., Deeks, JJ., Leeflang, M., Korevaar, DA., Whiting, P., Takwoingi, Y., Reitsma, JB., Cohen, JF., Frank, RA., Hunt, HA., Hooft, L., Rutjes, AWS., Willis, BH., Gatsonis, C., Levis, B., Moher, D., McInnes, MDF., Salameh, JP., Bossuyt, PM. Preferred reporting items for systematic review and meta-analysis of diagnostic test accuracy studies (prisma-dta): explanation, elaboration, and checklist. BMJ, vol-370:m2632, 2020.

26. Yali Liu Liang Yao Janne Estill Zhaoxiang Bian Taixiang Wu Hongcai Shang 8 Myeong Soo Lee Dang Wei Jinhui Tian Bin Ma Yongfeng Wang Guihua Tian Kehu Yang Xiaoqin Wang, Yaolong Chen. Reporting items for systematic reviews and meta-analyses of acupuncture: the prisma for acupuncture checklist. *BMC Complement Altern Med*, Vol-12, Issue- 19(1), 208, 2019

27. Page, M. J., McKenzie, J. E., Bossuyt, P. M. Boutron, I., Hoffmann, T., C., Mulrow, C. D. et al. The prisma 2020 statement: an updated guideline for reporting systematic reviews. *BMJ*; 372:n71, 2021.

28. Siw Waffenschmidt, Ana Patricia, Ayala David, Moher Matthew J. Page Jonathan B. Koffel PRISMA-S Group Melissa L. Rethlefsen, Shona Kirtley. Prisma-s: an extension to the prisma statement for reporting literature searches in systematic reviews. Syst Rev, vol-10, 39, 2021.
29. Borenstein, M., Hedges, L. V., Higgins, J. P., and Rothstein, H. R. (2009). Introduction to Meta-Analysis. Hoboken, NJ: Wiley.
30. J.P.T. Higgins, Tianjing Li, and Jonathan Deeks. Choosing effect measures and computing estimates of effect, ch6, p. 143–176. 2019.
31. Berben, L., Sereika, S. M., and Engberg, S. (2012). Effect size estimation: Methods and examples. *International Journal of Nursing Studies*, 49, 1039–1047.
32. Bajwa, S. J. (2015). Basics, common errors and essentials of statistical tools and techniques in anesthesiology research. *Journal of Anaesthesiology, Clinical Pharmacology*, vol. 31, 547–553.

40 A Fresh Approach using the Dijkstra and A-star Algorithms to Look up Dominant Failure Sequences in Structures

Minangshu Baidya[1,a], Baidurya Bhattacharya[2,b], Trisha Chakravorty[3,c], and Aritra Chatterjee[1,d]

[1]Assistant Professor, Department of Civil Engineering, Indian Institute of Technology, Kharagpur, Kharagpur, West Bengal

[2]Professor, Department of Civil Engineering, Indian Institute of Technology, Kharagpur, Kharagpur, West Bengal

[3]Graduate Student, Department of Civil Engineering, Indian Institute of Technology, Kharagpur, Kharagpur, West Bengal

Abstract

Numerous failure sequences may occur in a civil engineering structure. Local or global failure patterns are also possible. Still, the structures are more susceptible to local failure sequences with unreasonably long durations. It suggests using the Dijkstra and A-star search algorithms to develop effective search strategies to identify typical failure occurrences. System dependability aids in identifying probabilistically dominating sequences, which in turn helps in identifying major failure sequences. To test the accuracy and practicality of the technique, a complex structure is utilized, such as a truss structure with 31 members that is similar to a bridge. The probabilistically dominant sequence was effectively searched using the Dijkstra and A-star approach, coupled with special truncation criteria and heuristic functions, in a shrinking manner. A comparative study has done with existing methods. It is faster to identify the main components of the structure using this approach.

Keywords: A-star, dijkstra, redundancy, system reliability.

Introduction

The term "progressive collapse" generally refers to the process of partial or complete structural collapse following initial member failure Ellingwood et al. [1]. If a civil engineering structure breaks down, it could be the initial indication that the entire structure is damaged. Other elements of the structure become overworked and begin to fail in a domino fashion up until a certain point if there is a lack of redundancy in the system. Sequential failure describes the sequential behavior of a member failure from the original member failure to a specific region of the failure. Large redundant structures can have a very large number of sequences of failures. The structure is the most likely to experience a local failure among all the likely failure sequences. The most important sequences of failures need to be easily recognized. To determine the key failure sequence, the structure needs to be repeatedly analyzed with multiple members removed or restored. As such, determining the critical failure sequence is a challenging endeavor. Popular sequence search techniques

[a]aritra@civil.iitkgp.ac.in, [b]baidurya@civil.iitkgp.ernet.in, [c]trishachakravorty2000@gmail.com, [d]aritra@civil.iitkgp.ac.in

now in use are not credible. A system failure point of view states that a significant factor in raising the likelihood of a system failure is the critical failure sequence. A novel search method for locating the probabilistically dominating sequence is presented in this study. This method is based on Dijkstra and A-star search algorithm. This search technique is used to identify the dominant sequences efficiently for large complex structure.

System Reliability Formulation

In order to prevent structural collapse from occurring, it is very important to identify critical components. The most important components are the dominating sequences. Numerous sequences affect how likely it is for a system to fail. Probabilistically dominant sequences make up the major portion of them. System failure can be written as:

$$P_{sys} = P\left(\bigcup_{i=1}^{N_{id}} E_i\right)$$
(1)

where (i=1 to N_{id}) is the incidence of the ith detected system failure sequence (Ei) Lee and Song [4]. Each sequence can be described as a cut-set. Now, if each cut set is minimal, and if each minimal cut set is defined in terms of a sequence of member failures and system failure probability can be written as the sum:

$$P_{f,sys} = \sum_{i=1}^{n_I} S_I$$
(2)

The Ith minimal cut set denotes Equation Baidya and Bhattacharya [2]. The *I*th minimal cut set $S_I = \left\{ F^{n_I^1, n_I^2, ..., n_I^{m(I)}} \right\}$, signifies the sequence of member failures from left to write:

n_i^1 is the ID of the first member to fail, n_i^2 is the ID of the second member to fail, and so on up to $n_i^{m(I)}$ where $m(I)$ is the total number of members in the Ith cut set. Defined this way, it is easy to see that any two minimal cut sets S_i and S_j are disjoint $(i \neq j)$. Let the random capacity (strength) of the Ith member be C_i. The event S_I is the intersection of a sequence of member level limit states:

$$S_I = \left\{ F^{n_I^1, n_I^2, ..., n_I^{m(I)}} \right\} = \left\{ C_{n_I^1} < \sigma^K_{n_I^1}(\underline{L}), C_{n_I^2} > \sigma^K_{n_I^2}(\underline{L}), C_{n_I^3} > \sigma^K_{n_I^3}(\underline{L}), ..., C_{n_I^{m(I)}} > \sigma^K_{n_I^{m(I)}}(\underline{L}) \right\}$$
$$\cap \left\{ C_{n_I^2} < \sigma^{K,n_I^1}_{n_I^2}(\underline{L}), C_{n_I^3} > \sigma^{K,n_I^1}_{n_I^3}(\underline{L}), ..., C_{n_I^{m(I)}} > \sigma^{K,n_I^1}_{n_I^{m(I)}}(\underline{L}) \right\}$$
$$\cap \left\{ C_{n_I^3} < \sigma^{K,n_I^1,n_I^2}_{n_I^3}(\underline{L}), ..., C_{n_I^{m(I)}} > \sigma^{K,n_I^1,n_I^2}_{n_I^{m(I)}}(\underline{L}) \right\} .... \cap \left\{ C_{n_I^{m(I)}} < \sigma^{K,n_I^1,n_I^2,...,n_I^{m(I)-1}}_{n_I^{m(I)}}(\underline{L}) \right\}$$
(3)

where means the second member of cut set I's survival, denotes the first member of cut set I's failure, and so on. It is emphasized that the applied random load vector L determines the load impact in each member. We generate the load vector L periodically using Latin Hypercube sampling to calculate the likelihood of a system failure:

$$P_{f,sys} = \int\limits_{all,L} \sum_{i=1}^{n_C} P[S_I, L = l] f_L(l) dl \tag{4}$$

Applicability of Dijkstra Algorithm

This innovative approach uses an event tree diagram with different damage states as the search space Figure 40.1. We first employ Dijkstra shortest path algorithm to find dominant sequences first proposed by Dijkstra [5], later in transportation network Deng et al. [6] used this algorithm. Parent node '0' defines an intact structure, and it contains numerous child nodes that define varying damage states for its members, some of which are defined to fail. The goal nodes in this search space are dark coloured nodes with varying weighted distances from the parent node to the goal nodes. The goal nodes represent structural instability or collapse. Finding the shortest routes from parent nodes to goal nodes in progressively shorter order is the study's goal. In a progressive failure analysis, the various shortest paths represent equivalent load paths or failure sequences. Above algorithm is search process to find shortest path from source to goal vertex. The procedure is to carried out up to a certain truncation criterion for shortest path dominant sequences. Truncation criteria is:

$$\frac{P_n}{\sum\limits_{i=1}^{n} P_i} \times 100 = \varepsilon < \varepsilon_{tol} \tag{5}$$

P_i is the probability of occurrences of the ith sequence and n is the number of identified sequences so far. Generally, the value of ε_{tol} lies between 10^{-3} to 10^{-4} for best approximation to system failure.

Applicability of A-star Algorithm

A* algorithm was first published by Hart et al. [7]. The actual distances between distinct damage states are found in A* search. Using structure analysis, this is calculated. In terms of various intermediate failure occurrences, the sequence can be expressed Bhattacharya [3], Baidya and Bhattacharya [2].

$$F_{1,2,3}^0 = F_1^0 \cap F_2^1 \cap F_3^{1,2} \tag{6}$$

Where $F_{1,2,3}^0$, is the sequence that member 1, 2 and 3 fails sequentially starting from intact-structure which is similar to equation (4). Where n_I is = 1 to 3. F_1^0 is = 1st member to fail from intact structure. F_2^1 is = 2nd member to fail given that 1st member has failed. $F_3^{1,2}$ is = 3rd member to fail given that 1st and 2nd member has failed. Then equation (6) can be written as:

$$P(F_{1,2,3}^0) = P(F_{2,1}^3 \mid F_1^2) P(F_1^2 \mid F_1^0) P(F_1^0) \tag{7}$$

Next, equation (7) can be written as:

$$-\log(P(F^0_{1,2,3})) = -\log(P(F^3_{2,1} \mid F^2_1)) - \log(P(F^2_1 \mid F^0_1)) - \log(P(F^0_1)) \qquad (8)$$

In equation (8) negative logarithm of conditional failure probabilities are the distances between different damage states which can be taken as $g(n)$.

Heuristic Function

The heuristic value, which derives from the mechanics of the problem, is typically a quick estimate of the distances to objective. System failure is the objective node in this issue. In a progressive failure analysis, heuristic functions show how a search can approach a goal or a system failure quickly. System failure may happen in a number of ways. They are: 1) Member removal causing (det(K) = 0) unstable structure 2). Member removal causing local instability or local mechanism (condition number exceeds large due to removal of members) 3) Displacements become large due to removal of member(s).

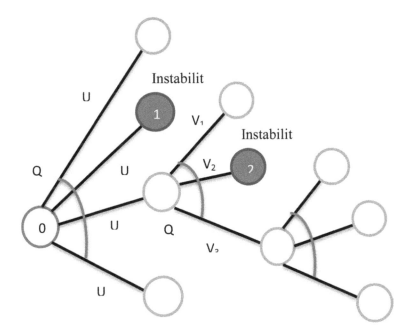

Figure 40.1 Event tree diagram of search space.

This several conditions are needed to check in the search process which is heuristic information. Next, heuristic value should provide bounding operation in dominant sequences. Thus, insignificant sequences should be excluded and unnecessary structure analysis function call should not be performed. Therefore, the heuristic value should be considered such that it will estimate the probable upper bound of probability of occurrences of sequence. Besides, the heuristic value should not overestimate the distances. This several conditions are needed to check in the search process, which is heuristic information. Next, heuristic value should provide bounding operation in dominant sequences. Thus, insignificant sequences should be excluded and unnecessary structure analysis function call should not be performed. Therefore, the heuristic value should be considered such that it will estimate the probable upper bound of probability of occurrences of sequence. Besides, the heuristic value should not overestimate the distances. At node '0' the heuristic value is:

$$h^0(n) = -\log(\max(\prod_{i=1}^{n} P_{f,i}(n)))$$ (9)

Where n is = the number of member failure in sequence. n is dependent on degree of indeterminacy. If the structure has degree of indeterminacy is = 2. Then n is = 3. Not only may that but structure fails in local instability. Then n may be less than D.O.I. In the next damage state heuristic value $h'(n) = h^0(n) - g^1(n)$. Where $g^1(n)$ are the distances in first damage state. In search process the heuristic value needs to change to higher value depending D.O.I (Static degree of indeterminacy) +1 to check whether any sequences are ignored and contribution in system failure probability up to minimum by the nth sequence.

Numerical Example

To test the applicability of proposed procedure a large and more complex structure is taken as in the Figure 40.2. Load P_1 and P_2 follows normal distribution with mean 250kN and COV 20%. The members Table 40.1 are brittle and follows normal distribution with mean 430 MPa and COV is= 20%. The members are statistically independent. According structure analysis the most stressful members are 24 and 4. The structure's global degree of indeterminacy is very high. But local mechanism may cause failure of the significant portion. The structure is more likely to fail in local failure.

Table 40.1: Member properties details.

Members	Area (m²)	Elastic modulus (GPa)
1–6	15×10^{-4}	200
7–12	14×10^{-4}	200
13–17	12×10^{-4}	200
18–31	13×10^{-4}	200

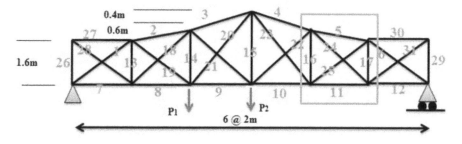

Figure 40.2 31-member truss structure.

A-star algorithm is applied here to find out dominant failure sequences. The truss system is combination of series and parallel system. The marked green section is parallel connection of members in system reliability point of view. Generally, the whole truss is series connection of small squared sections. In progressive collapse which elements are the key elements can be found out by dominant sequence analysis. In progressive failure members are suddenly removed.

Results

Initially 2 members' failure may cause system failure. Therefore, all combination of two members' failure is analyzed at intact structure. According to failure probability of the individual members the upper bound of probability of occurrence is = 7.655×10^{-6}. Hence heuristic value at start node is = 5.116. Based on that shortest distance branches are 24, 4 and 3. Expanding the branches system failure observed and the dominant sequences are 24-11 and 24-5. Similarly other dominant sequences found out 4-10, 4-22, 3-9 and 3-20. Other branches are not expanded. To discover other dominant sequences the heuristic value is increased because of three members 'failure. Thus, upper bound of probability of occurrence is 2.8862×10^{-10} and heuristic value is 21.202. Due to new heuristic value more numbers of dominant sequences discovered. Number of dominant sequences increased to 55 from 6. Time requirement to find out 6 sequences is 9.539 sec but time requirement to find out 55 sequences is 147.69 sec which is exponentially increased. Setting heuristic value to higher value increases number of dominant sequences but they hardly improve the system failure probability. Therefore, the upper bound of probability of occurrences should not be higher than acceptable bound. Numbers of possible global failure sequences are 1.32×10^{10} which is not considering local failure sequences for the bridge type truss structure. Considering local instability numbers of sequences are much less. In comparison with Branch and Bound algorithm Lee and Song [4] 6-31, 31-6, 12-31, 29-6 and 31-12 sequences are missed out. But in both the cases by Dijkstra and A-star search algorithm all failure sequences are identified which coincides with dominant sequences by enumeration approach. Among all sequences most likely to failure sequences are local failure and 70 sequences have significant non zero values up to order of 8×10^{-16}. But top 19 sequences are dominant whose probability of occurrences value up to 9×10^{-8}. In comparison with Dijkstra search top 19 sequences are identified with tolerance value 6.47×10^{-2} and tolerance value reach at 3×10^{-3} Figure 40.5 for top 36 sequences whose order of probability of occurrence is 4×10^{-9}. The system failure probability is 1.25×10^{-4}. Beyond this this tolerance value no more improvement is observed in system failure probability.

Conclusion

In this work A-star algorithm is implemented with new heuristic function and actual sequential failure formulation is used in progressive collapse analysis. The sequences are identified in gradually decreasing order without ignoring such significant sequences. So many significant sequences are missed out by conventional branch and bound (Lee and Song [4] and Xiao and Mahadevan [8]. Heuristic function helps to identify the critical sequence quickly and heuristic value estimates the upper bound of probability of occurrences of the sequences. The heuristic value is itself the bounding criteria for search strategy. In case of elastic modulus, a general structure made of same material or in some cases multiple material can be used. Their dominant sequences pattern changes as most likely to failure components differ. Similarly Cross-sectional area plays fundamental role in dominant sequences. Dominant sequences, majorly depends on cross sectional area. By analyzing structures, it is concluded that with increasing static degree of redundancy numbers of sequences are more to converge system failure probability. Number of structure analysis function call is less with respect to Branch and Bound method and truncated enumeration approach. Therefore, proposed methodology is efficient method.

References

1. Ellingwood, B. R., Smilowitz, R., Dusenberry, D. O., Duthinh, D., and Carino, N. J. (2007). Best practices for reducing the potential for progressive collapse in buildings, NIST Interagency or Internal Reports, 7396.
2. Baidya, M., Bhattacharya, B. (2022). System Reliability and New Measure of Robustness of Truss Structure in Progressive Collapse. In: Maiti, D.K., et al. Recent Advances in Computational and Experimental Mechanics, Vol II. *Lecture Notes in Mechanical Engineering*. Springer, Singapore.
3. Bhattacharya, B. (2021). A reliability based measure of structural robustness for coherent systems. *Structural Safety*, 89, 102050.
4. Lee, Y. J., and Song, J. (2011). Risk analysis of fatigue-induced sequential failures by branch-and-bound method employing system reliability bounds. *Journal of Engineering Mechanics*, 137(12), 807–821.
5. Dijkstra, E. W. (1959). A note on two problems in connexion with graphs. *Numerische Mathematik*, 1(1), 269–71.
6. Deng, Y., Chen, Y., Zhang, Y., and Mahadevan, S. (2012). Fuzzy dijkstra algorithm for shortest path problem under uncertain environment. *Applied Soft Computing*, 12, 1231–1237.
7. Hart, P. E., Nilsson, N. J., and Raphael, B. (1968). A formal basis for the heuristic determination of minimum cost paths. *IEEE Transactions on Systems Science and Cybernetics*, 4(2), 100–7.
8. Xiao, Q., and Mahadevan, S. (1994). Fast failure mode identification for ductile structural system reliability. *Structural Safety*, 13(4), 207–26.

41 Indoor Air Quality Monitoring System for Asthma Patients

Avishek Banerjee[1,a]*, Ramesh P Sah[2,b], Santanu Metia[3,c], Annesha Chakraborty[1,d], Arnab Mitra[4,e], and Tanmay Bhowmik[5,f]

[1]Department of CSBS, Asansol Engineering College, Asansol, India

[2]Departmentof ME, Asansol Engineering College, Asansol, India

[3]Faculty of Engineering & Information Technology, University of Technology Sydney, 15 Broadway, Ultimo, NSW 2007, Sydney, Australia

[4]Department of Computer Science and Engineering, SRM University-AP, Andhra Pradesh, India

[5]Department of Computer Engineering, Pandit Deendayal Energy University, Gandhinagar, Gujarat, India

Abstract

The major focus of the paper is to determine indoor air condition through different machine learning models. The indoor air quality is measured through IoT devices. The health of asthma sufferers is seriously threatened by the huge environmental problem that is indoor air pollution. For people with asthma, particulate matter (PM) can seriously harm their health. When these individuals are exposed to indoor air pollutants like particulate matter 10 (PM10), their health quality might deteriorate rapidly. Since most individuals spend their time inside, indoor air quality, or IAQ, is important for keeping track of asthma sufferers' health problems. For the benefit of asthma sufferers, safe indoor air quality may be monitored and polluting indoor air particles can be identified using IoT technology. Three key measures (PM 10) have been used in this work to measure the inside air quality. It is determined whether indoor air quality is harmful based on the human-acceptable threshold values for the key factors. The acceptable threshold value fluctuates within a specified range for asthma sufferers. Consequently, this research incorporates five machine learning models to assess the system's accuracy. The models used include Multiple Linear Regression (MLR), K-nearest neighbor (KNN), Linear Support Vector Machine (L-SVM), Logistic Regression (LR), and Decision Tree (DT). When these machine-learning models were compared, it was discovered that the L-SVM model had the highest level of accuracy in this case.

Keywords: Inside air quality, IoT (Internet of Things), machine learning models.

Introduction

Inside air quality is getting worse every day because of expanding transportation, industrialization, rising global heat, and abrupt climate change. Eye, lungs, nose, and throat discomfort result from poor indoor air quality. In addition to causing respiratory issues,

[a]avishek.csbs@aecwb.edu.in, [b]sramesh2031@gmail.com, [c]anneshac2002@gmail.com, [d]Santanu.Metia@uts.edu.au, [e]arnab.m@srmap.edu.in, [f]tanmaybhowmik@gmail.com

indoor air pollution makes asthma and other pre-existing illnesses worse. Long-term asthma sufferers are maintained in an area with the least amount of indoor airpollution. Patients with asthma who are subjected to indoor air pollution for an extended period have a higher chance of developing the condition. As a result, anapproach is required to monitor pollution levels in real-time and send out timely alerts about the general indoor air quality. In the proposed study, an inexpensive IoT-based indoor aircharacteristic monitoring system that would track the levels of PM10 in the inside air is constructed.

The succeeding steps are used to assess the accuracy of various threshold PM-10 levels that are detrimental for people with asthma at a specific location.

Step 1. To make it affordable for most people, low-cost sensors are being used. However, the input values are calibrated with the aid of locally accessible metrological data from the internet to lower the error rate.

Step 2. For various PM-10 levels, various threshold values are recognized.

Step 3. The various PM-10 values are subjected toexamination over the threshold value. This is done to rank the important model parameters.

Step 4. In the data model, normalization is done to both deciding and deciding parameters.

Step 5. Various models have been used to generate the desired accuracy levels.

The successivesubdivisions of the article are ordered as follows: Section 2 specifies an outline of the current research in the area. Section 3 delves into the research technique employed in this research. Section 4 exhibits the empirical results attained from the search. Finally, Section 5 offers a thoroughoutline of the paper's findings.

Literature Review

Over the past decade, the usage of low-cost sensingequipment for observinginside air-contamination has made significantimprovements. Some related works have focused on developing and adapting affordable devices to supervise and analyze indoor air quality in indoor conditions. These devices have proven useful in detecting indoor air pollutants and assessing their impact on public health, reducing related health risks [1–3]. Moreover, these portable and low-maintenance sensors have enabled broader participation, engaging citizens in various science projects and community-driven scientific research.Despite the advantages, these low-cost sensors often appeared with design negotiations and portability concerns that can affect data accuracy and reliability. As the number of studies and projects using such sensors increases, information tends to become scattered. In response, this study critically reviews recent research related to the modification of indoor air quality monitoring devices using low-cost sensors [4]. In a similar project, a wireless distributed portableinside air pollution observingapproachwas developed utilizing "General Packet Radio Service (GPRS) sensors" [5]. This exemplifies how advances in wireless sensors and communications technology are promptlyaltering the field of indoor air pollution monitoring. Internet of Things (IoT) technologies further enable the innovation of smart-tech environments where objects cooperate and conjoin with each other. Many improvements have been made to currentindoor air pollution monitoring systems. For instance, an IoT-based inside air pollution and noise monitoring system was developed to detect noise levels and indoor air pollution in the environment [5–7]. Additionally, researchers suggested an IoT-based indoor air pollution observing system incorporated with a web server that sparks an alert when the inside air quality dropslower the threshold level. The model displays indoor air quality on an LCD display and a webpage, enabling real-time monitoring. Furthermore, an IoT-based indoor air pollution technique employs Single Board Computers (SBC) to integrate IoT with wireless sensor networks (WSN), reducing processing complexity and making the alerting process smarter. Although its network coverage area was limited, the proposed system demonstrated a cost-effective and scalable implementation. Nevertheless, it proved to be a stable and effective solution, with potential applications for real-time monitoring in industrial settings, enhancing industrial output [8]. Another real-time inside air pollution observing system utilizes IoT with many sensors and employs neural net works to analyze the acquired data.In order to monitor, an IoT-based Indoor air Pollution Monitoring System is made in which we will monitor the Indoor air Quality over a web server using the Internet and it will create an alarm when the indoor air quality goes down beyond a certain level, which means when there are enough harmful gases are present in the indoor air like CO2, smoke, alcohol, benzene and NH3 [9–11]. The IoT project enables easy monitoring of indoor air quality by showing the contamination levels on both an LCD display and a webpage.

Methodology and Model Specifications

In this paper, the following machine learning models have been implemented.
A. Multiple Linear Regression (MLR)
B. Decision Tree (DT)
C. Logistic Regression (LR)
D. KNearest Neighbour (KNN)
E. L-SVM (Linear-Support Vector Machine)

A. Multiple Linear Regression (MLR)

1. **Dataset:** Gather a dataset using IoT devices that includes samples of indoor air quality measurements for different areas or patients. Each sample should contain PM10.0 values as well as the corresponding indoor air quality label ("Dangerous" or "Not Dangerous").
2. **Threshold Values:** Assign a specific threshold value for each patient or specific area. The threshold value determines when the indoor air quality is considered "Dangerous" based on PM10.0 values.
3. **Feature Scaling:** Perform feature scaling on the PM10.0 values to confirm they stay on a comparable scale. Conventional techniques include standardization (mean normalization) or normalization (min-max scaling).
4. **Model Training:** Splitting the dataset into the training andthe testing sets for evaluation of the model. Use the training set to train a Multiple Linear Regression model. Corresponding the model to the exercising data, where the PM10.0 values are the input features and indoor air quality is the target variable.
5. **Prediction:** Once the Multiple Linear Regression model is trained, use it to predict the indoor air quality for new, unseen samples. Calculate the combinational weighted sum value of PM10.0 for the test sample using the model's coefficients. Compare the sum value with the corresponding threshold to determine the indoor air quality label.

6. **Evaluation:** Assess the performance of a trained model applying the testing set. Calculate metrics such as the model's accuracy, recall,precession, f1-score, and support.

B. **Decision Tree (DT)**
1. **Dataset:** Same as the previous (MLR) model.
2. **Threshold Values:** Same as the previous (MLR) model.
3. **Decision Tree Structure:** Define a decision tree structure with inner nodes demonstrating feature tests and leaf nodes indicating the predicted indoor air quality labels.
   Each internal node corresponds to a decisive parameter and its threshold value, while each leaf node includes the predicted label.
4. **Recursive Splitting:** Develop the decision tree by recursively separating the dataset based on the decisive parameters and their threshold values. At each internal node, choose the decisive parameter and threshold value that best separates the samples based on indoor air quality.
5. **Prediction:** Once the decision tree is constructed, use it to predict the indoor air quality for new, unseen samples. Traverse the decision tree by comparing the PM10.0 values of the sample with the decisive parameter thresholds at each internal node. Follow the appropriate path until reaching a leaf node, which provides the predicted indoor air quality label.
6. **Classification:** Compare the combinational weighted sum value of PM10.0 for each sample with the corresponding patient's threshold value. If the combinational weighted sum value exceeds the threshold, classify the indoor air quality as "Dangerous"; otherwise, classify it as "Not Dangerous".
7. **Evaluation:** Evaluate the decision tree model's functioning by balancing the anticipatedinside air quality labels with the actual labels from the dataset. Calculate evaluation metrics, including accuracy, recall, accuracy, and F1-score, to gauge the model's efficiency.

C. **Logistic Regression (LR)**
1. **Dataset:** Same as the previous (MLR) model.
2. **Threshold Values:** Same as the previous (MLR) model.
3. **Feature Engineering:** Calculate the combinational weighted sum for each sample by multiplying PM10.0 values by their respective weights and summing them. The weights can be predetermined or derived from a training process.
4. **Data Preprocessing:** Breaking the dataset into training and testing sets for model evaluation. Perform any essential data preprocessing phases such as feature scaling or normalization.
5. **Model Training:** Train a Logistic Regression model using the training set. Fit the model to the training data, where the combinational weighted sum is the input feature and the indoor air quality is the target variable.
6. **Prediction:** Use the trained Logistic Regression template to forecast the indoor air quality for new, unseen samples. Calculate the combinational weighted sum for the new sample using the predefined weights. Compare the sum value with the corresponding threshold to determine the indoor air quality label.
7. **Evaluation:** Assess the performance of the trained model utilizing the testing set. Evaluate metrics such as accuracy, F1, or,recallgrade to consider the model's performance.

D. **K Nearest Neighbour (KNN)**
  1. **Dataset:** Same as the previous (MLR) model.
  2. **Threshold Values:** Same as the previous (MLR) model.
  3. **Distance Calculation:** Implement a distance metric, such as Euclidean distance, to measure the similarity between samples. Calculate the space between the test sample (unseen data point) and each sample in the training dataset.
  4. **K Nearest Neighbors:** Choose a value for K, which represents the number of nearest neighbors to consider. Select the K samples with the smallest distances to the test sample.
  5. **Majority Voting:** Determine the majority class (indoor air quality label) among the K nearest neighbors. If the majority class is "Dangerous," classify the indoor air quality as "Dangerous"; otherwise, classify it as "Not Dangerous."
  6. **Prediction:** Once the KNN model is trained, use it to predict the indoor air quality for new, unseen samples. Calculate the combinational weighted sum value of PM10.0 for the test sample. Find the K nearest neighbors based on the distance metric and assign the majority class label to the test sample.
  7. **Evaluation:** Measure the accomplishment of the KNN model by evaluating the predicted indoor air quality labels with the actual labels from the dataset. Calculate evaluation metrics such as accuracy, F1-score, and recall, to assess the model's success.

E. **L-SVM (Linear-Support Vector Machine)**
  1. **Dataset:** Same as the previous (MLR) model.
  2. **Threshold Values:** Same as the previous (MLR) model.
  3. **Feature Engineering:** Calculate the combinational weighted sum for each sample by multiplying the PM10.0 values by their respective weights and summing them. The weights can be predetermined or derived from a training process.
  4. **Data Preprocessing:** Partition the dataset into training and testing sets to facilitate model assessment. Perform any necessary data preprocessing steps such as feature scaling or normalization.
  5. **Model Training:** Train a SVM classifier utilizing the training set. Fit the model to the training data, where the combinational weighted sum is the input feature, and the indoor air quality is the target variable.
  6. **Prediction:** Use the trained SVM model to forecast the indoor air quality for new, unseen samples. Calculate the combinational weighted sum for the new sample using the predefined weights. Compare the sum value with the corresponding threshold to determine the indoor air quality label.
  7. **Evaluation:** Assess the functioning of the trained model utilizing the testing set and assess metrics like accuracy, F1, and recall, score to gauge its efficiency.

The proposed System Architecture is presented in following Figure 41.1.
  Creating an air monitoring system involves the following steps [12, 13]:

1. Utilizing PM sensors to measure environmental data.
2. Storing the calibrated data on a server.
3. Inputting the data into various machine learning models.
4. Selecting the most suitable machine learning model by evaluating accuracy.
5. Employing the optimal machine learning model to classify and determine air quality.

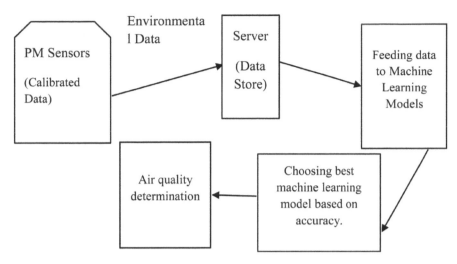

Figure 41.1 Proposed block diagram of the air monitoring system.

Empirical Results

Numerous ML models have been used in this research to test the system's accuracy. The table that follows shows how well the models performed:

Table 41.1: The performance of various ML models [7, 14].

Serial No	Machine Learning Model Name	Model Accuracy	Performance matrices			
			Precision	Recall	F1 score	support
1.	MLR-(Multiple Linear Regression)	0.7261	99.99%	75%	60%	16
2.	DT-(Decision Tree)	0.8881	98%	78%	66%	16
3.	LR-(Logistic Regression)	0.8965	97%	81%	69%	16
4.	KNN	0.9193	96%	82%	71%	16
5	L-SVM	0.9871	95%	88%	93%	16

In this research work, various machine learning models were evaluated for an indoor air monitoring system. The accuracy scores obtained for each model are as follows: MLR – 72.61%, DT – 88.81%, LR – 89.65%, KNN – 91.96%, and L-SVM – 98.71%, which exhibited the maximumaccurateness with respect to all parameters i.e., precision, f1-score, recall, and validation as compared to other models. Based on the outcomes, it can be inferred that L-SVM model outperformed the other models and proved to be the most suitable approach for the indoor air monitoring system in this study.

Conclusion and Future Research Direction

In this investigation, multiple Machine-Learning models were applied, and upon comparing their performances, it was observed that the L-SVM achieved maximum accuracy (0.9871). As a result, it can be assumed that for the indoor air monitoring system, Random

Forest is the most appropriate machine learning model, ensuring the highest precision. This finding specifies a valuable hypothesis for future researchers to develop even more accurate models.Additionally, the study explored a varied set of eight AIML models, opening the option for other researchers to employ alternative models and obtain significant results. With the aid of this low-cost indoor air monitoring system, IAQ (Inside air Quality) finding for asthma patients becomes more feasible in both urban and rural areas. Such areas may confrontvariable levels of particulate matter (PM) that could reach hazardous limits for asthma patients. The proposed project holds the potential to prevent high-risk situations by generating timely alarms.

In the future, we plan to measure the real-life performance of our proposed work.

References

1. Chojer, H., Branco, P. T. B. S., Martins, F. G., Alvim-Ferraz, M. C. M., and Sousa, S. I. V. (2020). Development of low-cost indoor air quality monitoring devices: recent advancements. *Science of the Total Environment*, 727, 138385.
2. Clements, A. L., Griswold, W. G., Rs, A., Johnston, J. E., Herting, M. M., Thorson, J., ... and Hannigan, M. (2017). Low-cost indoor air quality monitoring tools: from research to practice (a workshop summary). *Sensors*, 17(11), 2478.
3. Barot, V., Kapadia, V., and Pandya, S. (2020). QoS enabled IoT based low costindoor air quality monitoring system with power consumption optimization. *Cybernetics and Information Technologies*, 20(2), 122–140.
4. Kelechi, A. H., Alsharif, M. H., Agbaetuo, C., Ubadike, O., Aligbe, A., Uthansakul, P., ... and Aly, A. A. (2022). Design of a low-cost indoor air quality monitoring system using arduino and thingspeak. *Computers, Materials and Continua,* 70, 151–169.
5. Alshamsi, A., Anwar, Y., Almulla, M., Aldohoori, M., Hamad, N., and Awad, M. (2017). Monitoring pollution: applying IoT to create a smart environment. In 2017 International Conference on Electrical and Computing Technologies and Applications (ICECTA), (pp. 1–4). IEEE.
6. Shaban, K. B., Kadri, A., and Rezk, E. (2016). Urban indoor air pollution monitoring system with forecasting models. *IEEE Sensors Journal*, 16(8), 2598–2606.
7. Pushpam, V. E., and Kavitha, N. S. (2019). Iot enabled machine learning for vehicular indoor air pollution monitoring. In 2019 International Conference on Computer Communication and Informatics (ICCCI), (pp. 1–7). IEEE.
8. Shah, H. N., Khan, Z., Merchant, A. A., Moghal, M., Shaikh, A., and Rane, P. (2018). IOT based indoor air pollution monitoring system. *International Journal of Scientific and Engineering Research*, 9(2), 62–66.
9. Potbhare, P. D., Bhange, K., Tembhare, G., Agrawal, R., Sorte, S., and Lokulwar, P. (2022). IoT based Smart Indoor air Pollution Monitoring System. In 2022 International Conference on Applied Artificial Intelligence and Computing (ICAAIC), (pp. 1829–1834). IEEE.
10. Sai, K. B. K., Subbareddy, S. R., and Luhach, A. K. (2019). IOT based indoor air quality monitoring system using MQ135 and MQ7 with machine learning analysis. *Scalable Computing: Practice and Experience*, 20(4), 599–606.
11. Mabahwi, N. A. B., Leh, O. L. H., and Omar, D. (2014). Human health and wellbeing: Human health effect of indoor air pollution. *Procedia-Social and Behavioral Sciences*, 153, 221–229.
12. Mad Saad, S., Andrew, A. M., Md Shakaff, A. Y., Mat Dzahir, M. A., Hussein, M., Mohamad, M., and Ahmad, Z. A. (2017). Pollutant recognition based on supervised machine learning for indoor air quality monitoring systems. *Applied Sciences*, 7(8), 823.
13. Majdi, A., Alrubaie, A. J., Al-Wardy, A. H., Baili, J., and Panchal, H. (2022). A novel method for Indoor Air Quality Control of Smart Homes using a Machine learning model. *Advances in Engineering Software*, 173 (2022): 103253.
14. Patino, E. D. L., and Siegel, J. A. (2018). Indoor environmental quality in social housing: A literature review. *Building and Environment*, 131, 231–241.

42 WeeCare—secure Online Home and Car Maintenance Application using Flutter

Debankan Mandal[1,a], Sanjoy Maity[1,b], and Monalisa Banerjee[2,c]

[1]Alumni, MCA Department, Techno Main Salt Lake, Kolkata, India

[2]Professor, Techno Main Salt Lake, Kolkata, India

Abstract

In this modern era, time is the most crucial asset for a business to make profits. This paper proposes an application designed using Flutter SDK and Dart Programming language which provides cross-platform development using a single code base which saves a lot of time compared to developing the application for each platform individually. To show this, we developed a home and car maintenance service application. A user has a huge pool of services to choose from, which can be scheduled at any preferred date and time. This application uses Google Maps services for providing the location information of the place where the services are needed. Also, this application uses the Firebase Authentication services for OTP based validation which provides secure login. All the user data is securely stored on the Cloud Firestore which is a document-based NoSQL platform offered by Google.

Keywords: Android studio, cloud firestore, dart, firebase, flutter, google maps flutter plugin, visual studio code.

Introduction

This project is to facilitate people to maintain daily essentials like home services and car maintenance with our skilled experts. The user can access these facilities easily and maintain their work with no extra time. On the home page, there are options Home Services and Car Maintenance where users choose their required options according to their work. In our Home Services, there are many options like cleaning, plumbing, electricity, furniture, and decor and in our Car Maintenance, there are also many services like car repair, car paint, etc. All these types of work are done by our highly skilled professionals with care. In this modern era, every person has a busy lifestyle and has their hands full with day-to-day tasks. To ease all these processes "WeeCare" app comes to the rescue. Through our app users can do these types of essential daily work in an easy manner and very convenient way.

The main objective of the project on Online Home and Car Maintenance Application (WeeCare) is to maintain daily essential work like home services and car maintenance by our skilled and experienced professionals. Users can book this type of service online

[a]debankan2014@gmail.com, [b]sanjaymaity055@gmail.com, [c]head.mcabca.ti@gmail.com

from anywhere, anytime according to their preference. Manually searching for professional workers is much more difficult for the people but through our app, anyone can find skilled and experienced professionals for their required services. This proposed research is to make an expedient user-friendly application which provides the best User Interface offering top-quality services. This research paper offers a very straightforward and simple UI. Users have access to as much information as the need. The primary goal of the research paper is to deliver excellent services through an appropriate user interface.

Literature Survey

Cross-platform development has evolved rapidly over the years, with various frameworks and technologies emerging to cater to developers' needs. Among these, Flutter has gained considerable attention for its unique features and capabilities. This literature survey aims to explore and analyze the reasons why considering Flutter for mobile app development is a good choice.

The cross-platform capabilities aside, the hot reload feature and better performance are key elements to consider. Also, the disk usage and battery consumption much more optimized in Flutter than apps developed using React [1]. Flutter is also highly rated among the cross-platform development frameworks analysis conducted by Kewal Shah, Harsh Sinha and Prof. Payal Mishra [2].

There are different applications in market which either cater to automobile maintenance services or home maintenance services. Many a times an application provide only one of these services for example "Rapid Service" [3]. We can also refer to business like GoMechanic [4] which has a B2B business model, where they partner with different already established local businesses. This may give rise to inconsistent service provide by different partners, whereas in our case all the service experts will be trained highly and similarly thus increasing consistency. There are also businesses which provide home maintenance service such as Urban Company [5]. But this application provides the best of the both worlds in the same landing zone which provides more ease to the user to avail these services.

Proposed System

To overcome the difficulties faced while maintaining a house or a car, we propose this app called "WeeCare".

As shown in Figure 42.1, this application mainly focuses on two types of users mentioned below:

1. Customer
2. Service Expert

Customer

The user who is in need of the services provided by the application. This user, after successful sign-in process which uses OTP verification for security and validation, gets to choose the service required from mainly two different modules i.e. Home or Car. After choosing a service from either module, the user needs to specify a few more details related to the respective selected service along with the date, time and address. Also, the user needs to pay an amount of Rs. 100 for successfully booking a service request.

Service Expert

The user who will attend to the service requests made by the Customers. This user also goes through a sign-in process which uses OTP verification for security and validation. The user attends to the currently active services requests, and after successfully completing the service, he/she bills the Customer according to the service provided.

Now coming to the security and safety part. As discussed above the application has an OTP verification process, which ensures that the user base consists of valid and verified users. The verification itself is a product of Google's Firebase. The Phone Authentication is very secure and uses a 6-digit code which is better than a 4-digit code. Also, as soon as a user successfully gets past the OTP verification for the first time, the user is automatically assigned a User-Id which is unique to each mobile number. In case of any problems the user can be easily identified and blocked from further access.

As for the user's data, it is also in the very safe hands of Firebase's Firestore Database. Firestore lets the admin make certain read and write rules which make accessing of data from the data much more secure. Also, Firestore Database is a real-time database which means faster data access and better user experience.

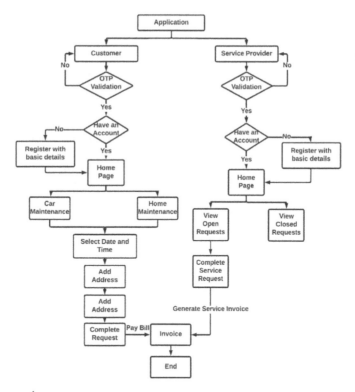

Figure 42.1 Proposed system.

Tools and Technologies

Android Studio Chipmunk 2021.2.1

Android Studio is the official IDE by Google and JetBrains for Android app development. It is purpose-built for Android to accelerate the development process and help build the best quality apps for every device based on Android Operating System. Android Studio provides the fastest possible turnabout on our coding and running workflow. The code editor enhances code quality, accelerates work, and boosts productivity through advanced completion, analysis, and refactoring. Android Studio suggests as you type in dropdowns list. Pressing the Tab key inserts the code. The "Apply Changes" feature lets us push code and resource changes to our currently running app without the need to restart our app also, in some cases, without even restarting the current activity. This flexibility provides us with how much control we have of our app when it is restarted to deploy and test small, incremental changes while preserving our device's current state.

Visual Studio Code (version 1.68)

Visual Studio Code is a concise code editor aiding tasks like debugging and version control. It prioritizes swift code cycles, deferring intricate workflows to richer IDEs like Visual Studio IDE. Compatible with macOS, Linux, and Windows, it merges Git control and terminal functions. Sporting a vast plugin system, it competes and exceeds Sublime's web language prowess. With essential features and plugins, it smoothly constructs complete Flutter apps.

Firebase

It can be described as a mixture of Google services in the cloud such as user authentication, hosting, analytics, storage, real-time databases, remote configuration, and performance monitoring. A Firebase project has an identification number (ID) automatically generated by Google. It is used in the integration process of certain services from Firebase or to call various APIs. In addition to that identification number automatically generated by Google, there is also a unique ID also defined by Google with the property of being modifiable by the user.

The user accessibility is covered by the Firebase Authentication service. It is available through a software development kit (SDK) that is characterized by an easy integration process in the mobile application. This service supports the following authentication methods: email and password, Facebook account, Google account, Twitter account, phone number, or anonymous authentication.

Cloud Firestore 3.1.17

Cloud Firestore stands as an adaptable, expandable database designed for mobile, web, and server development by Firebase and Google Cloud. Cloud Firestore ensures data synchronization across client apps using real-time listeners. Moreover, it provides offline support for responsive mobile and web apps, unaffected by network delays or availability of internet services. Moreover, Cloud Firestore seamlessly integrates with various Firebase and Google Cloud offerings, encompassing Cloud Functions.

Flutter SDK 3.0.2

Flutter can be defined as an ecosystem created by Google to build native compiled applications for mobile, web, and desktop using a single code base. Flutter is a cross-platform technology which offers developers the ability to build a mobile application, available on iOS and Android by writing a single code.

Flutter framework and the Dart programming language have been chosen to implement the mobile application for this paper. This choice has been made considering both costs and development complexity variables. From a financial point of view, Flutter is a good choice being an open-source platform. Also, the implementation process is relatively easy because it is enough to import certain packages and use them in the necessary context.

Dart 2.13

Dart, a language fine-tuned for efficient app creation across platforms, aims to be the pinnacle of productivity for multi-platform development. It is integrated with a flexible execution runtime designed for app frameworks. Geared towards client development, Dart excels in quick development using hot reload feature, alongside top-notch production quality for multiple compilation targets like web, mobile, and desktop. Dart serves as the bedrock for Flutter apps, while also addressing essential developer tasks such as code formatting, analysis, and testing.

Google Map

Google Maps stands as a web map service application developed by Google, catering to both desktop and mobile platforms. It delivers satellite imagery, 2D maps, and street view angles. Furthermore, it facilitates integration of maps into third-party websites through the Google Maps API.

Application Implementation

The following section shows the implementation of the WeeCare application from the customer's view point and not from the service expert's view point which is being run on an Android Emulator. Figure 42.2 is the splash screen of the application which greets the user when the application is opened. After the splash screen, the application moves to the next screen which is OTP Validation shown in Figure 42.3 which accepts a valid 10 digit Indian mobile number and awaits the input of a 6 digit OTP generated through Google Firebase's Phone Authentication feature as show in Figure 42.4.

Figure 42.2 Splash screen.

Figure 42.3 Mobile number input.

Figure 42.4 OTP input screen.

When the customer is successfully verified the application checks if the user is an old customer or a new customer. In case of a new user the customer is asked to fill out a simple form shown in Figure 42.5 after which the application takes the user to the home screen (in case of an old customer, the application directly moves to the home screen) shown in Figure 42.6.

Figure 42.5 New user form screen.   *Figure 42.6* Home screen.

After the customer makes the choice of service they require from the pool of services of Home Services (Figure 42.7) and Car Services (Figure 42.8), they need to select the date and time slot from the choices visible in Figure 42.9, they also need to provide the address where the service is required (Figure 42.10).

Figure 42.7 Home services.   *Figure 42.8* Car services.   *Figure 42.9* Date and time selection.   *Figure 42.10* Address page.

The address is fetched using the current location of the device's GPS (the user needs to allow location permission for this app). The user can change the address by interacting with a map present in the UI.

Once the address is provided the request is registered and the service expert will tend to it. After successfully resolving the request, the payment is made by the customer and received by the service expert.

Conclusion

This paper is only a humble venture to show the capabilities of Flutter cross-platform framework and Dart Programming Language with the help of an application names "WeeCare". "WeeCare" lets a user book a service request for any problems related to home maintenance or car maintenance. The user has a pool of services to choose from. After selecting the required service, user needs to choose a date and also a time slot most favorable to the individual along with the address where the service is to be provided.

References

1. Kishore, K., Khare, S., Uniyal, V., and Verma, S. (2022). Performance and stability comparison of react and flutter: cross-platform application development. In 2022 International Conference on Cyber Resilience (ICCR), Dubai, United Arab Emirates, 2022.
2. Shah, K., Sinha, H., and Mishra, P. (2019). Analysis of cross-platform mobile app development tools. In 2019 IEEE 5th International Conference for Convergence in Technology (I2CT), Bombay, India, March 2019.
3. Bhavani, K., Patel, M., Savaj, B., and Trada, A. (2021). Rapid service—mobile app for bike and car service. In 2021 5th International Conference on Electronics, Communication and Aerospace Technology (ICECA), Coimbatore, India, December 2021.
4. Karwa, K., Bhasin, A., Karwa, R., and Rana, N. GoMechanic (2023). https://gomechanic.in/
5. Bhal, A. S., Chandra, R., and Khaitan, V. UrbanClap Technologies India Pvt. Ltd (2023). https://www.urbancompany.com/
6. Bracha, G. (2015). The Dart Programming Language. Addison Wesley.
7. Bailey, T. (2021). Flutter for Beginners. Packt Publishing.
8. Flutter. (2023). Flutter documentation. Flutter. https://docs.flutter.dev/
9. Pub.dev. (2023). Dart package repository. pub.dev. https://pub.dev/

43 Image Segmentation And Medical Image Analysis Using Deep Learning And Convolution Neural Network: A Review

Pritusna Banik[1,a], Rupak Chakraborty[2,b], Arpita Dutta[2,c], Ratul Dey[3,d], and Md. Abdul Aziz Al Aman[2,e]

[1]Department of Computer Science and Engineering, Techno India University, Kolkata, India

[2]Department of Computer Science and Engineering-AI, Techno India University, Kolkata, India

[3]Department of Computer Science and Business Systems (CSBS), Techno India University, Kolkata, India

Abstract

Numerous crucial applications including medical image analysis (MIA), video surveillance, augmented reality, robotic perception, scene understanding, segmentation and compression of images are key tasks in image processing. There exists a number of segmentation algorithms are available in the literature. The novel image segmentation methods that manipulate DL models have been supported by Deep Learning (DL). When it is required to learn features of neural network automatically, deep learning is effectively employed as machine-learning tool. Deep convolution neural networks (CNNs), a deep learning method that is being employed for medical picture analysis. Various application areas such anomaly detection, computer assisted diagnosis, illness categorization, segmentation, and retrieval are included here. We look at the benefits, connections, and difficulties of CNN-based segmentation models, look into commonly available datasets, compare performance, highlight prospective future prospects. A comprehensive study of deep CNN-based image analysis has also been provided in this article. These strategies' potential and difficulties are also emphasized.

Keywords: Deep learning, image segmentation, medical images.

Introduction

A key element of many visual understanding systems is image segmentation. Images are divided into a variety of items or segments [1]. Image segmentation has a broad range of applications including MIA like identification of tumor, measurements of volumes tissues, automated cars like pedestrian and navigable surface recognition, video surveillance, and augmented reality. Each pixel in an image is given a name during the segmentation process so that pixels with the same label are connected by some visual or semantic property. In computer vision, image segmentation encompasses a sizable class of closely related issues. Many image segmentation algorithms, including histogram-based segmentation, K-means clustering, thresholding, conditional and Markov random fields, watershed methods, and more sophisticated algorithms like graph cuts, sparsity-based methods, and active contours approaches have been surveyed in the literature. Now a days, DL-based approaches have created a brandnew level of segmentation approaches with remarkable improved

[a]pritusna@gmail.com, [b]rupak.c@technoindiaeducation.com, [c]arpita.d@technoindiaeducation.com, [d]ratul.d@technoindiaeducation.com, [e]abdul.a@technoindiaeducation.com

performances, typically scoring the best accuracy rates on well-known datasets of benchmarks. Natural language processing (NLP), speech analysis, and computer vision areas all employ deep learning extensively. In the realm of MIA, using Deep Learning as a technique for pattern identification and machine learning is also becoming increasingly important. Clinical practices have used medical pictures as a diagnostic tool for classification, segmentation [2], and anomaly detection using images produced by a broad range of modalities of clinical imaging [3]. The cerebral cortex exhibits aberrant and excessive neuronal activity as a result of epileptic seizures, which may be detected by an electroencephalogram (EEG), a stereo-EEG, or an electrocorticography (ECoG). The number of studies utilizing long-term brain signal data to predict seizures by scalp EEG in their early phases has grown (Khan et al., 2023).

MIA Anwar et al. [4] seeks to help doctors and radiologists diagnose and treat patients more quickly. Effective MIA is necessary for Computer Aided Diagnosis (CAD) and Computer Aided Detection (CADx), making it crucial for performance. Precision, recall, accuracy, F-measure, sensitivity, and specificity are some crucial metrics that must perform well for MIA to produce high results. Deep learning techniques have the intrinsic capacity to learn complicated characteristics from the source of the data, which gives significant benefit. The most current developments in deep learning methodologies might be advantageous for future medical applications. There are several DL open source platforms available, including, but not limited totensorflow, caffe,keras, theano and torch. The difficulties result from the lack of clinical expertise among DL specialists and clinical expertise among DL experts. A new tutorial makes an attempt to close this gap by detailing how to apply DL to digital pathology photos step by step. By giving the code, a high-level introduction to the deep learning task of medical picture segmentation is offered. Choosing a suitable DL architecture based on the quantity of pictures and ground truth labels available continues to be difficult.

Literature Review

Some existing MIA methods that are based on deep CNN, have been surveyed in detail in this literature review section. The important performance indicators that were attained utilizing deep learning techniques and have therapeutic relevance are outlined and addressed. By performing several convolutional operations, the convolutional neural network may extract the COVID19 Wang et al. [17] feature information. The residual learning processes improve feature extraction while reducing gradient issues brought on by stacking convolutional layers. The capability also makes it possible for the suggested model to obtain useful feature information at a reasonable cost, allowing our model to maintain tiny weight parameters. Since deep learning algorithms learn from the provided data without supervision of human and hence support picture categorization, their application has significantly risen across all sectors of medical research [5, 6, 8]. Using the transfer learning approach, CNN variants have gained popularity over the past several decades and are being employed in a variety of classification tasks where a network trained on natural pictures is reused for medical images Kujur et al. [18].

The article "Medical image analysis" provides a succinct overview of the subject. Analysis of medical image using CNN gives an overview and sets an example that how deep CNN are preferred in this field. The most current developments in deep CNN techniques for analyzing of medical images are examined in "Discussion". The findings offered in "Conclusion" are then presented after this.

Medical Image Analysis (MIA)

The procedure that provides visual details of the human body is referred to as medical imaging. MIA Zhou et al. [7] seeks to help doctors and radiologists diagnose and treat patients more quickly. Processes that offer visual data about the human body are considered to be part of image analysis. Medical imaging is used for helping radiologists and clinicians to treat patients more quickly and effectively. MIA, a crucial matter of sickness treatment and diagnosis, uses a variety of imaging modalities. They also include X-rays, Magnetic Resonance Imaging (MRI), Positron Emission Tomography (PET), Computed Tomography (CT), and ultrasound of images. Figure 43.1 illustrates the essential features of medical imaging and associated technologies.

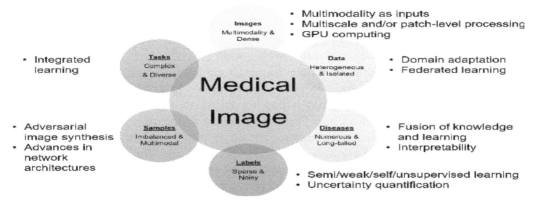

Figure 43.1 The traits in medical imaging along with its technical trends.

Figure 43.2 displays typical typology of image modalities for various sections that have been produced in the settings of radiology and laboratory. An essential tool of contemporary healthcare systems is medical imaging. Some of the applications like identification of tumor, detections of cancer, classification, guided therapy, annotation and retrieval of images are examples provided by deep learning in CADx.

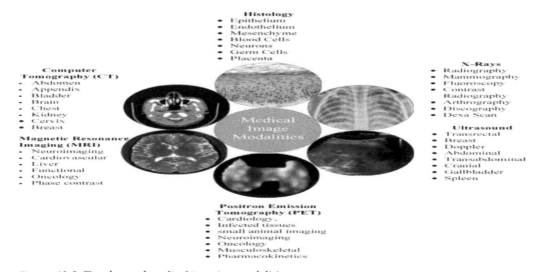

Figure 43.2 Typology of medical imaging modalities.

Content Based Image Retrieval (CbIR)

Figure 43.3 shows a schematic block of a generic system of CbIR. Based on the features, that are either extracted or generated from the pictures. CbIR retrieves images from large databases. Any CbIR system normally has two phases: the first is offline while the second is online. During the offline phase, a local features database is created by extracting features from sizable collections of photos which are used for training the system. Depending on how many training photos were used to teach the system, this phase might take a while. The same features are retrieved from the query picture in the online phase, and a similarity metric is produced by analyzing the features of query images and the features of the images in the database. The user is then shown the retrieval results that contain photos with a high degree of similarity or close proximity. Extraction of features and pre-processing follow the similar technique in both the phases.

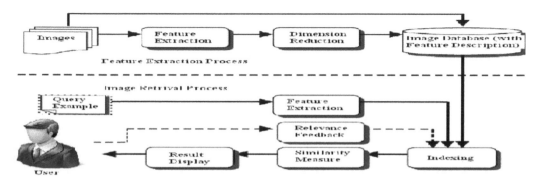

Figure 43.3 Block diagram of a generic CBIR system.

Detection and Classification Abnormality

Detecting an abnormality in a medical imaging entails locating a specific condition, such a tumor. Clinical professionals typically identify problems, but for that a lot of work and time are required. Therefore, the creation of automated systems for abnormality detection is becoming increasingly important. In the literature, many techniques for detecting abnormalities in medical photographs are discussed. Potential field segmentation is a method used in MRI segmentation, fusion-based brain tumor detection etc [9–13]. The effectiveness of this method has been evaluated using segmentation of brain tumor pictures, a publically accessible MRI standard. There are three categories of abnormality detection: contextual, collective, and point abnormalities. Finding a single anomalous point within a bigger dataset is known as point abnormality detection. The categorization of supervised abnormality detection and extremely imbalanced datasets are comparable. Semi-supervised modeling uses the model to both healthy and diseased data and seeks to train on only one class, usually the normal class. Unsupervised abnormality detection employs both normal and anomalous data, relies solely on the intrinsic qualities of the dataset, and does not use any labels at all [14]. The terms semi supervised and unsupervised abnormality detection are frequently applied to both semi supervised and unsupervised approaches in abnormality detection, leading to confusion [15, 16].

Computer Aided Diagnosis

In radiology, a Computer Aided Diagnosis (CAD) system is employed to help the radiologist and clinical practitioners analyze the diagnostic pictures. The system uses computer vision, machine learning-based techniques in the field of medical-image processing. A conventional CADx system acts as a second reader in clinical practice, helping to make judgments and providing more specific information about the aberrant location. Pre-processing, feature extraction, feature selection, and classification make up a typical CADx system. For diagnosing such as prostate cancer, fatty liver, dry eyes, breast cancer and Alzheimer's many approaches have been proposed.

Evaluation Metrics for MIA System

Different essential performance criteria, including precision, recall, accuracy, F1-score, sensitivity, specificity, and dice score, are used for evaluating a typical image processing system. These measurements are computed mathematically as

$$F1\text{-score} = 2 \times (\text{precision} \times \text{recall})/ (\text{precision} + \text{recall}), \tag{1}$$

where,

$$\text{precision} = t\_p/ (t\_p + f\_p), \tag{2}$$
$$\text{recall} = (t\_p)/ (t\_p + t\_n), \tag{3}$$
$$\text{accuracy} = (t\_p + t\_n)/(t\_p + t\_n + f\_p + f\_n), \tag{4}$$
$$\text{sensitivity} = t\_p/(t\_p + f\_n), \tag{5}$$
$$\text{specificity} = t\_n/ (t\_n + f\_p), \tag{6}$$
$$\text{dice Score} = 2 \times |p \cap gt |/ |p|+|gt |, \tag{7}$$

The no. of classes that were correctly identified as having a defect has been identified by True Positive (t_p), the no. of cases that were wrongly identified as having a defect is found by False Positive (f_p), the no. of objects which were correctly classified with no defect is indicated by True Negative (t_n), and the no. of cases which were wrongly classified as not having a defect is indicated by False Negative (f_n). p and gt in Equation 7 represent for the ground truth for the testing sample and the forecast produced by the system under assessment, respectively.

Convolutional Neural Networks (CNNs)

Multiple non-linear and linear processing approaches are placed in a deep CNN architecture as part of the ML technique known as deep learning to represent high-level abstraction contained in the data. Several deep learning methods are presently employed in a range of applications. These include CNNs, Deep Belief Networks (DBNs), Deep Restricted Boltzmann Machines (RBMs), auto-encoders, stack auto-encoders, and auto-encoders. In the MIA field as well as vision systems, CNN-based approaches have grown in favor recently.

Medical Image Analysis Using CNN

Figure 43.4 depicts a CNN architecture similar to LeNet-5 that can classify medical pictures into N different categories using a patch of 32 by 32 pixels taken from the original medical images. Convolutional (conv), max pooling, and fully connected layers form the network. The length of the feature maps that have been achieved by each conv layer and forwarded to the subsequent layers is reduced by the pooling layers. The necessary class prediction is generated by the fully-linked layers at the output nodes. The no. of layers, neurons (nodes) in each layer, and connections between nodes all affect how many parameters are needed to characterize a network. The network's training phase assures that the best weights are learned, resulting in great performance for the current issue. Deep CNN has been incorporated into medical image field to improve in deep learning techniques and computer power. Deep learning techniques are effectively employed for segmentation of medical images, computer assisted diagnosis, illness, medical image retrieval, classification and detection according to several recent research.

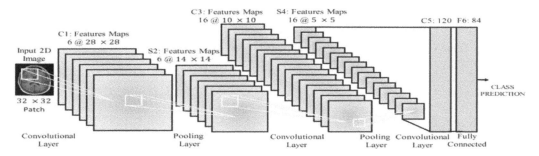

Figure 43.4 A typical CNN architecture for classification of medical images.

Figure 43.4 depicts a CNN architecture similar to LeNet-5 that can classify medical pictures into N different categories using a patch of 32 by 32 pixels taken from the original 2D medical image. Convolutional, maximum pooling, and fully linked layers make up the network. The size of the feature maps that are generated by each convolutional layer and passed to the subsequent layers is reduced by the pooling layers. The necessary class prediction is generated by the fully linked layers at the output. Neurons in each layer and the no. of layers, and connections between neurons all affect how many parameters are needed to characterize a network.

Empirical Results

We use several approaches, including InputCascadeCNN, Periera, Lisa, DeepMedic, SegNet, and 3DNet 3, as well as one widely used Dataset which is BRATS 2013. Applying these several approaches to the aforementioned dataset, BRATS 2013, yields values for Complete, Core, and Enhancing. Table 43.1 displays each of the pertinent data.

The applications of CNN for computer assisted diagnosis, medical image retrieval, and detection and classification tasks are highlighted in Table 43.2. It has been shown that CNN-based networks perform well in application domains involving several modalities for different kinds of jobs in medical image processing and provide encouraging outcomes at almost each situation. Then outcomes may change depending on the amount of classes,

Table 43.1: Results of various CNN-based approaches to obtain segmentation results.

Method	Dataset	Dice		
		Complete	Core	Enhancing
InputCascadeCNN Havaei et al. [19]	BRATS 2013	0.88	0.79	0.73
Periera, Pereira et al. [20]		0.84	0.72	0.62
Lisa, Jodoin et al. [22]		0.79	0.68	0.57
Deep medic, Kamnitsas et al. [23]	BRATS 2015	0.89	0.75	0.72
SegNet, Tseng et al. [21]		0.85	0.68	0.68
3DNet 3 (Hussain et al.)		0.91	0.83	0.76

Table 43.2: Some CNN based clinical applications.

Application	Method	Dataset	ImageNumber	Accuracy
Recognition of Body Part	2-stage CNN	CT slices of twelve body organs	6000 Synthetic 12 Classes	92.23%
Lung Pattern Classification	CNN	ILD CT scans	109 HR 7 classes CT scans	85.5%
Diagnosis of Thyroid Nodule	Pre-trained CNN	USG images	2 classes 15000 USG images	83%
Diagnosis of Breast Cancer	CNN with semi supervised learning	Mammographic images with ROIs	2 classes 3158 ROIs	82.43%
Diabetic Retinopathy	CNN	K Dataset	5 classes 80000 images	75%
Retrieval and classification of Medical Images	CNN	24 classes having Multimodal Data	24 classes 7200 multi-,modal images	99.77%
Retrieval of Radiographic Image	CNN	IRMA Database	31 classes 14410 images	97.79%

photos, and DCNN model employed. Based on CNN's achievements in the field of medicine, convolutional networks appear as crucial in the creation of future systems for medical image processing. Deep CNN have displayed exceptional performance in the medical field's photo analysis have been compared with other approaches utilized in related application fields.

Results of several methods for classifying lung patterns in ILD illness are summarized in Table 43.3. In terms of key performance measures, the CNN-based approach works much better than others. Table 43.4 compares the effectivity of a CNN-based technique with

Table 43.3: Method of comparison of classification of ILD.

Method	features	classifier	precision	recall	F1-Score
yan et al.	Bag of Words (BoW)+SIFT	Linear Regression (LR)	62.21	63.37	62.78
	BoW+SIFT	SVM	63.72	64.63	64.17
	Histogram of oriented gradients (HOG)	LR	67.74	68.71	68.22
	HOG	SVM	76.39	76.75	76.57
	CNN	CNN	92.25	92.21	92.23

Table 43.4: CNN-based methods comparison with other state of art approaches for recognition of body organs.

Methods	features	classifier	F_{avg}	Accuracy
Gangeh et. al.	Texon Intensity	RBF of SVM	0.7127	0.7152
Sorensen et. al.	Histogram+Local binary pattern	K-nearest neighbour	0.7322	0.7333
Anthimopoulous et. al.	Histogram+Local Discrete Cosine transgorm	Random forest	0.7786	0.7809
Anthimopoulous		CNN	0.8547	0.8561

other cutting-edge computer vision-based methods for identifying bodily organs. Evidently, the CNN-based approach improves key performance measures significantly.

Limitation of Deep Learning and Future Prospects

Voxel-wise classification, which is often used in picture segmentation, is one job for which deep learning approaches have significantly advanced in current years (3). By eliminating the time-consuming work of manually annotating cells, tissues, bones, and other structures, they are gradually assisting many scientists. Global thresholding, or separating the background from the bones using an intensity threshold, is a common technique for semi-automatically segmenting CT images. But this method is away from ideal and frequently manual modifications by a professional are required, which takes time and is prejudiced against the user and prone to mistake. The resulting CNN model can provide a significant potential to reduce the time- and effort-intensive process of segmentation of bones in CT scans, making manufacturing constructions more accessible and economical. Deep CNN, however, have significant drawbacks. The diversity in this data like resolution, contrast, and signal to noise is the key obstacles to the extensive use of CNN techniques in clinical practices.

Conclusion

A thorough overview of deep learning methods and how they are used to MIA is provided. In conclusion, segmentation and classification in all sub-parts of analysis of images are finding CNN-based deep learning approaches that will be widely accepted. Approaches like data augmentation and transfer learning are used to solve the issues with deep learning approaches caused by limited labels and scarce data. Larger datasets can perform better because to the avail more processing power and to improve DL architectures. Ultimately, this accomplishment would result in improved computer assisted detection and diagnosis techniques. More research is required before these techniques may be applied to imaging modalities where they are not already employed. Deep learning techniques would greatly improve medical picture analysis, given current developments.

References

1. Minaee, S., Boykov, Y., Porikli, F., Plaza, A., Kehtarnavaz, N., and Terzopoulos, D. (2022). Image segmentation using deep learning: a survey. *IEEE Transactions on Pattern Analysis and Machine Intelligence*, 44(7), 3523–3542. doi: 10.1109/TPAMI.2021.3059968.
2. Ghosh, S., Das, N., Das, I., and Maulik, U. (2019). Understanding deep learning techniques for image segmentation. *ACM Computing Surveys (CSUR)*, 52(4), 1–35.
3. Guo, Z., Li, X., Huang, H., Guo, N., and Li, Q. (2019). Deep learning-based image segmentation on multimodal medical imaging. *IEEE Transactions on Radiation and Plasma Medical Sciences*, 3(2), 162–169. doi: 10.1109/TRPMS.2018.2890359.
4. Anwar, S. M., Majid, M., Qayyum, A., Awais, M., Alnowami, M., and Khan, M. K. (2018). Medical image analysis using convolutional neural networks: a review. *Journal of medical systems*, 42, 1–13.
5. Tajbakhsh, N., Jeyaseelan, L., Li, Q., Chiang, J. N., Wu, Z., and Ding, X., (2020). Embracing imperfect datasets: a review of deep learning solutions for medical image segmentation. *Medical Image Analysis*, 63, 101693.
6. Singh, A., Sengupta, S., and Lakshminarayanan, V. (2020). Explainable deep learning models in medical image analysis. *Journal of Imaging*, 6(6), 52.
7. Zhou, S. K., Greenspan, H., and Shen, D. eds. (2017). Deep Learning for Medical Image Analysis. Academic Press.
8. Zhou, S. K., Greenspan, H., Davatzikos, C., Duncan, J. S., Van Ginneken, B., Madabhushi, A., and Summers, R. M. (2021). A review of deep learning in medical imaging: Imaging traits, technology trends, case studies with progress highlights, and future promises. *Proceedings of the IEEE*, 109(5), 820–838.
9. Hesamian, M. H., Jia, W., He, X., and Kennedy, P. (2019). Deep learning techniques for medical image segmentation: achievements and challenges. *Journal of digital imaging*, 32, 582–596.
10. Chen, L., Bentley, P., Mori, K., Misawa, K., Fujiwara, M., and Rueckert, D. (2018). DRINet for medical image segmentation. *IEEE Transactions on Medical Imaging*, 37(11), 2453–2462. doi: 10.1109/TMI.2018.2835303.
11. Qayyum, A., Anwar, S. M., Awais, M., and Majid, M. (2017). Medical image retrieval using deep convolutional neural network. *Neurocomputing*, 266, 8–20.
12. Chauhan, R., Ghanshala, K. K., and Joshi, R. C. (2018). Convolutional neural network (CNN) for image detection and recognition. In *2018 First International Conference on Secure Cyber Computing and Communication (ICSCCC)*, Jalandhar, India, 2018, (pp. 278–282). doi: 10.1109/ICSCCC.2018.8703316.
13. Albawi, S., Mohammed, T. A., and Al-Zawi, S. (2017). Understanding of a convolutional neural network. In 2017 International Conference on Engineering and Technology (ICET), Antalya, Turkey, 2017, (pp. 1–6). doi: 10.1109/ICEngTechnol.2017.8308186.
14. Hatamizadeh, A., Tang, Y., Nath, V., Yang, D., Myronenko, A., Landman, B., Roth, H. R., and Xu, D. (2022). Proceedings of the IEEE/CVF Winter Conference on Applications of Computer Vision (WACV), 2022, (pp. 574–584).

15. Chen, X., Wang, X., Zhang, K., Fung, K. M., Thai, T. C., Moore, K., Mannel, R. S., Liu, H., Zheng, B., and Qiu, Y. (2022). Recent advances and clinical applications of deep learning in medical image analysis. *Medical Image Analysis*, vol 79, 102444.

16. Perone, C. S., and Cohen-Adad, J. (2019). Promises and limitations of deep learning for medical image segmentation. *Journal of Medical Artificial Intelligence,* 2(1), 1–2.

17. Wang, B., Zhang, Y., Ji, S., Zhang, B., Wang, X., and Zhang, J. (2023). A COVID-19 detection model based on convolutional neural network and residual learning. *Computers, Materials and Continua*, 75(2), 3625–3642.

18. Kujur, A., Raza, Z., Khan, A. A., and Wechtaisong, C. (2022). Data complexity based evaluation of the model dependence of brain MRI images for classification of brain tumor and alzheimer's disease. *IEEE Access*, 10, 112117–112133.

19. Havaei, M., Davy, A., Warde-Farley, D., Biard, A., Courville, A., Bengio, Y., Pal, C., Jodoin, P. M., and Larochelle, H. (2017). Brain tumor segmentation with deep neural networks. *Medical Image Analysis*, 35, 18–31.

20. Pereira, S., Pinto, A., Alves, V., and Silva, C. A. (2016). Brain tumor segmentation using convolutional neural networks in mri images. *IEEE Transactions on Medical Imaging*, 35(5), 1240–1251.

21. Tseng, K. L., Lin, Y. L., Hsu, W., and Huang, C. Y. (2017). Joint sequence learning and cross-modality convolution for 3D biomedical segmentation. In Proceedings of the IEEE Conference on Computer Vision and Pattern Recognition, (pp. 6393–6400).

22. Jodoin, A. C., Larochelle, H., Pal, C., and Bengio, Y. (2017). Brain tumor segmentation with deep neural networks. *Medical image Analysis*, 35, (18–31).

23. Kamnitsas, K., Ledig, C., Newcombe, V. F., Simpson, J. P., Kane, A. D., Menon, D. K., Rueckert, D., and Glocker, B. (2017). Efficient multi-scale 3d cnn with fully connected crf for accurate brain lesion segmentation. *Medical Image Analysis,* 36, 61–78.

24. Khan, G. H., Khan, N. A., Altaf, M. A. B., and Abbasi, Q. (2023). A Shallow Autoencoder Framework for Epileptic Seizure Detection in EEG Signals. Sensors, 23(8), 4112.

25. Hussain, M. S., and Palit, R. (2020, July). Performance Analysis of Different 2D and 3D CNN Model for Liver Semantic Segmentation: A Review. In Medical Imaging and Computer-Aided Diagnosis: Proceeding of 2020 International Conference on Medical Imaging and Computer-Aided Diagnosis (MICAD 2020) (Vol. 633, p. 166). Springer Nature.

26. Yan, T., Du, Y., Sun, W., Chen, X., Wu, Q., Ye, Q., and Xue, J. (2023). Interstitial lung disease in adult patients with anti-NXP2 antibody positivity: a multicentre 18-month follow-up study. *Clin Exp Rheumatol*, 41, 247–253.

27. Gangeh, M. J., Sørensen, L., Shaker, S. B., Kamel, M. S., De Bruijne, M., and Loog, M. (2010). A texton-based approach for the classification of lung parenchyma in CT images. In Medical Image Computing and Computer-Assisted Intervention MICCAI 2010: 13th International Conference, Beijing, China, September 20–24, 2010, Proceedings, Part III 13 (pp. 595–602). Springer Berlin Heidelberg.

28. Sorensen, L., Nielsen, M., Lo, P., Ashraf, H., Pedersen, J. H., and De Bruijne, M. (2011). Texture-based analysis of COPD: a data-driven approach. *IEEE transactions on medical imaging*, 31(1), 70–78.

29. Anthimopoulos, M., Christodoulidis, S., Ebner, L., Christe, A., and Mougiakakou, S. (2016). Lung pattern classification for interstitial lung diseases using a deep convolutional neural network. *IEEE transactions on medical imaging*, 35(5), 1207–1216.

30. Anthimopoulos, M., Christodoulidis, S., Ebner, L., Geiser, T., Christe, A., and Mougiakakou, S. (2018). Semantic segmentation of pathological lung tissue with dilated fully convolutional networks. *IEEE journal of biomedical and health informatics*, 23(2), 714–722.

44 Parkinson's Disease Prediction from Voice Data using DNN and Bagging: An Integrated Approach

Aishwarya Barik[a], Kasturi Nandi, Rehana Mullick, Chandra Das[b], and Shilpi Bose

Dept. of CSE, NSEC, Kolkata, West Bengal, India

Abstract

One of the most prevalent neurodegenerative disorders is Parkinson's Disease (PD), which results in uncontrollable body movements. The patient's daily life may be severely impacted by this disorder's significant side effects. Approximately 10 million people worldwide are estimated to have experienced or are experiencing PD. Therefore, PD detection has been the subject of numerous prior investigations. Given that Parkinson's disease is unpredictable and incurable, automatic Parkinson's disease identification in voice recordings may be a breakthrough when compared to current expensive techniques of excluding tests. Already, a very limited number of works has been done in this field. In this regard, this study presented machine learning based a new feature selection-classification model named the Ensemble-Filters-DNN-BAGGING model for separating Parkinson's disease related individuals from healthy individuals based on voice-based data. For the feature selection part, we have ensemble results of different filter measures named Pearson correlation coefficient, mutual information, Chi-square, Annova, and Relief to select relevant features while in the classification part, a custom Deep neural network-based bagging model is introduced. Performance metrics, including accuracy, sensitivity, specificity, and precision, are employed to assess the models' effectiveness. The findings reveal that our proposed model exhibits promising results in terms of test accuracy and sensitivity.

Keywords: Classification, feature selection, machine learning, neurodegenerative disease.

Introduction

One of the most common neurodegenerative disorders, Parkinson's disease (PD), affects 1–2 people per 1,000 adults over 60 and has a prevalence rate of 1% [1]. The estimated number of people affected by PD worldwide grew by more than twice between 1990 and 2016 [2] due to an increase in the elderly population and age-standardized prevalence rates. The motor and non-motor components of movement, including planning, initiation, and execution, are all impacted by Parkinson's disease (PD), which is a degenerative neurological disorder, according to [3, 4]. Parkinson disease is often accompanied by the traditional symptoms and signs of Parkinson's disease, such as hypokinesia (lack of

[a]aishwarya25252@gmail.com, [b]chandra.das@nsec.ac.in

movement), bradykinesia (slowness of movement), rigidity (wrist, shoulder, and neck), and rest tremor (imbalance of neurotransmitters dopamine and acetylcholine). In addition to medications, unusual conditions like repeated brain infarctions and degenerative illnesses like progressive supranuclear palsy (PSP) and multiple system atrophy (MSA) can also cause Parkinson's disease [5]. Numerous research [6–8] have revealed that the quality of life (QoL), social interactions, and family relationships of PD patients are significantly impacted negatively. PD also creates a huge economic burden for the patient's family.

Historically, the diagnosis of Parkinson's disease has been made using motor symptoms. Despite the creation of cardinal symptoms of PD in clinical evaluations, the bulk of the rating scales used to assess the severity of the disease have not yet undergone extensive examination and verification [3]. Even though non-motor symptoms, such as cognitive changes, sleep problems, and sensory abnormalities like olfactory dysfunction, are common in many people before the onset of Parkinson's disease (PD) [3, 9], these symptoms are frequently ill-defined, difficult to diagnose, and/or vary from patient to patient. There is yet no reliable clinical test that reliably distinguishes Parkinson's disease from other disorders with comparable clinical symptoms. The clinical diagnosis is made mostly on the basis of the physical examination and history. Due to this reason machine learning and deep learning techniques [10–12], have emerged as a potential tool for anticipating Parkinson's disorders due to their ability to discover patterns via scanning different types of large amounts of complex data.

Machine learning algorithms give systems the capacity to autonomously learn from experience and make inferences. It is typically considered a sub-field of artificial intelligence. For the diagnosis of this disease, machine learning models have been applied to a variety of data modalities, such as handwriting patterns [13], movement [14], neuroimaging [15], speech [16], cerebrospinal fluid [17], cardiac scintigraphy [18], and optical coherence tomography (OCT) [19]. Apart from this machine learning techniques are applied via combining different data modalities [20] for prediction of this disease. We may find relevant features that are not frequently utilised in the clinical diagnosis of PD by utilising machine learning algorithms, allowing us to rely on these alternative measurements to detect PD in preclinical stages or a typical form.

Neurophysiologists can detect non-motor symptoms and provide early patient forecasts by thoroughly analyzing speech signals. So, in recent years, there has been growing interest in using voice data as a potential biomarker for PD prediction. Voice analysis offers a non-invasive and cost-effective approach for detecting subtle changes in vocal characteristics associated with PD. Several studies [21–26] have focused on developing machine learning models to predict PD using voice data. Although, some works have been done for predicting early-stage Parkinson's disease from voice data, still researchers have been continuing to explore and develop methods for improvement of early-stage Parkinson's prediction from voice data.

In this regard, here, we have proposed a Deep neural network (DNN) based bagging classification model to predict PD from voice data. Experimental findings demonstrate the superiority of the proposed approach.

Proposed Method

This paper presents a new feature selection-classification model named (Ensemble-Filters-DNN-BAGGING) for predicting Parkinson's Disease (PD) using voice data. For the feature selection part, we have used an ensemble of filters named the Ensemble-filters method, while for the classification part, we have proposed a custom Deep Neural Network (DNN) based ensemble bagging classification model named DNN-BAGGING. In this section, the Bagging Based Ensemble Classification model has been briefly introduced in the next subsection, and it is followed by a brief description of the DNN classifier. After that, the proposed Ensemble-Filters-DNN-BAGGING model is finally described.

Traditional Bagging Model

It is derived from the idea of bootstrapping and is referred to as bootstrap aggregating [27, 28]. It is among the first and most straightforward ensemble-based algorithms. Leo Breiman developed the idea of Bootstrap Aggregating and came up with the abbreviation "Bagging" in the late Nineties. Bagging was developed to improve classification by combining predictions of randomly generated training sets. In comparison to its base components, it primarily minimizes variance and creates a robust ensemble model. Here, a number of homogenous weak classifiers that have high variance are trained on various bootstraps that are produced by resampling the training dataset with replacement. Following training, the predictions of these weak classifiers are then pooled through an 'averaging' procedure. Here the weak learners are trained in parallel and independently. To further understand the working process of bagging, we have to first understand the concept of Bootstrapping.

Bootstrapping (random sampling with replacement) is a technique for creating a large number of random subsets from a training dataset. Many bootstrapped datasets are constructed by selecting random samples from the training dataset with replacement, a process known as bootstrapping in machine learning. In this case, sampling with replacement allows for the selection of samples more than once for each subset. By selecting various samples at random from the dataset with replacement, n sub-datasets are derived from the training dataset (where n is the size of the original dataset). After that, each homogenous weak learner has been trained using a bootstrapped dataset, and the predictions are finally merged to produce the output. There are numerous approaches to aggregation. If the issue is regression, the final output of the final ensemble learner is created by averaging the outputs of the separate base learners. When a classification issue arises, the ensemble learner's ultimate judgment will be based on the class with the highest number of votes.

Deep Neural Networks

Deep neural networks (DNNs) are artificial neural networks with many layers. These networks may be trained with a lot of data to automatically learn how to represent facts at different levels of abstraction. DNNs employ a backpropagation method to find intricate patterns in massive amounts of data. Using the error (the discrepancy between the system's output and the known expected output), the backpropagation technique adjusts the network's internal weights [11].

Proposed Model

The proposed Ensemble-Filters-DNN-BAGGING model is described here. The proposed study involves several crucial steps to ensure accurate and reliable predictions. Firstly, the dataset is carefully preprocessed, including data cleaning and normalization, to eliminate noise and inconsistencies that may impact the analysis. Additionally, feature variety and correlation analysis are conducted to gain insights into the distribution and relationships among the different features and their association with PD labels.

After preprocessing the training dataset is passed through the Ensemble-Filters-DNN-BAGGING model.

Ensemble-filters module: In the Ensemble-Filters module, different filter methods are applied independently on the voice dataset and p number of most relevant features is selected by every filter method. The union of these filters is taken and this features based reduce dataset is passed through a custom DNN-based bagging module. The filter methods which are used here are Pearson correlation coefficient, mutual information, Chi-square, Annova, and Relief [28].

Custom DNN-based bagging module: For the classification stage, we have proposed a custom Deep Neural Network (DNN) based ensemble bagging classification model. Here *q* numbers of bootstrapped datasets are created from the original training dataset and a custom DNN bagging classifier is trained using every bootstrapped dataset. In this way, *q* number custom DNN classifiers are trained. Finally, every test sample is passed through all *q* classifiers and the test sample's class is predicted using the majority voting technique of classification accuracy of all *q* classifiers. Hyper-parameter tuning is performed to find the optimal configuration for the models, exploring various combinations of layers, activation functions, and node numbers. The block diagram of the proposed Ensemble-Filters-DNN-BAGGING model is given in Figure 44.1.

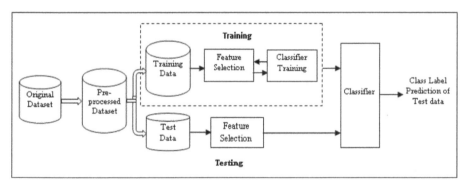

Figure 44.1 Block diagram of proposed ensemble-filters-DNN-BAGGING model.

Results

The proposed Ensemble-Filters-DNN-BAGGING model is compared with Baseline classifiers, including Support Vector Machine (SVM), Naive Bayes (NB), and Decision Tree (DT), and these models are also implemented to compare their performance with the proposed model. The hyper-parameters of these classifiers are tuned as well to ensure fair comparisons. Both train-test split and 10-fold cross-validation techniques are used here for evaluation.

Hyper-Parameters of Our Model

In our custom DNN model, we have used 5 layers. In the first layer there are 128 neuronal units where the RELU activation function is used, in 2nd layer 256 neuronal units are used with the activation function RELU, in the 3rd layer 128 neuronal units are used where also activation function is also RELU. In the 4th hidden layer, 64 neuronal units are used with the activation function RELU. The last layer is the output layer with 1 neuronal unit where a sigmoid activation function is used. In our Bagging model, we have created 50 bootstrapped samples.

Dataset Description

The dataset available on the UCI Machine Learning Repository consists of voice recordings from 31 individuals, with 23 individuals diagnosed with Parkinson's disease (PD) and 8 healthy control individuals [21, 22].

Each voice recording in the dataset is associated with a set of acoustic features extracted from sustained phonation tasks, where individuals were asked to sustain the vowel sound "/a/" as long as possible. The dataset includes a total of 22 acoustic features, which provide information about various aspects of vocal characteristics and speech production.

Comparative Performance Analysis

The objective of the comparison study done for this work was to evaluate the effectiveness of several machine learning models for Parkinson's disease identification utilising voice-based analysis. The evaluated models included Decision Tree, Naive Bayes, SVM Linear, SVM RBF, and our proposed Ensemble-Filters-DNN-BAGGING model. The proposed Ensemble-Filters-DNN-BAGGING model demonstrated favorable results in terms of accuracy, sensitivity, specificity, and precision as shown in Table 44.1. The findings highlight the potential of voice-based analysis in the early-stage diagnosis and continuous monitoring of this disease, offering a promising avenue for future research and development of diagnostic tools.

Table 44.1: Comparative performance analysis of our model with respect to other machine learning models in terms of 10 fold cross validation based on PD-related voice dataset.

Model	Accuracy	Sensitivity	Specificity	Precision
Proposed Ensemble-Filters-DNN-BAGGING model (10-fold)	0.923	0.959	0.822	0.947
Proposed Ensemble-Filters-DNN-BAGGING model (Training-Testing splitting)	0.949	1.000	0.714	0.941
Decision Tree	0.877	0.911	0.818	0.944
Naive Bayes	0.697	0.623	0.868	0.968
SVM Linear	0.866	0.993	0.454	0.854
SVM RBF	0.871	1.000	0.454	0.855
Decision Tree(Bagging)	0.867	0.901	0.768	0.924
Naive Bayes(Bagging)	0.707	0.649	0.834	0.942
SVM Linear(Bagging)	0.887	0.993	0.520	0.874
SVM RBF(Bagging)	0.871	1.000	0.454	0.855

Conclusion

In this work, a new feature selection-classification model named Ensemble-Filters-DNN-BAGGING is proposed for voice based prediction of Parkinson's disease using voice features.

The use of voice features is a significant aspect of this work. Voice analysis offers non-invasive and cost-effective means of detecting the disease, making it suitable for remote monitoring and screening. Voice-based detection can be particularly advantageous for early-stage diagnosis and continuous monitoring of patients. However, while voice-based detection of Parkinson's disease has merits, it also has limitations. Voice features alone may not capture the full complexity of the disease, and additional clinical data may be required for a comprehensive assessment. The models evaluated in this work relied solely on voice features and may not have utilized other relevant patient information. The caliber and diversity of the training and evaluation datasets may also have an impact on how well the machine learning models perform. The size, representativeness, and potential biases of the dataset can have an effect on how generalizable and applicable the models are in the actual world.

This proposed model is compared with other machine learning models named Decision Tree, Naive Bayes, SVM Linear, SVM RBF, and their corresponding bagging versions. The custom DNN model with bagging emerged as the top performer, achieving the highest test accuracy (94.87%) and sensitivity (100%) among the models. It demonstrated promising results in accurately classifying the data and detecting positive cases of Parkinson's disease. The model's higher specificity (71.43%) and precision (94.12%) compared to other models suggest its potential for reducing false positives and improving overall accuracy.

In conclusion, the custom DNN model with bagging demonstrated promising results in detecting Parkinson's disease using voice features. However, in future we will apply this model on other voice data with huge number of voice related features to enhance the models' performance, and also address the limitations of voice-based detection for Parkinson's disease.

References

1. Tysnes, O. B., and Storstein, A. (2017). Epidemiology of Parkinson's disease. *Journal of Neural Transmission*, 124, 901–905. doi: 10.1007/s00702-017-1686-y.
2. Dorsey, E. R., Elbaz, A., Nichols, E., Abd-Allah, F., Abdelalim, A., Adsuar, J. C. et al. (2018). Global, regional, and national burden of Parkinson's disease, 1990–2016: a systematic analysis for the global burden of disease study 2016. *Lancet Neurology*, 17, 939–953. doi: 10.1016/S1474-4422(18)30295-3.
3. Jankovic, J. (2008). Parkinson's disease: clinical features and diagnosis. *Journal of Neurology, Neurosurgery and Psychiatry*, 79, 368–376. doi: 10.1136/jnnp.2007.131045.
4. Contreras-Vidal, J., and Stelmach, G. E. (1995). Effects of parkinsonism on motor control. *Life Sciences*, 58, 165–176. doi: 10.1016/0024-3205(95)02237-6.
5. Hawkes, C. H., and Deeb, J. (2006). Predicting Parkinson's disease: worthwhile but are we there yet?. *Practical Neurology*, 6, 272–277.
6. Johnson, S. J., Diener, M. D., Kaltenboeck, A., Birnbaum, H. G., and Siderowf, A. D. (2013). An economic model of Parkinson's disease: implications for slowing progression in the United States. *Movement Disorders*, 28, 319–326. doi: 10.1002/mds.25328.
7. Kowal, S. L., Dall, T. M., Chakrabarti, R., Storm, M. V., and Jain, A. (2013). The current and projected economic burden of Parkinson's disease in the United States. *Movement Disorders*, 28, 311–318. doi: 10.1002/mds.25292.

8. Yang, J. X., and Chen, L. (2017). Economic burden analysis of Parkinson's disease patients in China. *Parkinson's Disease,* 2017, 8762939. doi: 10.1155/2017/8762939.

9. Tremblay, C., Martel, P. D., and Frasnelli, J. (2017). Trigeminal system in Parkinson's disease: a potential avenue to detect Parkinsonspecific olfactory dysfunction. *Parkinsonism and Related Disorders,* 44, 85–90. doi: 10.1016/j.parkreldis.2017.09.010.

10. AyonDey (2016). Machine learning algorithms: a review. *International Journal of Computer Science and Information Technologies,* 7(3), 1174–1179.

11. Sarker, I. H. (2021). Deep learning: a comprehensive overview on techniques, taxonomy, applications and research directions. *SN Computer Science,* 2, 420. https://doi.org/10.1007/s42979-021-00815-1.

12. Mei, J., Christian, D., and Johannes, F., (2021). Machine learning for the diagnosis of Parkinson's disease: a review of Literature, *Frontiers in Aging Neuroscience,* 13, 1–41.

13. Pereira, C. R., Pereira, D. R., Rosa, G. H., Albuquerque, V. H. C., Weber, S. A. T., Hook, C. et al. (2018). Handwritten dynamics assessment through convolutional neural networks: An application to Parkinson's disease identification. *Artificial Intelligence in Medicine,* 87, 67–77. doi: 10.1016/j.artmed.2018.04.001.

14. Wahid, F., Begg, R. K., Hass, C. J., Halgamuge, S., and Ackland, D. C. (2015). Classification of Parkinson's disease gait using spatial-temporal gait feature. *IEEE Journal of Biomedical and Health Informatics,* 19, 1794–1802. doi: 10.1109/JBHI.2015.2450232.

15. Cherubini, A., Nisticó, R., Novellino, F., Salsone, M., Nigro, S., Donzuso, G. et al. (2014b). Magnetic resonance support vector machine discriminates essential tremor with rest tremor from tremor-dominant Parkinson disease. *Movement Disorders,* 29, 1216–1219. doi: 10.1002/mds.25869.

16. Sakar, B. E., Isenkul, M. E., Sakar, C. O., Sertbas, A., Gurgen, F., Delil, S., et al. (2013). Collection and analysis of a Parkinson speech dataset with multiple types of sound recordings. *IEEE Journal of Biomedical and Health Informatics,* 17, 828–834. doi: 10.1109/JBHI.2013.2245674.

17. Maass, F., Michalke, B., Leha, A., Boerger, M., Zerr, I., Koch, J. C. et al. (2018). Elemental fingerprint as a cerebrospinal fluid biomarker for the diagnosis of Parkinson's disease. *Journal of Neurochemistry,* 145, 342–351. doi: 10.1111/jnc.14316.

18. Nuvoli, S., Spanu, A., Fravolini, M. L., Bianconi, F., Cascianelli, S., Madeddu, G. et al. (2019). [(123)I]Metaiodobenzylguanidine (MIBG) cardiac scintigraphy and automated classification techniques in Parkinsonian disorders. *Molecular Imaging and Biology,* 22, 703–710. doi: 10.1007/s11307-019-01406-6.

19. Nunes, A., Silva, G., Duque, C., Januário, C., Santana, I., Ambrósio, A. F. et al. (2019). Retinal texture biomarkers may help to discriminate between Alzheimer's, Parkinson's, and healthy controls. *PLoS One,* 14, e0218826. doi: 10.1371/journal.pone.0218826.

20. Wang, Z., Zhu, X., Adeli, E., Zhu, Y., Nie, F., Munsell, B. et al. (2017). Multi-modal classification of neurodegenerative disease by progressive graph-based transductive learning. *Medical Image Analysis,* 39, 218–230. doi: 10.1016/j.media.2017.05.003.

21. Tsanas, A., Little, M. A., McSharry, P. E., and Ramig, L. O. (2012). Nonlinear speech analysis algorithms mapped to a standard metric achieve clinically useful quantification of average Parkinson's disease symptom severity. *Journal of the Royal Society Interface,* 9(66), 2756–2770.

22. Little, M. A., McSharry, P. E., Hunter, E. J., Spielman, J., and Ramig, L. O. (2009). Suitability of dysphonia measurements for telemonitoring of Parkinson's disease. *IEEE Transactions on Biomedical Engineering,* 56(4), 1015–1022.

23. Arora, S., Baghai-Ravary, L., Tsanas, A., and Mackay, C. E. (2018). An ensemble method for predicting Parkinson's disease using deep convolutional neural networks. In Proceedings of the 15th International Conference on Machine Learning and Data Mining in Pattern Recognition, (pp. 237–250).

24. Dubey, H., Pandey, S. K., and Srivastava, R. (2020). Bagging ensemble learning-based approach for Parkinson's disease prediction using voice features. *Expert Systems with Applications,* 139, 112856.

25. Antonio, S., Giovanni, C., Francesco, A., Pietro, D. L., Mohammad, A. W., Giulia, L., Simona, S., Antonio, P., and Giovanni, S. (2022). Frontiers|voice in Parkinson's disease: a machine learning study. *Frontiers in Neurology*, 13, 831428.

26. Dao, S. V. T., Yu, Z., Tran, L. V., Phan, P. N. K., Huynh, T. T. M., and Le, T. M. (2022). An analysis of vocal features for Parkinson's disease classification using evolutionary algorithms. *Diagnostics (Basel)*, 12(8), 1980. doi: 10.3390/diagnostics12081980. PMID: 36010330; PMCID: PMC9406914.

27. Błaszczyński, J., and Stefanowski, J. (2017). Actively balanced bagging for imbalanced data. In Proceedings of International Symposium on Methodologies for Intelligent Systems, Cham, 2017, (pp. 271–281).

28. Bose, S., Das, C., Banerjee, A., Ghosh, K., Chattopadhyay, M., Chattopadhyay, S., and Barik, A. (2021). Multi-filtering and supervised gene clustering algorithm based ensemble classification model for microarray gene expression data. *PeerJ Comput. Sci.*, 1–40, DOI 10.7717/peerj-cs.671

45 Glaucoma Detection using Deep Learning Approach

Shubhajoy das[a] and Debashish das[b]

Techno India University, Kolkata, West Bengal, India

Abstract

Glaucoma is a disease that affects the optic nerve and affects people above 60 years of age. This disease causes blindness. The main causes of glaucoma are diabetes mellitus, family history, etc. Cases of glaucoma are very high in India compared to other countries. This disease may cause complete blindness in its severity.The pre-trained CNN model Efficient Net has been applied to the dataset to classify diseases into Glaucoma/Non-Glaucoma. An ophthalmologist uses metrics such as cup/disk ratio to predict glaucoma as a result if the images are segmented using deep neural network algorithms and it may be beneficial for detecting glaucoma. Several methodologies for performing segmentation on Glaucoma images have been surveyed and compared. Glaucoma is associated with interocular pressure which affects the vision drastically leading to permanent blindness.A precision close to 64% and 98% have been obtained on the RIMON and Drishti datasets. A convolutional neural network and an Autoencoder were used to predict the disease and we achieved an accuracy of above 95% on the metrics Precision, Recall, and F1-score, and in the RIMON dataset, a Precision, Recall of 64 % was achieved. Segmentation of the optic disc was also carried out using the Unet segmentation model and it achieved a Dice coefficient of above 86% on the RIMON dataset.

Keywords: Classification, deep neural networks, glaucoma detection, image processing, unet.

Introduction

Glaucoma is a disease which if untreated can lead to complete blindness. There are various methods for the detection of glaucoma such as increase in pressure of the optic nerve or the fluids associated with it. This disease causes blindness in people, More then 90% of glaucoma remains undiagnosed. There are special types of cameras which are used to detect glaucoma, because they require special vision sensors. These sensors emit blue light because the optic nerve reacts to the blue light differently, prolonged exposure to blue light is harmful for glaucoma many countries such as USA glaucoma is a dreaded disease like cancer because it affects a lot of people in that country. The optic disc comprises of rods and cons cells which results in distinguishing color from images. Glaucoma can be

---

[a]shubhajoydas@rocketmail.com, [b]debashis.d@technoindiaeducation.com

regarded as death of these rods and con cells. The detection of glaucoma requires complex imaging hardware [22]. Texture based features [Harlick features] have been used with support vector machines to detect glaucoma and an accuracy of above 85% was achieved [19]. Recently Optical Coherence tomography has been used for detecting glaucoma which are state of the art in classification tasks [23]. Optical Coherence tomography is a procedure which uses light waves to generate a cross-sectional view of the retina of the eye. The images produced are of high quality compared to other techniques.

Related Work

In Patil et al. [1] CNN and deep belief network has been used for classification of glaucoma images. The proposed model is compared with random forest, support vector machine and k-nearest neighbor.In Liu et al. [2] CNN has been used to classify the glaucoma images which uses a 18 layer convolution neural network to classify the images. An accuracy of 98% was achieved. In X. Chen et al [2015] a deep CNN and an attention guided neural network is used, To both of these networks an image is imputed and the output of the two neural networks are concatenated. The method improves upon the state of the art by about 6%. In Y. George [2020] a deep convolutional layer network is proposed which has 4 convolutional layers and two fully connected layers. In Bowd et al. [5] a 3d optical coherence tomography enabled deep learning model. The glaucoma detection model achieves an accuracy or AUC of 93.8% compared to a baseline model of 86%. In X. Chen et al [2015] Autoencoders have been used to detect the onset of glaucoma and optical coherence tomography is also used. In Nirmala et al. [7] A convolutional neural network has been designed which has 4 convolutional layers and two fully connected layers. The analysis was carried out on ORIGA and SCES dataset an AUC score of 0.831 and 0.887 was achieved on these two datasets. In Hemalatha et al. [8] The images were corrected using gamma correction with weighted distribution function. The blood vessels are removed using Gabor filter and morphological operators. The ROI that was obtained a contourlet wavelet transform was used to extract the features and then naïve bayes classifier was used to distinguish between Glaucoma and non Glaucoma images. In Zulfira and Suyanto [9] Snake model has been used to segment the image. The methodology used is active contour snake models. The optic cup to disk ratio is calculated to classify glaucoma from nonglaucoma images [11, 15]. In George Y et al [2020] a 3d convolution neural network has been proposed this model architecture has been used in a classification and a regression task. This is considered an advanced form of classification technique. In Jiang et al. [29] various techniques for detecting glaucoma has been summarized, to detect glaucoma ophthalmologist's measure the eye pressure, diameter of the optic disc etc to predict whether a person has glaucoma, Here the authors have tried to prepare a brief literature survey on how computers aid in detecting glaucoma.

Related Work on Segmentation of Optic Disk

In Obulesu et al. [12] morphological operators has been used for segmenting the optical disc from the diseased eye. In Almazroa et al. [13] template matching has been used to segment the image. All regions in the image is compared with a template image and the region where there is maximum match with the template is confirmed as glaucoma. In Mvoulana et al. [14] A RCNN based deep neural network has been proposed which shows good performance in classifying eyes having glaucoma disease. In Bajwa et al. [16] A RCNN based two stage deep neural network has been designed which localizes the optic disc and the retinal images. Later in the second stage a deep convolutional neural network has been used to classify glaucoma and non glaucoma diseases. In Nawaz et al. [17] Transfer Learning has been used to classify the glaucoma/non-glaucoma images. An Efficient net B0 has been used as the backbone to classify the disease. Classification accuracy has been measured on the RIM-ONE datasets. The Efficient Net b0 has better accuracy to classify the glaucoma/non glaucoma images. In Xiangyu Chen et al. [2015] Canny edge detector has been used to classify the glaucoma/non glaucoma disease. And then a support vector machine has been used to classify from glaucoma/non glaucoma diseases. Parashar and Agarwal [19] a 6 layered deep convolutional neural network is developed using six convolutional layers and two fully connected layers which has AUC score as 0.831 and 0.887 in two different datasets. Abdullah et al [20] a two layer architecture is used comprising of a classification network and a segmentation network which adopts the U-net architecture and another encoder path followed by two fully connected layers and a Faster RCNN network. Kass, et al [21] a hybrid deep learning neural network has been used to detect glaucoma. The validation accuracy was 1.0 and sensitivity of 1.0 as well.

Deep Learning Approach to Detect Glaucoma

Deep Learning approaches generally rely on huge datasets to perform inference regarding a disease. Image segmentation has also been performed using Encoder Decoder networks which are very robust to detect glaucoma disease such as U-net. An U-net comprises of two independent networks which are combined at different times to augment the loss of information that is introduced due to the max pooling operation. When the decoder works in the second stage The information from the previous layer at the same hierarchy is concatenated into the decoder layer, there have been approaches where an attention layer is introduced when the concatenation is done. This type of segmentation layer is known as U-net with attention Lv, et al. [24] which contains an attention layer with an bounding box, in the final layer of the U-net it is multiplied with that bounding box to achieve better segmentation accuracy. The accuracy that is obtained when this type of a Deep Neural Network is used is above 98% in terms of AUC SCORE.

Features Considered for Detecting Glaucoma Using Computer Vision

- Cup Disk ratio:
  If this ratio is very high then it may result in glaucoma disease of eye and if the ratio is less then it is a normal eye without glaucoma disease. The cup and discs measurements can be obtained from the segmented image. If the cup disk ratio is greater then 0.3 it is considered as glaucoma's.
- Harlick texture features:
  Harlick texture features are calculated from Gray level cooccurrence matrix. In [30] there was an attempt to detect glaucoma using harlick texture features, The popular K-nearest neighbour's was used to detect glaucoma and the accuracy achieved was above 98%.
- Optic disc diameter:
  Optic disc diameter is also used for detecting glaucoma. Most of the patients having glaucoma have a diameter in a particular range.

Datasets for Detecting Glaucoma Disease

- RIGA DATASET comprises of 3 different files (Retinal images for glaucoma analysis) The dataset comprises of 750 colour fundus images gained from three different sources. The images have varying sizes.
- MESSIDOR
  It contains 460 images of glaucoma/non glaucoma images. These images were gathered from a local hospital in spain
- RIM-ONE
  The RIM-ONE DL image dataset consists of 313 retinography images from normal subjects and 172 retinography images from patients with glaucoma. These images were collected from three hospitals in Spain.
  This dataset has been divided into training and test sets and they comprise of two different methodology of structuring them:
  Partitioned randomly: the Training/Test datasets are created randomly from the images of the dataset.
  Partitioned by hospital: the images from the HUC hospital are used in the training set, while the images from the HUMS hospital and HCSC hospital are used for test dataset.
- DRISHTI-GS1
  The dataset contains 101 eye fundus images with varying resolutions and ground truth labels for the optic disc and cup. This dataset was produced in a research carried out at IIIT Hyderabad India.
- DRIONS-DB
  The dataset contains 110 eye fundus images with a resolution of 600 × 400. There are two sets of ground-truth optic disc annotations are available. The 1st set is used training and testing purpose and the 2nd set serves a baseline for comparison.

Metrics used for Measuring Accuracy of Segmentation and Classification of Glaucoma

- Precision

$$Precision = {True\ Positive}\Big/{(True\ Positive + FalsePositive)}$$

- $$Recall = {True\ Positive}\Big/{(True\ Positive + FalseNegative)}$$

- $$F1\ score = {2*Precision*Recall}\Big/{(Precision + Recall)}$$

Interpretation of Classification Metrics

- True Positive (TP): The samples which are predicted as positive class during prediction and they actually belong to the positive class.
- True Negative (TN): The samples which are predicted as negative class during prediction and they actually belong to the negative class.

- False Positive (FP): The samples which are predicted as positive during prediction but actually belong to the negative class.
- False Negative (FN): The samples which are predicted as negative during prediction but actually belong to the positive class.

Image Segmentation Accuracy

$$Dice\ coefficient = 2 * Predicted * GroundTruth \Big/ \left(Predicted + GroundTruth\right)$$

Interpretation of Image Segmentation Accuracy

Dice coefficient estimates the area of overlap between Predicted class and Ground Truth annotation of of the image. The dice coefficient is having maximum value(1) when the images overlap each other completely. That is perfect segmentation is achieved.The area of intersection is the estimate of the accurate segmented area.

Results

Deep Learning Architectures used for Classification

A transfer Learning based approach using Efficient Nets have been used for performing classification. Efficient Nets are a family of deep neural networks which have been used and the accuracy achieved is very high compared to other deep neural network architectures [27]. There are even larger family of Efficient Net models such as EffcientNetb7 which performs even better because they have larger number of parameters to train. These models use transfer learning to classify images, they have some pretrained weights which have been obtained after training on the ImageNet dataset, and they are trained on the current domain with augmented examples. The images were augmented by left shifting, rotating right, to generate a larger set of training examples, which can improve the validation accuracy of the models.

Autoencoder's in Deep Learning

An Auto Encoder comprises of a block which comprises of an Convolution Block,MaxPooling Layer and finally an Up-sampling Block.The last layer uses mean squared error or SoftMax functions.

Transfer Learning for Classifying Images

Transfer Learning learns from a particular knowledge base and applies the knowledge gathered in a different domain to the current domain. Alex Net was the first Neural Network which was used in the ImageNet dataset for classifying more then a 100 categorical classes.

Attention Based Image Segmentation

In Attention Based image segmentation we incorporate an attention block in between the Unet Layers. There are variants of Unet such as Unet++ which adds more features to the concatenate block and also performs better in the segmentation tasks.Auxilary information has also been used with a independent convolution neural network which affects the results significantly.

Glaucoma Classification Using Vision Transformer

What is a Transformer?

A transformer is an improved variant of the attention neural networks. This type of a neural network uses Query, Key and Value vector to estimate the embedding. It also uses a positional encoding mapped to a different space. The last layer comprises of a Multi Layer perceptron and an Dense Layer containing the total Number of classes.

How a Vision Transformer Works?

In Natural Language processing applications we can use positional embeddings, But the type of embedding used by researchers are different. The most popular type of embedding used in vision transformers are learnable embeddings, which are strikingly similar to the positional embeddings used in Natural Language Processing applications. There have been variants by users who have used a modified form of vision transformer to perform classification. The positional embedding layer can be modified for various cases. Gabor Filters have been used by researchers to augment the positional embeddings. Edge level information have also been used by researcher's in the positional embeddings Jiang et al. [29].

Results and Discussions

Classification Results on Drishti Datasets

The EfficientNetb7 based deep neural network performs better than the Autoencoder based approach to classify Glaucoma/Non Glaucoma diseases in this dataset Table 45.1.

Table 45.1: Results of classification on drishti dataset.

Methodology	Precision	Recall	F1measure
Auto Encoder based	0.934	0.90	0.91
Efficient Net b7	0.984	0.94	0.97

Classification Results on RIM-ONE Datasets

The classification results on RIM-ONE database for different methods are provided in Table 45.2

Table 45.2: Results of classification on RIM-ONE dataset.

Methodology	Precision	Recall	F1 measure
Auto Encoder based	0.53	0.61	0.50
Efficient Net b7	0.76	0.74	0.74

Image Segmentation Accuracy Results

The segmentation results on different input images are depicted in Figure 45.1

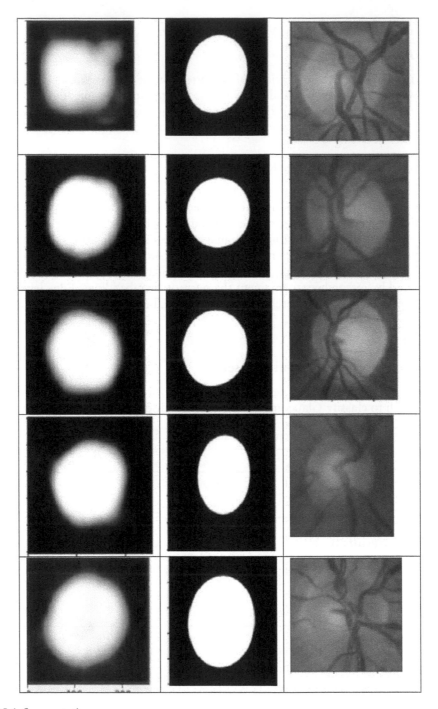

Figure 45.1 Segmentation accuracy.

Proposed Methodology of Segmentation of Optic Disc

The average segmentation accuracy on the optic disc image database is tabulated on Table 45.3

Table 45.3: Average dice coefficient of the dataset.

Average dice coefficient	Total images
0.85	455

How is Unet Image Segmentation Performed

The original Deep Learning architecture of U-net which was used in BioMedical image segmentation can be modified, The Unet architecture uses an auto-encoder to perform image segmentation, it comprises of Convolution Layers,,Maxpooling Layers,Upsampling Layers stacked on top of another. The final layer contains a SoftMax function for multiclass classification and sigmoid layers in case of binary classification. This type of image segmentation can classify the region of interest from the training samples. The loss function that is used is binary focal loss. The training runs for 100 epochs with learning rate 0.01,Google colab pro+ with 128 gb of Ram was used to perform deep learning using Unet. The proposed Unet architecture is shown in Figure 45.2

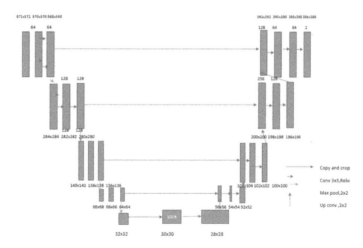

Figure 45.2 Unet architecture.

Binary Focal Loss

$$L(y,p) = \alpha y(1-P)^{\gamma}\log(P) - (1-\gamma)P\gamma\log(1-P) \tag{1}$$

The γ parameter penalizes heavily the misclassification. This type of a loss function is a variant of Binary Cross Entropy Loss which classifies the pixels in two different regions. The parameter can be changed based on the model under consideration.This affects how the neural network learns and the rate of convergence.

Table 45.4: Comparison with state of the art algorithms.

[25]	M-net,AUC Score of 0.8998,DENET, 0.9198 AUC score
Xinguua et al. (2015)	99.8% Sensitivity,99% Accuracy
[28]	0.88 % Accuracy

Comparison with Recent State of the Art

The benefit of M-net is that it uses 2d convolution on 3d images and so it is memory efficient. This architecture of Mnet is identical to Unet in all other respects.

Conclusion

Glaucoma detection approaches used was different for different researchers. Initial methods used edge detection to detect edges in the images and then classify it based on the content of the image. The approach used here makes use of autoencoders and transfer learning to detect the onset of glaucoma. The other approaches are summarized which uses the features from the images to classify from glaucoma/non glaucoma disease. The Auto Encoder based classification approach achieved an accuracy of above 90 % and the Efficient-Net based approach achieved an accuracy of 91% in the Drishti dataset and an Precision of 76% in the RIM-One dataset. The Unet approach can aid in diagnosing glaucoma by measuring the optic disc diameter. How vision transformers are used in classification tasks are summarized and a thorough analysis is done.

References

1. Patil, N., Patil, P. N., and Rao, P. V. (2021). Convolution neural network and deep-belief network (DBN) based automatic detection and diagnosis of glaucoma. *Multimedia Tools Application,* 80, 29481–29495. https://doi.org/10.1007/s11042-021-11087-5.
2. Liu, Y., Yip, L. W. L., Zheng, Y., and Wang, L. (2021). Glaucoma screening using an attention-guided stereo ensemble network. *Methods*, vol 202, pp. 14–21. ISSN 1046-2023, https://doi.org/10.1016/j.ymeth.2021.06.010. (https://www.sciencedirect.com/science/article/pii/S1046202321001626).
3. Chen, X., Xu, Y., Wong, D. W. K., Wong, T. Y., and Liu, J. (2015). Glaucoma detection based on deep convolutional neural network, In 2015 37th Annual International Conference of the IEEE Engineering in Medicine and Biology Society (EMBC), 2015, (pp. 715–718). doi: 10.1109/EMBC.2015.7318462.
4. George, Y., Antony, B. J., Ishikawa, H., Wollstein, G., Schuman, J. S., and Garnavi, R. (2020). Attention-guided 3D-CNN framework for glaucoma detection and structural-functional association using volumetric images. *IEEE Journal of Biomedical and Health Informatics*, 24(12), 3421–3430. doi: 10.1109/JBHI.2020.3001019.
5. Bowd, C., Belghith, A., Christopher, M., Goldbaum, M. H., Fazio, M. A., Girkin, C. A., Liebmann, J. M., de Moraes, C. G., Weinreb, R. N., and Zangwill, L. M. (2021). Individualized glaucoma change detection using deep learning auto encoder-based regions of interest. *Translational Vision Science and Technology*, 10(8), 19. https://doi.org/10.1167/tvst.10.8.19.
7. Nirmala, K., Venkateswaran, N., Kumar, C. V., and Christobel, J. S. (2017). Glaucoma detection using wavelet based contourlet transform. In 2017 International Conference on Intelligent Computing and Control (I2C2), 2017, (pp. 1–5). doi: 10.1109/I2C2.2017.8321875.

8. Hemalatha, R., Thamizhvani, T., Dhivya, A. J. A., Joseph, J. E., Babu, B., and Chandrasekaran, R. (2018). Active contour based segmentation techniques for medical image analysis. In Medical and Biological Image Analysis. London, United Kingdom: IntechOpen. 2018 [Online]. Available: https://www.intechopen.com/chapters/59741. doi: 10.5772/intechopen.74576.

9. Zulfira, F. Z., and Suyanto, S. (2020). Detection of multi-class glaucoma using active contour snakes and support vector machine. In 2020 3rd International Seminar on Research of Information Technology and Intelligent Systems (ISRITI), 2020, (pp. 650–654), doi: 10.1109/ISRITI51436.2020.9315372.

11. Wong, D. W. K., Liu, J., Lim, J. H., Jia, X., Yin, F., Li, H., and Wong, T. Y. (2008). Level-set based automatic cup-to-disc ratio determination using retinal fundus images in ARGALI. In Conference Proc IEEE Engineering in Medicine and Biology Society, 2008, (pp. 2266–2269). [PubMed] [Google Scholar].

12. Obulesu, G., Shaik, F., Sree Lakshmi, C., Vijay Vardhan Kumar Reddy, V., Nishanth, M., and Siva Shankar Reddy, L. (2021). Optic disk segmentation for glaucoma detection in retinal images. In Kumar, A., and Mozar, S. eds. ICCCE 2020. Lecture Notes in Electrical Engineering, (Vol. 698). Singapore: Springer. https://doi.org/10.1007/978-981-15-7961-5_41.

13. Almazroa, A., Burman, R., Raahemifar, K., and Lakshminarayanan, V. (2015). Optic disc and optic cup segmentation methodologies for glaucoma image detection: a survey. *Journal of Ophthalmology*, 2015, 180972. doi: 10.1155/2015/180972. Epub 2015 Nov 25. PMID: 26688751; PMCID: PMC4673359.

14. Mvoulana, A., Kachouri, R., and Akil, M. (2019). Fully automated method for glaucoma screening using robust optic nerve head detection and unsupervised segmentation based cup-to-disc ratio computation in retinal fundus images. *Computerized Medical Imaging and Graphics*, 77, 101643. doi: 10.1016/j.compmedimag.2019.101643. Epub 2019 Aug 14. PMID: 31541937.

15. Wang, Z., Dong, N., Rosario, S. D., Xu, M., Xie, P., and Xing, E. P. (2019). Ellipse detection of optic disc-and-cup boundary in fundus images. In 2019 IEEE 16th International Symposium on Biomedical Imaging (ISBI 2019), 2019, (pp. 601–604). doi: 10.1109/ISBI.2019.8759173.

16. Bajwa, M. N., Malik, M. I., Siddiqui, S. A., Dengel, A., Shafait, F., Neumeier, W., and Ahmed, S. (2019). Two-stage framework for optic disc localization and glaucoma classification in retinal fundus images using deep learning. *BMC Medical Informatics and Decision Making*, 19(1), 136. https://doi.org/10.1186/s12911-019-0842-8.

17. Nawaz, M., Nazir, T., Javed, A., Tariq, U., Yong, H. S., Khan, M. A., and Cha, J. (2022). An efficient deep learning approach to automatic glaucoma detection using optic disc and optic cup localization. *Sensors (Basel, Switzerland)*, 22(2), 434. https://doi.org/10.3390/s22020434.

19. Parashar, D. R., and Agarwal, D. K. (2021). SVM based supervised machine learning framework for glaucoma classification using retinal fundus images. In 2021 10th IEEE International Conference on Communication Systems and Network Technologies (CSNT), 2021, (pp. 660–663). doi: 10.1109/CSNT51715.2021.9509708.

20. Abdullah, F., Imtiaz, R., Madni, H. A., Khan, H. A., Khan, T. M., Khan, M. A., and Naqvi, S. S. (2021). A review on glaucoma disease detection using computerized techniques. *IEEE Access*, 9, 37311–37333. doi: 10.1109/ACCESS.2021.3061451.

21. Kass, M., Witkin, A., and Terzopoulos, D. (1988). Snakes: active contour models. *International Journal of Computer Vision*, 1, 321–331. https://doi.org/10.1007/BF00133570.

22. Wang, M., Yu, K., Zhu, W., Shi, F., and Chen, X. (2019). Multi-strategy deep learning method for glaucoma screening on fundus image. *Investigative Ophthalmology and Visual Science*, 60, 6148.

23. Chakrabarty, N., and Chatterjee, S. (2019). A novel approach to glaucoma screening using computer vision, In 2019 International Conference on Smart Systems and Inventive Technology (ICSSIT), 2019, (pp. 881–884). doi: 10.1109/ICSSIT46314.2019.8987803.

24. Lv, Y., Ma, H., Li, J., and Liu, S. (2020). Attention guided U-Net with atrous convolution for accurate retinal vessels segmentation. *IEEE Access*, 8, 32826–32839. doi: 10.1109/ACCESS.2020.2974027.

25. Fu, H., Cheng, J., Xu, Y., and Liu, J. (2019). Glaucoma detection based on deep learning network in fundus image. In Lu, L., Wang, X., Carneiro, G., and Yang, L. eds. Deep Learning and Convolutional Neural Networks for Medical Imaging and Clinical Informatics. Advances in Computer Vision and Pattern Recognition. Springer, Cham. https://doi.org/10.1007/978-3-030-13969-8_6.

27. Rehman, A. U., Taj, I. A., Sajid, M., and Karimov, K. S. (2021). An ensemble framework based on Deep CNNs architecture for glaucoma classification using fundus photography. *Mathematical Biosciences and Engineering,* 18(5), 5321–5346. doi: 10.3934/mbe.2021270. PMID: 34517490.

28. Saxena, A., Vyas, A., Parashar, L., and Singh, U. (2020). A glaucoma detection using convolutional neural network. In 2020 International Conference on Electronics and Sustainable Communication Systems (ICESC), 2020, (pp. 815–820). doi: 10.1109/ICESC48915.2020.9155930.

29. Jiang, K., Peng, P., Lian, Y., and Xu, W. (2022). The encoding method of position embeddings in vision transformer. *Journal of Visual Communication and Image Representation,* 89, 103664. ISSN 1047-3203. https://doi.org/10.1016/j.jvcir.2022.103664. (https://www.sciencedirect.com/science/article/pii/S1047320322001845).

30. Simonthomas, S., Thulasi, N., and Asharaf, P. (2014). Automated diagnosis of glaucoma using Haralick texture features. In International Conference on Information Communication and Embedded Systems (ICICES2014), 2014, (pp. 1–6). doi: 10.1109/ICICES.2014.7033743.

46 Bio Geography Based Optimization with Improved Initial Operator for Multiple Sequence Alignment

Rajeev Kumar Pathak[1,a], Ratan Mani Prasad[2,b,] and Rohit Kumar Yadav[3,c]

[1]Research Scholar, Department of Mathematics, Magadh University, Bodhgaya, Bihar, India.

[2]Assistant Professor, Department of Mathematics, SNS College Tekari, Magadh University, Bodhgaya, Bihar, India.

[3]Assistant Professor, Department of Computer Science and Engineering, Sarala Birla University Ranchi, India.

Abstract

NP-hard problems are computation intensive and Multiple sequence alignment (MSA) is one of them in bioinformatics science. Therefore, methods influenced by nature can more accurately estimate the solution. A novel biogeography-based optimization (NBBO) is suggested in the current study to resolve MSA problem. A novel paradigm for optimization is called biogeography-based optimization (BBO). However, there are certain shortcomings in the solution of complex issues, such as small population variety and a low rate of convergence. NBBO is an improved or upgraded version of BBO that proposes a new initialization process to get beyond BBO's drawbacks. A new initialization operators has been incorporated in place of random initial operator. This improved initialization operator protects exploitation potential, population diversity, and collect more information from different ecosystems. Performance analysis is conducted, and benchmark datasets made accessible to the public are used to test new and current approaches including VDGA, MOMSA, and GAPAM (i.e., Bali base). The proposed method appears to be competitive or superior to the methods already in use, according on observations.

Keywords: Biogeography-based optimization (BBO), CLUSTALW, habitat, multiple sequence alignment (MSA), suitability index variable (SIV), variables.

Introduction

Multiple sequence alignment refers to the simultaneous alignment of more than three protein sequences or amino acid sequences (MSA). The most crucial instrument for resolving biological issues is MSA. By employing MSA, we can solve a wide range of biological problems. The tertiary and secondary structures of Ribonucleic acid (RNA) and proteins can be predicted with the aid of MSA [1, 2]. Utilizing MSA, we can predicts the significance and role of an unknown amino acid which is helpful for sequencing i.e., in order (align) and replace with certain another known functions, we may reassemble phylogenetic trees. Utilizing MSA, we can also determine how similar the sequences are, which can be used to establish structural and functional similarity [3,4]. The whole sequences in the multiple alignments need to share a similar beginning or origin for an Multiple Sequence

[a]rajeevpathak2009@gmail.com, rajeevpathak_sap@yahoo.com, [b]ratanmaniprasad@gmail.com, [c]rohit.ism.123@gmail.com

Alignment to be considered legitimate. The main objective of MSA is to optimise the degree of amino acid or protein compatibility [5]. Consequently, MSA is a crucial issue in bio-informatics sciences to understand the interaction between the phylogeny and genetic systems. Numerous methods can be implemented to fix an MSA problem.

Dynamic programming can be used to overcome the MSA problem and produce an ideal alignment (DP). The scoring function used by DP has a broad domain. Needleman and Wunsch [6] suggested using the DP technique to address the issue of two sequence alignments in 1970. However, the drawback of using DP is that as sequence length and number rise, so does complexity in an exponential fashion.

The progressive method can similarly be used to tackle the MSA problem [22,23]. When addressing an MSA problem, the progressive technique requires less complexity in both space and time [7, 8]. The progressive alignment approach states that after first aligning more comparable sequences, more divergent sequences or groups of sequences should be incrementally aligned. CLUSTALW is the accepted exemplar of progressive techniques [9].

Since an iterative technique begins with starting alignment and enhances the answers each time till further enhancement is not possible, it is independent of the first alignment. Raising the overall quality of a sequence alignment is the primary goal of the iterative technique for MSA.

Population-based algorithms include evolutionary algorithms [10, 11]. In the first phase, we construct a random beginning population in accordance with these procedures. In the following phase, we use a few operators for modification the starting population in accordance with the next generation. Utilizing these operators repeatedly till we reach at the overall best solution. Utilizing evolutionary algorithms (EAs) to an MSA, the first generation is produced randomly, and then the EA's stages are used to increase the similarity between the sequences. For MSA, certain evolutionary calculations have been made [12–16]. SAGA, GAACO, RBT-GA, GAPAM, [17] VDGA, [18], MSA-EC, MSA-GA and MO-SAStrE are further genetic algorithm (GA)-based MSA techniques [21]. We provide the approach for an algorithm that uses GA to solve an MSA problem. The benchmark BaliBase 2.0 dataset and BaliBase 3.0 dataset's reference sets 1, 2, 3, 4, and 5 included 56 flaws that the authors [19] resolved. Due to the lack of variety in the solutions, these evolutionary approaches have the problem of producing local optima.

Contributions and drive in the field of biology, MSA is essential for resolving many common issues including predicting structure and phylogeny feature. The accessible literature indicates that the MSA continues to remain is an unsolved challenge. Therefore, we are motivated to tackle an MSA problem by utilising an enhanced BBO. In this paper our work highlights the following contribution furthermore.

A. To retain the variety of the solutions, we first presented a technique to enhance the migration operator in BBO.
B. In terms of the time factor, the outcomes of the experimental analysis are superior. Additionally, we offer a comparison table that argues that, in terms of matching score, our approach is superior to the currently available competitive options.

In this paper, BBO (Biogeography-Based Optimization) has been introduced and an improved method is proposed. Experimental results of the proposed methods are shown and the analysis of the results obtained has been done in this work. The conclusion section summaries the work done.

Biogeography-Based Optimization

The immigration and emigration of organisms from one environment to another served as the inspiration for BBO [20]. Candidate solutions in the BBO algorithm are referred to as habitats (or islands). The suitability index variable (SIV) denotes each characteristic of a solution symbolized by a habitat, and the habitat suitability index (HSI) gauges the habitat's quality (HSI). More species can survive in ecosystems with a high HSI than in settings with a low HSI. By absorbing new traits from greater desirable habitats during the evolution process, poor environments can raise their HSI.

Migration and mutation are the two major operators in BBO. On the basis of immigration rate α_i and emigration rate β_i values, the operator which is used for migration. It is a probabilistic operator can alter Suitability index variable (SIV) values. α_i and β_i depend on how many species are present in the ith environment (HB_i).

In the BBO method, immigration rate α_i and emigration rate β_i are supposed be linear therefore the α_i and β_i should exhibits linear correlations between the number of species and have identical maximum values for mathematical convenience.

$$\alpha_i = I * (1 - n_i)/n$$
$$\beta_i = F * n_i/n \tag{1}$$

Here, the ith habitat is represented by n_i which is the count of species and n symbolized as the max count of species. The F is the maximum feasible rate of emigration and I represents the greatest achievable rate of immigration.

By sharing characteristics among them, in BBO the migration operator which is a probabilistic operator is used to randomly alter all habitats HB_i. The likelihood that HB_i gets modified depends on its immigration rate α_i but the likelihood that HB_j is the source of the modification depends on its emigration rate β_i. The migration equation looks like this:

$$HB_i(SIV) = HB_j(SIV) \tag{2}$$

where the *i*th habitat's feature (SIV) is denoted by $HB_i(SIV)$.

As Simon noted, the migration operator does not really reproduce "children"; rather, it only moves SIVs from one solution to another [20].

Cataclysmic occurrences can alter a species count from its equilibrium value, abruptly altering the HSI of a habitat. This unexpected operation in BBO is modelled as a mutation. The mutation operator can randomly alter the SIVs of the ith habitat HB_i in accordance with the habitat's prior probability P_i. The expression for the ith habitat's mutation probability, S_i, is

$$S_{i=}m_{max}*(1 - T_i / T_{max}) \tag{3}$$

where $T_{max} = \max(T_i)$, $I = 1, 2$, and m_{max} is a user-defined value N. A new feature will takes places randomly for a SIV in each habitat, which is created in a probabilistic and random approach over the whole solution space in the BBO mutation operator. This process tends to boost population diversity.

Proposed Approach illustration of the habitat. Each solution in BBO is represented by a

$$Z_i = (Z_i^1, ..., Z_i^d, ..., Z_i^n) \; \forall 1 \leq i \geq N \tag{4}$$

where N tells about habitats number.

Put the gap in the specified MSA at random in the startup state. Figure 46.1 provides the initial solution.

Binary encoding strategy is used which is an encoding approach, place 1 at where there exist a gap and 0 at where protein sequences exist. An initial solution's encoding is shown in Figure 46.2.

Then, each column is read from bottom to top to get the decimal value of this binary-encoded value. Therefore, $Z_1 = (1, 0, 0, 8, 2, 4)$ is the habitat representation for this solution, and the column count of the MSA and the habitat's feature count are correlated. Now that the gap in the MSA has been created, we may produce 100 different solutions. Therefore, in initialization, we may locate 100 habitats.

Fitness Function

Fitness is a process. MSA fitness is evaluated using the sum of pairs. Here, the scores' product of each pair of symbols is added to get the score of each column in an alignment. The total alignment score is then calculated by summing the column scores using (4) and (5).

$$W = \sum_{i=1}^{P} W_{i,} \, where \, W_i = \sum_{i=1}^{N-1} \sum_{k=l+1}^{N} Cost\left(A_l, A_k\right) \tag{5}$$

In this equation, W defines the MSAs cost and P symbolized as the alignment's column-based length. W_i is used to define the ith column cost of length P and N symbolizes all possible sequences, and alignment between A_l and A_k which is calculated by Cost (A_l, A_k). While A_l "_" and A_k "_", the cost (A_l, Ak) is derived by proportion of acceptable mutations matrix. Additionally, Cost $(A_l, Ak) = 0$ when A_l and A_k are both equal to "_". The cost function comes last. When employing a model with linear gap penalties as indicated in (Eq. 5), Cost (A_l, A_k) is obtained by summing of the replacement insertion's cost and deletion's cost where A_l = "_" and $A_k \neq$ "_" or $A_l \neq$ "_" and A_k = "_".

$$G = Q + Ar \tag{6}$$

Here, G denotes the gap cost or penalty, Q for the opening or start gap cost, r for the extending gap cost, and A for the gap number. Gap penalties are used in this research (Gap opening carries a penalty of –5 and gap extension a penalty of –0.40.) creation of fresh solutions. Both migration and mutation are types of operators that are utilized in this procedure. In order to improve the solution, the species having the high HSI solution are accepted by the low HSI solution. Migration alludes to the process as a whole.

Proposed Method

Initialize the Habitat's position, update the Habitat's position using the position as well as velocity equation provided in (1) and (2), and apply the stopping criteria are the steps of our technique employing BBO. The steps for using this approach are described below.

Initial Generation

This step's goal is to come up with preliminary solutions. By putting the gap between the residues, we create the first solution in this stage. Below is an illustration of how initial population is generated:

ABCD
BCDE
CDEF
DEFG

Now, using the Needleman and Wunsch algorithm[6], For every pair, we need to determine how to figure out the pair-wise alignment. ABCD BCDE is the first pair (1, 2).

As a result, our alignment is ABCD BCDE.

In this instance, there are 4 sequences, accordingly we have 4*(4-2)/2=6 pairings. Similar to this, we can identify the alignment of each pair, pair by pair.

We can now produce random permutations between 1 and N using our method.

This specific example contains 4 sequences. Therefore, we must create a random permutation of 1 to 4.

Let's create a random permutation (3, 4, 1, and 2).

2nd pair (1, 2) and 1st pair (3, 4) are therefore

CDEF_
_DEFG
ABCD_
_BCDE.

In a similar manner, we can produce K times as many solutions. The output of a random permutation ranges from 1 to N. Our initial population has now been established.

Habitat Representation

Each solution in BBO is represented by a Habitat as shown in equation 4.

$$Z_i = (Z_i^1, \ldots\ldots, Z_i^d, \ldots\ldots, Z_i^n) \ \forall 1 \leq i \geq N$$

Here n tells the dimension's number within a habitats and N represent the habitats number. Out of 200 options, we are selecting one. Here, we outline the particle encoding for the MSA problem.

-	O	B	D	O	U
D	-	O	U	D	D
B	O	-	U	-	U
U	D	D	B	U	-

Figure 46.1 Initial solution.

Binary encoding strategy: Place 1 in the Gap location and 0 in the protein sequence location in the encoding scheme. Our MSA now changes.

1	0	0	0	0	0
0	1	0	0	0	0
0	0	1	0	1	0
0	0	0	0	0	1
8	**4**	**2**	**0**	**2**	**1**

Figure 46.2 Encoding scheme.

Habitat Representation

Then, each column is read from bottom to top to get the decimal value of this binary-encoded value. Therefore, Z1 is how we describe this answer using particles as $Z1 = (1, 2, 0, 2, 4,$ and $8)$. Therefore, In MSA both the number of dimensions and the number of columns must be equal in a habitat. We may now construct particles for any MSA that have been generated after the initials in accordance with the encoding scheme.

Migration

Migration is used to widen the solution space. The mutation is time intensive and computation intensive which makes the solution search space more intense. We are utilizing migration and mutation operators on the habitats on each and every iteration and share the characteristic of a high HSI environment during the migratory process.

Too little HSI to raise the quality of the solutions. The generated habitat differs greatly from the original habitat, and this operator is quite successful. Based on the rates of immigration and emigration, we selected two habitats. Then, one index in the emigration habitat was randomly selected, and this element/SIV moves to the same place in the immigration habitat. Figure 46.3 illustrates the procedure.

Mutation

The difference between the original habitat and the resulting habitat is hardly noticeable, and this operator is not significantly more efficient. This operator is uncommon yet

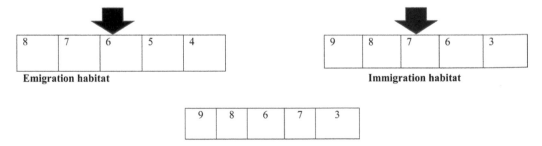

Figure 46.3 Graphical representation of migration process.

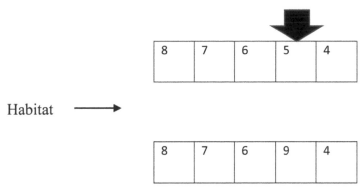

Figure 46.4 Graphical representation of mutation process.

accelerates the search space's solution. Based on the likelihood of mutation, one habitat is selected in this operator. After then, one index from this habitat is chosen at random, and a new element/SIV selected between 0 and 2N is substituted for this one. Figure 46.4 displays a visual illustration of this procedure.

Experimental Results and Analysis

To know the efficiency of proposed method, contrasting the suggested approach with the 5 best current approaches, including VDGA_Decomp_2 [21], VDGA_Decomp_3 [21], VDGA_Decomp_4 [21], GAPAM [17] and MO-SAStrE [21]. The fitness function, which is discussed in section 3.2, is used to calculate the fitness value. The proposed method was run 50 times on each data set, and the best value was recorded. In Table 46.1 provide the results in terms of Baliscore. These are well known recent methods for solving MSA problem. Table 46.1 shows the Results of proposed method and compare it with VDGA_ Decomp_2, VDGA_Decomp_3, and VDGA_Decomp_4. We have also consider GAPAM and MO-SAStrE for comparison and result analysis. In most of cases proposed methods performs better than the others methods in terms of Baliscore.

Table 46.1: Comparison table for proposed methods and state of art methods.

Datasets	GAPAM[17]	VDGA_ Decomp_2 [21]	VDGA_ Decomp_3 [21]	VDGA_ Decomp_4 [21]	MO-SAStrE [21]	PROPOSED METHOD
1idy	.565	.550	.651	.654	–	**.711**
1tvxA	.316	.316	.316	.310	–	**.325**
1uky	.402	.416	.459	.464	.403	**.484**
Kinase	.487	.531	.545	.548	**.808**	.792
1ped	.498	.443	.482	.451	.716	**.752**
2myr	.317	.347	.359	.282	.544	**.549**
1ycc	.845	.752	.839	.685	–	**.859**
3cyr	**.911**	.797	.898	.797	–	.703

(continued)

Table 46.1: Continued

Datasets	GAPAM[17]	VDGA_Decomp_2 [21]	VDGA_Decomp_3 [21]	VDGA_Decomp_4 [21]	MO-SAStrE [21]	PROPOSED METHOD
1ad2	.956	.950	.941		–	.878
1ldg	.963	.914	.946	.903	–	**.972**
1fieA	.963	.926	**.960**	.927	–	.956
1sesA	**.982**	.917	.962	.923	–	.932
1krn	.960	.942	.960	.892	–	**.966**
2fxb	.970	**.978**	.978	.978	–	.975
1amk	**.998**	.982	.984	.982	–	.932
1ar5A	.974	.942	.968	.954	–	**.988**
1gpb	.983	.976	**.984**	.983	–	.979
1taq	.945	.938	.959	.944	–	**.965**

Conclusions

An enhanced initial operator has been employed in this paper to effectively address the MSA problem. This improved operator help to maintain exploration and exploitation. A comparison between the new method and the original BBO algorithm was made. The matching Bali score of this solution was applied because previous methods of comparing MSA quality/accuracy did so using the Bali score. The experimental findings demonstrated that the suggested NBBO outperformed the alternative in the majority of test circumstances. Though not the optimal solution for all test circumstances, the proposed approach was picked since it was close.

References

1. Gusfield, D. (1997). Algorithms on Strings, Trees and Sequences Computer Science. Cambridge: Cambridge University Press.
2. Feng, D., Johnson, M., and Doolittle, R. (1985). Aligning amino acid sequences: comparison of commonly used methods. *J Mol Evol.* 21:112–25.
3. Bonizzoni, P., and Della Vedova G. (2001). The complexity of multiple sequence alignment with SP-score that is a metric. *Theor Comp Sci.* 259:63–79.
4. Carrillo, H., and Lipman, D. (1988). The multiple sequence alignment problem in biology. *SIAM J Appl Math.* 48:1073–82.
5. Dayhoff, MO., and Schwartz, RM. (1978). A model of evolutionary change in proteins. In: Dayhoff MO, ed. Atlas of Protein Sequence and Structure. Washington, DC: National Association for Biomedical Research. 345–52.
6. Needleman, SB., and Wunsch, CD. (1970). A general method applicable to the search for similarities in the amino acid sequence of two proteins. *J Mol Biol.* 48:443–53.
7. Taylor, WR. (1988). A flexible method to align large numbers of biological sequences. *J Mol Evol.* 28:161–9.
8. Feng, DF., and Doolittle, RF. (1987). Progressive sequence alignment as a prerequisite to correct phylogenetic trees. *J Mol Evol.* 25:351–60.

9. Thompson, JD., Higgins, DG., and Gibson, TJ. (1994). CLUSTAL W: improving the sensitivity of progressive multiple sequence alignment through sequence weighting, position-specific gap penalties and weight matrix choice. *Nucleic Acids Res.* 22:4673–80.

10. Yadav, RK., and Banka, H. (2015). Genetic Algorithm with Improved Mutation Operator for Multiple Sequence Alignment. *Springer AISC.* 340:515–524.

11. Yadav, RK, and Banka H. (2016). Genetic Algorithm using Guide Tree in Mutation Operator for solving Multiple Sequence Alignment. *Springer AISC.* 395:145–157.

12. Cai, L., Juedes, D., and Liakhovitch, E. (2000). Evolutionary computation techniques for multiple sequence alignment. IEEE In: Evolutionary Computation, Proceedings of the Congress on 2:829–835.

13. Chellapilla, K., and Fogel, GB. (1999). Multiple sequence alignment using evolutionary programming. IEEE In: Evolutionary Computation, CEC. Proceedings of the Congress on 1.

14. Horng, JT., Lin, CM., Liu, BJ., and Kao, CY. (2000). Using genetic algorithms to solve multiple sequence alignments. In Proceedings of the 2nd Annual Conference on Genetic and Evolutionary Computation, Morgan Kaufmann Publishers Inc. 883–890.

15. Ishikawa, M., Toya, T., Totoki, Y., and Konagaya, A. (1993). Parallel iterative aligner with genetic algorithm. *Genome Inform.* 4:84–93.

16. Notredame, C., and Higgins, DG. (1996). Saga: sequence alignment by genetic algorithm. *Nucleic Acids Res.* 24:1515–24.

17. Naznin, F., Sarker, R., and Essam D. (2012). Progressive alignment method using genetic algorithm for multiple sequence alignment. *IEEE Trans EvolComput.* 16:615–31.

18. Naznin, F., Sarker, R., and Essam, D. (2011). Vertical decomposition with genetic algorithm for multiple sequence alignment. *BMC Bioinformatics.* 12:353.

19. Zhu, H, He, Z, Jia, Y. (2015). A novel approach to multiple sequence alignment using multi-objective evolutionary algorithm based on decomposition. *IEEE J Biomed Health Inform.* 20(2):1–11.

20. Simon D. (2008). Biogeography-based optimization. *IEEE Trans EvolComput.* 12:702–13.

21. Chowdhury Biswanath, Garai Gautam. (2017). A review on multiple sequence alignment from the perspective of genetic algorithm, Genomics, 109:419–431.

22. Sofi, M. Y., Shafi, A., and Masoodi, K. Z.(2022). Chapter 6—Multiple sequence alignment, Academic Press, 47–53.

23. Reddy, Bharath and Fields, Richard. (2022). Multiple Sequence Alignment Algorithms in Bioinformatics. 10.1007/978-981-16-4016-2_9.